Volumes already published

Volume 1: Trees I (1986)
Volume 2: Crops I (1986)
Volume 3: Potato (1987)
Volume 4: Medicinal and Aromatic Plants I (1988)
Volume 5: Trees II (1988)
Volume 6: Crops II (1988)
Volume 7: Medicinal and Aromatic Plants II (1989)
Volume 8: Plant Protoplasts and Genetic Engineering I (1989)

Volumes in preparation

Volume 9: Plant Protoplasts and Genetic Engineering II (1989)
Volume 10: Legumes and Oilseed Crops I
Volume 11: Somaclonal Variation in Crop Improvement
Volume 12: Haploids in Crop Improvement
Volume 13: Wheat
Volume 14: Rice

Biotechnology in Agriculture and Forestry 8

Plant Protoplasts and Genetic Engineering I

Edited by Y. P. S. Bajaj

With 152 Figures

Springer-Verlag Berlin Heidelberg New York
London Paris Tokyo Hong Kong

Professor Dr. Y. P. S. Bajaj
A-137
New Friends Colony
New Delhi 110065, India

ISBN 3-540-19194-1 Springer-Verlag Berlin Heidelberg New York
ISBN 0-387-19194-1 Springer-Verlag New York Berlin Heidelberg

Plant protoplasts and genetic engineering / edited by Y. P. S. Bajaj. p. cm. – (Biotechnology in agriculture and forestry; 8–) Includes bibliographies and index. 1. Plant protoplasts. 2. Plant genetic engineering. I. Bajaj, Y. P. S., 1936– . II. Series. QK725.P578 1989 581.87'3 – dc 19 88-31197

Printed in Germany

Typesetting: K+V Fotosatz GmbH, Beerfelden.
Offsetprinting and bookbinding: Konrad Triltsch, Graphischer Betrieb, Würzburg.
2131/3145-543210 – Printed on acid-free paper

Dedicated to my sister Nirmal

Preface

Isolated protoplasts are a unique tool for genetic manipulation of plants. Since the discovery of a method for the enzymatic isolation of protoplasts by Professor E. C. Cocking in 1960, tremendous progress has been made in this very fascinating area of research. I have witnessed the struggle in the 1960's and early 1970's, when obtaining a clean preparation of protoplasts was considered an achievement. I also shared the pioneering excitement and enthusiasm in this field during the 2nd International Congress of Plant Tissue Culture held at Strasbourg in 1970, where Dr. I. Takebe of Japan presented his work on the induction of division in tobacco protoplasts. This was followed by my participation in three international conferences devoted to plant protoplasts held in 1972 in Salamanca (Spain) and Versailles (France), and then in 1975 in Nottingham (England). The enthusiasm shown by plant scientists at these meetings was ample proof of the bright future of protoplast technology, and it became evident that protoplasts would play a major role in plant biotechnology, especially in genetic engineering. Since then we have never looked back, and now the methods for isolation, fusion, and culture, as well as regeneration of somatic hybrids, have become routine laboratory procedures for most plant species. Currently the focus is on cereal and tree protoplasts.

In order to bring about any genetic manipulation through fusion, incorporation of DNA, and transformation, the regeneration of the entire plant through manipulation of protoplasts is a prerequisite. Today we are at the threshold of the breakthrough in genetic engineering through manipulation of protoplasts. This volume on *Plant Protoplasts and Genetic Engineering I* is the first in the series, containing 28 chapters dealing with the isolation, fusion, culture, immobilization, and cryopreservation of protoplasts, and the regeneration of somatic hybrid and cybrid plants. The technology and literature on the regeneration of complete plants from isolated protoplasts of rice, potato, citrus, soybean, linseed, cabbage, chicory, lettuce, butterbur, orchids, and some forest tree species are discussed. Various methods of fusion of protoplasts are also reviewed, and the work on the regeneration of interspecific and intergeneric somatic hybrids in *Lycopersicon, Petunia, Nicotiana, Hyoscyamus, Solanum, Citrus, Brassica, Medicago,* and *Trifolium* is discussed. In the next volume, *Plant Protoplasts and Genetic Engineering II*, some of the newer techniques such as microinjection, electrofusion, flow cytometry, uptake of organelles, and viruses are described and the work on recombinant DNA technology for direct gene transfer/*Agrobacterium*-mediated transformation in potato, rice, maize, cotton, tobacco, *Brassica*, poplars, *Vinca*, *Hyoscyamus* etc. is reviewed. These studies reflect the far-reaching impact of protoplast alterations for agriculture and forest biotechnology. The greatest use of these recent advances in protoplast technology is likely to be made for the induction of genetic variability and the development of new genotypes for specific plant improvement programs.

New Delhi, March 1989

Y. P. S. Bajaj
Series Editor

Contents

Section III Protoplast Fusion and Somatic Hybridization

List of Contributors

BAJAJ, Y. P. S., A-137, New Friends Colony, New Delhi 110065, India

BAKARAT, M. N., Tissue Culture and Plant Breeding Group, Crop Science Department, Faculty of Agriculture, Alexandria University, Alexandria, Egypt

BERGOUNIOUX, C., Laboratoire de Physiologie Végétale Moléculaire, Bât 430, Faculté des Sciences, F-91405 Orsay, France

BINDING, H., Botanisches Institut der Christian-Albrechts-Universität, Biologiezentrum, Olshausenstraße 40–60, D-2300 Kiel 1, FRG

BOYES, C. J., Plant Genetic Inc., 1930 Fifth Street, Davis CA 95414, USA

CAMPBELL, M. A., Department of Biological Sciences, Rutgers University, Newark, NJ 07102, USA

CHUPEAU, Y., Laboratoire de Biologie Cellulaire, I.N.R.A., Centre de Versailles, F-78025 Versailles, France

COCKING, E. C., Plant Genetic Manipulation Group, Botany Department, Nottingham University, Nottingham NG7 2RD, United Kingdom

COLIJN-HOOYMANS, C. M., Institute for Horticultural Plant Breeding, P.O. Box 16, NL-6700 AA Wageningen, The Netherlands

DAVEY, M. R., Plant Genetic Manipulation Group, Department of Botany, University of Nottingham, University Park, Nottingham, NG7 2RD, United Kingdom

DEMARLY, Y., Laboratoire d'Amélioration des Plantes, Université de Paris-Sud, F-91405 Orsay, France

DONS, J. J. M., Institute for Horticultural Plant Breeding, P.O. Box 16, NL-6700 AA Wageningen, The Netherlands

ENOMOTO, S., Laboratory of Morphogenesis, Department of Cell Biology, National Institute of Agrobiological Resources, Tsukuba, Ibaraki 305, Japan

FAMELAER, Y., Laboratorium voor Plantengenetica, Vrije Universiteit Brussel, Instituut voor Moleculaire Biologie, Paardenstraat 65, B-1640, Sint-Genesius-Rode, Belgium

FOWKE, L. C., Department of Biology, University of Saskatchewan, Saskatoon, SAS, S7N OW9 Canada

GAHAN, P. B., Plant Cell and Moleculaire Biology Group, King's College London, Campden Hill Road, London W8 7AH, United Kingdom

GALUN, E., Department of Plant Genetics, The Weizmann Institute of Science, Rehovet 76100, Israel

GILMOUR, D. M., Molecular Genetics Department, Plant Breeding Institute, Maris Lane, Trumpington, Cambridge, CB2 2LQ, United Kingdom

GLEBA, Y. Y., Academy of Sciences, Institute of Botany, Repina 2, Kiev, USSR

GOLDS, T. J., Plant Genetic Manipulation Group, Department of Botany, University of Nottingham, University Park, Nottingham NG7 2RD, United Kingdom

HANSON, M. R., Section of Genetics and Development, Biotechnology Building, Cornell University, Bradfield Hall, Ithaca, NY 14853, USA

HINATA, K., Laboratory of Plant Breeding, Faculty of Agriculture, Tohoku University, 1-1 Tsutsumidori Amamiyamachi, Sendai 981, Japan

HINNISDAELS, S., Laboratorium voor Plantengenetica, Vrije Universiteit Brussel, Instituut voor Moleculaire Biologie, Paardenstraat 65, B-1640 Sint-Genesius-Rode, Belgium

HONKANEN, J., Kemira OY, ESPOO Research Centre, P.O. Box 44, SF-02271 Espoo

ISHII, S., Bioscience Research Laboratory, Kikkoman Corporation, 399 Noda, Noda-shi, Chiba-ken 278, Japan

JACOBS, M., Laboratorium voor Plantengenetica, Vrije Universiteit Brussel, Instituut voor Moleculaire Biologie, Paardenstraat 65, B-1640 Sint-Genesius-Rode, Belgium

KAMEYA, T., Institute of Genetic Ecology, Tohoku University, Sendai 980, Japan

KAO, K. N., Plant Biotechnology Institute, National Research, 110 Gymnasium Road, Saskatoon, Saskatchewan, S7N OW9, Canada

KIRBY, E. G., Department of Biological Sciences, Rutgers University, Newark, NJ 07102, USA

KYOZUKA, J., Plantech Research Institute, 1000 Kamoshida, Midori-ku, Yokohama 227, Japan

LAZAR, G., Department of Molecular Biology, Massachusetts General Hospital and Department of Gentics, Harvard Medical School, Massachusetts General Hospital, Boston, MA 02114, USA

LUU, H. T., The Upjohn Company, 9612-50-1, 301 Henrietta Street, Kalamazoo, MI 49001, USA

MICHAYLUK, M. R., Plant Biotechnology Institute, National Research Council, 110 Gymnasium Road, Saskatoon, Saskatchewan, S7N OW9, Canada

Mouras, A., Laboratoire de Biologie Cellulaire, Université Bordeaux II, Avenue des Facultés, F-33405 Talence-Cedex

NEGRUTIU, I., Laboratorium voor Plantengenetica, Vrije Universiteit Brussel, Instituut voor Moleculaire Biologie, Paardenstraat 65, B-1640 Sint-Genesius-Rode, Belgium

NEWELL, C. A., Monsanto Agricultural Company-Technology, 700 Chesterfield Village Parkway, St. Louis, MO 63198, USA

NIIZEKI M., Plant Breeding Laboratory, Faculty of Agriculture, Hirosaki University, Hirosaki, Aomori-ken 036, Japan

NISHIO, T., National Research Institute of Vegetables, Ornamental Plants and Tea, Ano, Mie 514–23, Japan

O'CONNELL, M. A., Plant Genetic Engineering Laboratory, Department of Agronomy and Horticulture, Box 3GL, New Mexico State University, Las Cruces, New Mexico 88003, USA

OGURA, H., Department of Agricultural Sciences, Ishikawa College of Agriculture, Nonoichi, Ishikawa 921, Japan

OHYAMA, K., Department of Cell Biology, National Institute of Agrobiological Resources, Kannondai Tsukuba, Ibaraki 305, Japan

PENCHEL, R. M., Aracruz Florestal S. A., Departemento de Biotecnologia Florestal, 29190-Aracruz-ES, Brazil

RYÖPPY, P., Plant Molecular Biology Laboratory, University of Helsinki, Valimotie 7, SF-00380 Helsinki

SHIMAMOTO, K., Plantech Research Institute, 1000 Kamoshida, Midori-ku, Yokohama 227, Japan

SIDEROV, V., Academy of Sciences, Ukr. S.S.R., Institute of Botany, Repina 2, Kiev, 252601, USSR

SINK, K. C., Department of Horticulture, Michigan State University, East Lansing, MI 48824, USA

SCHNABL, H., Institut für Landwirtschaftliche Botanik der Universität Bonn, Meckenheimer Allee 176, D-5300 Bonn, FRG

TABAEIZADEH, Z., Department of Biology, University of Quebec in Montreal, C.P. 8888, Station "A", Montréal, H3C 3P8, Canada

TAKAYANAGI, K., National Research Institute of Vegetables, Ornamental Plants and Tea, Ano, Mie 514–23, Japan

TORIYAMA, K., Laboratory of Plant Breeding, Faculty of Agriculture, Tohoku University, 1-1 Tsutsumidori-Amamiyamachi, Sendai 981, Japan

VARDI, A., Institute of Horticulture, Agricultural Research Organization, The Volcani Center, P.O. Box 6, Bet Dagan 50250, Israel

VIDAIR, C., Department of Radiation Oncology, CED-200, University California of San Francisco, CA 94143, USA

YABE, K., Aichi-ken Agricultural Research Centre, Sagamine 1-1, Yazako, Nagakute-cho, Aichi-gun, Aichi 480-11, Japan

YAMASHITA, Y., Plantech Research Institute, 1000 Kamoshida, Midori-ku, Yokohama 227, Japan

YASUGI, S., Department of Horticulture, University of Minamikyushu, Hibarigoaka, Takanabe-cho, Miyazaki Prefecture, Japan 884

ZIMMERMANN, U., Lehrstuhl für Biotechnologie der Universität Würzburg, Röntgenring 11, D-8700 Würzburg, FRG

ZUBA, M., Norddeutsche Pflanzenzucht, Hans-Georg-Lembke KG, Hohenlieth, D-2331 Holtsee

Section I Isolation, Culture, and Preservation of Protoplasts

I.1 Recent Advances in the Isolation and Culture of Protoplasts and Their Implications in Crop Improvement

Y. P. S. Bajaj[1]

1 Introduction

Since the first report on the regeneration of complete plants from isolated protoplasts (Takebe et al. 1971), tremendous progress has been made and a number of crops of commercial importance such as potato, tomato, tobacco, rice, linseed, alfalfa, cucumber, eggplant, lettuce and *Brassica* species etc. have been routinely regenerated. This has paved the way for their genetic manipulation. Consequently considerable interest has generated in two areas, i.e., the culture of protoplasts of cereals and those of trees, and a large volume of information has accumulated recently (Table 1). This has been made possible for a number of reasons, primarily due to the upsurge of activity in plant biotechnology (see Bajaj 1986a), commercial interest, the tremendous increase in the number of workers during the last 5 years, advancement of in vitro technology and automation. Moreover, it has created awareness in agriculture, especially of plant breeders. There has been inflow of funds and finances, both from the private and the government agencies which has resulted in setting up of centres of biotechnology in various countries. This has contributed a great deal for the promotion of in vitro technology which has, in fact, now become a new trend in plant science research.

The list of plant species regenerated from isolated protoplasts has considerably increased (Table 1) and plants can, in fact, now be regenerated from protoplasts of a large number of species belonging to diverse genera, including cereals, which were once thought to be recalcitrants, This has helped to create new protoplast-derived somaclones; especially those of potato, rice and alfalfa have been tested in the field. The refinement in technology has also enabled the successful culture of somatic hybrids/cybrids at interspecific and intergeneric levels. This has far-reaching implications in crop improvement programmes for widening the pools of germplasm. Another point of considerable interest is the storage of protoplasts through immobilization and cryopreservation; both these aspects are of great importance, especially for the pharmaceutical industry.

Earlier, the author reviewed the subject, and pointed out the potentials of protoplast work in agricultural research and crop improvement (Bajaj 1974a, 1977; Bajaj et al. 1988). In this chapter only recent advances in the development and refinement of technology of protoplast isolation and culture are summarized and the literature reviewed.

[1] Former Professor of Tissue Culture, Punjab Agricultural University, Ludhiana, India. Present address: A-137 New Friends Colony, New Delhi 110065, India

Biotechnology in Agriculture and Forestry, Vol. 8
Plant Protoplasts and Genetic Engineering I (ed. by Y. P. S. Bajaj)

Table 1. Some recent references on the culture of protoplasts (for earlier references see Bajaj 1977; Bajaj et al. 1987)

Plant species	Mode of regeneration	Reference
Atropa belladonna	Plants from frozen (−196 °C) protoplasts	Bajaj (1988)
Brassica oleracea	P	Toriyama et al. (1987a)
Cucumis sativus	P	Orczyk and Malepszy (1985); Dons and Bouwer (1986)
Datura innoxia	Shoots from frozen protoplasts	Bajaj (1988)
Foeniculum vulgare	E, P	Miura and Tabata (1986)
Lactuca sativa	P	Ohyama et al. (1986)
Lactuca saligna	P	C. Brown et al. (1987)
Linum usitatissimum	P	Barakat and Cocking (1985)
Lycopersicon esculentum	P	Niedz et al. (1985); Tan et al. (1987a)
Medicago sp.	P	Gilmore et al. (1987)
Nicotiana tabacum	P	Firoozabady (1986)
	Plants from frozen protoplasts	(Bajaj 1988)
Petasites japonicus	P	Yabe et al. (1986)
Solanum lycopersicoides	P	Tan et al. (1987b)
Solanum melongena	P	Gleddie et al. (1986)
Solanum tuberosum	P	Bajaj (1986b); Tavazza and Ancora (1986); de Vries and Tempelaar (1987)

Species	Observations	Reference
Culture of cereal protoplasts		
Oryza sativa	Complete plants from cell suspension-derived protoplasts	Fujimara et al. (1985); Abdullah et al. (1986); Toriyama et al. (1986); Yamada et al. (1986); Coulibaly and Demarly (1986)
	Higher grain yield in field-grown plants	Ogura et al. (1987)
	Rice × pea fusion, occasional division	Bajaj (1983a)
	Rice × soybean fusion, callus formation	Niizeki et al. (1985)
	Rice × carrot cell lines	Sala et al. (1985)
Zea mays	Microcolonics, callus, embryos	Ludwig et al. (1985); Imbrie-Milligan and Hodges (1986); Vasil and Vasil (1987); Rhodes et al. (1988)
Culture of tree protoplasts		
Citrus sp.	E, P	Vardi et al. (1982)
Liriodendron tulipifera	E, P	Merkle and Sommer (1987)
Populus tremula	S, P	Russell and McCown (1988)
Pyrus communis	S, P	Ochatt and Caso (1986)
Santalum album	E, P	Rao and Ozias-Akins (1985)
Ulmus x Pioneer	S	Sticklen et al. (1986)

Table 1 (continued)

Species involved	Reference
Culture of somatic hybrid protoplasts	
Brassica oleracea × *Moricandia arvensis*	Toriyama et al. (1987 a)
Citrus sinensis × *Poncirus trifoliata*	Ohgawara et al. (1985)
Various *Citrus* sp.	Vardi et al. (1987)
Lycopersicon esculentum × *Solanum rickii*	O'Connell and Hanson (1986)
Lycopersicon esculentum × *Solanum lycopersicoides*	Handley et al. (1986)
Oryza sativa × *Echinochloa oryzicola*	Terada et al. (1987)
Sinapis turgida × *Brassica oleracea*	Toriyama et al. (1987 b)
Sinapis turgida × *Brassica nigra*	
Solanum melongena × *Solanum sisymbrifolium*	Gleddie et al. (1986)
Solanum tuberosum × *Solanum brevidens*	Austin et al. (1986)
Solanum tuberosum × *Solanum phureja*	Puite et al. (1986)

E = embryo; S = shoot; P = plant.

The successful culture of protoplasts and their subsequent regeneration into complete plants depends on a number of factors, the important ones being genotype, tissue from which protoplasts are isolated, the physiological conditions under which the plants cultures have been raised, purity of the enzymes and the plasmolyticum, period of incubation, culture media and growth regulators, milieu of protoplasts/plating density, method of culture (liquid/solid), incubation conditions etc. Some of these factors are discussed below.

2 Enzymes

The enzymatic isolation of protoplasts was first reported by Cocking (1960) from tomato seedling root tips by using enzyme preparation (cellulase) from *Myrothecium verrucaria.* This was followed by a period when all types of crude and impure enzyme preparations were tried and hence frequently the results could not be repeated. The easy availability of commercial purified enzymes has now enabled the raising of the percent viability of protoplasts and their subsequent response in cultures. The period of treatment is reduced and thus the deleterious effects on plasmamembrane are avoided. Enzymes such as cellulase, cellulysin, pectinase, macerozyme, driselase, rhozyme, helicase, hemicellulases etc. are commercially available (see Chap I.2, this Vol.).

The concentration of enzyme required depends on a number of factors. For instance, old cell suspensions yield very little or no protoplasts at all, whereas young cultures, especially embryogenic ones, are easily degraded, as has been recently demonstrated in rice (Abdullah et al. 1986). Likewise young in vitro-grown plants (Bajaj 1972), young tissues and explants such as root tips (Xu et al. 1982), hypocotyl (Glimelius 1984), cotyledons (Hammatt et al. 1987) and shoots (Russell and McCown 1986) require low concentrations of enzymes and a relatively short period of treatment as compared to large leaves from old or mature plants.

A manipulation of young tissue, pure enzyme and the reducation in the period of incubation often results in higher viability and plating efficiency.

3 Viability

Methods are now available which enable better, precise and rapid evaluation of viability of protoplasts. In the 1970's most of the workers based their results on staining procedures, which were not accurate, consequently the data were misleading. The merits and demerits of various methods have been dealt with by Gahan (see Chap I.3, this Vol.). The fluorescein diacetate (FDA) method although good, may sometimes yield a false number of counts for viability. Nitroblue tetrazolium can offer a good guide to the early state of damage and can help to distinguish the subpopulations of protoplasts which are viable with the FDA method, but which are irreversibly damaged and would not survive subsequent culturing. Scheurich et al. (1980) judged viability of the protoplasts (intactness of the plasmalemma of the immobilized protoplasts) by repeated studies of the shrinkage and swelling process in response to corresponding changes of the osmotic pressure in the external medium, and also by mesurement of hydraulic conductivity of the protoplast membrane in *Avena.*

The recent application of flow cytometry has yielded useful information (Galbraith et al. 1983; Glimelius 1987), which enables a large number of protoplasts to be screened for viability in a short time (300–1000 s). It has specially facilitated the sorting and enrichment of somatic hybrid protoplasts (Afonso et al. 1985).

4 Agarose Plating

Agarose is a purified agar from which sulphate esters have been removed, and thus the negative charge becomes much lower than in agar. It was first used for the culture of lettuce protoplasts (Engler and Grogan 1982) and since then numerous reports have appeared on its successful application in a variety of plant species (see Chap. I.4, this Vol.). Agarose has certain definite advantages over the traditional agar plating, i.e.

1. The melting temperature of agarose is lower than that of agar and therefore chances of damage to protoplasts are less.
2. Plating efficiency is higher.
3. Protoplasts remain in a fixed position, and are thus easier to observe.
4. The protoplasts do not aggregate much, and thus cloning is facilitated.

Shillito et al. (1983) compared different gelling agents and observed that plating in agarose stimulated colony formation as compared to agar or liquid culture, moreover protoplasts from *Hyoscyamus muticus* formed colonies only on agarose. Likewise, of the several gelling agents tested by Imbrie-Milligan and

Hodges (1986) for the culture of maize protoplasts, only agarose resulted in sustained division and formed microcalli. The plating efficiency of potato protoplasts was also stimulated by low melting temperature agarose (Tavazza and Ancora 1986). Recently Furze et al. (1987) reported that from various types of agar/agarose tested, Sea Plaque agarose was the most effective for hop protoplasts.

Two methods are generally followed for gelling; (1) *Agarose bead type culture* in which the protoplasts are first embedded in agarose beads, and then once the divisions have started, they are transferred to liquid medium. This has been successfully demonstrated for tobacco, crepis and petunia (Shillito et al. 1983). (2) *Agarose disc type culture.* Dons and Bouwer (1986) employed this method for cucumber protoplasts which normally could not be cultured otherwise in liquid or agar media. They plated the agarose-embedded protoplasts in small droplets (discs) surrounded by liquid medium and obtained a plating efficiency of over 50%.

In addition to plating of protoplasts, agarose has been put to a number of miscellaneous uses in research. Dons and Colijn-Hooymans (Chap. I.4, this Vol.) have listed the following applications of agarose:

1. *As an Aid in Fixing Protoplasts for Microinjections.* Lawrence and Davies (1985) first embedded protoplasts in soft agarose (0.3%) and then put them in microdroplets (1 μl) on top of a layer of hard agarose (0.9%). During microinjection the protoplasts were fixed firmly. This technique has also been applied for the microinjection of chromosomes into protoplasts (De Laat and Blaas 1987).
2. *Immobilization of Protoplasts.* For immobilization of protoplasts agarose has been employed as a gel to form beads. The protoplasts of carrot and *Catharanthus* mixed with 3% agarose at 35 °C remained viable and protected against osmotic and mechanical stress (Linsefors and Brodelius 1985). This method may be employed for the storage of protoplasts, as well as in reactor cultures for commercial exploitation.
3. *Facilitation of Cloning and Selection.* Agarose is also likely to facilitate selection and cloning of single protoplasts, somatic hybrid and transformant cells. For instance, Holbrook et al. (1985) reported that *Medicago* microcalli formed after embedding the protoplasts in 0.4% agarose could be separated by melting the agarose at 40 °C. The microcalli remained alive and were plated at three of four times lower density and obtained clones. Potrykus et al. (1985b) used the agarose method for the selection of transformed cells. They transformed tobacco protoplasts and subsequently embedded them in agarose for culturing. The beads were transferred to a kanamycin-containing medium and thus rare transformed cells could be selected efficiently. This method has also been employed for *Triticum monococcum* (Lörz et al. 1983), and can also be extended to the selection of mutants and hybrid cells.

A survey of the literature thus shows that agarose has not only considerably improved the culture ability of the protoplasts, but has been put to a number of other uses in plant protoplast research, specially for the immobilization and selection of transformed cells. In future, agarose is certain to be used more often as a method preferred over agar plating.

5 Regeneration of Plants from Protoplasts

The regeneration of complete plants from isolated protoplasts of herbaceous dicot species is an almost routine exercise now. Moreover, due to concerted efforts, cereals and tree species, which were once considered recalcitrant, have also recently yielded positive results. Likewise fusion products of protoplasts at interspecific and intergeneric level have been reared to form somatic hybrids/cybrids. This subject is discussed in detail in Section II of this Volume. Some recent examples of the culture of protoplasts are given in Table 1 and discussed below.

5.1 Culture of Cereal Protoplasts

The callus cultures especially obtained from the immature embryos show a high degree of differentiation and regeneration of plants in wheat, rice, maize and barley. Embryogenic cell suspensions have proved to be a good source of regenerable protoplasts, the best example being rice. Recently there have been several reports on the regeneration of complete rice plants from isolated protoplasts, from different laboratories (Fujimara et al. 1985; Abdullah et al. 1986; Yamada et al. 1986; Toriyama et al. 1986; Coulibaly and Demarly 1986; Ogura et al. 1987). Although these reports differ in details, one common point is the use of fast-growing young cell suspensions.

Kyozuka et al. (1987) raised cell suspensions established initially from seed-derived callus, and isolated protoplasts from suspension culture 3–5 days after subculture using enzyme solution consisting of 4% cellulase RS, 1% macerozyme R10, and 0.4 M mannitol at pH 5.6. They cultured protoplasts in the presence of nurse cells to induce high frequency of division. The use of agarose-bead-type culture was essential to obtain reproducible colony formation. Shoots as well roots were regenerated from protoplast-derived callus within 3–8 weeks after the transfer of colonies to N6R medium and plants were regenerated from all the cultivars tested, thus showing the dependency of the method. Ogura et al. (1987) studied growth behaviour of protoplast-derived plants of four Japanese cultivars under a submerged paddy field. The plants exhibited somaclonal variation (80% of the plants appearing normal). There were more stems which resulted in more panicles than in respective control plants. The seed fertility was slightly lower, but grain yield was higher.

Recently, by using the above technique, somatic hybrids between *Oryza sativa*×*Echinochloa oryzicola* have been obtained (Terada et al. 1987). Although work on protoplast culture of other cereals has also been done (Table 1), this dependent and highly reproducible plant regeneration paves way for its use in genetic manipulation studies of rice.

5.2 Culture of Tree Protoplasts

As compared to the successful culture of protoplasts of herbaceous species, the reports on the regeneration of shoots/complete plants from protoplasts of tree species are very few, i.e., in *Citrus* spp. (Vardi et al. 1982), *Santalum album* (Rao and Ozias-Akins 1985), *Pyrus communis* (Ochatt and Caso 1986), *Ulmus* (Sticklen et al. 1986), *Populus tremula* (Russell and McCown 1988), and recently in *Liriodendron tulipifera* (Merkle and Sommer 1987) plants were obtained via embryogenesis.

In *Liriodendron*, whereas the thin layer and hanging droplet method of culture failed, low melting point Sea Plaque agarose was successful. About 75% of the protoplasts divided at least once within a week. After the droplets solidified, the plates were floated with liquid regenerating medium. Embryogenesis occurred on a hormone-free medium.

Protoplasts of a number of tree species were isolated from various seedling material, leaves of juvenile and adult materials and callus. Hakman and Fowke (1987) and Russell and McCown (1986) successfully employed in vitro-grown shoots of white spruce (*Picea glauca*) and poplar respectively to obtain leaf protoplasts. In general, suspension cultures have proved better. In *Pinus contorta* (Hakman and von Arnold 1983) and *Pinus taeda* (Teasdale and Rugini 1983), cell suspension in the exponential phase were used. However, recently emphasis has been on the use of embryogenic cell cultures (Gupta and Durzan 1987).

The initial work on *Citrus* protoplast culture (Vardi et al. 1975) led to the formation of trees (Vardi et al. 1982). This has now been done with orange (*Citrus sinensis*), bitter orange (*C. aurantinum*), grapefruit (*C. paradisi*), mandarin (*C. reticulata*) and *Microcitrus* (Vardi et al. 1987b), and trovita oranges (Kobayashi et al. 1985).

The refinement in the techniques for isolation, culture and fusion of *Citrus* protoplasts has resulted in the formation of true somatic hybrids between *Citrus sinensis* × *Poncirus trifoliata* (Ohgawara et al. 1985) and cybrid trees (Vardi et al. 1987a). These studies are of special interest because of the limitations in sexual reproduction in *Citrus*.

Recently there has been increased interest in the genetic transformation of trees through *Agrobacterium*, and work has been reported on trees such as Douglas-fir (Dandekar et al. 1987), poplar (Parsons et al. 1986; Pythoud et al. 1987; Russell and McCown 1988) and loblolly pine (Sederoff et al. 1986), but protoplasts have not been employed. The experiments with microinjection, electroporation, somatic fusion, and *Agrobacterium* could prove to be highly rewarding.

5.3 Culture of Somatic Hybrid Protoplasts

The last decade has witnessed refinements in the techniques of protoplast fusion, and the culture of fused protoplasts. Electrofusion has greatly enhanced the fusion frequencies (Zimmermann and Scheurich 1981) and likewise flow cytometry (Galbraith et al. 1983; Glimelius 1987) has facilitated the isolation of heterokaryons. The computer-manipulated isolated protoplasts, and the culture of individually selected single protoplasts or fused products in micro-quantities of the media is another step likely to eliminate selection procedures (Spangenberg and Schweiger 1986; Schweiger et al. 1987). The list of somatic hybrids/cybrids has considerably increased, some recent examples are cited in Table 1 (for more information see Sect. II, this Vol.).

6 Culture of Haploid Protoplasts

Haploids are of great importance in plant breeding, and they can be obtained from cultured anthers, pollen and cells (Bajaj 1983b). The culture and potentials of haploid protoplasts has been fairly well emphasized (Bajaj 1983c). In addition to various other biotechnological potentials they may help in understanding many intricate problems of somatic cell genetics, especially in the induction of mutations and genetic variability. Perhaps the best use to which haploid protoplasts can be put is the induction of mutations. As most of the mutations are recessive, the haploid system, with one set of chromosomes, makes its detection easier. A number of mutants of various types, auxotrophic, autotrophic, and resistant ones have been induced through cell and protoplast culture (see King 1986). The protoplast system has several similarities with the microorganisms. They can be plated and exposed to mutagens in large numbers (10^6/ml) in a small Petri dish; and each protoplast is a potential plant. The advantage of protoplasts over callus cell suspensions is obvious in that they are genetically more stable.

In earlier experiments, haploid callus cultures were employed for the induction of mutations by Binding et al. (1970) and Carlson (1970). The mutants obtained by Carlson were leaky auxotrophs, and this perhaps was due to the amphidiploid nature (tobacco is not a true haploid) of *N. tabacum.* The auxotrophic cells incorporated lethal amounts of 5-bromodeoxyuridine into their DNA and were thus eliminated by light. In six cell lines growth was observed to increase two to six times on supplementing the medium with biotin, arginine, proline, or lysine. Later, realizing the limitations of the use of callus cells, Carlson (1973a, b) employed haploid protoplasts and obtained methionine sulfoximine-resistant callus. Such calli were diploidized and regenerated into plants. Two out of five plants thus regenerated from the resistant calli contained up to five times the amount of free methionine. Since the toxin produced by the bacterial pathogen (*Pseudomonas tabaci*) that causes the wild-fire disease of tobacco is an analogue of methionine, the resistance may be thus due to over-production of methionine. Later, a number of reports on the use of haploid protoplasts for the induction of mutations of various types appeared. Schieder (1976) irradiated mesophyll protoplasts of *Datura innoxia* with 1 krad of X-rays and obtained ten strains with altered pigments. Valine-resistant plants of tobacco have been obtained from protoplasts subjected to ultraviolet and then plated on a medium containing valine. The valine-induced inhibition of protoplasts was reversed by isoleucine (Bourgin 1976, 1978). Krumbiegel (1979) observed that *Datura* protoplasts treated with NG (10 µg/ml) showed 50% survival, alterations in pigmentations, and a mutation frequency of 2×10^{-4}. Likewise, isolation of temperature-sensitive (Gebhard et al. 1981) and chlorate-resistant (Straus et al. 1981) mutants have been obtained from protoplasts of *Hyoscyamus muticus* treated with N-methyl-N-nitro-N-nitrosoguanidine.

Some studies on the effect of irradiation and chemical mutagens have also been carried out. Galun and Raveh (1975) compared the radiosensitivity of haploid and diploid mesophyll protoplasts of *N. tabacum*, and reported haploid protoplasts with LD_{50} at 0.75 krad (X-irradiation). Later, Aviv and Galun (1977) made an attempt to isolate mutants, and developed a method for the isolation of auxotroph. The procedure was based on the selective killing of dividing cells, and the use of irradiated protoplasts as the feeder layer. Bajaj (1983c) subjected haploid mesophyll protoplasts of tobacco to 0.1 and 0.25% EMS, followed by plating and selection of colonies.

Marton et al. (1982) selected nitrate reductase-deficient (NR^-) cell lines by their resistance to 40 mM chlorate in haploid protoplasts of *N. plumbaginifolia*. The frequency of chlorate resistance which was only 5.8×10^{-5} in the non-mutagenized cultures could be increased up to 25 times by treatment with N-ethyl-N-nitrosourea (NEU) or gamma irradiation. Out of 136 chlorate-resistant clones, 29 were fully deficient in nitrate reductase. The plants were regenerated only in the clones which contained residual NR activity. Thus from the literature it is evident that haploid protoplasts, especially of mesophyll origin, are an excellent tool for the induction of mutations.

By using X-irradiated haploid mesophyll protoplasts from *Petunia hybrida*, Steffen and Schieder (1984) selected four nitrate reductase lines showing resistance to 100 mM chlorate. The complementation analysis via protoplast fusion confirmed that the mutants comprised three non-allelic groups. Likewise, Shimamoto and King (1983) isolated one histidine-requiring clone (PO3) of *Hyoscyamus muticus* by testing 375 surviving colonies in an experiment in which 5-bromodeoxyuridine (BUdR) was applied to cell cultures derived from N-methyl-N-nitro-nitrosoguanidine (MNNG)-treated haploid leaf protoplasts. In subsequent experiments inconsistent effects of BUdR on cell growth were observed and no further auxotrophs were obtained. Many more examples of the use of haploid protoplasts for the induction of mutants are available (see Maliga 1984; Negrutiu et al. 1985; King 1986).

Work needs to be done on pollen protoplasts. No substantial progress has been made since the early work (Bajaj 1974b; Bajaj and Davey 1974; Bajaj et al. 1975). Pollen protoplasts, being natural haploids, would be ideal for DNA uptake and genetic transformation. The foremost potential of haploid protoplasts lies in the early detection of mutants. The most rewarding investigations for plant improvement would be to use haploid systems for the induction of mutants that are resistant to pests, pathotoxins, herbicides, and tolerant to various salts and stresses.

Recently, Chuong et al. (1988) employed fusion between protoplasts from haploid cytoplasmic atrazine resistant (CATR) and haploid cytoplasmic male sterile (CMS) *Brassica napus* plants to produce a diploid (CMS/CATR) cybrid, and discussed the advantages of haploid protoplasts in such studies.

7 Protoplast Culture and Somaclonal Variation

About a decade ago, it was demonstrated that plants regenerated from protoplasts of *Atropa belladonna* showed genetic variability, chromosome number ranging from haploid to polyploids and aneuploids (Bajaj et al. 1978). Since then, numerous reports on the somaclonal variation in protoplast-derived plants of a number of crops such as potato (Shepard et al. 1980; Karp et al. 1982; Sree Ramulu et al. 1983; Creissen and Karp 1985; Bajaj 1986b; Jacobsen 1987), rice (Ogura et al. 1987); alfalfa (Latunde-Dada and Lucas 1983; Johnson et al. 1984), tobacco (Bajaj et al. 1978; Barbier and Dulieu 1983; Prat 1983), eggplant (Gleddie et al. 1986), sugarbeet (Steen et al. 1986), lettuce (Engler and Grogan 1982) etc. have been reported. The work on potato is of great significance and needs special attention.

Most cultivars of potato are tetraploids ($2n = 4\times = 48$) and are vegetatively propagated. The breeding usually involves crossing favourable plants followed by vegetative propagation of F_1 plants to form clones. The most important objectives in potato breeding are to develop resistance to late blight and to viral diseases. There is a great need for the induction of genetic variability and the protoplasts have proved to be a good source (Shepard et al. 1980). Various studies have enabled protoplast-derived somaclones which show a varied degree of variation. Shepard and his coworkers (Shepard et al. 1980; Secor and Shepard 1981) regenerated about 1000 clones from mesophyll protoplasts of cv. Russet Burbank which show extentive variation for tuber shape, size, skin colour, yield etc. Out of 800 somaclones tested, 20 showed some resistance to *Phytophthora infestance*. The frequency of resistant clones was later increased by isolating protoplasts from the selected resistant clones and regenerating a second generation of somaclones (A. R. Ayers and Shepard 1981). Similarly, various types of somaclonal variation have been observed in cv. Fortyfold (Karp et al. 1982), Bintje (sree Ramulu et al. 1983), Majestic (Creissen and Karp 1985), and other cultivars (see Bajaj 1987), and replicated field trials are being conducted (Thomson et al. 1986).

Reviewing the progress of work on regeneration of somaclonal variants in various crops, although there are some obvious demerits, it is likely to be increasingly used as an additional tool by plant breeders. The culture of protoplasts, especially of the fused products, may result in an array of gentic combinations not otherwise available in nature.

8 Immobilization of Protoplasts

Immobilization basically involves the entrapment of cells/organisms in a matrix. Considerable progress has been made in the immobilization of plant cells and tissues (see Hall et al. 1988), and recently the interest has shifted to protoplasts (see Chap. I.5, this Vol.). This is primarily due to great potential in the large-scale production of valuable secondary metabolites, and also for the biosynthesis/biotransformation of new products for the pharmaceutical industry. It has been demonstrated that ageing process of biomembranes of gel-immobilized protoplasts is considerably delayed (Schnabl et al. 1983a; Schnabl and Youngman 1985) but stimulated by low concentrations of ectoxicological substances (Schnabl et al. 1984). These observations may lead to new applications and according to Schnabl and Zimmermann (Chap. I.5, this Vol.) a simple biological detection and screening system for pollutants is being developed for commercial use.

The protoplasts, due to lack of cell wall, are more sensitive and prone to damage, thus they have to be handled rather carefully for the entrapment process and, additionally, mild and rapid immobilization methods have to be used to overcome the problem of fragility. Entrapment is usually done in a polymerase matrix by various methods (Brodelius 1984; Woodward 1985; Hartmeyer 1986), however, for protoplasts gel formation is carried out by ionic cross-linking of a charged polymer such as alginate. It is a block copolymer of D-mannuronate and L-gluronate joined by 1,4-linkages. An aqueous solution of this polysaccharide forms a relatively stable gel in the presence of divalent or polyvalent cations. Schnabl and Zimmermann have given a protocol for the immobilization of protoplasts – the gels are usually made by dripping a suspension of protoplasts in the alginate solution. Small spherical beads of uniform size are preferred, the larger ones can suffer from mass transfer limitations (Kennedy and Cabral 1983). For *Vicia* protoplasts an aliquot of the protoplast suspension is gently mixed with sodium alginate (7%) in 500 µl, spread onto a slide and covered with calcium chlorate (100 mM). After about 5 min, Ca concentration is lowered (5 mM), and the immobilized protoplasts are stored in dark at 4 °C.

The viability of the protoplasts/intactness of the plasma membrane of immobilized protoplasts can be tested by repeated studies of the shrinkage and swelling process in response to corresponding changes of the osmotic pressure in the external medium (Scheurich et al. 1980). Measurement of hydraulic conductivity of the membrane has also been applied in *Avena* protoplasts (Scheurich et al. 1980). In immobilized protoplasts the enzymic activity and pool sizes of the products of lip peroxidation remained unchanged (Schnabl and Youngman 1985), showing thereby that membrane integrity is maintained.

The regeneration of the cell wall in the immobilized protoplasts is a sure test of viability. Recently, Tricoli et al. (1986) successfully cultivated alginate bead-embedded soybean protoplasts in shake cultures containing soybean cell suspension. Other examples have been cited by Dons and Colijn-Hooymans (see Chap. I.4, this Vol.).

In addition to the many uses and implications of immobilized protoplasts in basic and fundamental research for the production of biochemicals etc., one important aspect is the storage and transport of germplasm. The protoplasts of carrot and *Catharanthus* mixed with 3% agarose at 35 °C remained viable and protected against osmotic and mechanical stress (Linsefors and Brodelius 1985). The immobilized protoplasts of guard cells of *Vicia faba* stored for 2–3 weeks in alginate matrix showed no change in morphology, membrane integrity or cellular functions. Similar examples from protoplasts of *Cleviceps* (Komel et al. 1985) and higher plants have been noted (Lörz et al. 1983).

This method has also been employed for the selection of transformed protoplasts (Potrykus et al. 1985 b). The beads were transferred to a medium containing kanamycin and the rare transformants could be efficiently selected. The method can be extended to selection of mutants and somatic hybrid protoplasts. Thus immobilization methods can be employed not only for storage, as in "artificial seed" from somatic embryos (Redenbough et al. 1986), but also for the conservation, transport and exchange of beads containing germplasm at national and international levels.

9 Cryopreservation of Protoplasts

Although much literature has accumulated on the cryopreservation of cell, tissue and organ cultures (Bajaj 1983a, 1984, 1986c) not sufficient information is available on frozen protoplasts. Steponkus (1985) has reviewed the cryobiological aspects of isolated protoplasts including intra-, and intercellular freezing, ice crystal formation, and the possible mechanisms involved. Though for all practical purposes freezing of protoplasts is similar to that of isolated cells and cell suspensions (Bajaj 1976) the protoplasts however differ from the cells in the lack of cell wall and the plasmodesmata, which makes them more sensitive to bursting. Thus they require a more appropriate concentration of osmoticum cum cryoprotectant to prevent them from bursting or causing an undue damage to plasma membrane. It has been suggested (Steponkus and Dowgert 1981) that the nucleation event is primarily the result of mechanical breakdown of the plasma membrane per se rather than a consequence of temperature dependent characteristics of the external ice. With high resolution video cryomicroscopy outward flow of the intracellular solution is observed before intracellular ice formation and thus effecting contact with extracellular ice matrix. Lysis occurs when the protoplasts are osmotically expanding. Infact freezing injury could be due to a number of interdependent factors i.e. intracellular freezing, dehydration, increase in the concentrations of solutes, formation of disulfide bonds, and also due to mechanical stress and rupture of the membrane.

Thorough understanding of the various cryobiological aspects of protoplasts will add to the present knowledge of handling them more efficiently and to increase their survival. Infact prevention of bursting of the protoplasts and the formation of intracellular ice formation and recrystallization are the main points which must be carefully studied and manipulated with the osmoticum, cryoprotectants, and the method and velocity of freezing-thawing (see Bajaj Chap. I.6, this Vol.).

In addition to various uses of cryopreservation of cells, i.e., conservation of germplasm, retention of biosynthetic potential, and the maintenance of genetic stability of clones, the isolated protoplasts offer the following potentials:

1. Protoplasts offer a system in which the effect of freezing on the plasma membrane can be studied directly in situ
2. Freeze-preservation of protoplasts reduces the number of steps in cloning mutant cells (Weber et al. 1983)
3. The cryomiscroscopy of protoplasts allows for a direct analysis of intracellular ice formation, and it also enables the precise control of temperature and cooling rate. With videoimage analysis the temperature at which intracellular nucleation occurs is observed directly (Steponkus and Dowgert 1981; Dowgert and Steponkus 1983).

The work so far done on the culture of frozen protoplasts has demonstrated that the retrieved protoplasts retain their morphogenetic potentials, and entire plants have been obtained. The carrot protoplasts (Takeuchi et al. 1980, 1982) frozen at $-196\,^{\circ}C$ showed totipotency. Later, Weber et al. (1983) observed that partial dehydration of soybean protoplasts frozen in 0.7 M sorbitol without any added cryoprotectant and stored in LN resulted in clones, whereas glucose was unsuitable as a cryoprotectant.

Recently, protoplasts of *Atropa belladonna*, *Datura innoxia* and *Nicotiana tabacum* have been successfully revived (Bajaj 1988). The frozen tobacco and *Datura* protoplasts in cultures after regeneration of the cell wall behaved like their frozen cell suspensions (Bajaj 1976). Tobacco callus eventually differentiated plants. The frequency of regeneration was 36% as compared to 53% control. The *Atropa* protoplasts likewise behaved similarly, but gave the best response in cultures, a maximum of 44% cultures underwent differentiation. The morphogenetic events were not impaired and the cultures showed both patterns of differentiation, i.e., embryogenesis and organogenesis. The plant thus regenerated, when transferred to pots, continued to grow to form normal flowers and fruit. The seeds obtained from these plants were fertile, and in turn showed normal germination.

Likewise, somatic hybrid protoplasts of wheat × pea, and rice × pea frozen in LN in the presence of mannitol and DMSO showed a survival rate varying between 10.8% and 17.9%. Budding of the protoplasts was a common phenomenon, but in one instance the retrieved hybrid cell underwent repeated division (Bajaj 1983a, b).

For better understanding of the mechanism of freezing, isolated protoplasts offer a unique tool because as it is at the cellular membranes that biological integrity appears to be insulted by freezing, it is at the membrane level that biochemical understanding of cryopreservation must be sought, and in this regard isolated protoplasts are an ideal tool. Thus for studies on basic aspects the use of protoplasts as a tool is to be encouraged.

10 Conclusions and Prospects

The list of species in which complete plants can now be obtained has considerably increased during the last decade. Recent studies especially on the use of embryogenic cell cultures for the isolation of protoplasts, the use of agarose, media manipulation, the refinement of technology and culture conditions have helped to increase percentage viability and the regeneration of complete plants, especially in cereals and tree species. Some of the newer techniques such as flow cytometry, electrofusion and electroporation are now ready for successful use for somatic hybridization, somaclonal variation and genetic manipulation, especially through direct gene transfer. Another important point worth mentioning is the storage of protoplasts through immobilization and cryopreservation. The immobilization of protoplasts in agarose beads and their subsequent storage and its use for biotransformation is likely to be taken up by the industry. The greatest impact of the recent advances in protoplast culture work is likely to be felt in the induction of genetic variability for plant improvement.

References

Abdullah R, Cocking EC, Thompson JA (1986) Efficient plant regeneration from rice protoplasts through somatic embryogenesis. Biotechnology 4:1087–1090

Afonso CL, Harkins KR, Thomas-Compton MA, Krejci AE, Galbraith DW (1985) Selection of somatic hybrid plants in *Nicotiana* through fluorescence-activated sorting of protoplasts. Biotechnology 3:811–816

Alexander RG, Cocking EC, Jackson PJ, Jett JH (1985) The characterization and isolation of plant heterokaryons by flow cytometry. Protoplasma 128:52–58

Anderson LK, Stack SM, Fox MH, Chuanshan Z (1985) The relationship between genome size and synaptonemal complex length in higher plants. Exp Cell Res 156:367–378

Ashcroft RG, Preston C, Cleland RE, Critchley C (1986) Flow cytometry of isolated chloroplasts and thyalkaloids. Photobiochem Photobiophys 13:1–14

Austin S, Ehlenfeld MK, Baer MA, Helgeson JP (1986) Somatic hybrids produced by protoplast fusion between *Solanum tuberosum* and *S. brevidens:* phenotypic variation under field conditions. Theor Appl Genet 71:682–690

Aviv D, Galun E (1977) An attempt at isolation of nutritional mutants from cultured tobacco protoplasts. Plant Sci Lett 8:299–304

Ayers AR, Shepard JF (1981) Potato variation. Environ Exp Bot 2:379–381

Ayers NM (1987) Inheritance of organelles in somatic cell hybrids and cybrids of *Nicotiana.* Ph D Thesis, University Nebraska, Lincoln

Baba A, Hasezawa S, Syono K (1986) Cultivation of rice protoplasts and their transformation mediated by *Agrobacterium* spheroplasts. Plant Cell Physiol 27:463–472

Bajaj YPS (1972) Protoplast culture and regeneration of haploid tobacco plants. Am J Bot 59 (6):647

Bajaj YPS (1974a) Potentials of protoplast culture work in agriculture. Euphytica 23:633–649

Bajaj YPS (1974b) Isolation and culture studies on pollen tetrad and pollen mother-cell protoplasts. Plant Sci Lett 3:93–99

Bajaj YPS (1976) Regeneration of plants from cell suspensions frozen at −20 °, −70 °, and −196 °C. Physiol Plant 37:263–268

Bajaj YPS (1977) Protoplast isolation, culture and somatic hybridization. In: Applied and fundamental aspects of plant cell, tissue, and organ culture. Reinert J, Bajaj YPS (eds). Springer, Berlin Heidelberg New York, pp 467–496

Bajaj YPS (1983a) Survival of somatic hybrid protoplasts of wheat x pea and rice x pea subjected to −196 °C. Indian J Exp Biol 21:120–122

Bajaj YPS (1983b) Somatic hybridization and cryopreservation studies on rice x pea and wheat x pea. In: Potrykus I, Harms CT, Hinnen A, Hutter R, King PJ, Shillito RD (eds) Protoplasts 1983. Birkhäuser, Basel, pp 248–249

Bajaj YPS (1983c) Haploid protoplasts. In: Giles KL (ed) Plant protoplasts. Int Rev Cytol, Suppl 16. Academic Press, New York, pp 113–141

Bajaj YPS (1983d) Cryopreservation and international exchange of germplasm. In: Sen SK, Giles KL (eds) Plant cell culture in crop improvement. Plenum, New York, pp 19–41

Bajaj YPS (1983e) In vitro production of haploids. In: Evans DA, Sharp WR, Ammirato PV, Yamada Y (eds) Handbook of plant cell culture, Vol 1. MacMillan, New York, pp 228–287

Bajaj YPS (1984) Cryopreservation of germplasm of crops – Present status and prospects. In: Novak FJ, Havel L, Dolezel J (eds) Plant tissue and cell culture application to crop improvement. Acad Sci, Prague, Czechoslovakia

Bajaj YPS (ed) (1986a) Biotchnology in agriculture and forestry 2. Crops I. Springer, Berlin Heidelberg New York Tokyo

Bajaj YPS (1986b) Cryopreservation of potato somaclones. In: Semal J (ed) Somaclonal variations in crop improvement. Nijhoff, Dordrecht Lancaster, pp 244–250

Bajaj YPS (1986c) In vitro preservation of genetic resources. Int Symp Nuclear techniques and in vitro culture for plant improvement. IAEA Vienna, pp 43–57

Bajaj YPS (1987) Biotechnology in agriculture and forestry 3. Potato. Springer, Berlin Heidelberg New York Tokyo

Bajaj YPS (1988) Regeneration of plants from frozen (−196 °C) protoplasts of *Atropa belladonna*, *Datura innoxia* and *Nicotiana tabacum*. Indian J Exp Biol 26:289–292

Bajaj YPS, Davey MR (1974) The isolation and ultrastructure of pollen protoplasts. In: Linskens HF (ed) Fertilization in higher plants. North Holland, Amsterdam, pp 73–80

Bajaj YPS, Davey MR, Grout BW (1975) Pollen tetrad protoplasts: A model system for the study of ontogeny of pollen. In: Mulcahy DL (ed) Gamete competition in plants and animals. North-Holland, Amsterdam, pp 7–18

Bajaj YPS, Gosch G, Ottma M, Weber A, Gröbler A (1978) Production of polyploid and aneuploid plants from anthers and mesophyll protoplasts of *Atropa belladonna* and *Nicotiana tabacum*. Indian J Exp Biol 16:947–953

Bajaj YPS, Gosal SS, Gill MS (1988) Somatic hybridization – a biotechnological approach to plant breeding. In: Ganguli PN, Mandal AK (eds) Plant breeding. Oxford & IBH Publ. New Delhi

Bajaj YPS, Reinert J (1977) Cryobiology of plant cell cultures and establishment of gene-banks. In: Reinert J, Bajaj YPS (ed) Applied and fundamental aspects of plant cell, tissue, and organ culture. Springer, Berlin Heidelberg New York, pp 757–777

Barakat MN, Cocking EC (1985) An assessment of the cultural capabilities of protoplasts of some wild species of *Linum*. Plant Cell Repts 4:164–167

Barbier M, Dulieu H (1983) Early occurrence of genetic variants in protoplasts cultures. Plant Sci Lett 29:201–206

Bergounioux C, Perennes C, Miege C, Gadal P (1986) The effect of male sterility on protoplast division in *Petunia hybrida*. Cell cycle comparison by flow cytometry. Protoplasma 130:138–144

Binding H, Binding K, Straus J (1970) Selektion in Gewebekulturen mit haploiden Zellen. Naturwiss 57:138–139

Bourgin JP (1976) Valine-induced inhibition of growth of haploid tobacco protoplasts and its reversal by isoleucine. Z Naturforsch 31c:337–338

Bourgin JP (1978) Valine-resistant plants from in vitro selected tobacco cells. Mol Gen Genet 161:225–230

Brodelius P (1984) Immobilization of cultured plant cells and protoplasts. In: Vasil IK (ed) Cell culture and somatic cell genetics of plants vol 1. Academic Press, Orlando, pp 535–546

Brown C, Lucas JA, Power JB (1987) Plant regeneration from protoplasts of a wild lettuce species (*Lactuca saligna* L.). Plant Cell Rep 6:180–182

Brown S, Renaudin JP, Prevot C, Guern J (1984) Flow cytometry and sorting of plant protoplasts: technical problems and physiological results from a study of pH and alkaloids in *Catharanthus roseus*. Physiol Veg 22:541–554

Brown S, Jullien M, Coutos-Thevenot P, Muller Ph, Renaudin JP (1986) Present developments of flow cytometry in plant biology. Biol Cell 58:173–178

Carlberg I, Glimelius K, Eriksson T (1984) Nuclear DNA content during the initiation of callus formation from isolated protoplasts of *Solanum tuberosum* L. Plant Sci Lett 35:225–230

Carlson PS (1970) Induction and isolation of auxotrophic mutants in somatic cultures of *Nicotiana tabacum*. Science 168:487–489

Carlson PS (1973a) The use of protoplasts for genetic research. Proc Natl Acad Sci USA 70:598–602

Carlson PS (1973b) Methionine sulfoximine-resistant mutants of tobacco. Science 180:1366

Chuong PV, Beversdorf WD, Powell AD, Pauls KP (1988) Somatic transfer of cytoplasmic traits in *Brassica napus* L. by haploid protoplast fusion. Mol Gen Genet 211:197–201

Cocking EC (1960) A method for the isolation of plant protoplasts and vacuoles. Nature 187:962–963

Coulibaly MY, Demarly Y (1986) Regeneration of plantlet from protoplasts of rice, *Oryza sativa* L. Z Pflanzenzücht 96:79–81

Creissen GP, Karp A (1985) Karyotopic changes in plants regenerated from protoplasts. Plant Cell Tissue Organ Cult 4:171–182

Dandekar AM, Gupta PK, Durzan DJ, Knauf V (1987) Transformation and foreign gene expression in micropropagated Douglas-fir. Biotechnology 5:587–590

De Laat AMM, Blaas J (1984) Flow cytometric characterization and sorting of plant chromosomes. Theor Appl Genet 67:463–467

De Laat AMM, Blaas J (1987) An improved method for protoplast micro-injection suitable for transfer of entire plant chromosomes. Plant Sci 50:161–169

De Vries SE, Tempelaar MJ (1987) Electrofusion and analysis of potato somatic hybrids. In: Bajaj YPS (ed) Biotechnology in agriculture and forestry 3. Potato. Springer, Berlin Heidelberg New York Tokyo, pp 211–222

Dons JJM, Bouwer R (1986) Improving the culture of cucumber protoplasts by using an agarose-disc procedure. In: Nuclear techniques and in vitro culture for plant improvement. IAEA, Vienna, pp 496–504

Dowgert MF, Steponkus PL (1983) Effect of cold acclimation on intracellular ice formation in isolated protoplasts. Plant Physiol 72:978

Engler DE, Grogan RG (1982) Isolation, culture and regeneration of lettuce leaf mesophyll protoplasts. Plant Sci Lett 28:223–229

Firoozabady E (1986) Rapid plant regeneration from *Nicotiana* mesophyll protoplasts. Plant Sci 46:127–131

Fujimura T, Sakurai M, Akagi H, Negishi T, Hirosa A (1985) Regeneration of rice plants from protoplasts. Plant Tissue Cult Lett 2 (2):74–75

Furze JM, Rhodes MJC, Robins RJ (1987) The use of agarose bead culture for the regeneration of single cell-derived colonies from protoplasts isolated from suspension culture of *Humulus lupulus*. Plant Cell Tissue Organ Cult 8:17–25

Galbraith DW (1984) Selection of somatic hybrid cells by fluorescence activated cell sorting. In: Vasil IK (ed) Cell culture and somatic cell genetics of plants, Vol 1. Academic Press, New York, pp 433–447

Galbraith DW (1989) Flow cytometric analysis and sorting of somatic hybrid and transformed protoplasts. In: Bajaj YPS (ed) Biotechnology in agriculture and forestry 9. Plant protoplasts and genetic engineering II. Springer, Berlin Heidelberg New York Tokyo

Galbraith DW, Shields BA (1982) The effect of inhibitors of cell wall synthesis on tobacco protoplast development. Physiol Plant 55:25–30

Galbraith DW, Harkins KR, Maddox JM, Ayres NM, Sharma DP, Firoozabady E (1983) Rapid flow cytometric analysis of the cell cycle in intact plant tissues. Science 220:1049–1051

Galun E, Raveh D (1975) In vitro culture of tobacco protoplasts: survival of haploid and diploid protoplasts exposed to X-ray radiation at different times after isolation. Rad Bot 15:79–82

Gebhardt C, Schnebli V, King PJ (1981) Isolation of biochemical mutants using haploid mesophyll protoplasts of *Hyoscyamus muticus*. II. Auxotrophic and temperature-sensitive clones. Planta 153:81–89

Giles KL (ed) (1983) Plant protoplasts. Academic Press, New York

Gilmore DM, Davey MR, Cocking EC (1987) Plant regeneration from cotyledon protoplasts of wild *Medicago* species. Plant Sci 48:107–112

Gleddie S, Keller WA, Setterfield G (1986) Production and characterization of somatic hybrids between *Solanum melongena* L. and *Solanum sisymbrifolium* Lam. Theor Appl Genet 71:613–621

Glimelius K (1984) High growth rate and regeneration capacity of hypocotyl protoplasts in some Brassicaceae. Physiol Plant 61:38–44

Glimelius K (1987) Enrichment of somatic hybrids with flow cytometry. IAPTC 52:2–10

Glimelius K, Djupsjobacka M, Fellner-Feldegg H (1986) Selection and enrichment of plant protoplasts heterokaryons of Brassicaceae by flow sorting. Plant Sci 45:133–141

Gupta PK, Durzan DJ (1987) Somatic embryos from protoplasts of loblolly pine proembryonal cells. Biotechnology 5:710–712

Hakman IC, von Arnold S (1983) Isolation and culture of protoplasts from cell suspensions of *Pinus contorta* Dougl. ex Loud. Plant Cell Rep 2:92–94

Hakman IC, Fowke LC (1987) An embryogenic cell suspension culture of *Picea glauca* (white spruce). Plant Cell Rep 6:20–21

Hall RD, Holden MA, Yeoman MM (1988) Immobilization of higher plant cells. In: Bajaj YPS (ed) Biotechnology in agriculture and forestry 4. Medicinal and aromatic plants I. Springer, Berlin Heidelberg New York Tokyo, pp 136–156

Hammatt N, Kim H-I, Davey MR, Nelson RS, Cocking EC (1987) Plant regeneration from cotyledon protoplasts of *Glycine canescens* and *G. glandestina*. Plant Sci 48:129–135

Handley LW, Nickels RS, Cameron MW, Moore PP, Sink KC (1986) Somatic hybrid plants between *Lycopersicon esculentum* and *Solanum lycopersicoides*. Theor Appl Genet 71:691–697

Hartmeyer W (1986) Immobilisierte Biokatalysatoren. Springer, Berlin Heidelberg New York

Herzenberg LA, Sweet RG (1976) Fluorescence activated cell sorting. Sci Am 234:108–117

Holbrook LA, Reich TJ, Iyer VN, Haffner M, Miki BL (1985) Induction of efficient cell division in alfalfa protoplasts. Plant Cell Rep 4:229–232

Imbrie-Milligan CW, Hodges TK (1986) Microcallus formation from maize protoplasts prepared from embryogenic callus. Planta 168:395–401

Jacobsen E (1987) Genetic diversity in protoplast- and cell-derived plants of potato. In: Bajaj YPS (ed) Biotechnology in agriculture and forestry 3. Potato. Springer, Berlin Heidelberg New York Tokyo, pp 358–374

Jia SR, Fu YY, Lin Y (1986) Embryogenesis and plant regeneration from cotyledon protoplast culture of cucumber (*Cucumis sativus* L.). J Plant Physiol 124:393–398

Johnson LB, Stuteville DL, Schlarbaum SE, Skinner DZ (1984) Variation in phenotype and chromosome number in alfalfa protoclones regenerated from nonmutagenized calli. Crop Sci 24:948–951

Karp A, Nelson RS, Thomas E, Bright SWJ (1982) Chromosome variation in protoplast-derived potato plants. Theor Appl Genet 63:265–272

Kennedy JF, Cabral JMS (1983) Immobilised living cells and their applications. Appl Biochem Bioeng 4:189–280

King J (1986) Plant cells and the isolation of conditional lethal mutants. Enzyme Microb Technol 9:514–522

Kobayashi S, Ikeda I, Uchimiya H (1985) Conditions for high frequency embryogenesis from orange (*Citrus sisnensis* Osb.) protoplasts. Plant Cell Tissue Organ Cult 4:249–259

Komel R, Rosman D, Puck A, Socic H (1985) Effect of immobilisation on the stability of *Cleviceps purpurea* protoplasts. Appl Microbiol Biotechnol 23:106–109

Krumbiegel G (1979) Response of haploid and diploid protoplasts of *Datura innoxia* Mill. and *Petunia hybrida* L. to treatment with X-rays and a chemical mutagen. Environ Exp Bot 19:99–103

Kumar A, Cocking EC (1987) Protoplast fusion: A novel approach to organelle genetics in higher plants. Am J Bot 74:1289–1303

Kyozuka J, Hayashi Y, Shimamoto K (1987) High frequency plant regeneration from rice protoplasts by novel culture methods. Mol Gen Genet 206:408–413

Landsmann PJ, Uhrig H (1985) Somaclonal variation in potato detected at the molecular level. Theor Appl Genet 71:500–505

Latunde-Dada AO, Lucas JA (1983) Somaclonal variation and reaction to *Verticillium* wilt in *Medicago sativa* L. plants regenerated from protoplasts. Plant Sci Lett 32:205–211

Lawrence WA, Davies DR (1985) A method for the microinjection and culture of protoplasts at very low densities. Plant Cell Rep 4:33–35

Linsefors L, Brodelius P (1985) Immobilization of plant protoplasts: viability studies. Plant Cell Rep 4:23–27

Lörz H, Larkin PJ, Thomson J, Scowcroft WR (1983) Improved protoplast culture and agarose media. Plant Cell Tissue Organ Cult 2:217–226

Lörz H, Baker B, Schell J (1985) Gene transfer to cereal cell mediated by protoplast transformation. Mol Gen Genet 199:178

Ludwig SR, Somers DA, Peterson WL, Pohlman RF, Zarowitz MA, Gengenbach BG, Messing J (1985) High frequency callus formation from maize protoplasts. Theor Appl Genet 71:344–350

Maliga P (1984) Isolation and characterization of mutants in plant cell culture. Annu Rev Plant Physiol 35:519–542

Marton L, Dung TM, Mendel RR, Maliga P (1982) Nitrate reductase deficient cell lines from haploid protoplast cultures of *Nicotiana plumbaginifolia.* Mol Gen Genet 186:301–304

Meadows MG (1982) Characterization of cells and protoplasts of the B73 maize cell line. Plant Sci Lett 28:337–348

Melamed MR, Mullaney PF, Mendelsohn ML (1979) Flow cytometry and sorting. John Wiley, New York

Merkle SA, Sommer HE (1987) Regeneration of *Liriodendron tulipifera* (family Magnoliaceae) from protoplast culture. Am J Bot 74:1317–1321

Miura Y, Tabata M (1986) Direct somatic embryogenesis from protoplasts of *Foeniculum vulgare.* Plant Cell Rep 5:310–313

Negrutiu I, De Brouwer D, Dirks R, Jacobs M (1985) Mol Gen Genet 199:330–337

Niedz RP, Rutter SM, Handley LW, Sink KC (1985) Plant regeneration from leaf protoplasts of six tomato cultivars. Plant Sci 39:199–204

Niizeki M, Tanaka M, Akada S, Hirai A, Saito K (1985) Callus formation of somatic hybrid of rice and soybean and characteristics of hybrid callus. Jpn J Genet 60:81–92

Ochatt SJ, Caso OH (1986) Shoot regeneration from leaf mesophyll protoplasts of wild pear (*Pyrus communis* var. Pyraster L.). J Plant Physiol 122:243–249

O'Connell MA, Hanson MR (1986) Regeneration of somatic hybrid plants formed between *Lycopersicon esculentum* and *Solanum rickii.* Theor Appl Genet 72:59–65

Ogura H, Kyozuka J, Hayashi Y, Koba T, Shimamoto K (1987) Field performance and cytology of protoplast derived rice (*Oryza sativa*); high yield and low degree of variation of four japonica cultivars. Theor Appl Genet 74:670–676

Ohgawara D, Kobayashi S, Ohgawara E, Uchimiya H, Ishii S (1985) Somatic hybrid plant obtained by protoplast fusion between *Citrus sisnesis* and *Poncirus trifoliata.* Theor Appl Genet 71:1–4

Ohyama K, Enomoto S, Suzuki H (1986) Protoplast regeneration system of lettuce (*Lactuca sativa* L.). In: Sommers DA, Gengenbach GB, Biesboer DD, Hackett WP, Green CE (eds) Proc VI Int Congr Plant Cell and Tissue Culture, Univ Minnesota, p 339

Orczyk W, Malepszy S (1985) In vitro culture of *Cucumis sativus* L. V. Stabilizing effect of glycine on leaf protoplasts. Plant Cell Rep 4:269–273

Parsons TJ, Sinker VP, Stettler RF, Nester EW, Gordon MP (1986) Transformation of poplar by *Agrobacterium tumefaciens.* Biotechnology 4:533–536

Petit P (1986) Flowcytometryanalysisofisolatedplantandanimalmitochondria. Biol Cell 58: suppl 15a

Pinkel D (1981) On the possibility of automated scoring of pollen mutants. Environ Health Perspect 37:133–136

Potrykus I, Saul MW, Petruska J, Paszkowski J, Shillito RD (1985a) Direct gene transfer to cells of a graminaceous monocot. Mol Gen Genet 199:183–188

Potrykus I, Shillito RD, Saul MW, Paszkowski J (1985b) Direct gene transfer; state of the art and future potential. Plant Mol Biol Rep 3:117–128

Prat D (1983) Genetic variability induced in *Nicotiana sylvestris* by protoplast culture. Theor Appl Genet 64:223–230

Puite KJ, Roest S, Pijnacker LP (1986) Somatic hybrid plants after electrofusion of diploid *Solanum tuberosum* and *S. phureja.* Plant Cell Rep 262–265

Pythoud F, Sinker VP, Nester EW, Gordon MP (1987) Increased virulence of *Agrobacterium rhizogenes* conferred by *vir* region of pTiBo542: Application to the genetic engineering of poplar. Biotechnology 5:1323–1327

Rao PS, Ozias-Akins P (1985) Plant regeneration through somatic embryogenesis in protoplast cultures of sandalwood (*Santalum album* L.). Protoplasma 124:80–86

Redenbough K, Paasch BD, Nichol JM, Kossler ME, Viss PR, Walker KA (1986) Somatic seeds: encapsulation of asexual plant embryos. Biotechnology 4:797–801

Rhodes CA, Pierce DA, Mettler IJ, Mascarenhas D, Detmer JL (1988) Genetically transformed maize plants from protoplasts. Science 240:204–207

Russell JA, McCown BH (1986) Culture and regeneration of *Populus* leaf protoplasts isolated from non-seedling tissue. Plant Sci 46:133–142

Russell JA, McCown BH (1988) Recovery of plants from leaf protoplasts of hybrid-poplar and aspen clones. Plant Cell Rep 7:59–62

Sala C, Biasini MG, Morandi C, Nielson B, Parisi B, Sala F (1985) J Plant Physiol 118:409

Scheurich P, Schnabl H, Zimmermann U, Klein JJ (1980) Immobilization and mechanical support of individual protoplasts. Biochim Biophys Acta 598:645–651

Schieder O (1976) Isolation of mutants with altered pigments after irradiating haploid protoplasts from *Datura innoxia* Mill. with X-rays. Mol Gen Genet 149:251–254

Schnabl H, Youngman RJ, Zimmermann U (1983) Maintenance of plant cell membrane integrity and function by the immobilisation of protoplasts in alginate matrices. Planta 158:392–397

Schnabl H, Kunne J, Florschutz K, Youngman RJ (1984) Differential effects of ecotoxicological substances on lipid peroxidation of plant cell protoplasts. Plant Cell Environ 7:499–505

Schnabl H, Youngman RJ (1985) Immobilisation of plant cell protoplasts inhibits enzymic lipid peroxidation. Plant Sci 40:65–69

Schweiger HG, Dirk J, Koop HU, Kranz E, Neuhaus G, Spangenberg G, Wolff D (1987) Individual selection, culture and manipulation of higher plant cells. Theor Appl Genet 73:769–783

Secor GA, Shepard JF (1981) Variability of protoplast-derived potato clones. Crop Sci 21:102–105

Sederoff R, Stomp AM, Chilton WS, Moore LW (1986) Gene transfer into loblolly pine by *Agrobacterium tumefaciens.* Biotechnology 4:647–649

Shepard JF, Bidney F, Shahin E (1980) Potato protoplasts in crop improvement. Science 208:17–24

Shillito RD, Paszkowski J, Potrykus I (1983) Agarose plating and a bead type culture technique enable and stimulate development of protoplast-derived colonies in a number of plant species. Plant Cell Rep 2:244–247

Shimamoto K, King PJ (1983) Isolation of a histidine auxotroph of *Hyoscyamus muticus* during attempts to apply BUdR enrichment. Mol Gen Genet 189:69–72

Spangenberg G, Schweiger H-G (1986) Controlled electrofusion of different types of protoplasts and subprotoplasts including cell reconstruction in *Brassica napus* L. Eur J Cell Biol 41:51–56

Sree-Ramulu K, Dijkhuis P, Roest S (1983) Phenotypic variation and ploidy level of plants regenerated from protoplasts of tetraploid potato (*Solanum tuberosum* L. cv. Bintje). Theor Appl Genet 65:329–338

Sree-Ramulu K, Dijkhuis P, Roest S, Bokelmann GS, De Groot B (1986) Variation in phenotype and chromosome number of plants regenerated from protoplasts of dihaploid and tetraploid potato. Z Pflanzenzücht 97:119–128

Steen P, Keimer B, D'Halluin K, Pedersen HC (1986) Variability in plants of sugar beet (*Beta vulgaris* L.) regenerated from callus, cell-suspension and protoplasts. In: Horn W (ed) Genetic manipulation in plant breeding. De Gruyter, Berlin, pp 633–634

Steffen A, Schieder O (1984) Biochemical and genetical characterization of nitrate reductase deficient mutants of *Petunia.* Plant Cell Rep 3:134–137

Steponkus PL (1985) Cryobiology of isolated protoplasts: Applications to plant cell preservation. In: Kartha KK (ed) Cryopreservation of plant cells and organs. CRC Press, Boca Raton, pp 49–60

Steponkus PL, Dowgert MF (1981) Phenomenology of intracellular ice nucleation in isolated protoplasts. Plant Physiol 67:S-58

Sticklen MB, Lineberger RD, Domir SC (1986) Shoot regeneration from protoplasts of *Ulmus* × Pioneer. Plant Sci 47:29–34

Strauss A, Bucher F, King PJ (1981) Isolation of biochemical mutants using haploid mesophyll protoplasts of *Hyoscyamus muticus.* I. A NO_3^- non-utilizing clone. Planta 153:75–80

Sundberg E, Glimelius K (1986) A method for production of interspecific hybrids within Brassiceae via somatic hybridization using resynthesis of *Brassica napus* as a model. Plant Sci 43:155–162

Tagu D, Bergounioux C, Perennes C, Brown S, Muller Ph, Gadal P (1987) Analysis by flow cytometry of the primary events of *Petunia hybrida* protoplast transformation. Plant Sci (in press)

Takebe I, Labib G, Melchers G (1971) Regeneration of whole plants from isolated mesophyll protoplasts of tobacco. Naturwissenschaften 58:318–320

Takeuchi M, Matasushima H, Sugawara Y (1980) Long-term freeze preservation of protoplasts of carrot and *Marchantia.* Cryo Lett 1:519

Takeuchi M, Matasushima H, Sugawara Y (1982) Totipotency and viability of protoplasts after long-term freeze preservation. In: Fujiwara A (ed) Plant tissue culture 1982. Maruzen, Tokyo, pp 797–798

Tan MMC, Rietveld EM, van Marrewijk GAM, Kool AJ (1987 a) Regeneration of leaf mesophyll protoplasts of tomato cultivars (*L. esculentum*): factors important for efficient protoplast culture and plant regeneration. Plant Cell Rep 6:172–175

Tan MMC, Boerrigter HS, Kool AJ (1987 b) A rapid procedure for plant regeneration from protoplasts isolated from suspension cultures and leaf mesophyll cells of wild *Solanum* species and *Lycopersicon pennellii.* Plant Sci 49:63–72

Tavazza R, Ancora G (1986) Plant regeneration from mesophyll protoplasts in commercial potato cultivars (Primura, Kennebec, Spunta, Desirée). Plant Cell Rep 5:243–246

Teasdale RC, Rugini R (1983) Preparation of viable protoplasts from suspension-cultured loblolly pine (*Pinus taeda*) cells and subsequent regeneration to callus. Plant Cell Tissue Organ Cult 2:253–261

Terada R, Kyozuka J, Nishibayashi S, Shimamoto K (1987) Plantlet regeneration from somatic hybrids of rice (*Oryza sativa* L.) and barnyard grass (*Echinochloa oryzicola* Vasing). Mol Gen Genet 210:39–43

Thomson AJ, Gunn RE, Jellis GJ, Boulton RE, Lacey CND (1986) The evaluation of potato somaclones. In: Semal J (ed) Somaclonal variations in crop improvement. Nijhoff, Dordrecht Lancaster, pp 236–243

Toriyama K, Hinata K, Sasaki T (1986) Haploid and diploid plant regeneration from protoplasts of anther callus in rice. Theor Appl Genet 73:16–19

Toriyama K, Hinata K, Kameya T (1987a) Production of somatic hybrid plants, "Brassicomoricandia", through protoplast fusion between *Moricandia arvensis* and *Brassica oleracea*. Plant Sci 48:123–128

Toriyama K, Kameya T, Hinata K (1987b) Selection of a universal hybridizar in *Sinapsis turgida* Del. and regeneration of plantlets from somatic hybrids with *Brassica* species. Planta 170:308–313

Tricoli DM, Hein MB, Carnes MG (1986) Culture of soybean mesophyll protoplasts in alginate beads. Plant Cell Rep 5:334–337

Vardi A, Spiegel-Roy P, Galun E (1975) *Citrus* cell culture: Isolation of protoplasts, plating densities, effect of mutagens and regeneration of embryos. Plant Sci Lett 4:231–236

Vardi A, Spiegel-Roy P, Galun E (1982) Plant regeneration from *Citrus* protoplasts: variability in methodological requirements among cultivars and species. Theor Appl Genet 62:171–176

Vardi A, Breiman A, Galun E (1987) Citrus cybrids: Production by donor recipient protoplast-fusion and verification by mitochondrial-DNA restriction profiles. Theor Appl Genet 75:51–58

Vasil V, Vasil IK (1987) Formation of callus and somatic embryos from protoplasts of a commercial hybrid of maize (*Zea mays* L.). Theor Appl Genet 73:793–798

Weber G, Roth EJ, Schweiger H-G (1983) Storage of cell suspensions and protoplasts of *Glycine max* (L.) Merr., *Brassica napus* (L.), *Datura innoxia* (Mill), and *Daucus carota* (L.) by freezing. Z Pflanzenphysiol 109:29–39

Woodward J (1985) Immobilized cells and enzymes; a practical approach. IRL Press, Oxford

Xu ZH, Davey MR, Cocking EC (1982) Organogenesis from root protoplasts of the forage legumes *Medicago sativa* and *Trigonella foenum-graecum*. Z Pflanzenphysiol 107:231–235

Yabe K, Nishio T, Takayanagi (1986) Plant regeneration from mesophyll protoplasts of Japanese butterbur (*Petasites japonicus* Fr. Schmit). Jpn J Breed 36:131–137

Yamada Y, Yang ZQ, Tang DT (1986) Plant regeneration from protoplast-derived callus of rice (*Oryza sativa* L.). Plant Cell Rep 5:85–88

Zimmermann U, Scheurich P (1981) High frequency fusion of plant protoplasts by electric fields. Planta 151:26–32

I.2 Enzymes for the Isolation of Protoplasts

S. Ishii[1]

1 Introduction

Cocking (1960) first used enzymes to isolate higher plant protoplasts. He applied culture filtrates of a cellulolytic fungus *Myrothecium verrucaria* to release protoplasts from tomato root tips. Later, Takebe and co-workers succeeded in applying commercially available cell wall-degrading enzymes, Macerozyme and Cellulase Onozuka, to isolate large quantities of protoplasts from a wide variety of plant tissues (Takebe et al. 1968; Otsuki and Takebe 1969). Thereafter, many enzmye preparations have been produced commercially and used successfully for the isolation of plant protoplasts. These enzmyes are divided into three categories: cellulases, hemicellulases and pectinases by their catalytic actions (Table 1). Combined use of these enzymes and development of isolation procedures have made

Table 1. Some cell wall-degrading enzymes in common use for protoplast isolation

Enzyme	Organism	Commercial source
Cellulases		
Cellulase Onozuka R-10	*Trichoderma viride*	Yakult Honsha Co., Tokyo, Japan
Cellulase Onozuka RS	*Trichoderma viride*	Yakult Honsha Co., Tokyo, Japan
Meicelase P-1	*Trichoderma viride*	Meiji Seika Kaisha, Tokyo, Japan
Driselase	*Irpex lacteus*	Kyowa Hakko Kogyo, Tokyo, Japan
Cellulase YC	*Trichoderma viride*	Seishin Pharmaceutical Co., Tokyo, Japan
Cellulysin	*Trichoderma viride*	Calbiochem, San Diego, CA, USA
Cellulase CEL	*Trichoderma viride*	Cooper Biomedical Inc., Malvern, PA, USA
Hemicellulases		
Helicase	*Helix pomatia*	Industrie Biologique Française, Gennevilliers, France
Hemicellulase	*Aspergillus niger*	Sigma Chemical Co. St. Louis, MO, USA
Rhozyme HP150	*Aspergillus niger*	Genencor Inc., South San Francisco, CA, USA
Pectinases		
Macerozyme R-10	*Rhizopus arrhizus*	Yakult Honsha Co., Tokyo, Japan
Pectolyase Y-23	*Aspergillus japonicus*	Seishin Pharmaceutical Co., Tokyo, Japan
Macerase	*Rhizopus arrhizus*	Calbiochem, San Diego, CA, USA
PATE	*Bacillus polymyxa*	Farbwerke-Hoechst AG, Frankfurt, FRG

[1] Research & Development Division, Kikkoman Corporation, 399 Noda, Noda-shi, Chiba-ken 278, Japan

Biotechnology in Agriculture and Forestry, Vol. 8
Plant Protoplasts and Genetic Engineering I (ed. by Y. P. S. Bajaj)

it possible to isolate protoplasts from almost all kinds of plant tissues. However, certain tissues such as pollen cells or pollen tetrad require specific enzyme combinations or special treatment to release protoplasts (Bajaj 1983). A number of reviews have dealt with the methods of isolation, stabilization, and culture of protoplasts in some detail (Cocking 1972; Bajaj 1977; Ruesink 1980; Davey 1983; Evans and Bravo 1983; Eriksson 1985; Bajaj et al. 1988).

Plant protoplasts isolated by enzymes, when cultured under appropriate conditions, regenerate cell walls and some of the cells thus formed have the potential for cell division, callus formation and regeneration of complete plants. In other cases, however, protoplasts lose their viability during the process of enzymic isolation. Although Pectolyase Y-23 and Cellulase CEL are relatively highly purified enzymes, commercial cell wall-degrading enzymes are all crude enzymes and therefore contain contaminants in their preparations. Impurities and other enzyme activities contaminated in commercial crude enzymes were suggested as one of the possible agents that damage protoplasts (Cocking 1972). The removal of low molecular weight impurities from commercial enzymes by gel filtration did enhance protoplast viability in several instances (Slabas et al. 1980; Santos et al. 1980; Patnaik et al. 1981; Arcioni et al. 1982), but such gel filtration alone could not remove high molecular weight contaminants such as non-desirable or detrimental enzymes to protoplasts from the preparations. Cocking (1972) suggested that nucleases, proteases, lipases and peroxidases may reduce protoplast viability. Indeed, Tseng and Mount (1974) observed that purified protease and phospholipase from culture filtrates of *Erwinia carotovora* could burst the protoplasts of cucumber. Thus, it is expected that highly purified cell wall-degrading enzymes may be useful for the isolation of more viable and intact protoplasts. On the other hand, however, several investigators have observed that highly purified enzyme preparations were less efficient for protoplast isolation (Vasil and Vasil 1980; Eriksson 1985).

Protoplasts are formed enzymically from plant tissues in the following steps: maceration or release of single cells from the tissues and the subsequent degradation of cell walls of the released cells. Up to recently it had not been known what kinds of enzymes are actually required for maceration and cell wall degradation of higher plant tissues. The answer to this question was a prerequisite for the utilization of highly purified enzymes for the isolation of plant protoplasts.

2 Cell Wall Compositions

In higher plant cell walls, cellulose microfibrils are embedded in the matrix of polysaccharides. The main components, although a detailed structure of these polysaccharides and their interaction in the cell walls was not yet known, were isolated and characterized (McNeil et al. 1984; York et al. 1986). It was demonstrated that cell wall components of monocots, especially Gramineae, were different from those of dicots (Table 2). In dicot tissues pectic polysaccharides are major cell wall components (~35%) and play an important role in cell wall cementing. Thus, pectinase as well as cellulase preparations were essential for the

Table 2. Comparison of cell wall components between monocots and dicots

Monocots	Dicots
Cellulose	Cellulose
Protein	Protein
Hyp-rich glycoprotein	Hyp-rich glycoprotein
Hemicelluloses	Hemicelluloses
Xyloglucan	Xyloglucan
Glucuronoarabinoxylan	
β-1,3:1,4-mixed glucan	
Pectic polysaccharides	Pectic polysaccharides
	Homogalacturonan
	Rhamnogalacturonan
	Arabinogalactan
Phenolic acids	
Ferulic acid	
Diferulic acid	
p-Coumaric acid	

isolation of protoplasts from dicot tissues. It has been demonstrated in many species, however, that pectic polysaccharides are very minor components (about 5%) in cell walls of Gramineae (Bishop et al. 1958; Wada and Ray 1978; Darvill et al. 1980). This suggested that the components do not play as important a role in cell walls of Gramineae as they do in cell walls of dicots. Indeed, cellulase preparation alone was effective for the isolation of protoplasts from mesophyll tissues of Gramineae (Kanai and Edwards 1973; Okuno and Furusawa 1977; Geballe and Galston 1982). However, the addition of Pectolyase Y-23, a pectinase preparation of *Aspergillus japonicus*, to the cellulase solution improved the yield of protoplasts from Gramineae tissues (Nagata and Ishii 1979; Wittenbach et al. 1982). Furthermore, cell walls of Gramineae contain glucuronoarabinoxylan, β-1,3:1,4 mixed glucan and phenolic acids as characteristic components. Thus, it is still unknown which combination of enzymes is most effective for the isolation of Gramineae protoplasts.

3 Enzymes that are Effective for Isolating Protoplasts from Gramineae Tissues

A cellulase preparation of *Trichoderma viride*, which by itself was able to release protoplasts from oat leaves, was fractionated into several fractions by chromatography on columns of DEAE-Sephadex, CM-Sephadex and SP-Sephadex and gel filtration. Each of the fractions isolated was assayed for protoplast-forming activity individually and in various combinations (Ishii and Mogi 1983a). No fraction by itself was active. It was found that the three enzyme fractions, D-9, P-3,

Table 3. Polysaccharase activity of purified P-3, P-4, and D-9. (Ishii and Mogi 1983a)

Polysaccharide	μ mol glucose eq/mg protein/min		
	P-3	P-4	D-9
Avicel cellulose	0.04	0	0.12
Carboxymethylcellulose	21.88	0	0
Xylan	0.07	16.70	0
Pectin	0	0	0
Arabinan	0	0	0
Galactan	0	0	0
Lichenan	23.67	0	0
Xyloglucan	0.04	0	0
Starch	0	0	0

and P-4 were effective for isolating protoplasts from oat leaves, D-9 and P-4 being essential fractions. These fractions were purified to homogeneity on SDS-polyacrylamide gel electrophoresis. Polysaccharase activities of purified P-3, P-4 and D-9 were determined by measuring the release of reducing sugars from polysaccharide solution (Table 3). It was demonstrated that P-4 could hydrolyze only xylan and is therefore xylanase. Purified D-9 could attack avicel cellulose but not soluble cellulose (carboxymethylcellulose), which suggests that it may be cellulase C_1. The evidence that D-9 is cellulase C_1 was further confirmed by its disintegrating activity of filter paper, especially when D-9 was combined with cellulase Cx. On the other hand, P-3 preferentially attacked soluble cellulose. P-3 also had a strong ability to hydrolyze lichenan, β-1,3:1,4 mixed glucan which is major cell wall components of Gramineae. Anderson and Stone (1975) demonstrated that the mixed glucan is hydrolyzed by β-1,4-glucanase, namely cellulase Cx. Cellulase Cx is also able to hydrolyze xyloglucan (Hayashi et al. 1984). The production of a small amount of reducing sugars from xylan by P-3 may be due to the contamination of cellulose in the substrate. The results indicated that cellulase C_1, cellulase Cx, and xylanase were the main enzyme components in the cellulase preparation which are responsible for the isolation of protoplasts from oat leaves.

To the mixture of cellulase C_1, cellulase Cx, and xylanase were added pectin lyase, polygalacturonase and a maceration-stimulating factor, the main enzyme components of Pectolyase Y-23 (Nagata and Ishii 1979). The addition of pectin lyase but not polygalacturonase was found to stimulate the isolation of protoplasts from oat leaves (Table 4). In as much as pectin lyase is specific for methylgalacturonide linkages, pectic polysaccharides of oat cell wall may have a high degree of methyl esterification. Although pectic polysaccharides are very minor components in cell walls of Gramineae as compared with cell walls of dicots, the result indicates that the components with a high degree of esterification are important constituents in cell walls of oat leaves.

The maceration-stimulating factor, in combination with pectin lyase and polygalacturonase, greatly stimulated the isolation of single cells from dicot

Table 4. Additional effect of pectin lyase (PL), polygalacturonase (PG), and maceration stimulating factor (MSF) on protoplast formation from oat leaves by cellulase C_1 (C_1), cellulase Cx (Cx) and xylanase (X). (Ishii and Mogi 1983a)

Enzymes	Protoplast-forming activity (Absorbance 660 nm)
C_1 + Cx + X	0.710
C_1 + Cx + X + PL	1.155
C_1 + Cx + X + PG	0.831
C_1 + Cx + X + PL + MSF	0.959
C_1 + Cx + X + PG + MSF	0.862
C_1 + Cx + X + PL + PG	1.140

tissues and onion bulbs (Ishii and Kiho 1976) and protoplasts from tobacco leaves (Nagata and Ishii 1979), but it did not stimulate protoplast formation from Gramineae (Table 4). It seems possible that this may be due to the difference in the structure of pectic polysaccharides in cell walls between Gramineae and other tissues.

Cell walls of Gramineae contained β-1,3:1,4 mixed glucan (Nevins et al. 1978; Stinard and Nevins 1980) and phenolic acids such as ferulic acid and p-coumaric acid (Harris and Hartley 1980) as characteristic components. The phenolic acids were bound to arabinosyl residues of glucuronoarabinoxylan through ester linkages (Smith and Hartley 1983). Then we examined the effect of β-glucanase and phenolic-ester hydrolase on protoplast isolation from oat leaves. β-Glucanase that is specific for the hydrolysis of β-1,3:1,4 mixed glucan was purified from a commercial amylase preparation of *Bacillus subtilis*, Novo BAN 120 (Novo Industri AA/S, Copenhagen, Denmark), by the method of Kato and Nevins (1984). Cellulase AP (Amano Pharmaceutical Co., Nagoya, Japan) was used as a starting material for the purification of phenolic-ester hydrolase, as it was the most active among the commercial enzyme preparations tested. The enzyme was purified using ammonium sulfate fractionation and ion exchange and adsorption chromatography (Ishii unpublished data). The addition of β-glucanase and phenolic-ester hydrolase to the reaction mixture consisting of cellulase C_1, xylanase and pectin lyase did not increase the yield of protoplasts from oat leaves. This indicated that neither of the enzymes is necessary for the isolation of protoplasts from grass leaves. This suggested that these components, despite their presence in significant amounts in Gramineae cell walls, may not contribute to cell wall integrity of Gramineae, at least in young tissues.

The mixture of cellulase C_1, cellulase Cx, xylanase, and pectin lyase was also effective in isolation of protoplasts from mesophyll tissues of wheat, barley, and corn (Ishii and Mogi 1983a) and suspension-cultured cells of rice (Ishii unpublished data). These results lead to the conclusion that cellulase C_1, cellulase Cx, xylanase and pectin lyase were primary enzymes for the isolation of protoplasts from Gramineae tissues.

Table 5. Comparison of enzymes between oat and petunia leaves that are effective for isolating protoplasts. (Ishii and Mogi 1983b)

Purified enzymes[a]	Protoplast formation (A 660 nm)	
	Oat leaves	Petunia leaves
$C_1+Cx+X+PL$	1.008	1.397
$-C_1$	0	0.185[b]
$-Cx$	0.331	0.705[c]
$-X$	0.240	1.365
$-PL$	0.737	0

[a] Cellulase C_1 (C_1), cellulase Cx (Cx), xylanase (X), and pectin lyase (PL).
[b] Single cells.
[c] Single cells and protoplasts.

4 Comparison of Enzymes Between Monocots and Dicots that are Effective for Isolating Protoplasts

A marked difference between oat (monocot) and petunia (dicot) was found in the requirement of pectin lyase for the isolation of protoplasts from these tissues (Ishii and Mogi 1983b). If this enzyme was removed from the reaction mixture, no protoplasts were formed from petunia leaves, while considerable amounts of protoplasts were released from oat leaves (Table 5). This indicated that pectinase was essential for the isolation of protoplasts from petunia leaves. Pectinases such as polygalacturonase, pectate lyase, and pectin lyase macerate dicot tissues (Chesson 1980). Liliaceae and Araceae, both belonging to the monocots, were similar in tissue maceration to dicots (Ishii 1976; unpublished data). The effectiveness of each pectinase on tissue maceration primarily depended on the degree of methylesterification of pectic polysaccharides, cell wall-cementing materials of dicot and some monocot tissues (Ishii 1976).

When xylanase was removed from the reaction mixture, the yield of protoplasts from oat leaves decreased. The yield from petunia leaves, however, was not affected by this removal. Xylanase was not an essential, but a quite significant enzyme for the isolation of protoplasts from oat leaves. Ishii (1984) demonstrated that both pectin lyase and xylanase could macerate grass leaves. A mixture of the two enzymes brought about more rapid isolation of single cells than did either of the two enzymes alone, indicating a synergistic effect. In dicots and Liliaceae tissues there was no synergistic effect between pectin lyase and polygalacturonase in tissue maceration (Ishii 1976), but the maceration of these tissues by pectinases was greatly stimulated by the addition of an enzyme-like stimulating factor produced by *Aspergillus japonicus* (Ishii and Kiho 1976).

Cellulase C_1 was an essential enzyme for the degradation of cell walls of both tissues (Table 5). Omission of cellulase C_1 resulted in the formation of single cells only. In the case of oat leaves, more vigorous shaking was necessary for the

Plant Tissues →	**Single Cells** → Maceration	**Protoplasts** Cell Wall Degradation
Dicots **Monocots** Liliaceae Araceae	Polygalacturonase Pectate Lyase Pectin Lyase Stimulating Factor	Cellulase C_1 Cellulase Cx
Monocots Gramineae	Xylanase Pectin Lyase	Cellulase C_1 Cellulase Cx

Fig. 1. Comparison between monocots and dicots of enzymes that are effective for maceration and cell wall degradation

release of single cells from the tissues (Ishii 1984). Cellulase Cx, however, was not essential, but facilitated the degradation of cell walls of both tissues. This was in contrast to the degradation of highly crystallized celluloses, in which both cellulase C_1 and cellulase Cx were essential for their degradation (Wood and McCrae 1978). Figure 1 summarizes enzymes that are effective for maceration and cell wall degradation among plant species.

5 Purification Procedures of Enzymes

Highly purified preparations of cellulase C_1, cellulase Cx and xylanase were obtained from Cellulase YC (Seishin Pharmaceutical Co.) and pectin lyase from Pectolyase Y-23 (Seishin Pharmaceutical Co.) by the following procedures. All procedures were carried out at 4 °C.

Pectolyase Y-23 (10 g) was dissolved in 200 ml of 0.1 M sodium acetate buffer (pH 5). The enzyme solution was treated with DEAE-Sephadex batchwise. About 25 g (wet weight) of DEAE-Sephadex A-50 equilibrated with 0.1 M sodium acetate buffer (pH 5) was added to the solution and the mixture was stirred for 30 min. The mixture was filtered and the residue was repeatedly washed with 0.1 M sodium acetate buffer (pH 5). The treated solution (about 250 ml) was adjusted to pH 4.5 with 1 N acetic acid. The solution was then subjected to chromatography on a SP-Sephadex C-50 column (2.5×50 cm) equilibrated beforehand with 0.1 M sodium acetate buffer (pH 4.5). After the column was washed with the same buffer, elution was carried out with a linear gradient of 1 l each of 0 and 0.34 M NaCl in the buffer. Pectin lyase was eluted between 0.11 and 0.13 M NaCl, while polygalacturonase was eluted at 0.25 M NaCl concentration (Ishii and Yokotsuka 1975). Active fractions were collected, dialyzed against distilled water, and concentrated using collodion bag (Sartorius GmbH, Göttingen, W Germany). The preparation of pectin lyase was highly purified and free from most other enzyme activities, but not homogeneous on disc electrophoresis. If further purification was needed, the preparation was subjected to a second SP-Sephadex column chromatography and gel filtration (Ishii and Yokotsuka 1975).

Cellulase C_1, cellulase Cx and xylanase were isolated and purified from Cellulase YC. Cellulase YC (100 g) was dissolved in 1 l of water and pH of the solution was adjusted to 5.5 with 1 N HCl. To the solution was added 420 g of ammonium sulfate with continuous stirring. After allowing to stand overnight, the precipitate was collected by centrifugation and dissolved in 100 ml of water. After the removal of undissolved materials by centrifugation, the solution (about 150 ml) was applied to a Sephadex G-25 column (2.8×100 cm) to remove ammonium sulfate. The protein fractions eluted at void volumes were collected and lyophilized. The lyophilized powder (about 10 g) was dissolved in 50 ml of 0.1 M sodium acetate buffer (pH 5). The solution was then subjected to chromatography on a DEAE-Sephadex column (2.5×50 cm) equilibrated beforehand with 0.1 M sodium acetate buffer (pH 5). After the column was washed with 300 ml of the same buffer, the elution was carried out with a linear gradient of 1 l each of 0 and 0.6 M NaCl in the same buffer. Non-adsorbed fraction (D-1) and D-9 which was eluted between 0.13 and 0.18 M NaCl as a major protein peak were obtained (Ishii and Mogi 1983a). Purified cellulase C_1 was obtained from fraction D-9 by DEAE-Sephadex column chromatography and Bio-gel P-200 gel filtration. Fraction D-1 was concentrated to 2 ml by ultrafiltration, and then applied to a Bio-gel P-200 column (1.9×100 cm). Cellulase Cx was purified from fraction P-3 (third protein peak in the order of molecular weight) by SP-Sephadex column chromatography and Bio-gel P-200 gel filtration. Purified xylanase was obtained from fraction P-4 (fourth protein peak) by CM-Sephadex column chromatography and Bio-gel P-200 gel filtration.

These purified enzyme preparations were concentrated and dialyzed against 0.5 M mannitol. They could be stored for at least 1 year in the frozen state without apparent loss in activity.

6 Protoplasts Isolated by Purified Enzymes

Usually purified enzymes were used as cellulase C_1 (0.6 mg/ml), cellulase Cx (40 μg/ml), xylanase (33 μg/ml), and pectin lyase (7.4 μg/ml). Sufficient amounts of protoplasts were isolated from petunia leaves only by the combination of cellulase C_1 and pectin lyase (Table 6). Although the same enzyme composition

Table 6. Isolation of protoplasts from petunia and oat leaves by the combination of purified enzymes

Leaf tissues	Purified enzymes[a]	Number of protoplasts from 1 g fresh wt of leaves
Petunia	C_1+PL	3.5×10^6
Oat	C_1+PL	0.4×10^6
	C_1+PL+X	1.7×10^6

[a] Cellulase C_1 (C_1), pectin lyase (PL), and xylanase (X).

Table. 7. Comparison of viability between protoplasts isolated by purified enzymes and by crude enzymes

Experiment	Plating efficiency (%)	
	Purified enzymes	Crude enzymes
I	30 ± 4	24 ± 4
II	48 ± 6	33 ± 6
III	26 ± 2	16 ± 2
IV	36 ± 2	25 ± 3

released protoplasts from oat leaves, further addition of xylanase was required for preparing large amounts of protoplasts from the tissues.

Petunia leaf protoplasts isolated by purified enzymes were more viable than those isolated by crude enzymes (Cellulase YC + Pectolyase Y-23). When these protoplasts were cultured in MS medium for a week, the frequency of cell division of the protoplasts isolated by purified enzymes were 10–15% higher than that of the protoplasts isolated by crude enzymes (Table 7). Oat leaf protoplasts isolated by purified enzymes, although they regenerated cell walls within a few days, did not divide under various conditions tested.

7 Summary and Conclusion

Cell walls of higher plants are composed of several polysaccharides and other constituents. Little is known about the precise function of these components in cell wall integrity, except for cellulose microfibrils. However, only the combination of pectin lyase and cellulase C_1 was found to be sufficient for the formation of protoplasts from petunia leaves, while further addition of xylanase was necessary for preparing large amounts of protoplasts from oat leaves. Although there are many differences in cell wall components between monocots and dicots, the principal difference between the two in enzymic isolation of protoplasts was observed in tissue maceration. In dicot and some monocot tissues only pectinases such as pectin lyase, pectate lyase and polygalacturonase are macerating enzymes. On the other hand, both pectin lyase and xylanase can macerate Gramineae tissues with a synergistic effect.

Highly purified enzymes which are available for protoplast isolation were easily obtained from commercial cell wall-degrading enzymes. Intact and more viable protoplasts prepared by these purified enzymes may be useful for the basic research of plant physiology (Pilet 1985) and for genetic manipulation of higher plants (Power et al. 1986) using protoplasts as experimental tools. The use of highly purified enzymes to obtain viable protoplasts is not always essential, however. In numerous cases plating efficiency higher than 90% could be obtained in protoplasts isolated by unpurified enzymes (Potrykus and Shillito 1986). Furthermore, even in most recalcitrant Gramineae protoplasts, recent developments have made it possible to regenerate whole plants from rice protoplasts isolated by unpurified commercial enzymes (Fujimura et al. 1985; Abdullah et al. 1986; Toriyama et al. 1986; Yamada et al. 1986; Kyozuka et al. 1987).

References

Abdullah R, Cocking EC, Thompson JA (1986) Efficient plant regeneration from rice protoplasts through somatic embryogenesis. Biotechnology 4:1087–1090

Anderson MA, Stone BA (1975) A new substrate for investigating the specificity of β-glucan hydrolases. FEBS Lett 52:202–207

Arcioni S, Davey MR, Santos AVP dos, Cocking EC (1982) Somatic embryogenesis in tissues from mesophyll and cell suspension protoplasts of *Medicago coerulea* and *M. glutinosa*. Z Pflanzenphysiol 106:105–110

Bajaj YPS (1977) Protoplasts isolation, culture and somatic hybridization. In: Reinert J, Bajaj YPS (eds) Applied and fundamental aspects of plant cell, tissue, and organ culture. Springer, Berlin Heidelberg New York, pp 467–577

Bajaj YPS (1983) Haploid protoplasts. In: Giles KL (ed) Plant Protoplasts. Academic Press, New York, pp 113–141

Bajaj YPS, Gosal SS, Gill MS (1988) Somatic hybridization – a biotechnological approach to plant breeding. In: Ganguli PN, Mandal AK (eds) Plant Breeding, Oxford + IBH Publ, New Delhi

Bishop CT, Bayley ST, Setterfield G (1958) Chemical constitution of the primary cell walls of *Avena* coleoptiles. Plant Physiol 33:283–289

Chesson A (1980) Maceration in relation to the post-harvest handling and processing of plant material. J Appl Bacteriol 48:1–45

Cocking EC (1960) A method for the isolation of plant protoplasts and vacuoles. Nature 187:962–963

Cocking EC (1972) Plant cell protoplasts. Isolation and development. Annu Rev Plant Physiol 23:29–50

Darvill A, McNeil M, Albersheim P, Delmer DP (1980) The primary cell walls of flowering plants. In: Tolbert NE (ed) The biochemistry of plants. Vol. 1. Academic Press, New York, pp 91–162

Davey MR (1983) Recent developments in the culture and regeneration of plant protoplasts. Experientia 46 (suppl.):19–29

Eriksson TR (1985) Protoplast isolation and culture. In: Fowke LC, Constabel F (eds) Plant protoplasts. CRC Press, Boca Raton, pp 1–20

Evans DA, Bravo JE (1983) Protoplast isolation and culture. In: Evans DA, Sharp WR, Ammirato PV, Yamada Y (eds) Handbook of plant cell culture. Vol. 1. Macmillian, New York, pp 124–176

Fujimura T, Sakurai M, Akagi H, Negishi T, Hirose A (1985) Regeneration of rice plants from protoplasts. Plant Tissue Cult Lett 2:74–75

Geballe GT, Galston AW (1982) Wound-induced resistance to cellulase in oat leaves. Plant Physiol 70:781–787

Harris PJ, Hartley RD (1980) Phenolic constituents of the cell walls of monocotyledons. Biochem Syst Ecol 8:153–160

Hayashi T, Wong YS, Maclachlan G (1984) Pea xyloglucan and cellulose. Hydrolysis by pea endo-1,4-β-glucanases. Plant Physiol 75:605–610

Ishii S (1976) Enzymic maceration of plant tissues by endo-pectin lyase and endopolygalacturonase from *Aspergillus japonicus*. Phytopathology 66:281–289

Ishii S (1984) Cell wall-cementing materials of grass leaves. Plant Physiol 76:959–961

Ishii S, Kiho K (1976) Evidence for a factor that stimulates tissue maceration by pectolytic enzymes. Phytopathology 66:1077–1081

Ishii S, Mogi Y (1983a) Identification of enzymes that are effective for isolating protoplasts from grass leaves. Plant Physiol 72:641–644

Ishii S, Mogi Y (1983b) Isolation and identification of enzymes that are responsible for the isolation of plant protoplasts. In: Potrykus I, Harms CT, Hinnen A, Hütter R, King PJ, Shillito RD (eds) Protoplasts 1983. Birkhäuser, Basel, pp 6–7

Ishii S, Yokotsuka T (1975) Purification and properties of pectin lyase from *Aspergillus japonicus*. Agric Biol Chem 39:313–321

Kanai R, Edwards GE (1973) Separation of mesophyll protoplasts and bundle sheath cells from maize leaves for photosynthetic studies. Plant Physiol 51:1133–1137

Kato Y, Nevins DJ (1984) Enzymic dissociation of *Zea* shoot cell wall polysaccharides. Dissociation of $(1\rightarrow3),(1\rightarrow4)$-$\beta$-D-glucan by purified $(1\rightarrow3),(1\rightarrow4)$-$\beta$-D-glucan 4-glucanohydrolase from *Bacillus subtilis*. Plant Physiol 75:745–752

Kyozuka J, Hayashi Y, Shimamoto K (1987) High frequency plant regeneration from rice protoplasts by novel nurse culture methods. Mol Gen Genet 206:408–413

McNeil M, Darvill AG, Fry SC, Albersheim P (1984) Structure and function of the primary cell walls of plants. Annu Rev Biochem 53:625–663

Nagata T, Ishii S (1979) A rapid method for isolation of mesophyll protoplasts. Can J Bot 57:1820–1823

Nevins DJ, Yamamoto R, Huber DJ (1978) Cell wall β-D-glucans of five grass species. Phytochemistry 17:1503–1505

Okuno T, Furusawa I (1977) A simple method for the isolation of intact mesophyll protoplasts from cereal plants. Plant Cell Physiol 18:1357–1362

Otsuki Y, Takebe I (1969) Isolation of intact mesophyll cells and their protoplasts from higher plants. Plant Cell Physiol 10:917–921

Patnaik G, Wilson D, Cocking EC (1981) Importance of enzyme purification for increased plating efficiency and plant regeneration from single protoplasts of *Petunia parodii*. Z Pflanzenphysiol 102:199–205

Pilet PE (1985) Introduction. The use of plant protoplasts in physiological research. In: Pilet PE (ed) The physiological properties of plant protoplasts. Springer, Berlin Heidelberg New York Tokyo, pp 1–5

Potrykus I, Shillito RD (1986) Protoplasts isolation, culture, plant regeneration. Methods Enzymol 118:549–578

Power JB, Davey MR, Freeman JP, Mulligan BJ, Cocking EC (1986) Fusion and transformation of plant protoplasts. Methods Enzymol 118:579–594

Ruesink A (1980) Protoplasts of plant cells. Methods Enzymol 69(C):69–84

Santos AVP dos, Outka DE, Cocking EC, Davey MR (1980) Organogenesis and somatic embryogenesis in tissues derived from leaf protoplasts and leaf explants of *Medicago sativa*. Z Pflanzenphysiol 99:261–270

Slabas AR, Powell AJ, Lloyd CW (1980) An improved procedure for the isolation and purification of protoplasts from carrot suspension culture. Planta 147:283–286

Smith MM, Hartley RD (1983) Occurrence and nature of ferulic acid substitution of cell-wall polysaccharides in graminaceous plants. Carbohydr Res 118:65–80

Stinard PS, Nevins DJ (1980) Distribution of noncellulosic β-D-glucans in grasses and other monocots. Phytochemistry 19:1467–1468

Takebe I, Otsuki Y, Aoki S (1968) Isolation of tobacco mesophyll cells in intact and active state. Plant Cell Physiol 9:115–124

Toriyama K, Hinata K, Sasaki T (1986) Haploid and diploid plant regeneration from protoplasts of anther callus in rice. Theor Appl Genet 73:16–19

Tseng TC, Mount MS (1974) Toxicity of endo-polygalacturonate transeliminase, phosphatidase and protease to potato and cucumber tissue. Phytopathology 64:229–236

Vasil IK, Vasil V (1980) Isolation and culture of protoplasts. In: Vasil IK (ed) Perspectives in plant cell and tissue culture. Academic Press, New York, pp 1–19

Wada S, Ray PM (1978) Matrix polysaccharides of oat coleoptile cell walls. Phytochemistry 17:923–931

Wittenbach VA, Lin W, Hebert RR (1982) Vacuolar localization of proteases and degradation of chloroplasts in mesophyll protoplasts from senescing primary wheat leaves. Plant Physiol 69:98–102

Wood TM, McCrae SI (1978) The cellulase of *Trichoderma koningii*. Purification and properties of some endoglucanase components with special references to their action on cellulose when acting alone and in synergism with the cellobiohydrolase. Biochem J 171:61–72

Yamada Y, Yang ZQ, Tang DT (1986) Plant regeneration from protoplast-derived callus of rice (*Oryza sativa* L.). Plant Cell Rep 5:85–88

York WS, Darvill AG, McNeil M, Stevenson TT, Albersheim P (1986) Isolation and characterization of plant cell walls and cell wall components. Methods Enzymol 118:3–40

I.3 Viability of Plant Protoplasts

P. B. GAHAN[1]

1 Introduction

In the context of work with plant protoplasts, viability may be defined as the capacity of an isolated protoplast to continue to grow in culture, to replace its lost cell wall and to form either suspension cells, callus or plantlets. However, for many practical situations, protoplasts need to be checked for viability at an early stage prior to use in a biochemical, genetical or cytological study. In spite of this, most workers often report neither on the viability levels of the protoplasts used in their experiments nor how long the protoplasts are left between preparation and use in an experiment (vide several articles in Pilet 1985 a). The process of removing the cell wall by enzyme digestion followed by washes and centrifugation of the isolated protoplasts must render them considerably changed from their stage in the cell. In addition, it is likely that the plasmalemma will be damaged or very modified from that of the intact cell. It is difficult to say precisely how long after isolation it takes for the plasmalemma to recover, since we are unable to know all of its features when a cell wall is present. However, it is clear that a proportion of the protoplasts in the preparation immediately after isolation will not survive in culture, this proportion depending upon the type of material employed as the protoplast source. Since the production of protoplasts became routine, a number of tests have been employed to check protoplast viability at an early stage after preparation, some having been used in viability tests for almost a century, others being developed for current needs. Unfortunately, the use of the classical test of reversible plasmolysis is not possible with protoplasts, although cyclosis does offer a quick and simple test for viability.

The following text has three aims. Firstly, to provide a list of methods available for testing protoplast viability and their mechanisms of action, secondly to comment on the efficiency of each method, and finally to try to suggest a relatively simple test system for reliable viability prediction.

2 Methodology

The methods currently used in different laboratories fall into four main categories, namely markers of change in the plasmalemma, the vacuolar system,

[1] Department of Biology, Plant Cell & Moleculaire Biology Group, King's College, London, Campden Hill Road, London W8 7AH, UK

Biotechnology in Agriculture and Forestry, Vol. 8
Plant Protoplasts and Genetic Engineering I (ed. by Y. P. S. Bajaj)

organelles and cell walls. In turn it is important to cross-reference those markers to check to the validity of their use – either independently or in tandem, and to understand the overall biology so described. These methods are basically cytochemical or cytophysical and can be applied either to single protoplasts or to large populations of protoplasts by automated cytological methods (Brown 1984; Gahan and McLellan 1985). A number of these methods depend upon the use of vital dyes, many of which are toxic to plant cells in high concentration. However, they are molecules which easily enter cells and are non-toxic to them in the very low concentrations employed and, on modification by the cell, colour either selective cell inclusions or organelles or the whole cell (Baker 1958).

Such dyes have been used with plant cells for over a century (Pfeffer 1886; Baker 1958; Gahan 1981, 1984a). Only a few of the very broad spectrum of such dyes are in use currently, and these include Janus green B, Evans blue, trypan blue, neutral red, brilliant cresyl blue and the fluorochromes tetracycline, chlorotetracycline, fluorescein diacetate, phenosafranin and 3,4-benzpyrene.

A number of workers have tried to use ethidium bromide, acridine orange or 4,6-diamino-2-phenylindole (DAPI) (Bornman and Bornman 1985), but these are essentially markers of RNA and DNA when used under precise conditions (Paolillo 1964; McCully 1976; James and Jope 1978; Gahan 1984a), and are not very effective as viability test molecules.

2.1 Plasmalemma-Dependent Tests

Dye Exclusion Methods. These can indicate the viability of the protoplast by virtue of its intactness and primarily normal physiological function. Thus, one test relies on the ability of the plasmalemma to block the entry of dye molecules, i.e. the dye exclusion test. This is normally performed with Evans blue [which differs from the more classically used trypan blue by a change in the position of the sulphate groups (Conn 1969)]. If the protoplasts are exposed to a 1% solution of Evans blue for 5 min, the dead or dying protoplasts will have stained blue, whilst those with an intact outer membrane will remain unstained (Gaff and Okang'o-Ogola 1971). Percentage viability can then be assessed by counting.

A second dye exclusion test employs the fluorochrome phenosafranin which is used as a 1% solution in 0.05 M phosphate buffer at pH 5.8 (Widholm 1972). On examining the protoplasts by fluorescence microscopy, only dead or damaged protoplasts will fluoresce.

Membrane Enzyme. One of the most popular viability tests depends upon plasmalemma integrity and an associated esterase activity. Fluorescein diacetate (FDA) is the substrate employed by mixing the protoplast with an equal volume of 0.1% FDA in 0.05 M phosphate buffer at pH 5.8 for 5–15 min (by dilution from a stock solution of 5 mg/ml in acetone) (Heslop-Harrison and Heslop-Harrison 1970; Widholm 1972). The esterase cleaves the acetate from the FDA as it passes into the protoplast, and on irradiation with ultraviolet or blue light the fluorescein fluoresces, unlike the FDA. The reaction product is very soluble and can freely move out of the protoplast so that only a protoplast with the intact enzymes-plasmalemma combination can ensure a surfeit of fluorescein in the protoplast – and hence maintain the high levels of fluorescence. Thus, in contrast to the dye exclusion methods, only viable protoplasts will fluoresce. Some workers prefer to add ethidium bromide (EB) to the FDA solution, the EB actively staining the protoplasm of dead or damaged protoplasts which fluoresce red when excited by blue or green light.

Membrane Lipids. Normally, the unfixed membrane is not very responsive to lipid stains, due apparently to the way in which proteins are dispersed within the membrane creating "a masking" effect. However, membrane changes result in exposure of larger clusters of lipid molecules which are amenable to lipid staining. Three methods found to be helpful in this respect are 3,4-benzpyrene [3,4-BP] (Berg 1951) a lipophilic fluorochrome, Sudan black B, a general lipid stain (Gahan 1984a), and the acid haematein reaction for phospholipids (Baker 1946; Gahan 1965) and choline-containing phospolipids in particular (Adams and Bayliss 1975). Of these, the most useful is, perhaps, 3,4-BP (Fig. 1) whilst there can be problems in the survival of the protoplast during the rather long acid haematein procedure.

Ca-Calmodulin-Binding Sites. The fact that chlorotetracycline (CTC) fluoresces when excited by ultraviolet light permitted Chandler and Williams (1978a, b) and Wolniak et al. (1980) to develop a cytochemical method for the detection of Ca-calmodulin-binding sites on membranes (Fig. 1). Protoplasts should be immersed for 20 min in an equal volume of unbuffered 3.5% glucose solution containing 200 μM CTC and examined by blue light fluorescence microscopy.

Zeta Potential Measurements. A micro-electrophoresis apparatus (Rank Brothers, Cambridge, UK) has been used successfully for the determination of the zeta potential of protoplasts (Grout et al. 1972; Senn and Pilet 1981; Pilet et al. 1984, 1985a; Gahan 1984b; Onyia et al. 1984), which measurement is theoretically close to the surface potential of the protoplast. Thus, the surface charge on the plasmalemma of a protoplast can be determined by electrophoresis when the rate of migration of the protoplast is related to the net negative surface charge. The slower the rate of migration, the less negatively charged is the protoplast surface. The zeta potential is calculated by using the Smoluchowski equation (Nagata and Melchers 1978)

$$\text{zeta potential} = \frac{4\pi\eta \mathrm{m}}{\varepsilon} \mathrm{V} ,$$

where U = electrophoretic mobility (velocity/V/cm)
η = viscosity in centipoise and
ε = the dielectric constant.

Similar studies have been made with laboratory-built electrophoresis equipment (Iverson 1985).

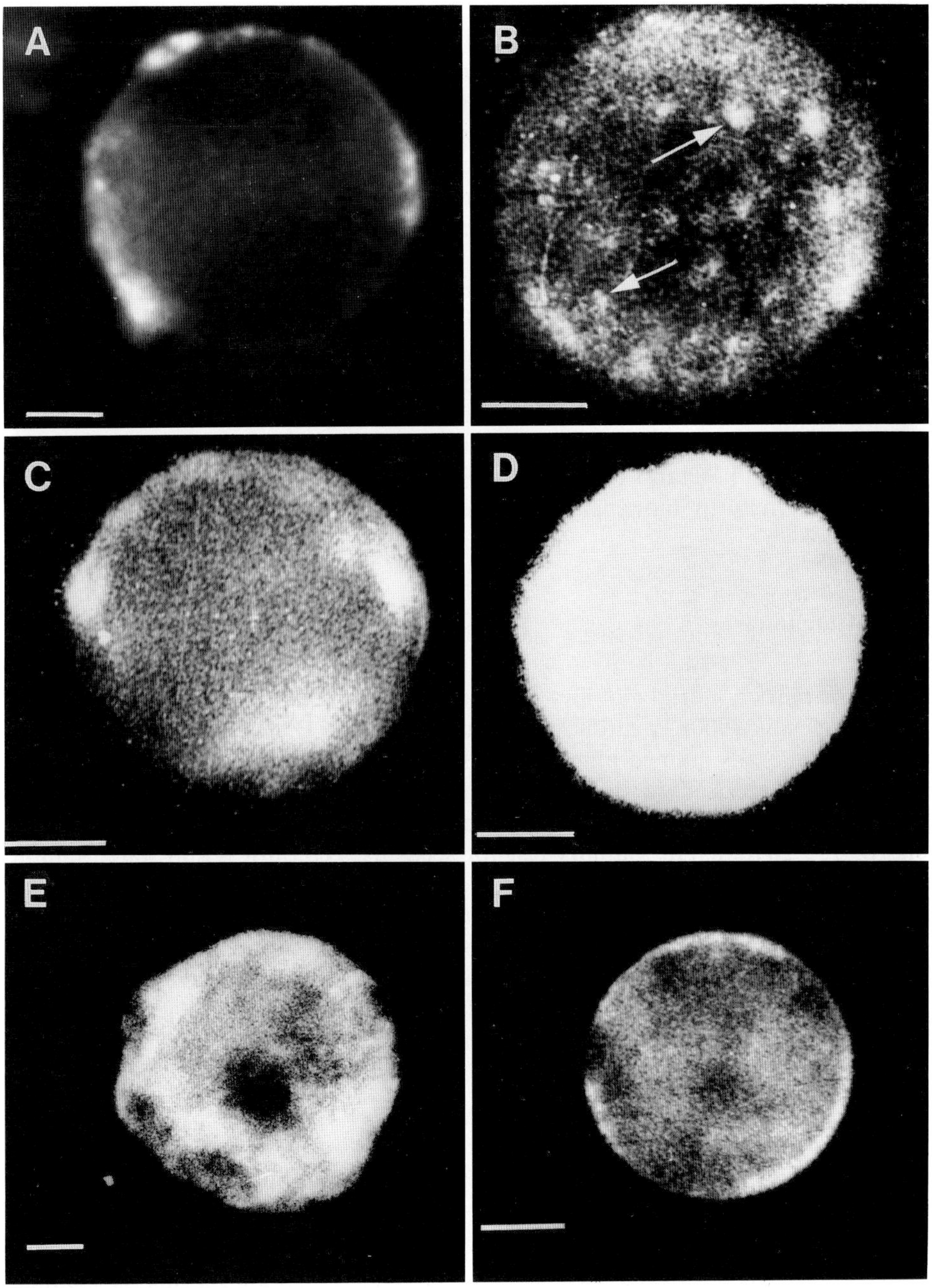

Fig. 1A–F. Protoplasts of meristem cells from tubers of *Dioscorea rotundata* Poir. treated with either 3,4-benzpyrene or chlorotetracycline after various chilling regimes to show the increase in binding of the fluorochromes with increasing chilling injury. Bar = 10 μm. **A** Healthy protoplast with low CTC fluorescence. **B** Protoplast held at 3 °C for 2 h – CTC fluorescence. **C** Protoplast held at 3 °C for 3 h – CTC fluorescence. **D** Protoplast frozen-and-thawed; high non-specific fluorescence with CTC. **E** Protoplast frozen-and thawed; high 3,4-BP fluorescence. **F** Protoplast held at 3 °C for 2 h – 3,4-BP fluorescence

2.2 Vacuolar System

One assay for the intactness of the vacuole concerns the vital dye, neutral red (Guilliermond 1941). If protoplasts are bathed in a 0.1% 20 mM phosphate buffer at pH 7.5 and containing 600 mM potassium nitrate and 1 mM calcium chloride (Basham and Bateman 1975), the dye molecules will traverse the intact plasmalemma and be concentrated in the vacuoles. Dead or damaged cells will stain completely. Another vital dye, brilliant cresyl blue, can be employed similarly as a 0.005% solution in Schoen's medium. However, a word of caution is necessary (see later), for Robards (1970) found that cells in which the plasmalemma was damaged still accumulated neutral red in the vacuoles, the tonoplast being apparently tougher (thicker) than the plasmalemma. Although as yet untested on protoplasts, a new vital dye with specificity for the Golgi apparatus was recently described for use with animal cultured cells, and employs the fluorescent lipid C-NBD-ceramide-(NL7-[4-nitrobenzo-2-oxa-1,3-diazole]-6-aminocaproyl sphingosine) (Lipsky and Pagano 1985). Staining patterns are upset by interference with the Golgi activity by monensin and colcemid.

2.3 Mitochondria

A test for mitochondrial viability used Janus green B as a 0.001% aqueous solution (Michaelis 1900; Lazarow and Cooperstein 1953). Dead or damaged protoplasts are totally stained blue-green. However, when the dye enters viable protoplasts it is reduced to its colourless form to be reoxidized by the electron pathway of the mitochondria, so rendering them blue-green in colour after reacting for 15–20 min. The viability test for seeds uses colourless and water soluble triphenyltetrazolium chloride which is reduced to a coloured, water-insoluble formazan in viable embryos of maize, wheat, barley, oats and rye (Lakon 1939, 1942a, b). This may be extended for use with protoplasts. Thus, Onyia and Gahan (1985) reported that during an examination of chilling damage in tubers of *Dioscorea rotundata* there was an increase in some dehydrogenase activities as adjudged by quantitative enzyme cytochemical assay (Gahan 1984a). The test used is a two-step reaction:

Substrate + NAD(P)	$\xrightarrow{A}$	Product + NAD(P)H	$\xrightarrow{B}$	Product + NAD(P)
+ tetrazolium salt		+ tetrazolium salt		+ formazan
(colourless)		(colourless)		(coloured)
I		II		III

where A = a dehydrogenase, and B = a tetrazole reductase. If the substrate was omitted from reaction medium I, nevertheless, when using nitroblue tetrazolium (NBT), a formazan was still produced presumably through the presence of endogenous substrates (Fig. 1; Table 1).

Whilst this reaction was marked, it was less than that seen when a dehydrogenase substrate was added (see later). However, both of these changes occurred in advance of any other signs of chilling damage, and in consequence the reaction

Table 1. NADP-tetrazolium reductase activity in yam tuber meristem cells either after chilling for 9 days at 0 °C and returned to 20 °C for 1 – 4 days, or kept for 9 days plus 1 – 4 days at 20 °C

Treatment	10 of days at 20 °C			
	1	2	3	4
After 9 days at 0 °C	0.28 ± 0.05 [a] (15)	0.29 ± 0.05 (24)	0.40 ± 0.07 (14)	0.36 ± 0.05 (20)
After 9 days at 20 °C	0.21 ± 0.04 (15)	0.19 ± 0.03 (27)	0.24 ± 0.05 (21)	0.24 ± 0.04 (20)

[a] Mean absolute integrated optical density units per μm of reacting cytoplasm + standard deviation (vide Gahan 1984a for details) with numbers of tubers in brackets below.

$$\mathrm{NAD(P)} + \mathrm{NBT} \xrightarrow{A} \mathrm{NAD(P)H} + \mathrm{NBT} \xrightarrow{B} \mathrm{NAD(P)} + \text{formazan},$$

where A involves endogenous substrates and dehydrogenases and B the tissue-bound, tetrazolium-reductase system is a good indication of irreversible damage to the protoplasts. There is no reaction in the absence of added NAD(P). The tetrazolium reductase system may reside in part in the mitochondria and in part with the endoplasmic reticulum, where it may be a useful marker of endoplasmic reticulum-bound cytochrome P_{450} activity.

Studies on damage in animal cells have shown that glutamate dehydrogenase may be readily lost from mitochondria. However, studies on chilling injury in yam tubers showed that it was not possible to detect a loss of this enzyme from mitochondria, but that the level of glutamate dehydrogenase activity was increased with damage (Onyia and Gahan 1985; Onyia 1984) (Fig. 2).

Two more recent approaches may also prove useful in studying the integrity of mitochondria. They employ fluorescent dye molecules to identify both functional mitochondria and viable protoplasts, and they depend upon the high transmembrane potential maintained by intact mitochondria. One such probe is merocyanine 540, the success of the method depending upon a potential-dependent partitioning of charged lipophilic molecules across biological membranes, the mitochondria appearing as fluorescent structures on irradiation with blue light (Johnson et al. 1981). The second type of molecules are the permeant cationic fluorescent probes, e.g. rhodamine 123, which are selectively accumulated by mitochondria in viable protoplasts. This mitochondria-specific interaction with such molecules is apparently dependent upon the high trans-membrane potential (inside is negative) maintained by functional mitochondria. The reaction may be controlled by the use of ionophores or inhibitors of electron transport which eliminate the selective mitochondrial association of those compounds (Johnson et al. 1981).

2.4 Chloroplasts

Photosynthetic fluorescence was considered as a means of assessing the possible damage to chloroplasts (Bornman and Bornman 1985). Variable fluorescence of

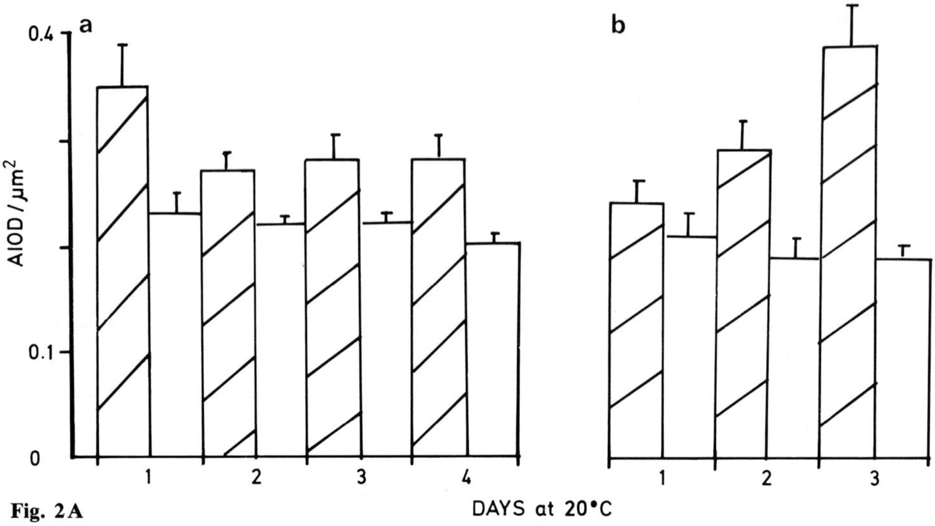

Fig. 2A

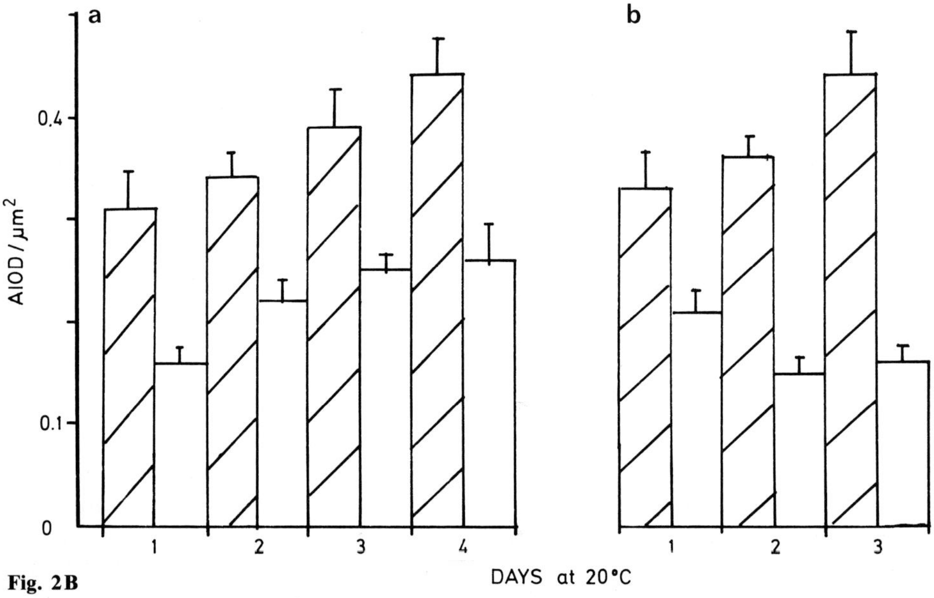

Fig. 2B

Fig. 2A–C. Quantitative tetrazolium reaction with meristem cells from tubers of *Dioscorea rotundata* Poir. Control tubers (*open*) maintained at 18 °C and those chilled at 3 °C (*hatched*) for either 4 days (**a**) of 7 days (**b**) are compared. Rates of activity determined by integrating microdensitometry and expressed as absolute integrated optical density units per μm^2 of reacting surface (AIOD/μm^2) (Gahan 1984a). **A** NAD(P)-tetrazolium reductase activity. **B** Glucose-6-phosphate dehydrogenase activity. **C** Glutamate dehydrogenase activity

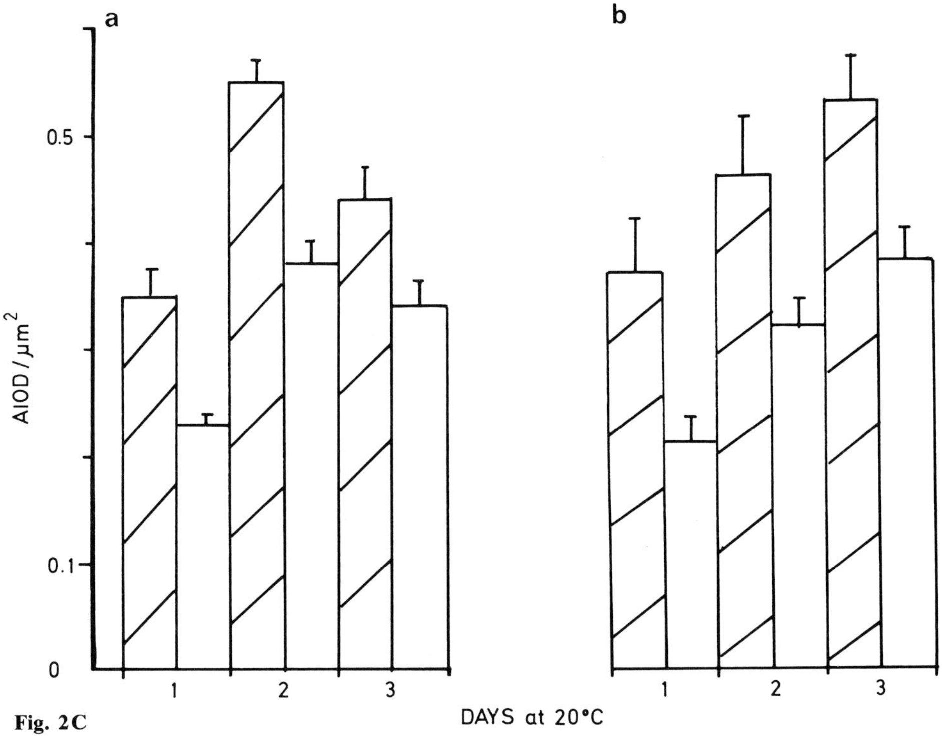

Fig. 2C

dark-adapted chloroplasts were measured in a modified Aminco-Bowman spectrofluorimeter and calculated according to

$$F_v = F_{max} - F_0 ,$$

where F_v = variable
F_{max} = maximum
F_0 = steady state of initial fluorescence (Bornman et al. 1985).

The time taken for the system to reach $F_0 + 0.75\ F_v$ was plotted against total time in order to check for damage to the photosynthetic system. Results with protoplasts prepared from *Brassica napus* showed that whilst FDA indicated a high percentage of viable protoplasts during the first 16 h with only a 9% decrease from time 0 h, the fluorescence rise-time measurements showed a deterioration of 17% by 16 h in the relatively functional photosynthetic ability.

2.5 Cell Walls

These are involved for two reasons, namely (a), the production of a new cell wall is indicative of a viable protoplast, but (b) there is a need to check that old cell wall material has been lost completely after enzyme hydrolysis. Brighteners have proved useful in that they can be employed to detect the first of the newly synthesized cell wall components. In particular calcofluor M2R new (Hughes and McCully 1975), calcofluor white (Hahne et al. 1983) or photine HV (Tampion et al. 1973) have proved useful. They may be added to the culture medium at a concentration of 10 μg/ml and fluoresce when excited by ultraviolet light.

The absence of a cell wall usually means that a variety of ions and molecules will be lost from the protoplasts and it is possible to check for such losses. Thus, leakage of ions can be assessed by measuring the level of conductivity in the medium in which the protoplasts have been placed (Dexter et al. 1932). Enzyme losses can also be monitored, e.g. acetylesterases and carboxylesterases, two cell wall enzymes which can be found in the protoplast culture medium during the first 18–24 h after the isolation of protoplasts from *Daucus carota* suspension cells. This is achieved by reacting with the substrate naphthol AS-D acetate in the

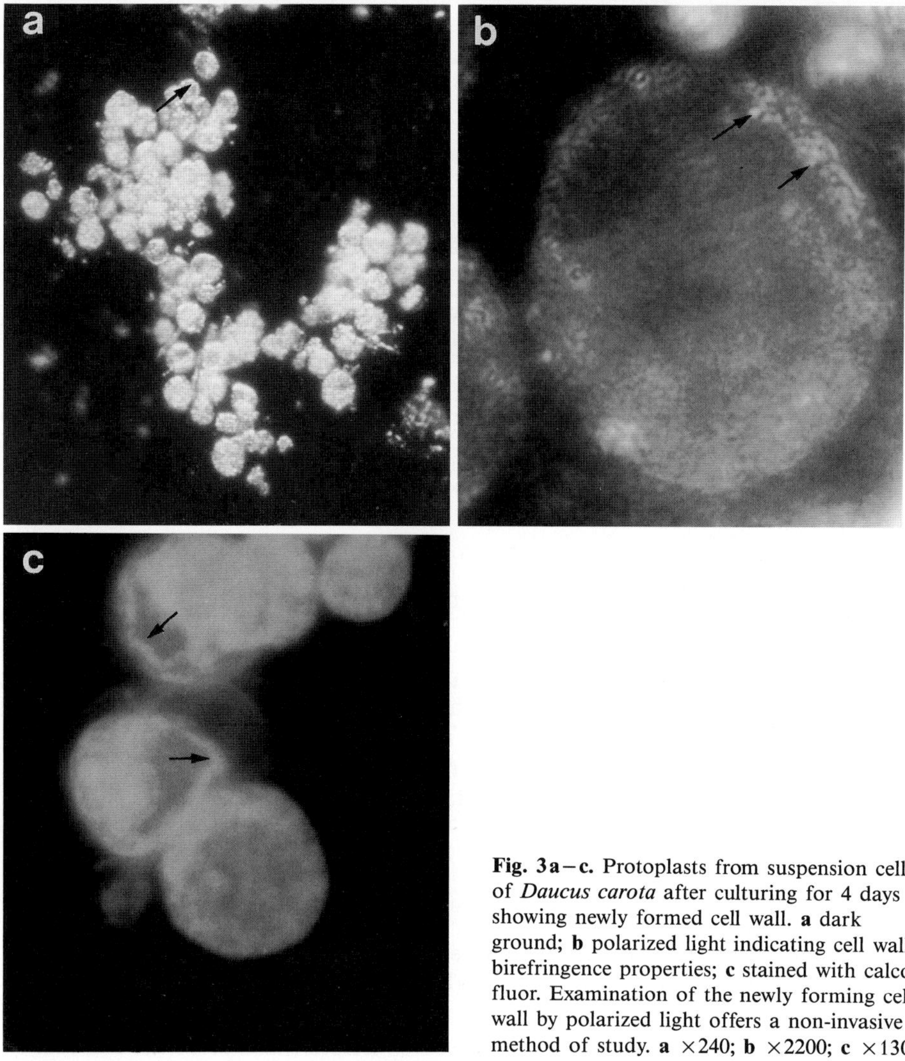

Fig. 3a–c. Protoplasts from suspension cells of *Daucus carota* after culturing for 4 days showing newly formed cell wall. **a** dark ground; **b** polarized light indicating cell wall birefringence properties; **c** stained with calcofluor. Examination of the newly forming cell wall by polarized light offers a non-invasive method of study. **a** ×240; **b** ×2200; **c** ×1300

presence of fast blue BB when a blue, water-insoluble end product is formed (Gahan 1984a; Pearce, O'Brien, Gahan and Morre, in preparation). Cessation of this leakage would indicate that the plasmalemma is no longer "permeable" and that cell wall components are being formed as assessed by polarized light microscopy (Fig. 3) (Gahan 1984a) and staining with calcofluor white (Fig. 3).

Protoplasts will also secrete polysaccharides into the extra-cellular milieu (Hanke and Northcote 1974; Takeuchi and Komamine 1978; Blaschek et al. 1981) during the early stages of wall formation when much of the cell wall polysaccharides disintegrate from the surface of the cells. As with the esterases above, the amount of extra-cellular polysaccharides decreases when compared to the amount of regenerated wall of, e.g., tobacco protoplasts (Blaschek et al. 1981).

It is possible to exploit these two examples of secretion to test for protoplast viability in the sense that the protoplasts are synthesizing and exporting these molecules to the external milieu which they would not be expected to do if dead or dying.

3 Efficacy of the Methods

The methods discussed above are considered because they may enable us to define when a protoplast is or is not viable. Most researchers who have published work on protoplasts have tended to rely on just one of these methods. Essentially we are trying to find out if the protoplasts are in the state before or after the point-of-no-return (P) (Fig. 4).

However, we need to be sure on two counts, namely, that (a) all of the tests for viability are equally good, and (b) the protoplasts under investigation represent a homogeneous population. Two examples may serve to demonstrate that it is not necessarily wise to rely on a single test or on the homogeneity of the protoplast population.

The first example concerns a study of the effects of chilling injury on protoplasts prepared from the meristem cells of the tubers of *Dioscorea rotundata* (Onyia et al. 1984). In this experiment, the viability of the protoplasts was studied

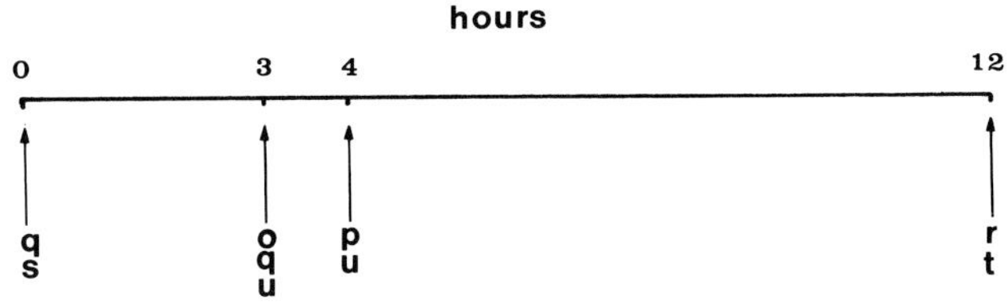

Fig. 4. Identification of point of no return in protoplasts from tuber of *Dioscorea rotundata* Poir. chilled at 3 °C. *o* maximum fluorescence with 3,4-benzpyrene; *p* maximum fluorescence with chlorotetracycline; *q* protoplasts viable by neutral red and Evans blue methods; *r* all protoplasts non-viable by tests in q; *s* maximum mean electrophoretic velocity (5.06 ± 0.13 µm/s); *t* minimum mean electrophoretic velocity (3.16 ± 0.21 µm/s); *u* mean velocity of protoplasts chilled for 3 h at 3 °C (4.13 ± 0.14 µm/s)

Table 2. Response of protoplasts prepared from meristematic tissue of tubers of *Dioscorea rotundata* Poir. exposed to 3 °C for varying periods of time. Point of no return appears to fall between 3 and 4 h. (After Gahan 1984b)

Time (h)	Zeta potential (μm/s)	3,4-Benzpyrene	Chlorotetracycline	Neutral red	Evans blue
0	5.06 ± 0.13	Little[a]	Weak[a]	Viable	Viable
3	4.13 ± 0.14	Maximum[a]	Increased[a]	Viable	Viable
4	4.13 ± 0.14	Reduced[a]	Maximum[a]	Viable	Viable
12	3.16 ± 0.21	Little[a]	Maximum[a]	Non-viable	Non-viable

[a] Fluorescence.

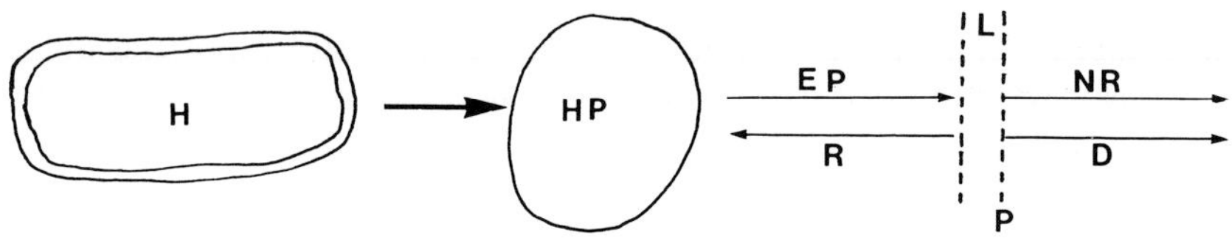

Fig. 5. Relationship between recoverable (*R*) and non-recoverable (*NR*) damage to protoplasts due to environmental factors (*EP*). A number of protoplasts may appear to be viable with the FDA test and yet be sufficiently close to the point of no return (*P*) and to loss of compartmentation (*L*) as to be non-viable (*D*). Healthy cell (*H*) and healthy protoplast (*HP*)

by the use of neutral red, Evans blue, 3,4-benzpyrene, chlorotetracycline and zeta potential (Table 2; Fig. 5).

These data may be interpreted to show that between 3 h and 4 h of chilling at 3 °C the protoplasts have reached a point of no return with respect to damage at that temperature, i.e., prior to this point the damage is reversible but after that point it is not (Fig. 5). The increase of 3,4-BP staining from virtually none at 0 h to marked staining at 3 h may be due to a re-arrangement of the lipid and protein molecules within the membranes to yield larger clusters of phospholipids from within the membranes which stain readily. However, the subsequent reduction in lipid staining after 3 h is due most likely to a real loss of phospholipids, as has been described for other tissues (Yamaki and Uritani 1972, 1973a, 1974; Wright and Simon 1973). This is accompanied by a reduction in the surface net negative charge as evidenced by zeta potential measurements. The binding of calcium/calmodulin to the plasmalemma also increases by 3–4 h as adjudged by CTC binding, so further contributing to the reduced negative charge on the plasmalemma. In spite of these results, neutral red and Evans blue tests continue to demonstrate the viability of these protoplasts, these tests not showing loss of viability until much later, so raising doubts as to the efficiency of these two tests of viability under all circumstances. This is in keeping with the observation of Robards (1970) that the vacuole can continue to accumulate neutral red after the plasmalemma has been damaged. Thus, only a partial breakdown of compartmentation has occurred in these protoplasts, yet this would appear to mean irreversible damage to the protoplast.

The second experiment questions essentially the homogeneity of protoplast populations. Since leaf mesophyll cells are supposed to form a homogeneous pop-

Fig. 6A–C. Protoplasts from mesophyll cells of *Lycopersicon esculentum* chilled at 2 °C for 24 h or 96 h and compared to those retained at 25 °C. Analysis using FDA and flow cytometry to indicate both viability levels and population heterogeneity. Zones in each graph represent the field of scatter of records of individual protoplasts. Graph **A** 25 °C; Graph **B** 24 h at 2 °C; Graph **C** 96 h at 2 °C. Plot of number of protoplasts (y axis) versus intensity of fluorescence (x axis)

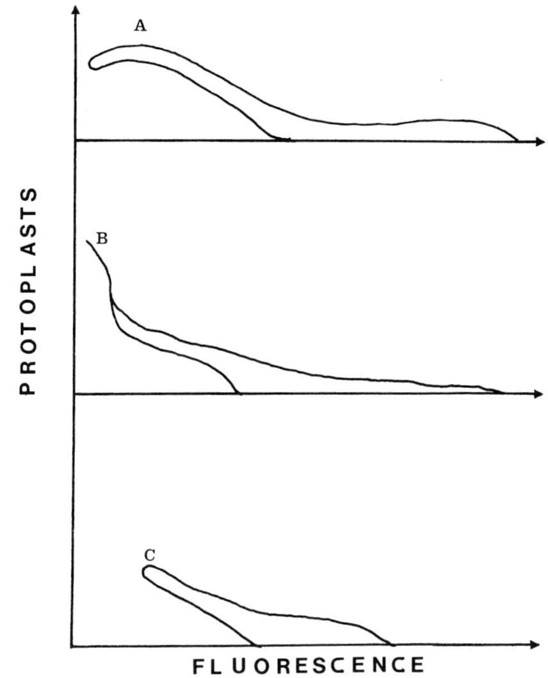

Table 3. Response (% viability) to FDA and 3,4-benzpyrene reactions by leaf mesophyll protoplasts of *Lycopersicon esculentum* on chilling at 2 °C for various periods of time (having been cultured for 6 h prior to being used) (McLellan and Gahan in preparation)

Chilling time (h)	0	12	24	48	72	96
FDA [a]	87	82	64	47	26	6
FDA [b]	81	67	38	–	19	9
3,4-BP	11	14	31	39	79	86

[a] Measured immediately after chilling.
[b] Measured after 8 h at 25 °C between chilling and assessment.

ulation of cells (Mayer 1985), it seems to follow that the protoplast population obtained from these mesophyll cells will also be homogeneous.

In consequence, data are presented from an experiment in which protoplasts prepared from leaf mesophyll cells of a chill-sensitive cultivar of *Lycopersicon esculentum* (Gahan and McLellan 1985; McLellan and Gahan 1986; McLellan and Gahan in preparation; Table 2; Fig. 6). Populations of about 5000 protoplasts were examined for each timing (Table 3) using a Leitz MPV3 flow cytometer. Thus, on examining protoplasts by the FDA method 6 h after preparation, 87% were indicated as viable (time 0 h in Table 3). However, examination of the plot of number of protoplasts versus amount of fluorescence does not show the expected single peak of fluorescence intensity for the viable protoplasts, but a continuous curve showing decreasing numbers of protoplasts with increasing intensity of fluorescence. This behaviour was seen also for protoplasts chilled at 2 °C for 24 h and having a viability count of 64% when a population of weakly fluorescing protoplasts shows almost a peak, the rest being spread over a broad range of high levels of fluorescence. Finally, after 96 h at 2 °C, only 6% of the protoplasts show fluroescence and these are essentially in the higher fluorescence population. A similar situation occurred when reacting the protoplasts with 3,4-BP (Table 3), although on this occasion the response was the inverse of that seen with FDA, as would be expected.

This unequal reaction pattern within a population of leaf mesophyll protoplasts demonstrates a striking heterogeneity of staining behaviour both in the control (0 h) and the chilled samples, and so may not represent the homogeneity implied by the statement of Mayer (1985). In this context, the capacity of only part of the total population of mesophyll cells or protoplasts from leaves of *Zinnia elegans* to be transformed into xylem elements (Kohlenbach and Schmidt 1975; Fukuda and Komamine 1980) would seem to be pertinent in confirming the heterogeneity of mesophyll cell physiology rather than a homogeneity.

A further point emanating from the FDA data is that on estimating the viability with the FDA test immediately after chilling the protoplasts, there was a consistently high reading when compared to repeat measurements after having transferred the chilled protoplasts to culture at 25 °C for 8 h (Table 3). This is noticeable particularly after 12 h and 24 h of chilling and implies that the FDA is signalling viability when the protoplasts are not viable.

4 Summary

The following main points can be made:

1. Neutral red and Evans blue are not necessarily good indicators of protoplast viability.
2. FDA can be used, although under certain conditions it may yield a false high number of counts for viability.
3. CTC and 3,4-BP can be useful adjuncts to FDA.
4. Zeta potential measurements can give a good indication of viability, but this is a time-consuming method and requires experience in the use of the apparatus, whether home-made or commercial in origin.
5. When feasible, flow cytometry should be employed to check on the levels of viability since large numbers of protoplasts can be handled rapidly (300–500/m) so working with a more reliable sample size, and identifying subpopulations, some of which may be close to non-viability.
6. Dyes normally used for assaying levels of nucleic acids, e.g. ethidium bromide, acridine orange and DAPI, cannot be used for assaying viability because (a) the dyes cannot be used under the conditions normally employed for nucleic acid studies, and (b) protoplasts can be non-viable but still contain the same amounts and distribution of DNA and RNA as their viable counterparts.
7. NBT with NAD(P) can offer a good guide to the early stages of damage and

can help to distinguish the subpopulations of protoplasts which are indicated as viable with the FDA method, but which are irreversibly damaged and will not survive in subsequent culture.
8. FITC is not a very specific marker for viability, and moreover, leaks very easily from one protoplast to another.

Acknowledgements. I would like to thank Central Reseach Fund of the University of London and the Queen Elizabeth College Research Scholarship for partial financial support for some of the work reported here and to Leitz Instruments Ltd. for the loan of the MPV3 Flow Cytometer.

References

Adams CWM, Bayliss OB (1975) Lipid histochemistry. Tech Biochem Biophys Morphol 2:99–156

Baker JR (1946) The histochemical recognition of lipine. Q J Microscop Sci 87:441–471

Baker JR (1948) The cell theory – a restatement, history and critique 1. Q J Microsc Sci 89:103–125

Baker JR (1958) Principles of biological microtechnique. Methuen, London

Basham HG, Bateman DF (1975) Relationship of cell death in plant tissue treated with a homogeneous endopectate lyase to cell wall degradation. Physiol Plant Pathol 5:249–262

Berg O (1951) A histological study of masked lipids. Stainability, distribution and functional variations. Acta Pathol Microbiol Scand 90(Suppl):1–192

Blaschek W, Haass D, Koehler H, Franz G (1981) Cell wall regeneration by *Nicotiana tabacum* protoplasts: chemical and biochemical aspects. Plant Sci Lett 22:47–57

Bornman JF, Bornman CH (1985) Plant protoplast viability. In: Pilet PE (ed) The physiological properties of plant protoplasts. Springer, Berlin Heidelberg New York Tokyo, pp 29–36

Bornman JF, Bjorn LO, Akerlund H-E (1985) Action spectrum for photoinhibition by ultraviolet radiation of photosystem II activity in spinach thylakoids. Photobiochem 8:305–313

Brown S (1984) Analysis and sorting of plant material by flow cytometry. Physiol Veg 22:341–349

Chandler DE, Williams JA (1978a) Intracellular divalent cation release in pancreatic acinar cells during stimulus-secretion coupling. I. Use of chlorotetracycline as a fluorescent probe. J Cell Biol 76:371–385

Chandler DE, Williams JA (1978b) Intracellular divalent cation release in pancreatic acinar cells during stimulus-secretion coupling. II. Subcellular localization of the fluorescent probe chlorotetracycline. J Cell Biol 76:386–399

Conn HJ (1969) Biological stains. 8th edn, revised by RD Lillie. Williams and Watkins, Baltimore

Dexter ST, Tottingham WE, Graber LF (1932) Investigations on the hardiness of plants by measurements of electrical conductivity. Plant Physiol 7:63–78

Fukuda H, Komamine A (1980) Direct evidence for cytodifferentiation to tracheary elements without intervening mitosis in a culture of single cells isolated from the mesophyll of *Zinnea elegans*. Plant Physiol 65:61–64

Gaff DF, Okang'o-Ogola O (1971) The use of non-permeating pigments for testing the survival of cells. J Exp Bot 22:756–758

Gahan PB (1965) Histochemical evidence for the presence of lipids on the chromosomes of animal cells. Exp Cell Res 39:136–144

Gahan PB (1981) Cell senescence and death in plants. In: Bowen ID, Lockshin RA (eds) Cell death in biology and pathology. Chapman and Hall, London, pp 145–169

Gahan PB (1984a) Plant histochemistry and cytochemistry: an introduction. Academic Press, London

Gahan PB (1984b) Reversible and irreversible damage in plant cells of different ages. In: Davies I, Sigee DC (eds) Cell ageing and cell death. Cambridge Univ Press, New York

Gahan PB, McLellan MR (1985) Flow cytometry and fluorescence activated sorting of plant suspension culture cells and isolated protoplasts. Proc R Microsc Soc 20:7–8

Grout BWW, Willison JHM, Cocking EC (1972) Interaction at the surface of plant cell protoplasts. Bioenergetics 4:585–602

Guilliermond A (1941) The cytoplasm of the plant cell. (LR Atkinson, trans). Chronica Bot Co. Waltham, Massachusetts

Hahne G, Herth W, Hoffman F (1983) Wall formation and cell division in fluorescence-labelled plant protoplasts. Protoplasma 115:217–221

Hanke DE, Northcote DH (1974) Cell-wall formation by soybean callus protoplasts. J Cell Sci 14:29–50

Heslop-Harrison J, Heslop-Harrison Y (1970) Evaluation of pollen viability by enzymatically induced fluorescence: Intracellular hydrolysis of fluorescein diacetate. Stain Technol 45:115–120

Hughes J, McCully ME (1975) The use of an optical brightener in the study of plant structure. Stain Technol 50:319–329

Iverson TH (1985) Protoplasts and gravireactivity. In: Pilet PE (ed) The physiological properties of plant protoplasts. Springer, Berlin Heidelberg New York Tokyo, pp 236–249

James TW, Jope C (1978) Visualization by fluorescence of chloroplast DNA in higher plants by means of the DNA-specific probe 4′6-diamidino-2-phenylindole. J Cell Biol 79:623–630

Johnson LV, Walsh ML, Bockus BJ, Chen LB (1981) Monitoring of relative mitochondrial membrane potential in living cells by fluorescence microscopy. J Cell Biol 88:526–535

Kohlenbach HW, Schmidt B (1975) Cytodifferentiation in the mode of a direct transformation of isolated mesophyll cells to tracheids. Z Pflanzenphysiol 75:369–374

Lakon G (1939) Das Schwinden der Keimfähigkeit der Samen, insbesondere der Getreidefrüchte. Ber Dtsch Bot Ges 57:191–203

Lakon G (1942a) Topographischer Nachweis der Keimfähigkeit der Getreidefrüchte durch Tetrazoliumsalze. Ber Dtsch Bot Ges 60:299–305

Lakon G (1942b) Topographischer Nachweis der Keimfähigkeit des Maises durch Tetrazoliumsalze. Ber Dtsch Bot Ges 60:434–444

Lazarow A, Cooperstein SJ (1953) Studies on the enzymatic basis for the Janus green B staining reaction. J Histochem Cytochem 1:234–241

Lipsky NG, Pagano RE (1985) A vital staining for Golgi apparatus. Science 228:745–747

Mayer Y (1985) Mitotic cycle of mesophyll protoplasts. In: Pilet PE (ed) The physiological properties of plant protoplasts. Springer, Berlin Heidelberg New York Tokyo, pp 143–151

McCully ME (1976) The use of fluorescence microscopy in the study of fresh plant material. Proc Microscop Soc Can 3:172–173

McLellan MR, Gahan PB (1986) The behaviour of isolated plant vacuoles during low temperature stress. In: Dorffling K (ed) Proc 5th Congr Fed Eur Soc Plant Physiol Hamburg, 31 Aug–4 Sept

Michaelis L (1900) Die vitale Färbung, eine Darstellungsmethode der Zellgranula. Arch Mikr Anat 55:558

Nagata T, Melchers G (1978) Surface charge of protoplasts and their significance on cell interaction. Planta (Berl) 153:447–452

Onyia GOC (1984) Quantitative cytochemical studies of chilling injury in tubers of white yam (*Dioscorea rotundata*, Poir). Ph D Thesis. University of London

Onyia GOC, Gahan PB (1985) A quantitative cytochemical study of glutamate and glucose-6-phosphate dehydrogenase activities during chilling injury in tubers of *Dioscorea rotundata*, Poir. J Exp Bot 169:1249–1257

Onyia GOC, Gahan PB, Norman H (1984) The use of new probes for protoplast integrity following isolation and purification of protoplasts from tubers of white yam (*Dioscorea rotundata*, Poir). Plant Sci Lett 33:231–238

Paolillo DJ (1964) Acridine orange fluorescence in shoots of Ephedra. Acta Histochem 18:276–282

Pfeffer W (1886) cited in Baker JR (1948) The cell theory – a restatement, history and critique. I. Q J Microsc Sci 89:103–125

Pilet PE (1985a) Some physiological properties of protoplasts from gravireacting maize roots. In: Pilet PE (ed) The physiological properties of plant protoplasts. Springer, Berlin Heidelberg New York Tokyo, pp 226–235

Pilet PE (ed) (1985b) The physiological properties of plant protoplasts. Springer, Berlin Heidelberg New York Tokyo, pp 1–5

Pilet PE, Herve R, Senn A (1984) Zeta potential of protoplasts from gravireacting maize roots. Planta (Berl) 162:17–22

Robards AW (1970) Electron microscopy and plant ultrastructure. McGraw-Hill, London, p 81

Senn A, Pilet PE (1981) Electrophoretic mobility, zeta potential and surface charge of maize root protoplasts. Z Pflanzenphysiol 102:19–32

Takeuchi Y, Komamine A (1978) Composition of the cell-wall formed by protoplasts isolated from cell suspension of *Vinca rosea* protoplasts. Planta (Berl) 140:227–232

Tampion J, McKendrick ME, Holt G (1973) Use of fluorescent brighteners to visualize the sites of cellulose synthesis in root hairs of species of *Peperonia*. Physiol Plant 29:440–441

Widholm JM (1972) The use of fluorescein diacetate and phenosafranin for determining the viability of cultured cells. Stain Technol 47:189–194

Wolniak SM, Hepler PK, Jackson WT (1980) Detection of the membrane-calcium distribution during mitosis in *Haemanthus* endosperm with chlorotetracycline. J Cell Biol 87:23–32

Wright M, Simon EM (1973) Chilling injury in cucumber leaves. J Exp Bot 24:440–411

Yamaki S, Uritani I (1972) The mechanism of chilling injury in sweet potato. VI. Changes of lipid composition in the mitochondrial membrane during chilling storage. Plant Cell Physiol 13:67–79

Yamaki S, Uritani I (1973a) Mechanism of chilling injury in sweet potato. VIII. Morphological changes in chilling injured sweet potato. Agric Biol Chem 37:183–186

Yamaki S, Uritani I (1973b) Mechanism of chilling injury in sweet potato. X. Change in lipid-protein interaction in mitochondria from cold-stored tissue. Plant Cell Physiol 51:883–888

Yamaki S, Uritani I (1974) Mechanism of chilling injury in sweet potato. XII. Temperature dependency of succinoxidase activity and lipo-protein interaction in mitochondria from healthy or chill-stored tissue. Plant Cell Physiol 15:669–680

I.4 Agarose Plating of Protoplasts and its Applications

J. J. M. DONS and C. M. COLIJN-HOOYMANS[1]

1 Introduction

Successful regeneration of plants from protoplasts involves a series of important events. Several steps in the process can be discriminated which all have their own, often empirically determined, requirements. These different steps and the parameters that have been shown to be important in the process are described in Table 1.

There is a lack of knowledge on the parameters which are important in these steps. Therefore, this type of research remains mainly empirical and based on trial and error. Different aspects of protoplast culture in general have been reviewed (Davey 1983; Dale 1983; Evans and Bravo 1983; Gleba and Sytnik 1984; Reinert and Binding 1986). In this chapter we will focus on the culture of protoplasts, and especially on the application of culture systems in which agarose is used as gelling agent.

2 Embedding of Protoplasts in Agarose

2.1 Properties of Agarose

Embedding of protoplasts into solid or semi-solid medium was described for the first time in 1971 by Nagata and Takebe, who showed that tobacco mesophyll protoplasts were able to divide and to form visible colonies in a medium solidified with 0.6% (w/v) agar. Because of the high gelling temperature of the agar that was used, the melted agar had to be mixed with the protoplast suspension at 45 °C. This relatively high temperature might be deleterious for protoplasts. Moreover, contaminating components in the agar may be toxic for protoplasts. This problem can be overcome by using higher quality agar, by purification of agar or by using agarose. Agarose is purified agar from which sulphate esters have been removed, by which the negative charge is much lower than in agar. By introduction of hydroxyethyl groups to the chains, the melting and gelling temperatures of agarose are lowered, as is the case in so-called low melting temperature agarose (LMT agarose), of which the Sea Plaque agarose from FMC Bioproducts (Marine Colloids, Rockland, USA), is the most frequently used. Some properties of the different agar types are summarized in Table 2.

[1] Institute for Horticultural Plant Breeding, P.O. Box 16, 6700 AA Wageningen, The Netherlands

Biotechnology in Agriculture and Forestry, Vol. 8
Plant Protoplasts and Genetic Engineering I (ed. by Y. P. S. Bajaj)

Table 1. Important stages in protoplast isolation, culture and regeneration

Starting material	Genotype/species Leaves/seedling material Immature embryos Callus/cell suspension
Conditioning of the donor material	Physiological condition Age, cell stage Pretreatment, (dark, cold)
Protoplast isolation	Preplasmolysis Osmotic pressure Enzyme mixture Plasmolyticum
Protoplast purification	Filtration Centrifugation Floating on sucrose
Protoplast culture	Plating density Nutrient composition of the medium Phytohormones Osmotic pressure Light/temperature *Culture system:* liquid medium agarose medium Use of feeder cells
Induction of embryogenesis or organogenesis and plant regeneration	Medium regeneration Decrease in osmolarity Changing hormone and nutrient composition

Table 2. Some properties of agar and agarose. Data obtained from manufacturers product information

Type	Manufacturer	Gelling temperature °C (conc. %)	Melting temperature °C (conc. %)	Sulphate content %	Electro endosmosis
Agar no. 1	Oxoid	32 – 39 (1.5)	>85 (1.5)	0.9	0.32
Agarose VII	Sigma	<30 (1.0)	>65 (1.0)	<0.15	<0.15
Sea Plaque agarose	FMC	28 (1.0)	>65 (1.0)	<0.15	<0.15
Sea Prep agarose	FMC	8 – 17 (0.8)	>50 (1.0)	<0.10	<0.05

The use of Sea Plaque agarose in protoplast culture was described for the first time by Engler and Grogan (1982) for study of the regeneration of lettuce protoplasts. Since the gelling temperature of this agarose type is about 28 °C, the medium and protoplasts can be mixed at 32 °C. This procedure resulted in higher plating efficiencies of the lettuce protoplasts used.

2.2 Comparison of Agar and Agarose as Gelling Agent

A comparison of a series of agar and agarose types as gelling agent in protoplast culture (Lörz et al. 1983) consistently showed the superiority of agarose over agar. In agreement with the first observation by Engler and Grogan (1982), especially agarose with a low melting point proved to be good. Higher cell division frequencies were obtained with protoplasts from a number of species, such as *Daucus carota, Hyoscyamus muticus, Nicotiana tabacum* and *Nicotiana plumbaginifolia*. A comparison of different gelling agents has also been made by Shillito et al. (1983), who showed that plating in agarose stimulated colony formation from protoplasts when compared with liquid culture or cultures solidified with agar. It was demonstrated, for instance, that mesophyll protoplasts from *Nicotiana tabacum* developed colonies at a lower plating density than required when cultured in liquid medium. Protoplasts from *Hyoscyamus muticus* which did not divide in agar-containing medium formed colonies when agarose was used.

The favourable effects of agarose plating have been described in other papers as well. For instance, Furze et al. (1987) showed that from three types of agar/agarose tested, Sea Plaque agarose was the most effective for callus formation of hop protoplasts. Plating of potato protoplasts in LMT agarose stimulated cell division (Tavazza and Ancora 1986). Several gelling agents were tested for the culture of maize protoplasts by Imbrie-Milligan and Hodges (1986). Only agarose plating resulted in sustained divisions leading to the formation of microcalli. Thus summarizing, it can be concluded that agarose and especially LMT agarose is a good gelling agent for the culture of protoplasts. The reason for this stimulation is not known. As mentioned before, a major advantage is, of course, the ability to reduce the temperature at which the protoplasts are mixed and plated. This will improve the viability of the protoplasts, especially of those species from which only fragile protoplasts can be isolated. The low toxicity and neutral character of purified agarose will further improve the viability of the protoplasts. Finally, by entrapment of the protoplasts in the gel matrix, the diffusion of cell metabolites with a high molecular weight will be reduced, which might improve the plating efficiencies compared to liquid cultures.

2.3 Advantages of Agarose Embedding

The embedding of protoplasts in media solidified by low melting point agarose has a number of advantages over culture in liquid media:

1. Plating of protoplasts in agarose solidified media was shown to improve plating efficiency.
2. In agarose, protoplasts remain in a fixed position in the gel matrix, which allows investigation of single protoplasts throughout the culture period.
3. Protoplasts embedded in agarose cannot aggregate as often observed in liquid culture, where sometimes large aggregates are formed. Therefore, when protoplasts are plated in low density the developing microcalli will be derived from single protoplasts, which facilitates cloning.

3 Methods for the Culture of Agarose-Embedded Protoplasts

Protoplasts embedded in agarose-solidified media can be cultured for different periods of time. In general, cell division starts about 4 days after initiation of the culture, subsequently microcalli develop. Different procedures may be used for the further culture and regeneration of calli. Sometimes the original agarose culture is maintained until the calli are large enough to be transferred directly to solid agar medium. However, it has been shown to be advantageous to culture the embedded protoplasts in combination with liquid medium (Shillito et al. 1983; Dons and Bouwer 1986).

3.1 Agarose Bead-Type Culture

If protoplasts are plated in agarose-containing media, it is easy to transfer parts of the solidified medium with the divided protoplasts to liquid medium. It was shown by Shillito et al. (1983) that this "Bead-Type Culture" procedure further improved the plating efficiency of the protoplasts. The procedure is outlined in a short protocol in Table 3 A and presented in Fig. 1. As indicated by Shillito, it was advantageous for certain species, e.g. petunia, tobacco and crepis, to culture the embedded protoplasts for a certain period without liquid medium before transferring them to liquid culture medium. However, protoplasts from *Brassica napus* had to be transferred to liquid medium as soon as the agarose was solidified.

Table 3

A Agarose bead type culture procedure (Shillito et al. 1983)

1. Isolate and purify the protoplasts. Finally resuspend the protoplasts in liquid culture medium at a concentration twice the final plating density (usually 1×10^5 per ml).
2. Mix equal volumes of the protoplast suspension and a solution of Sea Plaque agarose with twice the final concentration (usually 0.4% or 0.6% w/v). Keep both solutions at 37 °C before mixing.
3. Plate the protoplasts by pouring 3 ml of the mixture in a 6-cm Petri dish. Seal with parafilm and incubate in the dark at 25 °C. The culture period is dependent on the type of protoplasts used.
4. Start bead cultures by cutting the agarose in parts and transfer them to a 9-cm Petri dish with 4 ml of liquid culture medium. Incubate the dishes on a gyrotory shaker at about 30 rpm in the dark at 25 °C.
5. During subsequent culture replace the liquid medium at regular intervals (e.g. 14 days, but dependent on the division rate of the cells). By exchanging the medium, the osmotic pressure can be decreased steadily and regeneration induced by changing the phytohormone concentration.

B Agarose disc procedure (Dons and Bouwer 1986)

1. and 2. as described above.
3. Pipette five small droplets (each 100 µl) of the protoplast suspension into a 9-cm Petri dish. After gelling of the agarose, 4 ml of liquid medium is added. Culture at 25 °C in the dark on a gyrotory shaker at 30 rpm.
4. After 14 days take the medium from the culture and add 4 ml of the new medium.
5. Subsequent culture as described above.

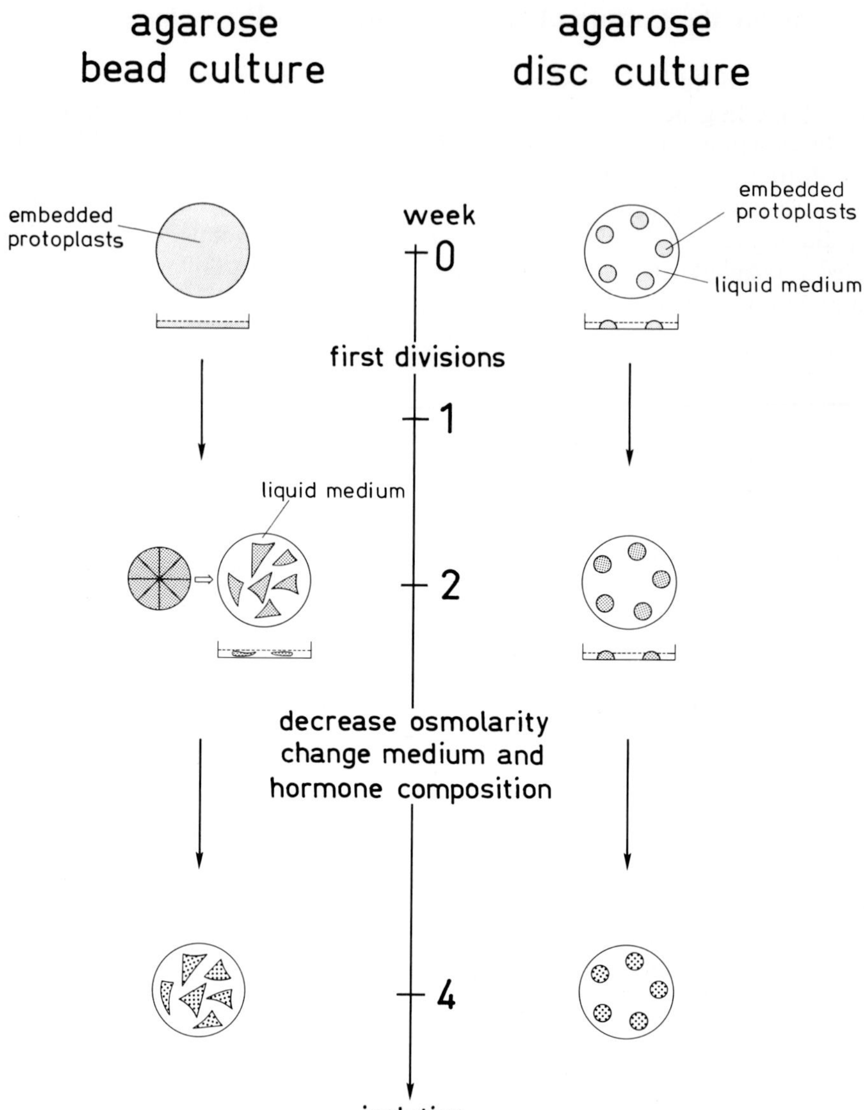

Fig. 1. A scheme of the agarose bead and agarose disc culture procedures for the culture of protoplasts. The sequence of steps as well as the time schedule is only indicative and is dependent on the protoplasts used, the time of first division and the division rate

3.2 Agarose Disc-Type Culture

In our experiments with protoplasts from cucumber, the importance of agarose plating was demonstrated (Dons and Bouwer 1986). The regeneration of cucumber cells is difficult and up to now no reproducible method for the regeneration of cucumber protoplasts is available, although plant regeneration has been achieved recently (Orczyk and Malepszy 1985; Trulson and Shahin 1986). However, we were not successful in culturing cucumber protoplasts in a liquid medium. Protoplasts from neither leaf mesophyll nor from cotyledons divided under these conditions (plating efficiencies less than 2%) and they tended to form large aggregates. These problems have been circumvented by the use of agarose plating, which resulted in high plating efficiencies up to 50% or more (Table 4). The plating efficiency appeared to be strongly dependent on the concen-

Table 4. Plating efficiency of cucumber protoplasts. Protoplasts isolated from cucumber cotyledons were plated at different densities. After 14 d of culture the plating efficiency was determined and the average number of cells in microcalli was estimated by microscopic observation. (Dons and Bouwer 1986)

Plating density (pps/ml)	Bead culture		Disc culture	
	Plating efficiency (%)	Estimated callus size (No. of cells)	Plating efficiency (%)	Estimated callus size (No. of cells)
1×10^5	79	30	84	50–100
0.5×10^5	53	10	6	5
0.25×10^5	0	–	0	–
0.10×10^5	0	–	0	–

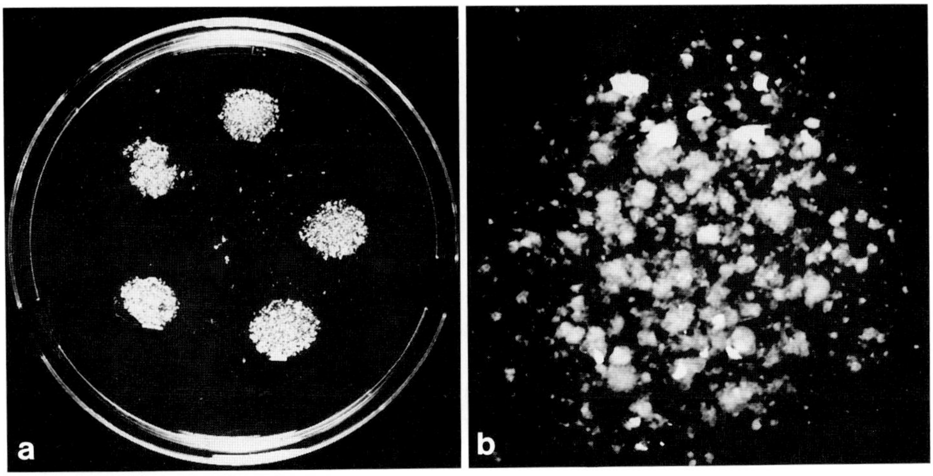

Fig. 2. a The agarose-disc culture. Cucumber protoplasts were plated at a density of 10^5 pps/ml in 5 droplets (100 µl) and surrounded by 4 ml of liquid medium. **b** An agarose disc in detail, showing developing microcalli

tration of the cells. No divisions were observed in cultures with densities lower than 0.5×10^5 pps/ml. Sometimes severe budding of protoplasts was observed instead of regular divisions. To prevent this, as an alternative to agarose bead culture, we plated the agarose-embedded protoplasts in small droplets (discs) surrounded by liquid medium (Fig. 2). For a number of protoplast isolates, this procedure improved not only the plating efficiency, but the sustained growth of the microcalli as well (Table 4) (Dons and Bouwer 1986).

The main difference between this agarose disc and the original agarose bead method is that the embedded protoplasts were diluted immediately with the surrounding medium, by which the final protoplast concentration is about ten times lower than at the start of the bead-type culture. This may be advantageous, especially when protoplasts produce toxic products. The agarose disc procedure, which is an alternative to the bead procedure, is outlined in Table 3B and both methods are compared in the scheme presented in Fig. 1.

3.3 Advantages of Agarose Bead and Disc Culture

The embedding of protoplasts in media solidified by low melting point agarose and the culture according to the bead- and disc-type method has a number of advantages over culture in liquid or solid media:

1. The cultures are easy to handle. The medium can be replaced or diluted without interfering with the cells. The exchange of medium takes place slowly by diffusion through the agarose gel.
2. Compared to liquid medium, smaller volumes can be used. Therefore, the number of protoplasts needed to obtain sufficiently high concentrations (approximately 1×10^5) to induce cell division is lower.
3. Embedded protoplasts can be cultured in liquid medium (as in the agarose bead or agarose disc procedure). In this way, high concentrations of protoplasts inside the gel are combined with large volumes of liquid medium. The methods combine the advantages of high and low density culture.
4. The agarose bead and disc culture systems are amenable for special applications, e.g. the controlled addition of drugs for the selection of cells.

4 A Survey of Literature on the Application of Agarose Plating in Protoplast Culture

4.1 Protoplast Regeneration Using Agarose Culture

Since the findings of Lörz et al. (1983) and Shillito et al. (1983), agarose plating has been successfully used for culture of protoplasts from a number of plant species, and survey of literature is given in Table 5.

In most cases the protoplasts were plated directly in medium solidified with Sea Plaque agarose in concentrations ranging from 0.4 to 1.2%. In some cases the culture was continued in the original agarose gel, but mostly the agarose embed-

Table 5. Agarose plating for protoplast culture. A survey of recent papers

Plant species	Plating method	Culture conditions	Response		Reference
			Division	Regeneration	
Cucumis sativus	0.4% Agarose (Serva)	Bead culture	+	Plants	Orczyk and Malepszy (1985)
	0.6% Agarose (Sea Plaque)	Disc culture	+	Plants	Dons and Bouwer (1986)
Foeniculum vulgare	0.5% Agarose (Sigma Type VII)	Solid[a]	+	Embryos	Miura and Tabata (1986)
Glycine canescens	0.6% Agarose	Disc culture	+	Plants	Hammatt et al. (1987)
Helianthus annuus	0.4% Agarose (Sea Plaque)	Bead culture	+	–	Bohorova et al. (1986)
Humulus lupulus	0.8% Agarose (Sea Plaque)	Bead culture	+	–	Furze et al. (1987)
Hyoscyamus muticus	0.6% Agarose (Sea Plaque)	Solid[a]	+	Not given	Lörz et al. (1983)
H. muticus	0.4% Agarose (Sea Plaque)	Bead culture	+	Not given	Shillito et al. (1983)
Lactuca sativa	0.6% Agarose (Sea Plaque)	Solid[a]	+	Plants	Engler and Grogan (1982)
L. saligna	1.0% Agarose (Sea Plaque)	Bead culture	+	Plants	Brown et al. (1987)
Lycopersicon esculentum	0.6% Agarose (Sea Plaque)	Disc culture	+	Plants	Tan et al. (1987a)
Medicago sativa	0.4% Agarose (Sea Prep)	Melting and plating on solid medium	+	–	Holbrook et al. (1985)
Oryza sativa	1.2% Agarose (Sea Plaque)	Bead culture	+	Plants	Abdullah et al. (1986)
Solanum tuberosum	Liquid	Softened (after 7 days) with 0.2% agarose (Sea Plaque)	+	Plants	Tavazza et al. (1986)
S. tuberosum	Liquid	After several weeks Softened with 0.4% agarose (Sea Plaque)	+	Plants	de Vries et al. (1987)
S. lycopersicoides	Liquid	Softened after 14 days with 0.2% agarose (Sea Plaque)	+	Plants	Tan et al. (1987b)
Zea mays	0.4% Agarose (LMP Agarose BRL)	Solid[a]	+	–	Imbrie-Milligan and Hodges (1986)
Z. mays	0.3% Agarose (Sea Plaque)	Solid[a]	+	Embryos	Vasil and Vasil (1987)

[a] Solid: no further treatment of the agarose embedded protoplasts.

ding was combined with liquid culture, according to either the disc-type procedure or the bead-type procedure. However, for *Solanum* species it appeared to be better to start the protoplast culture in liquid medium. After 7 to 14 days the cultures must be softened with 0.2–0.4% agarose (Tavazza and Ancora 1986; Tan et al. 1987b; de Vries et al. 1987). This special procedure for *Solanum* species was developed originally for *Solanum tuberosum* by Bokelmann and Roest (1983), who introduced this softening technique with 0.4% agar.

4.2 Comparison of Agarose Culture and Liquid Culture

One might expect that during the development of a new procedure for protoplast regeneration from a specific plant species different culture methods have been compared. Only in a few papers has such a comparison been described. *Medicago, Helianthus* and *Glycine* protoplasts only divide when they are embedded in agarose (Holbrook et al. 1985; Bohorova et al. 1986; Hammatt et al. 1987). Protoplasts plated in liquid medium did not continue to divide and in the case of *Glycine* resulted in browning of the culture. Results obtained at our institute showed that in a recalcitrant crop such as cucumber, culture of protoplasts in agarose solidified medium was essential for induction and continuation of cell divisions (Dons and Bouwer 1986). However, a comparison of different culture methods for tomato protoplasts did not reveal significant differences (Tan et al. 1987a). It is obvious that for plant species which are recalcitrant in tissue culture, agarose embedding may be advantageous for protoplast culture. For plants like tomato and tobacco no differences in plating efficiency were obtained between liquid and agarose-embedded protoplasts. As far as we know, however, an advantage of liquid medium over agarose-embedded protoplasts has never been described. This suggests that when a protoplast culture method has to be developed for a new plant genotype, it is worthwhile to start with the agarose-embedded protoplast culture procedure.

4.3 Protoplast Regeneration Using Liquid Cultures

Although the results presented show progress in protoplast technology, the importance of the new culture procedures should not be overestimated. A survey of literature showed that successful protoplast division in liquid medium is described several times, even in tissue culture recalcitrant plant species such as *Helianthus annuus* (Lenee and Chupeau 1986), *Brassica napus* (Chuong et al. 1985), *Brassica oleracea* (Glimelius 1984), *Populus* (Russell and McCown 1986), *Cucumis sativus* (Jia et al. 1986; Trulson and Shahin 1986) and *Oryza sativa* (Yamada et al. 1986). If this short list is compared with Table 5, it can be seen that for some species liquid culture as well as agarose culture have been described. This is, for instance, the case for *Helianthus*. Bohorova et al. (1986) showed that embedding protoplasts is essential for division, whereas Lenee and Chupeau (1986) described division in liquid medium. This means that other factors important for successful protoplast division (summarized in Table 1) may be more essential for protoplast division than embedding of the protoplasts in agarose. Nevertheless, when the optimal conditions for protoplast division are not yet known, embedding in agarose can be very helpful in the development of a procedure for protoplast culture.

5 Special Applications of Agarose Culture

5.1 Agarose Plating as an Aid in the Fixation of Protoplasts for Microinjection

In recent years, the method for microinjection has been improved, resulting in transformed tobacco plants (Crossway et al. 1986). An important aspect of the procedure is the immobilization of the protoplasts during the injection. Different methods have been used, such as suction on glass micropipettes (Crossway et al. 1986) or attachment to glass by chemical means, e.g. poly L-lysine (Steinbiss and Stabel 1983). An interesting method is the embedding of protoplasts in low melting point agarose. Lawrence and Davies (1985) embedded protoplasts at a concentration of 10^4 per ml in 0.3% (w/v) agarose and these were put in microdroplets (1 μl) on top of a layer of 0.9% (w/v) agarose. During microinjection, the protoplasts are fixed in the agarose firmly enough to allow needle penetration. Advantages of this method are its simplicity and the ease of further culturing of the protoplasts (for which they use a feeder system of agarose beads containing protoplasts at high density). Moreover the system avoids toxicity problems sometimes encountered in chemical fixation of protoplasts. Recently microinjection of metaphase-chromosomes into agarose-embedded protoplasts from *Nicotiana plumbaginifolia* was described by De Laat and Blaas (1987).

5.2 Entrapment of Protoplasts in Spherical Beads

Linsefors and Brodelius (1985) described a procedure for the immobilization of protoplasts in small spherical beads by using different gel-forming polysaccharides including Agarose type VII (Sigma). The protoplast suspension was gently mixed with 3% (w/v) agarose at 35 °C. The mixture was added under continuous stirring to vegetable oil, and small droplets (< 1 mm) were formed, which were stabilized by cooling to 10 °C, resulting in the solidification of the agarose. Protoplasts from *Catharanthus roseus* and *Daucus carota* were viable and protected against mechanical and osmotic stress. Such spherical beads might be very useful in reactor culture of cells and protoplasts.

5.3 Isolation of Microcalli by Melting of the Agarose

In most cases the protoplasts embedded in agarose bead or disc culture are plated in relatively high densities (1×10^5/ml), which is necessary for protoplast division. However, under these conditions the microcalli grow close to one another.

Dilution of the medium is not possible because the temperature needed for liquifaction of agarose is high. Even for low melting point agarose a temperature above 45 °C is necessary (e.g. Sea Plaque agarose, Table 2). In attempts to solve this problem, Adams and Townsend (1983) cultured tomato and potato protoplasts for 14 days in Sea Prep agarose medium, which has very low gelling and melting temperatures (Table 2). Thereafter the cultures were incubated for 2 h at 40 °C and diluted with warm medium and subsequently the microcalli were isolated by centrifugation and replated at a lower density. This method has also been applied by Holbrook et al. (1985), who showed that alfalfa microcalli formed after embedding the protoplasts in 0.4% (w/v) agarose could be separated by melting the agarose at 40 °C for 1 h. The microcalli remained alive and were plated at three- to fourfold lower density to ensure that the microcalli grow separately. This method facilitates cloning of microcalli derived from single protoplasts.

5.4 Selection in Agarose Cultures

The bead-type culture technique has been applied for the selection of protoplasts transformed with the selectable marker gene for neomycin phosphotransferase (NPT II gene), which confers resistance of the cells to kanamycin. Protoplasts of *Nicotiana tabacum* were transformed using the direct gene transfer method (Potrykus et al. 1985), subsequently embedded in agarose and cultured according to the bead-type culture procedure (Shillito et al. 1983). The agarose beads were transferred to a medium containing kanamycin. In this way an efficient selection was obtained by which the rare transformed cells could be identified and isolated. This procedure has been described for tobacco (Hain et al. 1985) and *Triticum monococcum* (Lörz et al. 1985). Also for the selection of hybrids and mutants obtained after somatic hybridization or mutagenic treatment, the agarose bead and disc culture can be helpful, as the culture medium can be replaced easily for the selective medium.

6 Concluding Remarks

Difficulties encountered in the regeneration of whole plants from isolated protoplasts are still a major obstacle in the use of protoplasts for the improvement of crops. One technical aspect of the culture of protoplasts, i.e. the use of agarose is reviewed. Since the discovery of Shillito et al. (1983) and Lörz et al. (1983) that the culture of protoplasts in agarose solidified media stimulate cell division and callus formation, this procedure has been applied with success. Protoplast culture is strongly dependent on the concentration of stimulating and inhibiting substances released from the protoplasts. Therefore dilution and/or refreshment of the media should be done in time. Agarose culture according to the described procedures (agarose bead and agarose disc) offers the possibility of doing this in a controlled way. For the development of a protoplast culture system for a new plant species it is therefore worthwhile to investigate agarose embedding.

References

Abdullah R, Cocking EC, Thompson JA (1986) Efficient plant regeneration from rice protoplast through somatic embryogenesis. Biotechnology 4:1087–1090

Adams TL, Townsend JA (1983) A new procedure for increasing efficiency of protoplast plating and clone selection. Plant Cell Rep 2:165–168

Bohorova NE, Cocking EC, Power JB (1986) Isolation, culture and callus regeneration of protoplasts of wild and cultivated *Helianthus* species. Plant Cell Rep 5:256–258

Bokelmann GS, Roest S (1983) Plant regeneration from protoplasts of potato (*Solanum tuberosum* cv. Bintje). Z Pflanzenphysiol 109:259–265

Brown C, Lucas JA, Power JB (1987) Plant regeneration from protoplasts of a wild lettuce species (*Lactuca saligna* L.). Plant Cell Rep 6:180–182

Chuong VP, Pauls KP, Beversdorf WD (1985) A simple culture method for *Brassica* hypocotyl protoplasts. Plant Cell Rep 4:4–6

Crossway A, Oaks JV, Irvine JM, Ward B, Knauf VC, Shewmaker CK (1986) Integration of foreign DNA following microinjection of tobacco mesophyll protoplasts. Mol Gen Genet 202:179–185

Dale PJ (1983) Protoplast culture and plant regeneration of cereals and other recalcitrant crops. In: Potrykus I, Harms CT, Hinnen A, Hütter R, King PJ, Shillito RD (eds) Protoplasts 1983. Birkhäuser, Basel, pp 31–43

Davey MR (1983) Recent developments in the culture and regeneration of plant protoplasts. In: Potrykus I, Harms CT, Hinnen A, Hütter R, King PJ, Shillito RD (eds) Protoplasts 1983. Birkhäuser, Basel, pp 19–31

Dons JJM, Bouwer R (1986) Improving the culture of cucumber protoplasts by using an agarose-disc procedure. In: Symp. Nuclear techniques and in vitro culture for plant improvement. IAEA Vienna, pp 497–504

Engler DE, Grogan RG (1982) Isolation, culture and regeneration of lettuce leaf mesophyll protoplasts. Plant Sci Lett 28:223–229

Evans DA, Bravo JE (1983) Protoplast isolation and culture. In: Evans AD, Sharp RW, Ammirato PV, Yamada Y (eds) Handbook of plant cell culture. Vol I. Macmillan, New York, pp 124–176

Furze JM, Rhodes MJC, Robins JR (1987) The use of agarose bead culture for the regeneration of single cell-derived colonies from protoplasts isolated from suspension culture of *Humulus lupulus*. Plant Cell Tissue Organ Cult 8:17–25

Gleba YY, Sytnik KM (1984) Protoplasts fusion – genetic engineering in higher plants. In: Shoeman R (ed) Monogr Theor Appl Genet 8: Springer, Berlin Heidelberg New York Tokyo

Glimelius K (1984) High growth rate and regeneration capacity of hypocotyl protoplasts in some Brassicaceae. Physiol Plant 61:38–44

Hain R, Stabel P, Czernilofsky AP, Steinbiss HH, Herrera-Estrella L, Schell J (1985) Uptake, integration, expression and genetic transmission of a selectable chimaeric gene by plant protoplasts. Mol Gen Genet 199:161–168

Hammatt N, Kim H-I, Davey MR, Nelson RS, Cocking EC (1987) Plant regeneration from cotyledon protoplasts of *Glycine canescens* and *G. glandestina*. Plant Sci 48:129–135

Holbrook LA, Reich TJ, Iyer VN, Haffner M, Miki BL (1985) Induction of efficient cell division in alfalfa protoplasts. Plant Cell Rep 4:229–232

Imbrie-Milligan CW, Hodges TK (1986) Microcallus formation from maize protoplasts prepared from embryogenic callus. Planta 168:395–401

Jia SR, Fu YY, Lin Y (1986) Embryogenesis and plant regeneration from cotyledon protoplast culture of Cucumber (*Cucumis sativus* L.). J Plant Physiol 124:393–398

Laat AMM de, Blaas J (1987) An improved method for protoplast microinjection suitable for transfer of entire plant chromosomes. Plant Sci 50:161–169

Lawrence WA, Davies DR (1985) A method for the microinjection and culture of protoplasts at very low densities. Plant Cell Rep 4:33–35

Lenee P, Chupeau Y (1986) Isolation and culture of sunflower protoplasts (*Helianthus annuus* L.): Factors influencing the viability of cell colonies derived from protoplasts. Plant Sci 43:69–75

Linsefors L, Brodelius P (1985) Immobilization of plant protoplasts: Viability studies. Plant Cell Rep 4:23–27

Lörz H, Larkin PJ, Thomson J, Scowcroft WR (1983) Improved protoplast culture and agarose media. Plant Cell Tissue Organ Cult 2:217–226

Lörz H, Baker B, Schell J (1985) Gene transfer to cereal cells mediated by protoplast transformation. Mol Gen Genet 199:178–182

Miura Y, Tabata M (1986) Direct somatic embryogenesis from protoplasts of *Foeniculum vulgare*. Plant Cell Rep 5:310–313

Nagata T, Takebe I (1971) Plating of isolated tobacco mesophyll protoplasts on agar medium. Planta 99:12–20

Orczyk W, Malepszy S (1985) In vitro culture of *Cucumis sativus* L. V. Stabilizing effect of glycine on leaf protoplasts. Plant Cell Rep 4:269–273

Potrykus I, Shillito RD, Saul MW, Paszkowski J (1985) Direct gene transfer; State of the art and future potential. Plant Mol Biol Rep 3:117–128

Reinert J, Binding H (eds) (1986) Differentiation of protoplasts and of transformed plant cells. Springer, Berlin Heidelberg New York Tokyo

Russell JA, McCown BH (1986) Culture and regeneration of *Populus* leaf protoplasts isolated from non-seedling tissue. Plant Sci 46:133–142

Shillito RD, Paszkowski J, Potrykus I (1983) Agarose plating and a bead type culture technique enable and stimulate development of protoplast-derived colonies in a number of plant species. Plant Cell Rep 2:244–247

Steinbiss HH, Stabel P (1983) Protoplast from tobacco cells can survive capillary microinjection of the fluorescent dye Lucifer Yellow. Protoplasma 116:223–227

Tan MMC, Rietveld EM, Marrewijk van GAM, Kool AJ (1987 a) Regeneration of leaf mesophyll protoplasts of tomato cultivars (*L. esculentum*): factors important for efficient protoplast culture and plant regeneration. Plant Cell Rep 6:172–175

Tan MMC, Boerrigter HS, Kool AJ (1987 b) A rapid procedure for plant regeneration from protoplasts isolated from suspension cultures and leaf mesophyll cells of wild *Solanum* species and *Lycopersicon pennellii*. Plant Sci 49:63–72

Tavazza R, Ancora G (1986) Plant regeneration from mesophyll protoplasts in commercial potato cultivars (Primura, Kennebec, Spunta, Desirée). Plant Cell Rep 5:243–246

Trulson AJ, Shahin EA (1986) In vitro plant regeneration in the genus *Cucumis*. Plant Sci 47:35–43

Vasil V, Vasil IK (1987) Formation of callus and somatic embryos from protoplasts of a commercial hybrid of maize (*Zea mays* L.). Theor Appl Genet 73:793–798

Vries SE de, Jacobsen E, Jones MGK, Loonen AEHM, Tempelaar MJ, Wijbrandi J, Feenstra WJ (1987) Somatic hybridization of amino acid analogue-resistant cell lines of potato (*Solanum tuberosum* L.) by electrofusion. Theor Appl Genet 73:451–458

Yamada Y, Zhi-Qi Y, Ding-Tai T (1986) Plant regeneration from protoplast-derived callus of rice (*Oryza sativa* L.). Plant Cell Rep 5:85–88

I.5 Immobilization of Plant Protoplasts

H. SCHNABL[1] and U. ZIMMERMANN[2]

1 Introduction

The potential of immobilized plant cells and protoplasts in biotechnology has been emphasized by several groups (Brodelius et al. 1979; Brodelius and Mosbach 1982; Brodelius 1983, 1984, 1985 a, b, c; Lindsey and Yeoman 1984 a, b; Rosevaer and Lambe 1985; Rhodes 1985; Hall et al. 1988). The large-scale production of valuable compounds (secondary metabolites) for clinical, pharmaceutical and agricultural purposes have so far been at the focus of interest.

The most important characteristic of higher plants is that individual cells within a particular cell line are capable of diverse expression of their inherent genetic information. Plant cells are, therefore, able to perform almost any bioconversion or biosynthesis. However, the enormous size of plant cells, their internal compartmentation and their nutrient demands preclude their use for production of primary and simple secondary metabolites on an industrial level. Conventional chemical reactions or simpler biocatalysts such as immobilized enzymes and microbial cells (Chibata 1978; Kennedy and Cabral 1983; Mosbach 1984; Woodward 1985; Hartmeyer 1986) are superior to immobilized plant cells. On the other hand, it is generally believed (Rosevaer and Lambe 1985) that immobilized plant cells and protoplasts can compete with other technologies in two fields. The first of these is the production of high value secondary products; the second is the biosynthesis of new complex chemical analogues, particularly when the synthetic pathways are only found in eukaryotes. The establishment of genetically stable plant cell lines by somatic hybridisation (i.e. by protoplast-to-protoplast fusion and plasmid transfer) may give further impetus in research in these directions (Zimmermann et al. 1981 a; Zimmermann 1982, 1986).

The development of new techniques for the repeated, reversible release of products stored intracellularly is another further operational condition for the exploitation of immobilized plant cells and protoplasts on a large scale. Encouraging methods used so far successfully in release of intracellularly located compounds from mammalian cells were very recently also applied to plant cells (Zimmermann et al. 1981 a; Zimmermann 1982, 1986; Knight and Scrutton 1986). Even though the longer-term prospects of immobilized plant cells in biotechnology are quite promising they should not be overestimated – at least

[1] Institut für Landwirtschaftliche Botanik der Universität Bonn, Meckenheimer Allee 176, 5300 Bonn 1, FRG
[2] Lehrstuhl für Biotechnologie der Universität Würzburg, Röntgenring 11, 8700 Würzburg, FRG

Biotechnology in Agriculture and Forestry, Vol. 8
Plant Protoplasts and Genetic Engineering I (ed. by Y. P. S. Bajaj)

for the time being. However, immobilized plant cells offer other exciting applications.

Immobilization of somatic embryos combined with the production of seedling-quality plants from individual artificial seeds obtained in this way has potential as a novel, universal delivery system by overcoming the inherent problems of vegetative propagation (Redenbaugh et al. 1986).

The physical similarity between the immobilized state and the cell-to-cell contact in a tissue of a whole plant allows the design of various model systems to reveal biophysical and biochemical problems of mass transfer and interactions between cells in close proximity to each other in high plant tissues (Büchner and Zimmermann 1982).

Furthermore, the entrapment of plant protoplasts in (partly) charged matrices results in changes of the surface potential of membrane which control – among other things – membrane structure and trigger membrane transport (see Zimmermann 1986). Studies in this direction may open new areas in membrane research. The finding that the ageing process of biomembranes of gel-immobilized plant protoplasts is considerably delayed (Schnabl et al. 1983a; Schnabl and Youngman 1985), but stimulated by very small concentrations of ecotoxicological substances (Schnabl et al. 1983b; Schnabl and Youngman 1987), is a striking example of how basic research on immobilized plant protoplasts can lead to new applications. A simple biological detection and screening system for pollutants based on this phenomenon is now approaching commercial use.

In this chapter we have focussed primarily on immobilized protoplasts. However, because many established processes are based on intact immobilized cells, information on these systems is included when necessary. It is to be expected that the techniques developed for cells can be successfully transferred to plant protoplasts because of their potential for biotechnology.

2 Immobilization in Gel Matrices

Compared to microorganisms, plant cells are relatively large and therefore sensitive to mechanical stress, in particular to shear forces. From this aspect, plant protoplasts are even more fragile. Further, their immobilization (as that of animal cells, Pilwat et al. 1980) presents severe problems, since the chemicals normally used for entrapment are membrane-reactive. Because of the lack of the protective cell wall, the membranes of protoplasts (or animal cells) are directly accessible to reactive chemicals, resulting in membrane damage and cell deterioration.

Mild and rapid immobilization procedures have to be used to overcome these problems. Therefore a covalent binding of plant cells and protoplasts to a solid matrix can be excluded. There is only one report on the successful covalent binding of plant cells to an activated carrier with preserved viability of the cells (Jirku et al. 1981). These authors have covalently bound *Solanum* cells to porous, high surface area beads of polyphenylene oxide which had been pre-activated with glutardialdehyde.

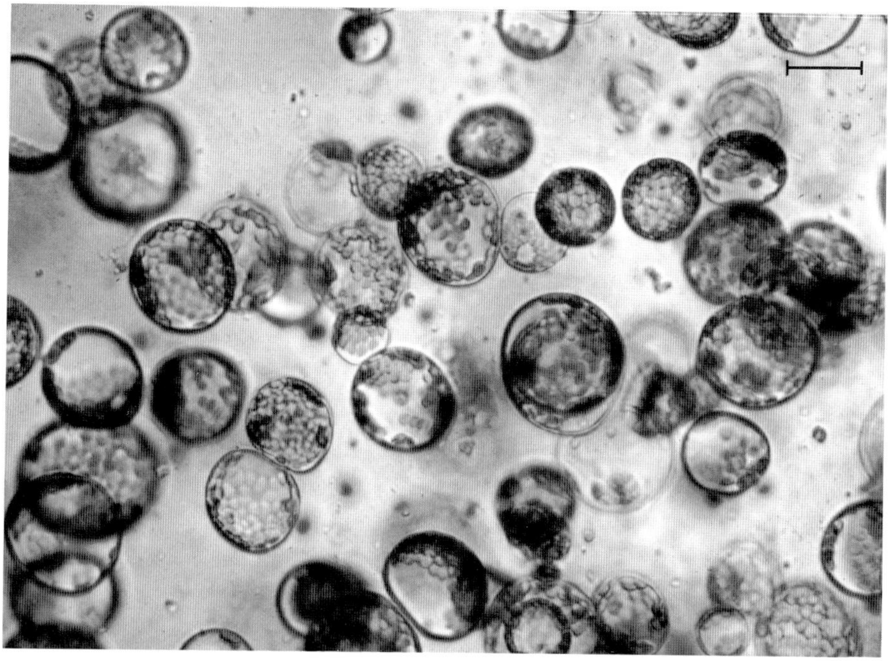

Fig. 1. Mesophyll cell protoplasts of *Vicia faba* suspended in 0.4 M mannitol solution were immobilized in a calcium-alginate matrix (10% alginate, Ca^{2+}-ion concentration 100 mM). Cross-linking was performed by spreading the alginate-protoplast suspension on a microslide followed by the addition of Ca^{2+}-ions. The immobilized protoplasts were kept for 2 weeks in a 0.6 M mannitol solution containing 5 mM $CaCl_2$ and antibiotics (in order to prevent growth of bacteria). For further details, see text. Scale: 1 cm = 20 μm. (Zimmermann, unpubl)

So far entrapment in a polymeric matrix under sterile conditions is the most convenient and efficient way. Three possibilities exist: (1) gel formation by ionic cross-linking of a charged polymer such as alginate and carragean, (2) gel formation by cooling of a heated polymer such as agar and agarose and (3) gel formation by chemical reactions, e.g. by use of glutardialdehyde cross-linked gelatine and polyacrylamide. These three methods are described in detail elsewhere (Brodelius 1984; Woodward 1985; Hartmeyer 1986). Here, we would like to elaborate only on the method of ionic cross-linked alginate gels because this is usually the procedure of choice for the immobilization of wall-less cells (Scheurich et al. 1980; Schnabl et al. 1980; Pilwat et al. 1980) (Fig. 1 and Table 1).

Alginate is a block copolymer of D-mannuronate and L-guluronate joined by 1,4-linkages. An aqueous solution of this polysaccharide forms a relatively stable gel in the presence of divalent or poylvalent cations. These gels are made by dripping a suspension of the plant cells or protoplasts in the alginate solution into an isotonic medium containing the appropriate concentration of the multivalent cations required for cross-linking. Droplet formation is normally obtained by using a dropping pipette or by pumping the biocatalyst/presupport mixture through a needle at a speed low enough to prevent a jet. An outer skin of gel forms around the droplet immediately on its entering the salt solution. However, complete gela-

Table 1. Plant species used for immobilization of isolated protoplasts, the kind of immobilization method and the application

Plant species	Family	Immobilization technique	Application	Reference
Vicia faba *Avena sativa*	Leguminosae Gramineae	Ionic cross-linked alginate gel; spreading	Long-term storage; built up of a pressure gradient	Scheurich et al. (1980)
Guard cell protoplasts of *Vicia faba*	Leguminosae	Alginate gel; spreading	Long-term storage; built up of a pressure gradient	Schnabl et al. (1980)
Datura innoxia	Solanaceae	Anchoring of protoplasts to microcarriers	Application of a new technique	Bornman and Zachrisson (1982)
Vicia faba	Leguminosae	Alginate gel; spreading	Delay of aging process; maintenance of membrane integrity	Schnabl et al. (1983a)
Hyoscyamus muticus *Nicotiana tabacum* *Petunia hybrida* *Brassica rapa* *Lycopersicon esculentum* *Crepis capillaris*	Solanaceae Solanaceae Solanaceae Brassicaceae Solanaceae Asteraceae	Agarose plating and bead culture	Improvement of division and colony formation of protoplasts	Shillito et al. (1983)
Nicotiana tabacum *N. plumbaginifolia* *Daucus carota* *Hyoscyamus muticus*	 Solanaceae Apiaceae Solanaceae	Agarose	Improved culture efficiency	Lörz et al. (1983)
Galium mollugo	Rubiaceae	Alginate gel; tablets	Application for ecotoxicological substances	Lüttge et al. (1984)
Daucus carota	Apiaceae	Alginate, agarose	Protoplast culture	Brodelius (1984)
Claviceps purpurea	Clavicipitaceae	Alginate	High efficiency from vegetative to productive hyphae	Komel et al. (1985)
Vicia faba	Leguminosae	Alginate gel; spreading	Delay of ageing process; inhibition of enzymic lipid peroxidation	Schnabl and Youngman (1985)
Glycine max	Leguminosae	Alginate gel; beads	Formation of calli in the presence of feeder cells	Tricoli et al. (1986)
Oryza sativa	Gramineae	Agarose	Protoplast culture in agarose media	Thompson et al. (1986)

tion, particularly if trivalent ions are used for cross-linking, takes from several minutes to 2 h, depending on the concentration of the cross-linking ions and on the alginate concentration. The viscosity of the alginate limits the practical range of the polymer to concentrations of up to 10% w/w.

Normally, the cells are evenly distributed in the gel matrix. The cell concentration within the gel may be as high as 50% (w/w), but it is normally 5 to 20% (w/w). The viscosity of the cell/alginate solution can result in large droplets and, in turn, in beads of wide size distribution. The average diameter may be as large as several millimeters (Rhodes 1985).

Small spherical biocatalyst beads of uniform size, however, are desirable for several reasons. Small beads will allow more effective substrate(s) and product(s) transfer than larger ones. The larger beads can suffer from mass transfer limitations (Kennedy and Cabral 1983). Considerable gradients of nutrients or released metabolites from the immobilized plant cells can exist, resulting in the limitation of the reaction rate by diffusion. (However, reduced diffusion of substrates to the immobilized cells may be of minor importance for resting cells and for plant protoplast with a decreased capacity for cell wall regeneration, see below.) Diffusion coefficients have been measured by several investigators for some substrates in metal ion cross-linked gel matrices using various methods (Hannoun and Stephanopoulos 1986; Hiemstra et al. 1983; Tanaka et al. 1984). Diffusivities decreased – as expected – with alginate concentration and with increasing bead size.

In contrast, beads that were too small, irregularly shaped or exhibited a wide size distribution gave rise to problems such as large pressure drops over, channeling in, or even clogging of packed-bed reactors. Therefore, at least for large-scale applications in industry – but also for the development of a screening method for ecotoxicological substances on the basis of immobilized plant protoplast – an immobilization method should satisfy at least the following two demands: high production capacity and formation of beads with a uniform and optimum size. A method which fulfils these requirements was recently developed by Hulst et al. (1985) and is based on the break-up of a jet of the biocatalyst/presupport mixture in uniform droplets with the aid of mechanical vibration (Fig. 2). The vibration nozzle technique is based on the physical principle that a liquid jet which is mechanically vibrated in a certain frequency range disrupts into drops of uniform size. The theory for the prediction of the optimum frequency at which break-up of the jet occurs was developed by Rayleigh (quoted in Hulst et al. 1985) for Newtonian liquids like water, and was later extended by Weber (quoted in Hulst et al. 1985) to include the effect of liquid viscosity. Schümmer and Tebel (1983) also included in the theory the break-up of jets consisting of viscoelastic liquids such as polymer melts and polymer solutions. Sodium alginate solutions are pseudoplastic and the theory of Weber can be roughly applied.

In the method of Hulst et al. (1985), the suspension (cells and alginate solution) is pushed through a nozzle into the attached capillary by the application of air pressure. Vibration is generated by a sinus generator, the signal of which is transformed by a vibrator into a mechanical movement of a pin which is connected to the rubber membrane of the nozzle (see Fig. 2). The rubber membrane transmits the signal to the liquid, resulting in the vibration of the jet in the

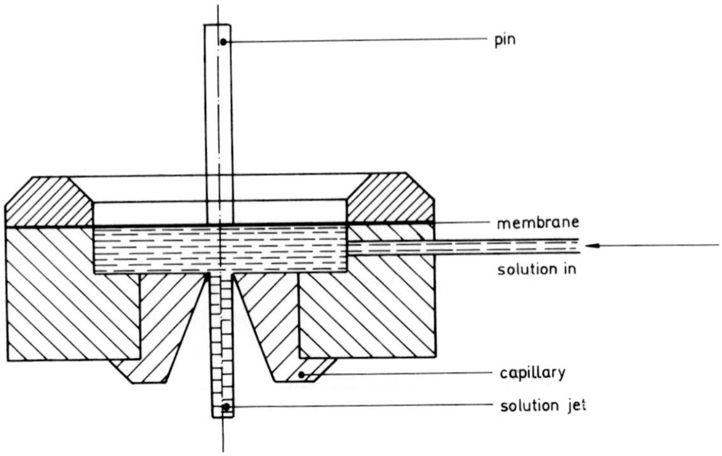

Fig. 2. Schematic diagram of the vibration nozzle used for the production of immobilized biocatalysts in large quantities. (After Hulst et al. 1985)

capillary. In accordance with the theory a suitable frequency range for breakup was observed between 300 and 450 Hz. At a frequency of 400 Hz the average diameter of (unloaded) beads was 1.87 mm with a standard deviation of 3.7%. Depending on frequency, nozzle diameter and volumetric flow rate, the bead size could be varied between 1 and 2 mm, with standard deviations of only 3 to 5% for yeast cells and 10 to 15% for plant cells (*Haplopappus gracilis*). There was no significant difference in uniformity between yeast loaded beads and unloaded beads. Immobilization of unsieved plant cells was not possible and the formation of plant cell aggregates sometimes led to clogging of the nozzle and to less uniform beads.

In this context it is interesting to note that respiration measurements (see below) of the immobilized plant protoplasts showed that the activity of the immobilized cells was higher than that of the free cells, provided that the cells were passed through the vibration nozzle a number of times before the droplets were collected in calcium chloride solution. The authors explained this finding by the assumption that the aggregates have been broken up by means of the actual shear forces in the nozzle, so that the cells occurred as single units in the cross-linked alginate matrix. In general, the disactivation of both yeast and plant protoplasts was negligible and comparable to that observed in the corresponding conventional methods. The production capacity of the vibration nozzle method was two magnitudes larger than that of the dripping procedure.

Another simple way of gel formation, which is particularly suitable for laboratory studies, is to spread the alginate-protoplast/cell suspension on a microslide (Schnabl et al. 1980; Scheurich et al. 1980; Rosevaer and Lambe 1985). This layer is then covered for a couple of minutes – depending on the concentration of bi- or polyvalent ions – with the solution containing the cross-linking ions. This is followed by rinsing the cross-linked layer with physiologically more appropriate solutions in order to remove excessive cross-linking ions. Under these conditions sheets of immobilized cells are obtained which can be handled as a single unit.

The strength and compressibility of the gel varies with the alginate concentration and with the amount and nature of the polyvalent metal ion used for cross-linking (Dainty et al. 1986). Increasing the concentration of the alginate solution and to a certain extent that of the polyvalent ion will result in more tightly "cross-linked" gels. Because of the toxicity of most of the metal ions – at elevated concentrations – it is difficult to vary one independently of the other in an alginate matrix.

The type of alginate, including its guluronic acid (G) to mannuronic acid (M) residue ratio, will also affect a gel's physical characteristics, with high gel strength alginates possessing a high G:M ratio (McDowell 1977).

Ca^{2+}, Ba^{2+}, La^{3+}, Al^{3+}, Fe^{3+} and other polyvalent metal ions can be used for entrapment of plant protoplasts and cells (Dainty et al. 1986).

Calcium cross-linked alginate complexes are plastic and fairly mechanically stable, whereas cross-linking with La^{3+}, Al^{3+} and Fe^{3+} ions leads to very elastic and much more stable sheets or beads. The Ca^{2+}, Al^{3+} and La^{3+} cross-linked matrices are transparent in contrast to gels obtained by the addition of Fe^{3+}-ions. Gels cross-linked with these ions can be suspended for some time (3 weeks to 3 months or more), particularly at low temperatures in physiological appropriate solutions without significant dissolution, provided that Ca^{2+} in the concentration range of 1 to 5 mM is present. Since calcium is a constituent of nutrient media it will inevitably replace any other polyvalent cation (with the exception of barium) in time and so is routinely chosen for most purposes. The alginate complex cross-linked by these ions can be easily dissolved by the addition of citrate or other chelating agents such as EDTA or polyphosphates and the entrapped cells can be recovered in a viable form (see below).

On the other hand, if long-term stability of the gel is required, phosphate and other chelating agents have to be avoided in the nutrition medium, since otherwise decalcification of the matrix would occur within a short time interval. Phosphate in the medium can be replaced by MES (Jones and Veliky 1981). This provides additional buffering capacity, but as phosphate is an essential nutrient, omission may prejudice the long-term stability of the matrix. Some flexibility in the phosphate levels may, however, be desirable because secondary metabolism is very often enhanced under limited growth conditions (Knobloch and Berlin 1980; Lindsey and Yeoman 1984a, b; Lindsey 1985; Watts and Collin 1985, see also below). The susceptibility of alginate beads cross-linked by various metal ions to phosphate chelation was recently studied by Dainty et al. (1986) in detail using alginate immobilized algal cells. These authors found that tolerance to phosphate chelation involves a complex interaction of cation type, concentration and pH of the phosphate solution.

Nearly completely stable alginate gels are obtained if barium ions are used instead of calcium ions or the other cations mentioned above (Dainty et al. 1986; Klenke and Zimmermann, unpublished data). Such cross-linked gels can also not be dissolved by long term treatment of EDTA and other chelating agents. To our knowledge there are no reports on plant cells immobilized in barium/alginate matrices. However, experiments with mammalian cells in our laboratory showed, that viability (and growth) of these entrapped cells was surprisingly good. The porosity and, therefore, the permeability of the gels could be adjusted by inclusion of additives into the cross-linked matrix (Klenke and Zimmermann, unpublished data).

3 Entrapment of Cell Protoplasts in Different Matrices

In the past few years, attempts have been made to replace the potentially toxic entrapment processes on gel basis by various other procedures.

3.1 Adsorption

Adsorption to a solid support – although very simple and mild – has not been used for immobilizing whole plant cells. The limitation to this technique is the relatively weak binding forces involved, which are insufficient to prevent release of plant cells from the support. There is only one report on adsorption of plant protoplasts (Bornmann and Zachrisson 1982). The protoplasts were absorbed on microcarriers (Cytodex) which are often used for growth of anchorage-dependent animal cells. However, it should be noted that the immobilization of cells in foam particles, as described below, may also arise partly from adsorption.

3.2 Foam Particles

Lindsey et al. (1983) reported that liquid-suspended plant cells invade – and were subsequently strongly retained in – porous polyurethane foam particles (supplied by Declon, Corby, Northants, UK) over a 21-day culture period. The plant species used were *Capsicum frutescens* Mill. (which produce the pungent flavour capsaicin) and *Daucus carota* L. The average pore size of foam particles was 0.85 mm. The foam particles, submerged in water, were autoclaved before the immobilization process to remove any chemicals which might have otherwise leached out into the culture medium. A few foam particles were added to newly subcultured flask cultures. During the first 6–9 days of the carrot cell culture period, there was a lag phase in which the total cell fresh and dry weight increased only very slowly. However, after the third day, cells became entrapped in the foam particles, followed by a continued entrapment of cells until about day 15. After this time, increase of the dry weight of the suspended and immobilized cells decreased or almost ceased. On the other hand, the fresh weight of the immobilized and suspended cells continued to increase to day 21, probably due to cell enlargement. After 3 weeks cells had grown out through the surface of the foam particles, were extremely compact, displayed a variety of shapes and sizes and were occasionally organized into globular structures. In contrast to the carrot cells, the growth and immobilization process of the pepper cells had no lag phase. During the first 2 weeks the fresh and dry weights, as well as the number of immobilized cells, increased approximately linearly and – unlike in the case of the carrot cultures – there was no indication of a stationary phase of growth after 3 weeks.

The viability of the immobilized carrot and pepper cells was high (70 to 80% were alive after 3 weeks). On agitation of the loaded foam particles in fresh liquid medium, at least 95% of the cells remained entrapped after 3 to 4 days.

Mavituna and Park (1985) recently improved the physical entrapment of cells into foam particles. The foam particles were held stationary in the flasks by threading them onto a stiff stainless steel wire loop so that they could entrap the suspended cells, which were aseptically introduced into the flask as the inoculum, through filtration.

The relative velocity between the foam and the plant cell aggregates was sufficient to filter out most of the inoculum into the foams within a few hours. The fine aggregates were entrapped only after the narrowing of the pores by other immobilized aggregates. Even those which were larger than the pores of the foam matrix became immobilized by adsorbing to the outside of the particles. After about 2 weeks there were no free cells in suspension.

Cells remained 90% viable for over 12 weeks in the flasks agitated and maintained at 25 °C without addition of nutrients. In contrast, suspension cultures maintained under the same conditions were almost dead after 8 weeks from time of inoculation. Mavituna and Park (1985) claimed that this improved physical method of immobilization has several advantages over others: the immobilization is harmless and can be carried out in situ in a bioreactor; the foam particles have good mechanical strength and durability and the scaling-up process of the immobilization process is more easily compared to others. Furthermore, the immobilized cells are in direct contact with the nutrient medium, there is no permeability barrier to nutrients and metabolites such as could be created by a gel.

In this context similar experiments performed by Felix and Mosbach (1982) should also be mentioned. These authors immobilized cells of *Catharanthus roseus* in a polyurethane foam block formed by mixing wet cells with a Hypol 3000 polymer. Glutardialdehyde hardening reduced initial activity significantly, but increased long-term stability.

3.3 Modified Gel Entrapment

Felix and Mosbach (1982) also tried to entrap cells of *C. roseus* by cross-linking agarose gels with hexamethylene diamine and glutardialdehyde. However, this procedure led to adverse side-effects on key enzymes. Galun et al. (1983) immobilized plant cells in cross-linked polyacrylamide-hydrazide. According to this method, the cells to be immobilized are suspended under mild conditions in an aqueous solution of prepolymerized, linear, water-soluble polyacrylamide, partially substituted with acylhydrazide functional groups. Gelation is achieved by addition of dialdehyde (e.g. glyoxal). This procedure resulted in a very high retention of activity of the entrapped cells (see also below).

3.4 Membranes

Hollow fiber reactors were used by Schuler (1981) and Jose et al. (1983) for immobilization of plant cells. The cells are simply entrapped outside the fibers. The medium flows inside the fibers at a relatively high rate and provides the necessary nutrients and the substrate. The reactants and products of low molecular weight are freely exchanged. The diffusional limitation imposed by the support is minimal and the pressure drop across the reactor is small. With this set-up, phenolic compounds could be produced using cells of *Glycine max* (see below). Operational problems which may generally arise in the use of immobilized cells in a hollow fiber system are described by Tharakan and Chau (1986).

4 Viability of Immobilized Cells and Protoplasts

Preserved viability and biosynthetic capacity of the immobilized cells are both essential. The term "viability" is not simple to define because different viability tests may measure different aspects of membrane integrity and metabolitic activity. Examination of the immobilized cells and protoplasts under the light microscope give an initial rapid test of whether or not gross damage to the cells had occurred during the entrapment process.

Measurements of the intactness of the plasmalemma membrane of immobilized cells and protoplasts, however, is a much more sensitive criterion for retained viability. Various methods were used. Plasmolysis of walled cells is certainly the simplest approach to study viability. Brodelius and Nilsson (1980) exposed cells of *Catharanthus roseus* to a plasmolyzing agent such as glycerol and used the dye phenolsaffranin for detection of plasmolysis. This dye does not penetrate the intact membrane. Such dyes, which are excluded from cells with intact membranes, are frequently employed in viability studies under isotonic conditions. An elegant alternative is to supply cells with a non-fluorescent compound, fluorescein diacetate, which penetrates the intact membrane. After uptake, this compound is hydrolyzed by cellular esterases to release the highly fluorescent fluorescein, which cannot pass through the membrane and thus accumulates within the cell (Widholm 1972). In consequence, only intact cells fluoresce in ultraviolet light when exposed to this substrate.

The intactness of the plasmalemma of immobilized protoplasts can easily be tested by repeated studies of the shrinking and swelling process of the immobilized protoplasts in response to corresponding changes of the osmotic pressure in the external medium. As shown by Scheurich et al. (1980) and Schnabl et al. (1980), cell volume changes induced by hypertonic stress can readily be monitored under the microscope (Fig. 3) because of the separation of the protoplasts from the alginate matrix.

Büchner et al. (1981) and Büchner and Zimmermann (1982) measured the hydraulic conductivity of the whole membrane barrier of immobilized giant algal cells (*Valonia utricularis*, *Halicystis parvula* and *Acetabularia mediterranea*) and of immobilized suspension-cultured cells of *Chenopodium rubrum*, by insertion of a pressure probe into the cell (Zimmermann 1978; Zimmermann and Steudle 1978). The hydraulic conductivity was calculated from pressure relaxation curves (Fig. 4) generated by the injection of a negative or positive pressure pulse into the cell (for details of this technique, see Zimmermann and Steudle 1978). Although

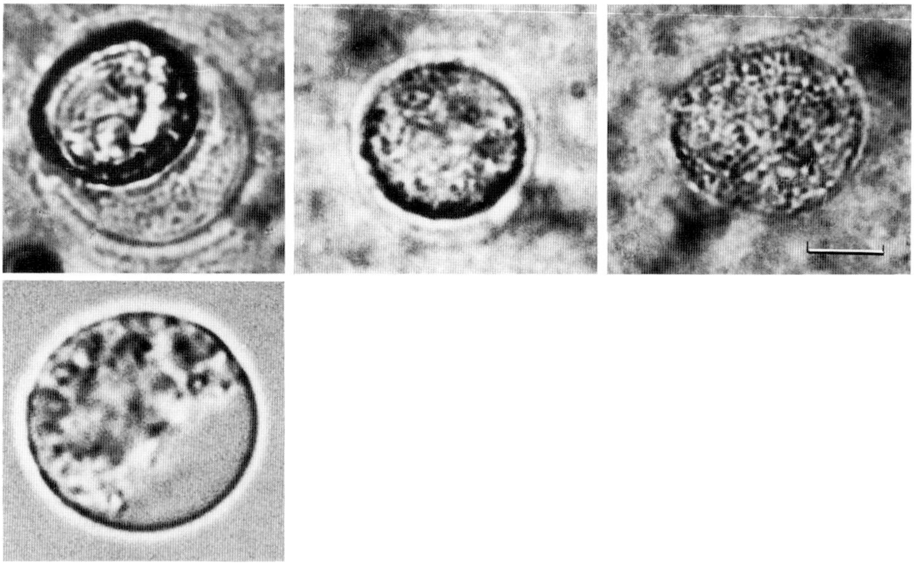

Fig. 3. *Above.* A guard cell protoplast of *Vicia faba* entrapped in a calcium-alginate matrix for 2 weeks. Immobilization was performed as described in Fig. 1. The immobilized cells were kept in a solution containing 0.5 M mannitol, 5 mM $CaCl_2$, 10 mM KCl, 20 mM NaCl, 5 mM $KHCO_3$, 10 mM glucose and a mixture of antibiotics. Scale: 1 cm = 6 μm. Osmolarity of the mannitol solution to which the immobilized cells were transferred (*from right to left*): 500 mOsmol, 640 mOsmol and 800 mOsmol, respectively. *Below:* A guard cell protoplast of *Vicia faba* released from the calcium-alginate matrix after 2 weeks storage. Release was achieved by addition of sodium citrate buffer (20 mM, mannitol 0.8 M, pH 7.4) for 2 h at room temperature. (After Schnabl et al. 1980)

this technique allows the detection of very small leakages in the tonoplast and plasmalemma membrane, the authors did not find an increase of the values of the hydraulic conductivity of the membranes of immobilized cells, and could therefore exclude cell damage on the molecular level. Application of this technique to protoplasts of *Avena sativa* immobilized in a calcium alginate matrix yielded similar results, although only small pressure gradients of the order of 0.5 bar could be built up (Scheurich et al. 1980).

Parr et al. (1984) developed a dual label isotope dilution method to measure the volume of intact cells under immobilized conditions and to follow subsequent growth and maintenance of cell integrity. In this rapid, non-destructive method, ^3H-labelled water and ^{14}C-labelled sorbitol or mannitol is added to the immobilized system. Since the sugar cannot penetrate the membrane, the difference in the volume occupied by ^3H- and ^{14}C-labelled compounds gives a measure of the intact cell volume at diffusion equilibrium.

It is clear that the proof of the existence of an intact plasmalemma does not present evidence for cell ability to undergo growth and division or to synthesize a specific metabolite in a rigid support.

In combination with such studies, additional methods are required to evaluate other specific cellular and membrane parameters which characterize the physiological state of the immobilized cells.

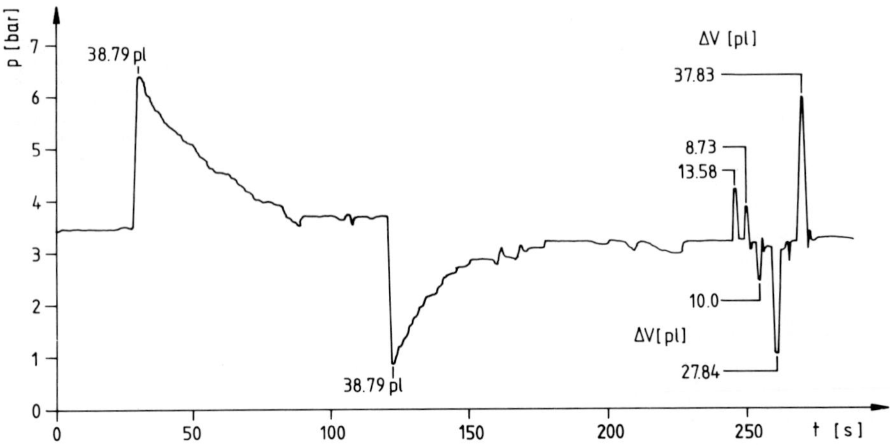

Fig. 4. Time course of cell turgor pressure measured in an individual cell of *Chenopodium rubrum* immobilized in a calcium alginate matrix. The pressure relaxations were induced by increasing or decreasing the turgor pressure by means of the pressure probe (Zimmermann and Steudle 1978). From the half-time of the relaxation the hydraulic conductivity of the cell membrane can be calculated as shown by Büchner et al. (1981). The following determination of the elastic properties of the cell wall was performed by injection of pressure pulses of different amplitudes into the cell and by simultaneous recording of the corresponding changes of the cell volume. Calculations showed that the hydraulic conductivity of the membrane and the elastic properties of the cell wall did not change under conditions of immobilization. (After Büchner et al. 1981).

Respiration, phosphor-metabolism and photosynthesis are vital parts of cell metabolism and were therefore frequently used as indicators of cell viability (e.g. Schnabl et al. 1980; Brodelius and Nilsson 1980; Jones and Veliky 1981; Brodelius and Mosbach 1982; Nilsson et al. 1983; Brodelius 1985a, b, c; Rhodes 1985; Rosevaer and Lambe 1985). Before measurement, the beads or sheets should be cut into small pieces in order to eliminate diffusion problems.

Brodelius and Vogel (1985) (see also Brodelius 1984, 1985b) studied metabolism of cells of *Catharanthus roseus* immobilized in agarose and calcium-alginate with the aid of non-invasive ^{31}P NMR. By using this technique, the uptake, storage and metabolism of inorganic phosphate can be followed within the cell. This method allows the identification of phosphorylated metabolites in different compartments. On comparison of ^{31}P NMR spectra of immobilized cells with freely suspended ones, these authors found no changes in oxygenation and cell metabolism.

A highly sensitive assay for the detection of both membrane deterioration and cellular decompartmentalization has been described by Schnabl et al. (1983a) and Schnabl and Youngman (1985). These authors studied calcium-alginate immobilized protoplasts of *Vicia faba* which were stored in isotonic mannitol solution under sterile conditions (in the dark at 4 °C). Several cellular and membrane processes were significantly delayed compared to freely suspended protoplasts. In non-immobilized protoplasts, a continuous decrease of light-dependent oxygen evolution (from 30 μM oxygen per mg chlorophyll and per hour to zero after 2 days) and of CO_2 uptake, a degradation of the chlorophyll a and b pigments, as

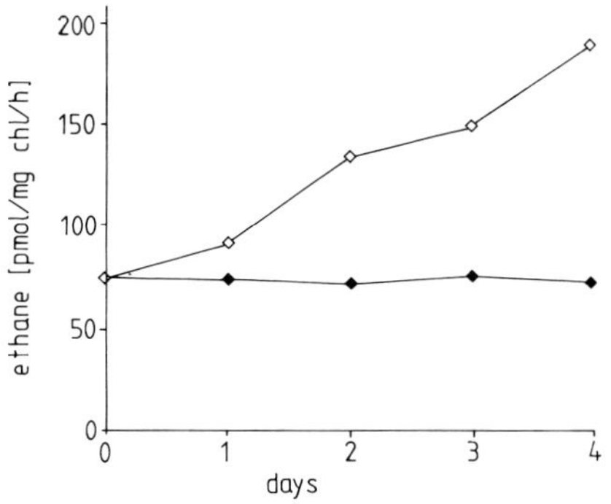

Fig. 5. Ethane production by suspended (○) and immobilized (●) protoplasts of *Vicia faba* stored in calcium-alginate strips for 1 to 4 days under 'semi-sterile' conditions in the dark at 4 °C. The nutrient solution consisted of mannitol (0.5 M), $CaCl_2$ (5 mM), NaCl (20 mM), KCl (10 mM), $KHCO_3$ (5 mM), glucose (10 mM) and antibiotics. After storage the protoplasts were transferred to Fernbach flasks fitted with serum rubber stoppers for the determination of ethane, and incubated in a phosphate buffer at pH 7 and at 22 °C for 30 min under light treatment (200 W/m^2). Headspace samples (1 ml) were withdrawn from the flasks and analyzed for ethane using gas chromatography. (After Schnabl and Youngman 1985)

well as a change in the activity of key enzymes such as ribulose bisphosphate carboxylase (RuBPcase, from 1500 to 480 μM $^{14}CO_2$ per mg chlorophyll per hour after 3 days) and neutral protease (from 4 to 5.5 pM per mg chlorophyll per hour after 3 days) has been monitored. In immobilized protoplasts, however, it was found that these changes did not occur within the monitored time period of 7 to 10 days.

Of most interest was the finding that the production of ethane, which increased in a time-dependent manner in freely suspended protoplasts (from 65 to 175 pM ethane per mg chlorophyll per hour within 4 days), was inhibited in the solid support (Fig. 5). Since the ethane evolution is known to be coupled with cellular disorder and the loss of compartmentation, it serves as a very sensitive indicator for membrane destruction (Riely et al. 1974; Elstner and Konze 1976). There is a body of evidence (e.g. Lieberman and Mapson 1962; Konze and Elstner 1978) that production of ethane results from free- radical-mediated peroxidation of membrane linolenic acid, apparently because free-radical scavenging mechanisms are lost when cells are decompartmentalized. In recent work, Schnabl and Youngman (1985) studied this phenomenon in more detail by investigation of four parameters linked with enzymic lipid peroxidation. These measurements included the determination of the activities of lipid acyl hydrolase (LAH) and lipoxygenase (LOX) (both are involved in the process of releasing and oxidizing fatty acids from membranes), of the concentration of hydroperoxides (the oxidized intermediate product of fatty acids) and of the amount of released ethane, which is one of the final products of lipid peroxidation. In protoplasts freely suspended in mannitol solution for 4 days, an increase in the activity of these enzymes (LAH from 20 to 35 μM per mg chlorophyll per hour; LOX from 10 to 30 μM per mg chlorophyll per hour) associated with a production of hydroperoxides (from 0.6 to 2 nM per μg chlorophyll) and ethane (described above) was recorded. In contrast, in immobilized protoplasts, the enzymic activities and pool sizes of the products of lipid peroxidation remained unchanged (Schnabl and Youngman 1985).

These data strongly support the assumption mentioned before that long-term stability of immobilized plant protoplasts (and presumably immobilized plant cells) arise from the maintenance of membrane integrity. These results also present very clearly for the first time that interactions between the membrane and the solid support occur probably via the charges of the outer membrane surface. It is well known that membrane processes are controlled via surface charges and correspondingly via the surface potential. Experiments have to be designed in future to reveal these relationships on a molecular level, because the outcome of such experiments is of great relevance for commercial applications of immobilized plant cells and protoplasts in bio-reactors.

When incubated in a complete medium, cell growth and cell division are the ultimate proof of cell viability in a solid support. Usually the increase of cell number (dry weight) and/or the mitotic index which reflects the number of cells in mitosis is determined (e.g. Brodelius and Nilsson 1980; Brodelius and Mosbach 1982; Brodelius 1983, 1984; Nilsson et al. 1983; Rhodes 1985).

If protoplastss which show a capacity for cell wall regeneration are immobilized, regeneration of the cell wall and subsequent division is the ultimate evidence for cell viability. Cell wall synthesis can be easily detected by staining the cell wall with the specific fluorescent dye Calcofluor White S (Zimmermann et al. 1981a; Thompson et al. 1986), (see Fig. 6).

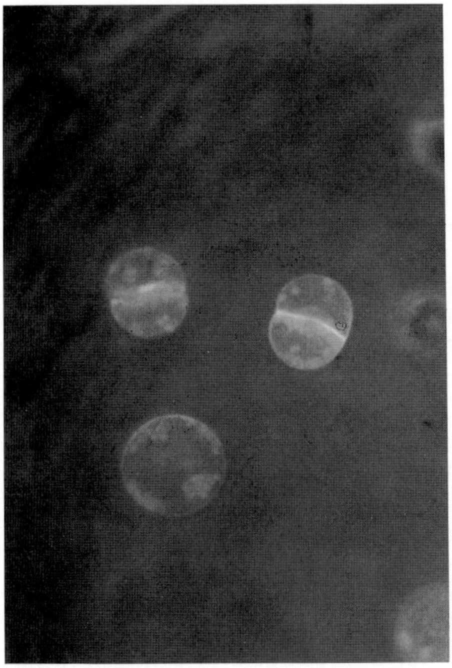

Fig. 6. *Petunia* protoplasts were immobilized in a calcium-alginate matrix as described in Fig. 1. Cell wall synthesis and cell division after 6 days is demonstrated by staining with Calcifluor White S. Scale: 1 cm = 10 μm. (Zimmermann unpubl)

Tricoli et al. (1986) recently reported on the successful culture of soybean (*Glycine max*) protoplasts in calcium-alginate beads. Immobilized protoplasts were induced to divide by culturing on shaker flasks containing an actively growing soybean cell suspension. In the presence of this feeder, layer-entrapped protoplasts first divided 3 to 5 days after encapsulation. After 18 days, the beads were dissolved and cell colonies placed on feeder-plates, giving reproducible formation of callus. Since the protoplasts were immobilized at low density, the cell colonies were derived from single cells.

In this context it is interesting to note that immobilized cells and protoplasts often – but not always – show reduced respiration and reduced cell growth (dry weight increase) compared to non-immobilized plant cells when complete growth medium is added. From a biotechnological standpoint this may, however, be an advantage because there is evidence (Lindsey and Yeoman 1983; see also below) that plant secondary metabolites are produced at higher concentrations in slow-growing or non-growing culture.

One of the reasons for reduced growth in rapid supports made up of alginate, agarose or agar may be the additional diffusion barriers within the aggregated immobilized cells. As mentioned above, diffusional barriers may play an important role in gels, such as calcium-alginate or agarose, which form a distinct boundary between medium and support. Brodelius and Nilsson (1980) reported that growth rate of *C. roseus* immobilized in alginate was lower than in suspension, whereas the reduction in growth rate was only marginal when agarose was used for immobilization. Since agarose has a more open porous structure than alginate, diffusion limitations may explain the effect of growth. Further support for this explanation is derived from experiments in which plant cells were immobilized in porous polyurethane foam particles (Lindsey et al. 1983; Mavituna and Park 1985). The maximum specific growth rate of non-immobilized cells of *C. frutescens* was only slightly higher than that of immobilized cells, but the overall growth pattern and final cell yields were similar. Similar results were obtained by Rhodes (1985) with beetroot cells immobilized on very porous nylon sheets.

Diffusion problems are certainly involved in growth rate reduction of immobilized cells supplied with complete medium. However, in the light of the results of Schnabl et al. (1980, 1983a) and Schnabl and Youngman (1985), it seems quite conceivable that the delay of the ageing process under immobilized conditions is linked with these growth phenomena. In any case, there is a body of evidence that reduced growth rate is not the result of cell damage during the entrapment process.

Several tests were also developed to prove the viability of plant cells and protoplasts after release from the solid support by addition of cation chelating agents (Fig. 3). These include – among other things – measurements of respiration and photosynthesis, as well as growth studies. The characteristic feature of all these methods is the demand for relatively large quantities of biological material, which is not always available. Furthermore, these methods average over a large number of cells. They do not allow – when decreased values of the measured parameters are observed – to decide whether the release process has partly reduced viability of all cells or has damaged completely only a few individual cells.

Single-cell methods have been recently developed to overcome these drawbacks. Changes in membrane and cytoplasmic conductivities that are expected to result from the matrix dissolution can be readily detectable using electrorotation (Zimmermann et al. 1981 b; Arnold and Zimmermann 1982; Arnold et al. 1985, 1986). Rotation of cells is observed in linear alternating electric fields at characteristic frequencies in the kHz range (20 to 100 kHz for plant protoplasts depending on medium conductivity) provided that at least two cells are present in close contact with each other. Single cell rotation can be induced by a planar, homogeneous rotating field in the same frequency range. Rotation direction is opposite to that of the field and can be reversed at will by reversing the field vector. The maximum speed of cell spinning is a few cycles per second and is proportional to the square of the amplitude of the fields.

The theory (Sauer and Schlögl 1985; Fuhr 1985) shows – at least for low medium conductivity – that from the relationship between the characteristic frequency and the external conductivity the values of membrane resistance and capacitance, as well as of the internal conductivity, can be deduced. Small changes in one of these parameters can be easily detected by a shift in the characteristic frequency/conductivity spectrum. Small changes in membrane integrity can be particularly sensitively picked up if the novel electrorotation-generator developed by Arnold and Zimmermann (1984) is used. This new equipment manipulates the frequency and the direction of the rotating field so that, when the frequency displayed matches the characteristic frequency of the cell under observation, a sharp null instead of a broad maximum in rotation rate is seen.

A modified, but also very sensitive technique is the so-called co-field rotation method (Arnold et al. 1986). This method makes use of the phenomenon that cells also show rotation in the MHz range, which is, however, in the same direction as the field. Whereas in the kHz range the charging process of the membrane is primarily responsible for the rotation, it is the ratio between the electrical properties of the cell interior and the medium in the MHz range that is important in determining the rotation speed and direction (Sauer and Schlögl 1985; Fuhr 1985). Between these two oppositely directed maxima in rotation, a frequency exists at which no rotation occurs. This frequency is selected for the given untreated cells and used for the assay of the treated cells. Damage to the cell integrity can then be easily monitored by the increase of number of cells which rotated in the same direction as the field, because the "null frequency" is shifted towards higher frequencies. Applications of these two methods to plant protoplasts released from a calcium-alginate matrix showed that most of the cells survive this procedure if the process is performed gently.

Another interesting approach was recently published by Hampp et al. (1986). These authors described a semi-quantitative assay for light-dependent oxygen evolution by a single mesophyll protoplast. The indicator in this assay is the density of aerotactic bacteria attracted to the protoplast (Fig. 7). This accumulation did not occur if plant protoplasts were not viable. Quantification of the accumulation effect of aerotactic bacteria in close proximity to the plant cell can be achieved by dark field microphotometry. The sensitivity was as low as 50 femtomoles oxygen per protoplast per minute. Preliminary experiments in our laboratory showed that this is also a very sensitive test to investigate all-or-none reac-

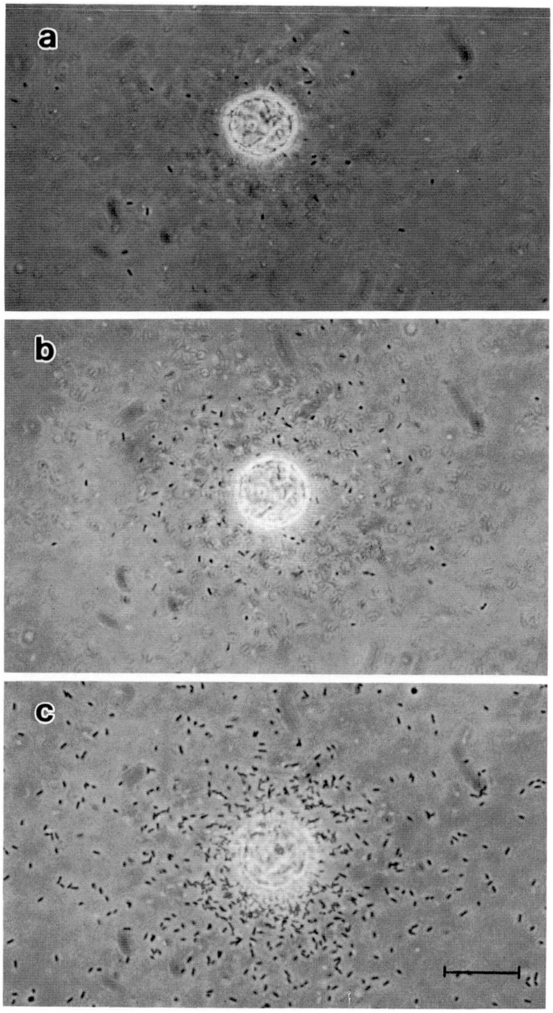

Fig. 7a–c. A protoplast of *Nicotiana tabacum* and *Pseudomonas aeruginosa* (aerotactic bacteria) viewed in the dark under the microscope **a**. Note that the bacteria are randomly distributed. **b** Illumination for 1 min initiates accumulation of the aerotactic bacteria in the neighbourhood of the oxygen-producing plant protoplast. **c** Accumulation is completed after 5 min illumination. Scale: 1 cm = 10 μm. (Photograph by courtesy of W. Mehrle, University of Tübingen)

tions occurring on the level of a single immobilized and subsequently released plant cell.

5 Applications of Immobilized Plant Cells and Protoplasts

The potential of immobilized plant cells is believed to be in biotransformation (i.e. conversions of one complex molecule into a related product) and in biosynthesis (Brodelius and Mosbach 1982; Rosevaer and Lambe 1985; Rhodes 1985; Brodelius 1985a, b, c). However, immobilized plant cells and protoplasts may also be valuable tools in the regeneration of whole plants from the protoplast stage

(Zimmermann et al. 1981 a; Thompson et al. 1986), in the production of artificial seeds (Redenbaugh et al. 1986) and in basic (biophysical) research (Büchner and Zimmermann 1982). A new application area of these biocatalysts was also recently introduced by Schnabl et al. (1983 b, 1984) and Schnabl and Youngman (1987). This is the development of a simple screening system for ecotoxicological substances.

5.1 Basic Research

Immobilized cells and protoplasts are interesting model systems for the elucidation of water transport parameters in tissue cells of higher plants.

Changes in turgor pressure (and in internal osmotic pressure) of tissue cells are normally achieved by immersing the intact or cut plant organ (e.g. the leaf) into solutions of different osmotic pressure. Because of the restriction of diffusion in the unstirred layers of the apoplasmic space, the actual osmotic pressure in the proximity of the cells in deeper-located layers of the tissue is significantly different to that in the bulk solution. Therefore, large errors arise when the water permeability and related parameters of tissue cells are calculated from such experiments.

A further uncertainty of the data derived in this way is introduced by the lack of methods to determine the reflection coefficient of the osmoticum used in the bulk solution. The reflection coefficient is a measure for the degree of coupling between water flow and solute (osmoticum) flow in a biological membrane and is, beside the solute permeability, the most important parameter for the appropriate description of the permeability of the osmoticum in respect to the cell membranes. This term is defined by the linear equations of the thermodynamics of irreversible processes which have to be used for the correct mathematical treatment of hydrostatic and/or osmotic pressure induced water flow (Zimmermann and Steudle 1978). As discussed below, knowledge of the reflection coefficients of various compounds for immobilized cells of different species might improve the development of techniques for the release of secondary products from storage vacuoles in the cells into the external medium. Some of the problems involved in the determination of the water relation parameters have been overcome by the pressure probe technique already mentioned above (Zimmermann 1978; Zimmermann and Steudle 1978). The pressure probe can be inserted even into the tiny tissue cells allowing the measurement of the hydraulic conductivity of the membranes, the related elastic properties of the cell wall (or of a surrounding gel matrix) and the reflection coefficients of various solutes. For calculation of these parameters, the phenomenological equations of the thermodynamics of irreversible processes have been used (e.g. Tomos et al. 1981). However, these were derived for single cells surrounded by a large, stirred volume of bulk solution. It was doubtful whether these equations were valid when applied to water exchange between an individual tissue cell and the apoplasmic space, because changes in the osmotic pressure and the occurrence of back pressures in the apoplasmic space during net water flow could not be ruled out a priori.

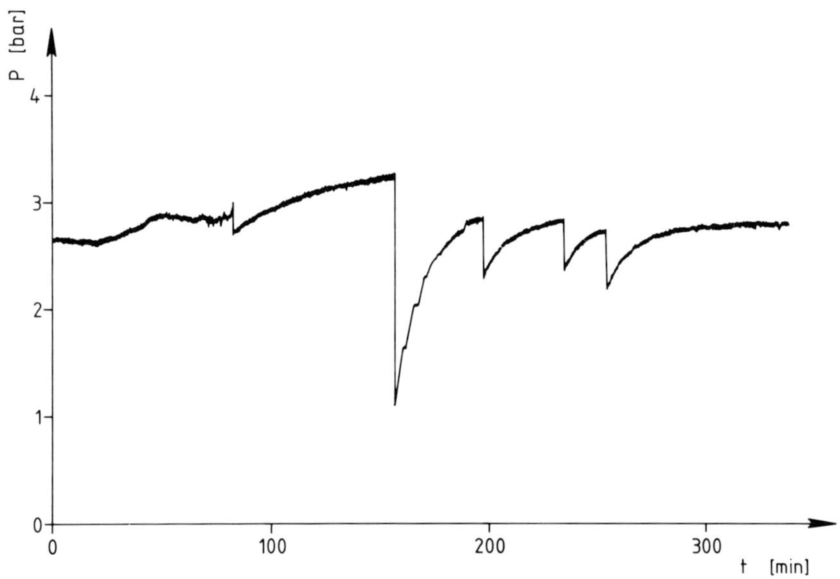

Fig. 8. The time course of turgor pressure in a cell of the marine giant alga *Valonia utricularis* which had been immobilized in a calcium-alginate matrix (after insertion of the pressure probe) and which had been subjected to lowered external osmolarity. The increase of turgor pressure was caused by the addition of distilled water to the sea water. At a maximum pressure of about 0.3 MPa reversible leakages occurred, leading to sudden decreases in turgor pressure followed by subsequent increases in turgor pressure up to the maximum value. (After Büchner and Zimmermann 1982)

The high packing density of immobilized cells, the presence of unstirred layers in a gel matrix and, therefore, the resultant similarity between the immobilized state and the situation in a tissue suggested that pressure relaxation measurements should be performed on immobilized cells as models of cells in tissue.

Using giant algal cells and suspension cells of *Chenopodium rubrum*, Büchner et al. (1981) and Büchner and Zimmermann (1982) demonstrated that the correct values of the hydraulic conductivity are only measured provided that the turgor pressure within the given cell is changed hydrostatically (Fig. 4). Under this condition unstirred layers and back pressure problems in such matrices can be neglected and, therefore, very likely also in the apoplasmic space. On the other hand, experiments in which water flow was osmotically induced showed that calculation of the hydraulic conductivity from such osmotically induced pressure relaxation curves was only possible if the half-time of water exchange through the cell membrane was long enough in comparison to the permeability and the thickness-variable diffusion pathway of the surrounding gel layer.

Experiments of this kind also allow in principle the estimation of the diffusion coefficient of solutes in gel matrices.

It is interesting to note that the calcium-alginate entrapment of these cells did not lead to any change in the elastic properties of the cell walls (measured by the parameter of the elastic modulus). However, for some species the rigid gel layer protected the cell against critical turgor pressure (in hypotonic solutions) which would normally result in the bursting of the cells. As shown in Fig. 8 leakages occurred in the membranes of *V. utricularis* if the turgor pressure exceeded a value of about 3 bar. No rupture occurred if the cells were immobilized. Moreover, the cells sealed up again after a rapid water (and solute) loss. The turgor pressure increased repeatedly up to the critical value.

Similar stabilization effects were observed by Scheurich et al. (1980) for immobilized plant protoplasts in which small pressure gradients of the order of 0.5 bar could be built up due to the mechanical support of the rigid gel.

Since the target product for biotechnological applications is often stored within the vacuoles of cells of suspension cultures, it seems possible to use the phenomenon described here for extraction of the product. This aspect is discussed in detail below.

The mechanical stabilization of protoplasts in a cross-linked alginate gels can also be used for storage of very sensitive plant protoplasts. Schnabl et al. (1980) have shown that immobilized protoplasts of guard cells of *V. faba* can be stored for 2 to 3 weeks in a calcium-alginate matrix without any apparent changes in morphology, membrane integrity or cellular functions. Such immobilized guard cells can also be sent over long distances and released from the matrix by the addition of chelating agents. Since free suspended guard cells are very difficult to prepare and usually die within a few days, immobilized guard cells are an elegant way to re-establish cell suspensions (or cultures) in other laboratories for research. Further work must be carried out to see if other very fragile and shear-sensitive cells can be handled in the same way. Similar results were also reported for immobilized protoplasts from *Claviceps purpurea* (Komel et al. 1985) and from other species (Lörz et al. 1983) immobilized either in alginate or in agarose.

A further aspect of the increased mechanical stability of immobilized protoplasts is the potential of this technology for crop improvement (Lörz et al. 1983; Thompson et al. 1986). There is currently considerable interest in somatic hybridization of plant protoplasts and the subsequent regeneration of a whole plant. Efficient methods are available for fusion of plant protoplasts of different origin or for injection of plasmids (foreign DNA) into protoplasts followed by expression of the respective genes. These methods are based on the application of field pulses of high intensity (Zimmermann et al. 1981a; Zimmermann 1982, 1986; Knight and Scrutton 1986). Fundamental to this objective is the need to be able to reproduce the culture of protoplasts from a range of genotypes. Thompson et al. (1986) recently reported that protoplasts isolated from cell suspensions of four rice cultivars and maintained in an amino acid-based culture medium showed a marked improvement in survival, division and plating efficiency when immobilized in agarose at higher concentration (about 1% w/w) (see also Lörz et al. 1983; Shillito et al. 1983). The protoplast-derived calli were similar in appearance to explant-derived morphogenic callus and produced distinct embryo-like structures. Such a procedure may enable further research in this field.

5.2 Production of Somatic Seeds

Many plant species are vegetatively propagated in vitro for commercial use; however, some disadvantages are involved, and the immobilization of somatic embryos to form artificial seeds has been experimented. Redenbaugh et al. (1986) recently reported on the successful encapsulation of somatic embryos of alfalfa, celery and cauliflorwer in a calcium-alginate matrix. The single-embryo beads were 4 mm in diameter. Somatic embryos of alfalfa and celery showed in vitro germination frequencies with production of seedling-quality plants of the order of 30 and 65%, respectively. In addition, these authors could demonstrate that immobilized somatic embryos developed into plants in sand trays or in transplant plugs at frequencies of about 10%. Artificial seeds of this kind may, therefore, have a great potential in agriculture.

5.3 A Simple Screening System for Monitoring Ecotoxicological Substances

Protoplasts have several advantages for a system suitable for monitoring ecotoxicological substances: they are highly sensitive for the determination of the phytotoxicity of substances and the use of the immobilization system makes them readily available even after long conservation. A new application technique was developed from the observation that the ageing process of membranes, which is reduced in the entrapped state, can be increased again by low concentrations of certain chemicals (Schnabl et al. 1983b). The possible ecotoxicity of test substances is evaluated on the basis of their effects on total photosynthesis (oxygen evolution and CO_2 fixation) (Schnabl and Youngman 1987) and on lipid peroxidation reactions induced by radicals or enzymes (Schnabl et al. 1983b, 1984).

The basic process of photosynthesis, determined as CO_2 fixation activity and oxygen production, was inhibited to a varying extent by the poisons tested with, however, marked concentration differences. The correlation between the concentration of chemicals and the degree of membranous and/or metabolic damage provides a method for monitoring the presence of the environmental poisons in a fast (20 min), exact, and reproducible way which can be undertaken even in a mobile laboratory.

The effect of chemicals on photosynthesis was determined by their action on oxygen evolution as an indirect or net indicator of photosynthesis and by examining RuBPcase (ribulose 1,5-bisphosphate carboxylase) activity in the absence or presence of the test substance. As can be seen from Figs. 9–11, pentachlorphenol (PCP) led to a concentration-dependent inhibition of light-dependent oxygen development (Fig. 9; concentration required for 50% inhibition (IC_{50} value) is 100 nM) and of RuBPcase (Fig. 10; IC_{50} value is 7.5 μM), in contrast to the effects on lipid peroxidation measured as ethane evolution in which PCP led to an increase in ethane production with higher concentrations (Fig. 11; concentration required for 50% stimulation is 1.2 μM). Tables 2–4 show the concentrations of a range of exotoxicological substances required for 20% inhibition (IC_{20}) of oxygen evolution and RuBPcase activity, such as Cl- and NO_2-substituted phenols (Table 2), anilins (Table 3) and hydrocarbons (Table 4). It appears that the oxygen production is a particularly sensitive indicator of toxicity induced by compounds of ecological significance. It proved to be more sensitive (about 1000 times) than that of RuBPcase. Comparing the IC_{20} values of oxygen evolution from the dif-

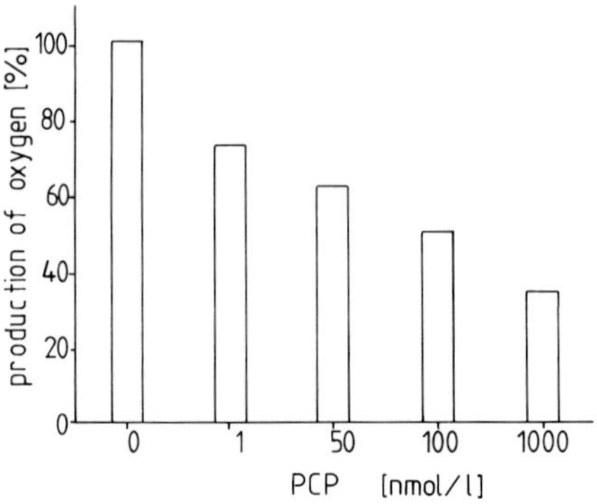

Fig. 9. Light-dependent oxygen production (in %) by protoplasts of *Vicia faba* immobilized in calcium-alginate strips for 1 day. The protoplasts were exposed to different concentrations of pentachlorphenol (PCP, nM) without any preincubation time. Each experiment took about 20 min; n = 3. The oxygen production of the control was 90 μM/mg/Chl/h. (After Schnabl and Youngman 1987)

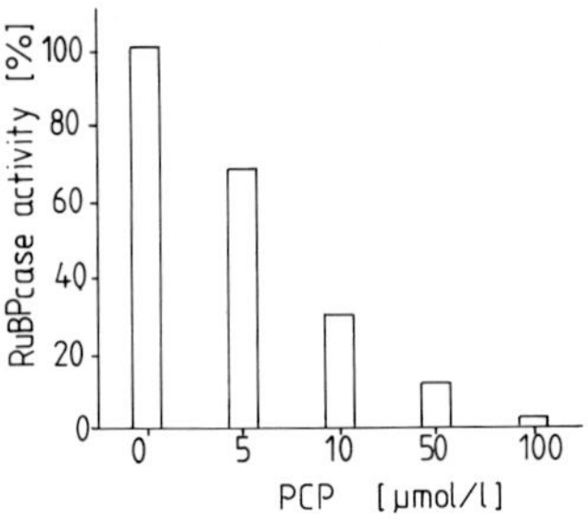

Fig. 10. CO_2 fixation measured as $^{14}CO_2$ incorporation (in %) to ribulosebisphosphate of *Vica faba* protoplasts which were immobilized for 1 day in calcium-alginate strips and released from the matrix by complexing the Ca^{2+}-ions with citrate buffer (20 mM; pH 7.4). The purified protoplasts were incubated at different concentrations of PCP (μM) without any preincubation time; n = 3. The CO_2 fixation of the control was 80 nM/μg Chl/h. The ^{14}C-uptake into acid stable products was determined by liquid scintillation counting. (After Schnabl and Youngman 1987)

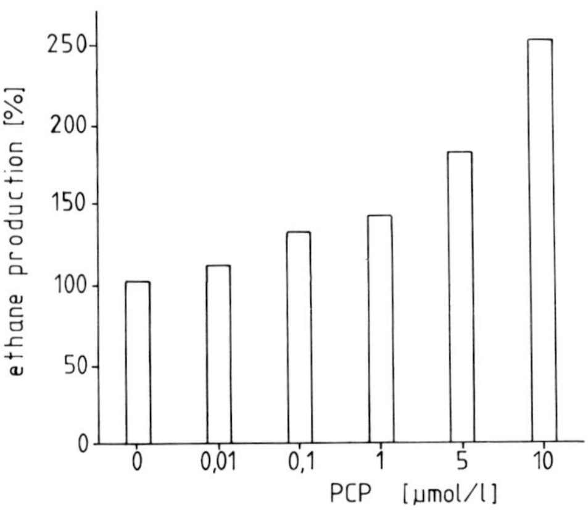

Fig. 11. Ethane development (in %) by protoplasts of *Vicia faba* immobilized in calcium-alginate strips for 1 day at different concentrations of PCP (μM). The measurements were performed as described in Fig. 5. The ethane value of the control was 50 pM/mg Chl/h. (After Schnabl et al. 1984)

ferent groups of substances, the phenols seem to damage the photosynthetic rate in a higher degree (between 1 nM and 1 μM, Table 2) than the anilins (1 – 200 μM, Table 3). The damage of oxygen evolution induced by hydrocarbons could be observed only with higher concentrations (1 – 70 mM, Table 4).

Table 2. IC_{20} values for inhibition of RuBPcase and O_2 evolution induced by a range of phenols. (Data taken from the final report of H. Schnabl, 1987 on the research programme supported by UBA, Berlin)

	RuBPcase	O_2-evolution
Phenol	2.9 mM	1 nM
4-Cl-phenol	560 μM	10 nM
2,3-Cl_2-phenol	180 μM	100 nM
2,4-Cl_2-phenol	12 μM	1 nM
2,6-Cl_2-phenol	2.1 mM	1 nM
2,4,6-Cl_2-phenol	300 μM	10 μM
Pentachlorphenol	2.5 μM	5 nM
Br-phenol	560 μM	8 μM
3-dimethylamino-phenol	86 μM	0.1 μM
2-NO_2-phenol	90 μM	1 nM
4-NO_2-phenol	130 μM	1 μM
2,4,6-$(NO_2)_3$-phenol	2 mM	0.1 μM
3,5-dimethoxy-phenol	1.3 mM	0.7 mM
4-OH-benzene acid	–	0.7 mM
4-acetamido-phenol	7 mM	0.5 nM
4-Cl-3,5-dimethyl-phenol	110 μM	0.1 μM
2-amino-4-methyl-phenol	22 μM	0.03 μM
4-CH_3-2-nitro-phenol	250 μM	0.5 μM
2-CH_3-4,6-dinitro-phenol	19 μM	10 μM
3-OH-phenylurea	–	8 nM
3-(trifluormethyl)-phenol	370 μM	0.05 μM

Table 3. IC_{20} values for inhibition of RuBPcase and O_2 evolution induced by a range of anilins. (Data taken from the final report of H. Schnabl 1987 on the research programme supported by UBA, Berlin)

	RuBPcase	O_2-evolution
Anilin	0.2 mM	5 μM
2-Cl-anilin	1.6 mM	0.3 mM
3-Cl-anilin	0.35 mM	8 μM
4-Cl-anilin	0.3 mM	50 μM
2,4-Cl-anilin	1.5 mM	0.8 μM
2,6-Cl-anilin	3.0 mM	200 μM
2-Br-anilin	1.8 mM	–
2,4-$(NO_2)_2$-anilin	7 μM	2 μM
2,4-$(CH_3)_2$-anilin	1.3 mM	50 μM
2-ethyl-anilin	1.5 mM	100 μM
4-ethyl-anilin	0.6 mM	10 μM
Benzidin	100 μM	0.3 μM
3-(trifluormethyl)-anilin	0.1 mM	100 μM
5-Cl,2,4-dimethoxy-anilin	6.0 mM	6 μM
4-Cl-2-NO_2-anilin	60 μM	20 μM
3-Cl-4-CH_3-anilin	0.5 mM	100 μM
2-OH-5-CH_3-anilin	22 μM	20 mM
3-NH_2-benzene acid	18 mM	1 mM
4 CH_3-anilin	0.5 mM	0.7 mM
4-methoxy-anilin	0.37 mM	–

Table 4. IC_{20} values for inhibition of RuBPcase and O_2 evolution induced by a range of hydrocarbons. (Data taken from the final report of H. Schnabl 1987 on the research programme supported by UBA, Berlin)

	RuBPcase	O_2-evolution
Methanol	0.8 M	0.4 M
Ethanol	0.38 M	0.15 M
1-propanol	0.15 M	30 mM
1-butanol	52 mM	10 mM
2-butanol	0.1 M	40 mM
iso-butanol	60 mM	20 mM
tert. butanol	0.16 M	70 mM
2-chlorethanol	74 mM	30 mM
2,3-dibrompropanol	6 mM	4 mM
2,2,2-trichlorethanol	70 mM	7 mM
Pentanol	22 mM	10 mM

The oxygen produced by immobilized protoplasts is measured on the basis of a definite chlorophyll amount per strip or bead (100 µg) and monitored using an oxygen electrode (Rank Brothers, Cambridge, UK), which is coupled with a recorder. The time required for one sample is about 20 min. Reduced activity of RuBPcase is assayed as $^{14}CO_2$ incorporation into acid-stable products and their determination by liquid scintillation (1.5 days are required for the measurement). The determination of ethane was followed, using the alginate strips containing the protoplasts with about 100 µg chlorophyll, which were incubated for 30 min in a phosphate buffer (0.2 M, pH 6) in Fernbach flasks. Incubations were performed at 22 °C with a light irradiance of 200 W/m^2. Samples (1 ml) of the gaseous headspace were withdrawn from flasks using a gas-tight syringe and analyzed for ethane by gas chromatography. This screening system with immobilized protoplasts was used for determining the degree of contamination of a sample of waste water derived from a sewage plant. An inhibition of oxygen production of 33% was measured indicating that the system responded well to typical levels of contamination.

Another simple screening system on the basis of immobilization techniques was developed by Lüttge et al. (1984) with immobilized yellow cells of *Galium mollugo* which turned their colour to orange when incubated in solution of ecotoxins. The influence of PCP, $HgCl_2$ and dodecylbenzosulphonic acid on the cells of *G. mollugo* entrapped in tablets of calcium-alginate matrix was investigated after a treatment of 24 h. The threshold concentrations to elicit a change of colour are: $HgCl_2$: 0.005 – 0.01 mM, PCP: 0.01 – 0.025 mM and dodecyl-benzenesulphonic acid: 0.05 – 0.1 mM.

Thus, immobilized protoplasts and cells may provide the basis for a suitable, sensitive and rapid assay to determine the effect of potential ecotoxins because of the high sensitivity and the simple ability to store the rest cells. As experiments have shown, this technique also permits a rapid screening of environmental water and soil samples.

6 Biotransformation

Biotransformations have been carried out with immobilized plant cells on a laboratory scale (Brodelius and Mosbach 1982; Rhodes 1985; Rosevear and Lambe 1985; Woodward 1985) and only a few model studies are presented here.

Brodelius et al. (1979) investigated the stereospecific conversion of the steroid digitoxin to digoxin, using alginate-immobilized cells of *Digitalis lanata*. The biotransformation involves hydroxylation at position 12. The biotransformation capacity of the immobilized cells was found to be unchanged compared to that of freely suspended cells, and increasing amounts of digoxin could be detected over 1 month in the medium. In the absence of digitoxin, no digoxin (or any other derivative of it) was produced, excluding the possibility of a de novo synthesis.

Using similar beads, Alfermann et al. (1980) found that β-methyl digitoxin could be converted into the clinically more useful β-methyl digoxin over a period of 2 months, whereby the cell productivity of the immobilized cells was higher than that of the free cells. However, a 50% loss of activity was measured on immobilization.

Using alginate-entrapped cells of *Daucus carota*, Jones and Veliky (1981) and Veliky and Jones (1981) reported transformation of digitoxeginin and gitoxigenin into the 5-hydroxy-derivative. Some loss of activity due to immobilization (in the order of 30%) and due to accumulation of inhibitory levels of the toxic substrates after about 2 weeks were observed.

The stereospecific reduction of monoterpenes by entrapped *Mentha* cells was studied by Galun et al. (1983) and Freeman (1984). Two cell lines capable of reducing (−) menthone to (+) neomenthol, or (+) pulegone to (+) isomenthone were immobilized in polyacrylamide-hydrazide cross-linked by glyoxal.

Transformation efficiency of the entrapped cells was as high as that recorded for the freely suspended cells. Furthermore, the distribution of both the precursor and product in the medium in relation to their content in the cells showed that less monoterpenes were retained in entrapped cells compared to the non-immobilized ones. In both biotransformations the total monoterpene recovery from the reaction medium after 24 h incubation was significantly higher in the case of entrapped cells. Repeated biotransformations by replacing the (−) menthone or (+) pulegone containing medium by fresh medium after 12 h resulted in the same yield.

Even though these experiments are promising, the potential of immobilized plant cells for biotransformation should not be overestimated. Less complex biocatalysts, such as immobilized enzymes and microbial cells (exhibiting only one membrane barrier which has to be passed by the secondary metabolite), may be much more effective for performance of such reactions.

7 Biosynthesis

Biosynthesis from precursors and de novo synthesis by immobilized plant cells have been investigated by several laboratories. This area has also been reviewed by Rhodes (1985), Rosevear and Lambe (1985) as well as by Brodelius and Mosbach (1982) and Brodelius (1985 a, b, c). Therefore, we would like to restrict ourselves to the discussion of particular results obtained.

Synthesis of various indole alkaloids from the common precursors tryptamine and secolaganin were studied by several authors using immobilized cells of *C. roseus*. Cells immobilized in various gels produced ajmalicine (which accumulated in the vacuoles) at approximately the same rate as the corresponding free cells, with the interesting exception of cells entrapped in an alginate matrix. These cells showed a significant increase in synthesis (up to 160% of that measured for freely suspended cells), which was explained by growth limitations. Brodelius and Nilsson (1980) also showed that the ability for the de novo synthesis of ajmalicine (with sucrose as the sole carbon source) is preserved in the immobilized cells. Synthesis of the related indole alkaloid serpentine was also achieved using alginate-entrapped cells of *C. roseus*. Enhanced synthesis could also be obtained by growth restrictions of the cells in the mechanical support. Continuous production of ajmalicine and serpentine for over 220 days was reported by Rosevaer and Lambe (1985) when entrapment of the cells was performed in a mixture of polyacrylamide and a viscous polysaccharide.

Alginate-immobilized cells of *Morinda citrifolia* were used for the production of anthraquinone from sucrose (Brodelius et al. 1979). These cells also showed an increased efficiency compared to the corresponding free cells under growth – (i.e. hormone) – limiting conditions. Entrapment of cells in hollow fibre reactors yielded similar results (Schuler 1981). The production of phenol and phenolic derivatives by cells of *Glycine max* and *Daucus carota* could be greatly enhanced over a period of 110 days, particularly when a membrane reactor with relatively large pores (125 μm) was used.

Lindsey and Yeoman (1984 a, b) using the foam immobilization technique described above, could demonstrate an increased productivity of capsacain in immobilized cells of *Capsicum frutescens*. In the absence of specific precursors, entrapped cells produced between two and three orders of magnitude higher yields than did freely suspended cells over 5-day (log-linear growth phase) or 10-day (earlier stationary growth phase) culture periods. In agreement with the hypothesis of an inverse relationship between culture growth rate (and correspondingly limited primary metabolic activity) and secondary metabolite accumulation, the authors found that the rate and extent of incorporation of radioactivity from ^{14}C phenylalanine was relatively greater into capsaicin than into protein in slow-growing, immobilized cells after 5 days than in rapidly growing freely suspended cells of the same age.

Further support for the growth/secondary metabolite production hypothesis was obtained from stimulation of accumulation of capsaicin under reduced growth conditions and from corresponding studies of the synthesis and accumulation of protein and capsaicin in immobilized cells cultured in media containing reduced concentrations of essential nutrients such as phosphate and/or nitrate. The data demonstrated that, in general, under those conditions in which the incorporation of radioactivity from ^{14}C-labelled phenylalanine into protein is greatest (at high intracellular nitrate concentrations), the incorporation of the label into capsaicin was lowest. The correlation between the growth index of the cultures and the incorporation of the precursor into protein was also apparent. The antagonism between primary and secondary metabolism broke down only when the intracellular concentrations of essential nutrients reached relatively low levels, presumably too low to maintain secondary biosynthetic activity. This was observed when the cells were kept for about 2 weeks in media lacking phosphate and nitrate or for 3 weeks in media containing either no nitrate or very low amounts.

In other experiments the authors showed that relatively high levels of capsaicin were measured in the medium within 5 h of culture (in growth medium and in the absence of specific precursors), and the levels detected thereafter fluctuated by a factor of 4 over a period of 12 weeks. In these experiments, medium was sampled for analysis of capsaicin and spent medium was replaced by fresh one at intervals of 2 weeks. A similar phenomenon has been reported for the extracellular accumulation of steroidal glycoalkaloids by cells of *Solanum aviculare* immobilized on polyphenyleneoxide (Jirku et al. 1981). Lindsey and Yeoman (1984a) explained these results in terms of a cyclical passive uptake form and presumably rapid release back into the medium. Since high yields were not accumulated intracellularly and the cumulative yield of capsaicin did not increase over the culture period, there may be a feedback inhibition of synthesis of this specific compound. The authors therefore suggested that, in a large-scale production system, capsaicin and perhaps other secondary products should be extracted continuously if the yield is to be maximized. Production of capsaicin could be further increased by supplementing the medium with precursors such as phenylalanine, isocapric acid and ferulic acid 11. Accumulation of capsaicin was also retained during prolonged culture periods (of up to 12 weeks' duration).

8 Product Release

As mentioned above, the desired products are often retained in storage vacuoles. The release of such products from plant cells would involve permeabilization of both the tonoplast and plasmalemma. Different approaches were developed to overcome this problem. Spontaneous release of normally intracellular stored products from a few immobilized plant species is reported (see review Brodelius 1985a) indicating that the state of immobilization affects membrane permeability under some circumstances. The underlaying mechanism is not understood.

In terms of the thermodynamics of irreversible processes, spontaneous release can be explained by the assumption that the reflection coefficients of some of the intracellular and/or extracellular compounds have been decreased due to the immobilization process and/or due to the storage in the solid support. The reason for this may be either an irreversible slight increase in membrane permeability on immobilization or an accumulation of toxic compounds (e.g. secondary products under growth limitation conditions). From our work on the delay of the ageing process in immobilized plant protoplasts, we know that the processes linked with senescence of plant protoplasts proceed if membrane integrity is affected by chemical (e.g. very small concentrations of heavy metal ions, see above) or physical means. Contamination of the alginate gel by such heavy metal ions might be one of the reasons because the quality of the commercial products can change considerably (Klenke and Zimmermann, unpubl. results). It is also well known that the toxicity of some secondary products, e.g. menthol (quoted in Rosevaer and Lambe 1985) can limit the operation range of immobilized *Mentha* cells. In this context, it is worth noting that even a specific increase in membrane permeability (due to environmental factors or released intracellular compounds) can dramatically change the osmotic balance provided that the compound, the reflection coefficient of which is affected, contributes significantly to the osmotic pressure. In any case, independently of which of these explanations is appropriate for the particular immobilized species, a decrease of the values of the reflection coefficients of some osmotically relevant extracellular compounds will lead to a decrease in the effective external osmotic pressure and, therefore to a net flow of water into the cells. On the other hand, if the critical turgor pressure and the critical volume, respectively of the cells (Zimmermann 1978) is reached (see above and Fig. 8) repeated spontaneous release of compounds can take place. Such processes are also observed in many biological cells, e.g. in *Acetabularia mediterranea* (Wendler et al. 1983). In this species the turgor pressure is not constant, but increases steadily due to potential-dependent uptake of K^+-ions up to a critical value of about 2.5 bar (at room temperature). Above this critical value electrically silent chloride burst occurred, resulting in a reduction of the turgor pressure followed repeatedly by an increase in turgor pressure associated with monovalent salt extrusion. In the light of this concept, the following result of Lambe et al. (1982) can also be easily explained. These authors found that cells of *C. roseus* immobilized in early stationary or late phases continued to release product at high levels for prolonged period, while very old cells rapidly released their alkaloid and died. It is well known (Zimmermann et al. 1981 a) that cells harvested from the stationary phase exhibit much lower values of the reflection coefficients for many extracellular compounds than those harvested from the log phase. Use of the same medium composition for both growth stages leads therefore to water uptake in the cells harvested from the stationary phase and, in turn, to an irreversible burst because of the very high hydrostatic gradients set up under these conditions. It is evident that there must exist a continuous transition from reversible bursts to cell lysis, depending on the magnitude of the pressure gradients built up across the membrane and cell wall barrier.

Lambe et al. (1982) also found that repeated changes of the medium in batch operation promotes release of ajmalicine and serpentine from *C. roseus* cells immobilized in alginate/acrylamide. This finding can be explained in the light of the considerations outlined above if we assume that the secondary products or other toxic compounds result in continuous increase of membrane permeability (and correspondingly in a decrease of the reflection coefficients of the respective compounds), provided that these compounds are not removed. The present authors believe strongly that the concept of reflection coefficients has undoubtedly to be taken into account in further research in the area of immobilized cells. In particular, measurements of the reflection coefficients of cells under various immobilized conditions over long periods of use are required to arrive at an advanced stage of this technique. This statement also holds for chemical treatment of the membranes of immobilized cells. Such treatments were used to artificially increase membrane permeability in order to enhance release of intracellularly stored products into the external medium.

Felix et al. (1981) showed that dimethyl sulphoxide (DMSO, which is used in cryoprotection) increased the extrusion of ajmalicine from storage vacuoles to the culture medium. The concentration of DMSO required for permeabilization depended on the plant species (5% in the case of *C. roseus*, 10 and 25% in the case of *Daucus carota* and *Datura innoxia*, respectively) as is expected in the concept of reflection coefficients introduced here. Other substances, such as nystatin, ether and lysolecithine, were also used (Hartmeyer 1986). Even though the viability of the cells was not affected under some conditions by this chemical treatment, lysis of the cells can occur very often.

Hartmeyer (1986) proposes the use of advanced methods, which allow a reversible permeabilization of the tonoplast and plasmalemma of plant cells and protoplasts. With such procedures, alternating changes between production and harvesting of the desired intracellular product would be conceivable, but also facilitated uptake of substrates precursors into the cell.

One approach is certainly controlled osmotic- and tugor-pressure-induced release of compounds as discussed above. This method requires the knowledge of the reflection coefficients of the compounds under various conditions. Another very promising way to change reflection coefficients of compounds in a reversible manner is the application of the electrical breakdown technique introduced by Zimmermann and coworkers (Zimmermann et al. 1981 a, b; Zimmermann 1982, 1986). In this technique cells are exposed to field pulses of high intensity (kV/cm range, depending on the radius of the cell) for very short durations (nano- to a couple of microseconds). If the critical breakdown voltage of the membrane in the order of 1 V (depending on temperature and pulse duration) is exceeded, break-down of the membrane occurred at membrane sites oriented in field direction. Application of supracritical pulses or application of a train of critical or supracritical pulses to the cell suspension lead to breakdown in larger membrane areas. Breakdown results in a dramatic increase in membrane permeability. Using the appropriate field strength, the permeability increase of the membrane can be matched to the size of the normally membrane-impermeable compound which must be made to cross the membrane. The field-induced high permeability stage of the membrane is reversible and the original membrane impermeability is restored (particularly at elevated temperature) provided that the pulse duration was short enough and the field intensity of appropriate magnitude. Otherwise – as with pressure gradient application – the reversible electrical breakdown goes over into an irreversible destruction of the cell membrane associated with cell lysis.

The electric field pulse technique was successfully used for entrapment of various compounds of low molecular and high molecular weight (such as enzymes and plasmid DNA) (Stopper et al. 1985, 1987, 1988), as well as of organelles and whole cells into a diversity of microbial, plant and animal cells (see review article, Zimmermann 1986). There is now a body of evidence that this procedure is very mild and harmless to the treated cells. Viability, growth and division of the cells is not affected. There are also some reports that this electropermeabilization technique can be used for the repeated release of compounds from cells (see review article, Knight and Scrutton 1986). Most of this work was performed on animal cells; however, recent studies in our laboratory (Zimmermann and Klenke, unpubl.) showed, that extrusion of secondary metabolites can be enhanced by field pulse application. In particular, when using immobilized plant cells, it could be demonstrated that release of these products could be further enhanced by repeated application of pulse trains over a period of 2 weeks. From these results it was obvious that – in analogy to the reversible turgor pressure-induced release of the substances mentioned above – the gel support protected the cells against burst once water has been taken up due to the (non-specific) field-induced decrease of the reflection coefficients of the osmotic relevant solutes in the external medium. From these preliminary and – as we feel – encouraging experiments we conclude that the longer-term prospects of this field pulse technique are quite promising for commercial applications of immobilized cells. However, more extensive studies are required to evaluate the potential of this technique for use in immobilized cell systems.

9 Conclusions

Immobilized plant cells and protoplasts have shown several promising advantages over freely suspended cells. The most promising areas are in basic research, in the production of biochemicals and artificial seeds, as well as in the development of bioassays. Other applications using somatic hybridization can be expected, corresponding to progress in genetic engineering and plant regeneration. Despite the large number of publications to date, the processes and applications based on the use of immobilized plant cells are so far only operative on a laboratory scale. However, research in this field is still in its infancy despite considerable progress in the last 5 years. As should be clear from the range of topics discussed in this chapter, the search must continue for a more thorough understanding of the biochemistry and biophysics of immobilized plant cells and for further improvements of the immobilization techniques, as well as for the development of reversible permeabilization methods. Suggestions for methods of solving these problems have been given. It is particularly fascinating to see ways in which the combination of different fields and methods may immediately provide a key for the solution of problems and a spring board into useful technology.

10 Protocol for the Immobilization of Protoplasts

In our studies on *Vicia faba* (Weisskernige Hangdown), the enzymatically isolated protoplasts are purified on a sucrose/mannitol gradient, pelleted and resuspended in mannitol (0.5 M). An aliquot of the protoplast suspension (containing 100 µg chlorophyll) is gently mixed with sodium alginate (7%; Manucol LD, Alginate Ind. Ltd., London) in a volume of 500 µl, spread onto a slide and covered with a solution of $CaCl_2$ (100 mM; 0.4 M mannitol). After 5 min the Ca^{2+}-concentration is lowered to 5 mM. The immobilized protoplasts are stored at 4 °C in the dark.

Acknowledgements. This work was supported by grants of the Bundesministerium für Forschung und Technologie to H. Sch. (no. 037269) and to U. Z. (no. 0373509) and of the Umweltbundesamt, Berlin to H. Sch. (no. 10603045) and of SFB 176 (B4 to U.Z.).

References

Alfermann AW, Schuller I, Reinhard E (1980) Biotransformation of cardiac glycosides by immobilized cells of *Digitalis lanata*. Planta Med 40:218–223

Arnold WM, Zimmermann U (1982) Rotating-field-induced rotation and measurement of the membrane capacitance of single mesophyll cells of *Avena sativa*. Z Naturforsch 37c:908–915

Arnold WM, Zimmermann U (1984) Electric field-induced fusion and rotation of cells. In: Chapman D (ed) Biological membranes, vol 5. Academic Press, London, pp 389–454

Arnold WM, Wendt B, Zimmermann U, Korenstein R (1985) Rotation of a single swollen thylakoid vesicle in a rotating electric field. Electrical properties of the photosynthetic membrane and their modification by ionophores, lipophilic ions and pH. Biochim Biophys Acta 813:117–131

Arnold WM, Geier BM, Wendt B, Zimmermann U (1986) The change in electrorotation of yeast cells effected by silver ions. Biochim Biophys Acta 889:35–48

Bornman CH, Zachrisson A (1982) Immobilisation of protoplasts by anchoring to microcarriers. Plant Cell Rep 1:151–153

Brodelius P (1983) Production of biochemicals with immobilized plant cells: possibilities and problems. Ann NY Acad Sci 413:383–393

Brodelius P (1984) Immobilization of cultured plant cells and protoplasts. In: Vasil IK (ed) Cell culture and somatic cell genetics of plants, vol 1. Academic Press, Orlando, Fla, pp 535–546

Brodelius P (1985a) Immobilized plant cells. In: Laskin AI (ed) Applications of isolated enzymes and immobilized cells to biotechnology. Addison-Wesley, Reading, pp 109–148

Brodelius P (1985b) The potential role of immobilization in plant cell biotechnology. Trends Biotechnol 3:280–285

Brodelius P (1985c) Immobilized plant cells: preparation and biosynthetic capacity. In: Woodward J (ed) Immobilized cells and enzymes: a practical approach. IRL Press, Oxford, pp 127–145

Brodelius P, Mosbach K (1982) Immobilised plant cells: general aspects. J Chem Biotechnol 32:330–337

Brodelius P, Nilsson K (1980) Entrapment of plant cells in different matrices. FEBS Lett 122:312–316

Brodelius P, Vogel HJ (1985) A phosphorus-31 nuclear magnetic resonance study of phosphate uptake and storage in cultured *Catharanthus roseus* and *Daucus carota* plant cells. J Biol Chem 260:3356–3560

Brodelius P, Deus B, Mosbach K, Zenk H (1979) Immobilised plant cells for the production and transformation of natural products. FEBS Lett 103:93–97

Büchner K-H, Zimmermann U (1982) Water relations of immobilized giant algal cells. Planta 154:318–325

Büchner K-H, Zimmermann U, Bentrup FW (1981) Turgor pressure and water transport properties of suspension-cultured cells of *Chenopodium rubrum* L. Planta 151:95–102

Chibata I (1978) Immobilized enzymes. John Wiley, New York

Dainty AL, Goulding KH, Robinson PK, Simpkins I, Trevan MD (1986) Stability of alginate-immobilised algal cells. Biotechnol Bioeng 28:210–216

Elstner EF, Konze JR (1976) Effect of point freezing on ethylene and ethane production by sugar leaf disks. Nature 263:351–352

Felix HR, Mosbach K (1982) Enhanced stability of enzymes in permeabilised and immobilised cells. Biotechnol Lett 4:181–186

Felix HR, Brodelius P, Mosbach K (1981) Enzyme activities of the primary and secondary metabolism of simultaneously permeabilized and immobilized plant cells. Anal Biochem 116:462–470

Freeman A (1984) Gel entrapment of whole cells and enzymes in cross-linked, prepolymerized polyacrylamide-hydrazide. Ann NY Acad Sci 434:418–426

Fuhr GR (1985) Über die Rotation dielektrischer Körper in rotierenden Feldern. Dissertation, Humboldt Universität, Berlin

Galun E, Aviv D, Dantes A, Freeman A (1983) Biotransformation by plant cells immobilized in cross-linked polyacrylamide-hydrazide. Planta Med 49:9–13

Hall RD, Holden MA, Yeoman MM (1988) Immobilization of higher plant cells. In: Bajaj YPS (ed) Biotechnology in agriculture and forestry 4. Medicinal and aromatic plants I. Springer, Berlin Heidelberg New York Tokyo, pp 136–156

Hampp R, Mehrle W, Zimmermann U (1986) Assay of photosynthetic oxygen evolution from single protoplasts. Plant Physiol 81:854–858

Hannoun BJM, Stephanopoulos G (1986) Diffusion coefficients of glucose and ethanol in cell-free and cell-occupied calcium alginate membranes. Biotechnol Bioeng 28:829–835

Hartmeyer W (1986) Immobilisierte Biokatalysatoren. Springer, Berlin Heidelberg New York Tokyo

Hiemstra H, Dykhuizen L, Harder W (1983) Diffusion of oxygen in alginate gels related to the kinetics of methanol oxidation by immobilized *Hansenula polymorpha* cells. Eur J Appl Microbiol Biotechnol 18:189–196

Hulst AC, Tramper J, Riet K van't, Westerbeck JMM (1985) A new technique for the production of immobilized biocatalysts in large quantities. Biotechnol Bioeng 27:870–876

Jones A, Veliky IA (1981) Effects of medium constituents on the viability of immobilised plant cells. Can J Bot 59:2095–2101

Jose W, Pedersen H, Chin Ch (1983) Immobilisation of plant cells in a hollow-fiber reactor. Ann NY Acad Sci 413:409–280

Jirku VT, Macek T, Vanerr T, Krumphanzl V, Kubanek V (1981) Continuous production of steroid glycoalcaloid by immobilized plant cells. Biotechnol Lett 3:447–450

Kennedy JF, Cabral JMS (1983) Immobilised living cells and their applications. Appl Biochem Bioeng 4:189–280

Knight DE, Scrutton C (1986) Gaining access to the cytosol: the technique and some applications of electropermeabilization. Biochem J 234:497–506

Knobloch K-H, Berlin J (1980) Influence of medium composition on the formation of secondary compounds in cell suspension cultures of *Catharanthus roseus* (L.) G Don. Z Naturforsch 35c:551–556

Komel R, Rosman D, Puck A, Sočič H (1985) Effect of immobilisation on the stability of *Claviceps purpurea* protoplasts. Appl Microbiol Biotechnol 23:106–109

Konze JR, Elstner EF (1978) Ethane and ethylene formation by mitochondria as indication of aerobic lipid degradation in response to wounding of plant tissue. Biochim Biophys Acta 528:213–221

Lambe CA, Reading A, Roe S, Rosevaer A, Thomson AR (1982) Production of alkaloids with immobilized cells of *Catharanthus roseus*. Enzyme Eng 6:137–138

Lieberman M, Mapson LW (1962) Fatty acid control of ethane production by sub-cellular particles from apples and its possible relationship to ethylene biosynthesis. Nature 195:1016–1017

Lindsey K (1985) Manipulation, by nutrient limitation, of the biosynthetic activity of immobilised cells of *Capsicum frutescens* Mill cv. *annuum*. Planta 165:126–133

Lindsey K, Yeoman MM (1983) The relationship between growth rate, differentiation and alkaloid accumulation in cell cultures. J Exp Bot 34:1055–1065

Lindsey K, Yeoman MM (1984a) The synthetic potential of immobilised cells of *Capsicum frutescens* Mill. cv. *annuum*. Planta 162:495–501

Lindsey K, Yeoman MM (1984b) The viability and biosynthetic activity of cells of *Capsicum frutescens* Mill. cv. *annuum* immobilized in reticulate polyurethane. J Exp Bot 35:1684–1696

Lindsey K, Yeoman MM, Black GM, Mavituna F (1983) A novel method for the immobilization and culture of plant cells. FEBS Lett 155:143–149

Lörz H, Larkin PJ, Thomson J, Scowcroft WR (1983) Improved protoplast culture in agarose media. Plant Cell Tissue Organ Cult 2:217–226

Lüttge U, Sellner M, Schnabl H, Zimmermann U (1984) Ökotoxikologische Bewertung von Umweltchemikalien und Entwicklung von biologischen Testsystemen aufgrund der Eigenschaften pflanzlicher Membranen. GIT-Suppl 4:36–42

Mavituna F, Park JM (1985) Growth of immobilised plant cells in reticulate polyurethane foam matrices. Biotechnol Lett 7:637–640

McDowell RH (1977) Properties of alginates. Alginate Industries, London

Mosbach K (1984) New immobilization techniques and examples of their applications. Ann NY Acad Sci 434:239–248

Nilsson K, Birnbaum St, Flygare S, Linse L, Schröder U, Jeppson U, Larsson P-O, Mosbach K, Brodelius P (1983) A general method for the immobilisation of cells with preserved viability. Eur J Appl Microbiol Biotechnol 17:319–326

Parr AJ, Smith J, Robins RJ, Rhodes MJ (1984) Apparent free space and cell volume estimation: a non-destructive method for assessing the growth and membrane integrity/viability of immobilised plant cells. Plant Cell Rep 3:161–164

Pilwat G, Washausen P, Klein J, Zimmermann U (1980) Immobilisation of human red blood cells. Z. Naturforsch 35c:352–356

Redenbaugh K, Paasch BD, Nichol JM, Kossler ME, Viss PR, Walker KA (1986) Somatic seeds: encapsulation of asexual plant embryos. Biotechnology 4:797–801

Rhodes MJC (1985) Immobilized plant cells cultures. Top Enzyme Ferment Biotechnol 10:51–87

Riely LA, Cohen G, Lieberman M (1974) Ethane evolution: a new index of lipid peroxidation. Science 183:208–210

Rosevear A, Lambe CA (1985) Immobilized plant cells. Adv Biochem Eng Biotechnol 31:37–58

Sauer FA, Schlögl RW (1985) Torques exerted on cylinders and spheres by external electromagnetic fields: a contribution to the theory of field induced cell rotation. In: Chiabrera A, Nicolini C, Schwan HP (eds) Interactions between electromagnetic fields and cells. Plenum, New York, pp 203–251

Scheurich P, Schnabl H, Zimmermann U, Klein J (1980) Immobilisation and mechanical support of individual protoplasts. Biochim Biophys Acta 598:645–651

Schnabl H, Youngman RJ (1985) Immobilisation of plant cell protoplasts inhibits enzymic lipid peroxidation. Plant Sci 40:65–69

Schnabl H, Youngman RJ (1987) A characterization of the effects of ecotoxicological substances on plant cell metabolic reactions. Angew Bot 61:493–504

Schnabl H, Scheurich P, Zimmermann U (1980) Mechanical stabilisation of guard cell protoplasts of *Vicia faba*. Planta 149:280–182

Schnabl H, Youngman RJ, Zimmermann U (1983a) Maintenance of plant cell membrane integrity and function by the immobilisation of protoplasts in alginate matrices. Planta 158:392–397

Schnabl H, Elbert Chr, Youngman RJ (1983b) Release of ethane from immobilised plant cell protoplasts in response of chemical treatment. Physiol Plant 59:46–49

Schnabl H, Künne J, Florschütz F, Youngman RJ (1984) Differential effects of ecotoxicological substances on lipid peroxidation of plant cell protoplasts. Plant Cell Environ 7:499–505

Schümmer P, Tebel KH (1983) A new elongational rheometer for polymer solutions. J Non-Newtonian Fluid Mech 12:331–347

Schuler ML (1981) Production of secondary metabolites from plant tissue culture – problems and prospects. Ann NY Acad Sci 369:65–79

Shillito RD, Paszkowski J, Potrykus I (1983) Agarose plating and a bead type culture technique enable and stimulate development of protoplast-derived colonies in a number of plant species. Plant Cell Rep 2:244–247

Stopper H, Zimmermann U, Wecker E (1985) High yields of DNA transfer into mouse L-cells by electropermeabilization. Z Naturforsch 402c:133–139

Stopper H, Jones H, Zimmermann U (1987) Improved DNA-transfection by electropermeabilization. Biochim Biophys Acta 900:38–44

Tanaka H, Matsumara M, Veliky IA (1984) Diffusion characteristics of substrates in calcium-alginate gel beads. Biotechnol Bioeng 26:53–58

Tharakan JP, Chau PC (1986) Operation and pressure distribution of immobilised cell hollow fiber bioreactors. Biotechnol Bioeng 28:1064–1071

Thompson JA, Abdullah R, Cocking EC (1986) Protoplast culture of rice (*Oryza sativa* L.) using media solidified with agarose. Plant Sci 47:123–133

Tomos AD, Steudle E, Zimmermann U, Schulze E-D (1981) Water relations of leaf epidermal cells of *Tradescantia virginiana*. Plant Physiol 68:1135–1143

Tricoli DM, Hein MB, Carnes MG (1986) Culture of soybean mesophyll protoplasts in alginate beads. Plant Cell Rep 5:334–337

Veliky IA, Jones A (1981) Bioconversion of gitoxigenin by immobilised plant cells in a column bioreactor. Biotechnol Lett 3:551–554

Watts MJ, Collin HA (1985) Growth and nutrient uptake by immobilized tissue culture cells of celery (*Apium graveolens*). Plant Sci 42:67–72

Wendler S, Zimmermann U, Bentrup F-W (1983) Relationship between cell turgor pressure, electrical membrane potential and chloride efflux in *Acetabularia mediterranea*. J Membr Biol 72:75–84

Widholm JM (1972) The use of fluorescein diacetate and phenosafranine for determining viability of cultured plant cells. Stain Technol 47:189–194

Woodward J (1985) Immobilized cells and enzymes; a practical approach. IRL-Press, Oxford

Zimmermann U (1978) Physics of turgor- and osmoregulation. Ann Rev Plant Physiol 29:121–148

Zimmermann U (1982) Electric field mediated fusion and related electrical phenomena. Biochim Biophys Acta 694:227–277

Zimmermann U (1986) Electrical breakdown, electropermeabilization and electrofusion. Rev Physiol Biochim Pharmacol 105:175–256

Zimmermann U, Steudle E (1978) Physical aspects of water relations of plant cells. Adv Bot Res 6:45–117

Zimmermann U, Scheurich P, Pilwat G, Benz R (1981 a) Cells with manipulated functions: new perspectives for cell biology, medicine, and technology. Angew Chem Int Ed Engl 20:325–344

Zimmermann U, Vienken J, Pilwat G (1981 b) Rotation of cells in an alternating electric field: the occurrence of a resonance frequency. Z Naturforsch 36c:173–177

I.6 Cryopreservation of Plant Protoplasts

Y. P. S. BAJAJ[1]

1 Introduction

During the last decade, plenty of literature has accumulated on the successful cryopreservation of cells, tissues and organ cultures of a large number of plant species (Bajaj 1986, 1988a). These cultures, freeze-preserved for various lengths of time, retain their morphogenetic potential and regenerate complete plants. However, not much information is available on the culture of frozen protoplasts (Takeuchi et al. 1982; Bajaj 1987). Although for all practical purposes freezing of protoplasts is similar to that of isolated cells and cell suspension (Bajaj 1976), they differ from the cells in their lack of a cell wall and plasmodesmata, which makes them more sensitive to bursting. Thus they require a critical concentration of an appropriate osmoticum cum cryoprotectant to prevent them from bursting or causing undue damage to the plasma membrane.

Steponkus (1985) has reviewed the cryobiological aspects of isolated protoplasts including intra-, and intercellular freezing, ice crystal formation, and the possible mechanisms involved. It would suffice to summarize his paper by pointing out that primarily there are three types of cryoinjury in the isolated protoplasts, (i) intracellular ice formation, (ii) expansion-induced lysis, and (iii) loss of osmotic responsiveness. It has been suggested (Steponkus and Dowgert 1981; Steponkus et al. 1982) that the nucleation event is primarily the result of mechanical breakdown of the plasma membrane per se rather than a consequence of temperature-dependent characteristics of the external ice. With high resolution video cryomicroscopy, outward flow of the intracellular solution is observed before intracellular ice formation, thus effecting contact with the extracellular ice matrix. Lysis occurs when the protoplasts are osmotically expanding and is hence termed "expansion-induced lysis". Lysis is a consequence of large area deformations incurred by the plasma membrane as a result of osmotically induced contraction-expansion of the protoplast during a freeze-thaw cycle rather than alterations in the osmotic characteristics (Steponkus and Wiest 1979; Wiest and Steponkus 1978). In fact, freezing injury could be due to a number of interdependent factors, i.e. intracellular freezing, dehydration, increase in the concentrations of solutes, formation of disulfide bonds, and also due to mechanical stress and rupture of the membrane.

[1] Former Professor of Tissue Culture, Punjab Agricultural University, Ludhiana, India.
Present address: A-137, New Friends Colony, New Delhi 110065, India

Biotechnology in Agriculture and Forestry, Vol. 8
Plant Protoplasts and Genetic Engineering I (ed. by Y. P. S. Bajaj)

Thorough understanding of various cryobiological aspects of protoplasts will add to the present knowledge of handling then more efficiently and increasing their survival. In fact, prevention of bursting of the protoplasts and the formation of intracellular ice formation and recrystallization are the main points to be carefully studied and manipulated with the osmoticum, cryoprotectants, and the method and velocity of freezing-thawing. The basic and fundamental aspects of the cryobiology of cells and protoplasts and the mechanisms involved have already been discussed (see Bajaj and Reinert 1977; Steponkus 1985). In this chapter literature on the culture of frozen protoplasts is reviewed and the significance of cryopreservation of isolated protoplasts is pointed out.

2 Significance of Cryopreservation of Protoplasts

In addition to the various uses of cryopreservation of cells already discussed (Bajaj 1988a), i.e. conservation of germplasm, retention of biosynthetic potentials of cells, and the maintenance of genetic stability of clones, the isolated protoplasts offer the following potentials.

(1) The protoplasts differ from cell suspensions in two important aspects, the lack of cell wall and the cytoplasmic connections with the adjacent cells. Thus protoplasts, in not being interdependent, are in fact a better system to study the mechanism of cryopreservation at a cellular level. Isolated protoplasts provide a system in which the effect of freezing on plasma membrane can be studied directly in situ.

(2) Isolated protoplasts provide an ideal system for quantitative cryomicroscopic studies (Steponkus et al. 1983). The cryomicroscopy of isolated protoplasts allows for a direct analysis of intracellular ice formation (Steponkus and Dowgert 1981; Dowgert and Steponkus 1983) and it also enables the precise control of temperature and cooling rate. With video image analysis the temperature at which intracellular nucleation occurs is observed directly.

(3) For genetic manipulation studies, since suspension cultures contain clumps of cells, the "clones" can often be from different cells. The cryopreservation of protoplasts would thus eliminate a cloning step after thawing, since all regrown colonies are the off-spring of single cells. Freeze-preservation of protoplasts reduces the number of steps in cloning mutant cells (Weber et al. 1983).

3 **Review of Work** (Table 1)

The freeze-preserved protoplasts, when brought to cultures, divide, grow and retain their morphogenetic potentials. The work done, although limited, is reviewed in Table 1.

Table 1. Summary of the work done on the cryopreservation of isolated plant protoplasts

Plant species	Viability and growth response	Reference
Atropa belladonna	44% cultures underwent differentiation compared to 71% control; plant regeneration	Bajaj (1988b)
Bromis inermis	Some protoplasts were viable, as shown by trypan blue dye	Mazur and Hartmann (1979)
Daucus carota	30% survival, plant regeneration	Takeuchi et al. (1980), Withers (1980), Hauptmann and Widholm (1982)
Datura innoxia	Colony formation in 5–6 weeks, occasional regeneration	Bajaj (1988b)
Glycine max	60% viability, callus	Takeuchi et al. (1982)
	Cell suspension	Weber et al. (1983)
Hordeum vulgare	68% viability, callus	Takeuchi et al. (1982)
Marchantia polymorpha	40% viability, normal thalli and rhizoids	Takeuchi et al. (1980, 1982)
Nicotiana tabacum	36% cultures underwent differentiation as compared to 53% control, plant regeneration	Bajaj (1988b)
Oryza sativa × *Pisum sativum*	10.8–13.9% survival	Bajaj (1983a)
Oryza sativa × *Sorghum bicolor*	10.3 to 13.8% survival	Bajaj (1983c)
Triticum aestivum	33% viability, callus	Takeuchi et al. (1982)
Triticum aestivum × *Pisum sativum*	12.2–17.9% survival	Bajaj (1983a)

4 Isolation, Freezing and Culture of Protoplasts

The isolation, freezing and culture of retrieved protoplasts of *Atropa belladonna, Datura innoxia* and *Nicotiana tabacum* (Bajaj 1988b) is summarized here. The details regarding isolation, media and culture conditions have been previously described for *Atropa* (Gosch et al. 1975), *Datura* (Furner et al. 1978) and *Nicotiana* (Bajaj et al. 1978).

For freezing studies, to 0.5 ml of protoplast suspension (10^4/ml) in 2 ml ampoules, an equal amount of double strength cryoprotectant solution (10% DMSO, 1 M mannitol, 4% sucrose) was gradually added drop by drop over a period of about 30 min (sudden mixing resulted in bursting or damage to the protoplasts). The ampoules were subjected to vapours of liquid nitrogen, and then gradually lowered into the LN cylinder (Bajaj 1983a) and kept for 5 min. Thereafter the ampoules were immediately thawed in a beaker containing warm water (35 °C) and centrifuged. Of the liquid 0.5 ml was pipetted out and the protoplasts were cultured in their respective media in Petri dishes. The cultures were periodically observed; visible colony formation and the regeneration of shoots/plants was recorded. The in vitro-produced plants were transferred to pots, and their further growth, flowering, and seed set was observed.

4.1 Culture of Control Protoplasts

Tobacco and *Datura* protoplasts were directly plated on the agar medium, while *Atropa* protoplasts were first grown in drop culture in the liquid medium and incubated in the dark. When colonies of cells were formed they were poured onto the top of an agar medium, the callus thus formed being used for differentiation studies.

First division was noticed after 2–4 days in tobacco, 4–5 days in *Atropa*, and 6–7 days in *Datura*. The protoplasts continued to divide to form colonies and eventually underwent differentiation; the frequency of colony formation and regeneration varied a great deal with the plant species.

In *Atropa* there were three modes of morphogenesis, i.e. rhizogenesis, shoot bud differentiation and embryogenesis (Bajaj et al. 1978), whereas in tobacco and *Datura* only organogenesis was observed. In *Atropa* and tobacco invariably complete plants were recovered, in *Datura* it was only occasional, and mostly bunches of shoots were formed.

Experiments were also conducted to see if the cryoprotectant solutions had any undesirable effects on protoplasts. For this purpose the protoplasts were treated with a mixture of mannitol, dimethyl sulfoxide and sucrose and cultured (Table 2). Initially the cryoprotectant-treated protoplasts looked unhealthy as compared to the control, and had a tendency to show a delay of 3–5 days in first division. Subsequently they continued to divide and formed colonies. There was no apparent difference between the colonies, except that they appeared smaller. However, with the passage of time, the untreated and the cryoprotectant-treated ones behaved similarly as far as the growth behaviour of the callus and its subsequent differentiation is concerned (Table 2).

4.2 Culture and Frozen Protoplasts

Earlier attempts to freeze protoplasts of *Datura innoxia* and *Nicotiana tabacum* were unsuccessful (Hauptmann and Widholm 1982); however, carrot protoplasts subjected to -30 and $-40\,°C$ before immersion in LN and stored for 4 months revived (Takeuchi et al. 1980, 1982). Later Weber et al. (1983) observed that partial dehydration of soybean protoplasts frozen in 0.7 M sorbitol without any added cryoprotectant and stored in LN resulted in clones with an efficiency of 10^{-3} as compared to non-frozen controls, whereas glucose was unsuitable as a cryoprotectant.

The observations on growth frequency, colony formation and regeneration of *Atropa, Datura* and tobacco protoplasts are summarized in Table 3. Initially in general the protoplasts lay quiescent for 3–5 days as compared to controls (Table 3). After that they started to elongate, the chloroplasts became scanty and yellowish, and reorganized themselves at the periphery. The first division and the subsequent colony formation were invariably delayed in all three species; however, once they regenerated walls they looked like the control protoplasts (Fig. 1 C). The frozen tobacco and *Datura* protoplasts in culture after the regeneration of wall behaved like their frozen cell suspensions (Bajaj 1976). Tobacco callus eventually

Table 2. Effect of cryoprotectant solution (DMSO + mannitol + sucrose) on the culture behaviour and regeneration in protoplasts of *Atropa belladonna, Datura innoxia* and *Nicotiana tabacum*. (Bajaj 1988b)

Species	Division						Colony formation (weeks)		Regeneration %	
	4 days		7 days		14 days					
	Control	Treated	Control	Treated	Control	Treated	Control	Treated	Control	Treated
Atropa belladonna	+	–	+ +	+	+ +	+ +	3	3	69	66
Datura innoxia	+	–	+	+	+ +	+ +	4	4 – 5	Occasional	
Nicotiana tabacum	+	+	+ +	+	+ +	+ +	3	4	56	55

Table 3. In vitro response of isolated protoplasts of *Atropa belladonna*, *Datura innoxia* and *Nicotiana tabacum* subjected to freezing (– 196 °C) in the presence of dimethyl sulphoxide, mannitol and sucrose and thawed at 35 °C. (Bajaj 1988b)

Plant species	Origin of protoplasts	Culture medium	Growth frequency		Colony formation weeks		Regeneration %	
			Control	Frozen	Control	Frozen	Control	Frozen
Atropa belladonna	Leaf mesophyll	Gosch et al. (1975)	+ + +	+ +	3	4	71	44
Datura innoxia	Cell suspension	Furner et al. (1978)	+ +	+	4	5 – 6	Occasional	
Nicotiana tabacum	Leaf mesophyll	Bajaj et al. (1978)	+ + +	+ +	3	4	53	36

Fig. 1 A–G. Regeneration of plants from protoplasts of *Atropa belladonna, Datura innoxia* and *Nicotiana tabacum* frozen in liquid nitrogen. **A** Mesophyll protoplasts of *Atropa* after freezing and thawing. **B** Same, 4 weeks after culture in a drop of liquid medium. **C** Retrieved callus cell suspension-derived protoplasts of *Datura* 3 weeks after culture. **D** A callus cell colony obtained from frozen mesophyll protoplast of tobacco. **E** A tobacco plantlet regenerated from frozen protoplast-derived callus (**D**). **F, G** Frozen protoplast-derived plant of *Atropa* 4 weeks after transfer to soil (**F**), and undergoing normal flowering (**G**). (Bajaj 1988 b)

differentiated plants (Fig. 1 D, E). The frequency of regeneration was 36% as compared to 53% control (Table 3). The *Atropa* protoplasts likewise behaved similarly; however, they gave the best response to cultures. The frequency of regeneration was higher than in tobacco, and a maximum of 44% cultures showed differentiation of callus. The morphogenetic events were not impaired and cultures showed both patterns, i.e. embryogenesis as well as organogenesis. The plants thus regenerated, when transferred to pots (Fig. 1 F), continued to grow to form normal flowers (Fig. 1 G) and fruits. The seeds obtained from these plants were fertile, and in turn showed normal germination.

5 Survival of Somatic Hybrid Protoplasts

Somatic hybrid protoplasts obtained as a result of fusion between wheat × pea, rice × pea and rice × sorghum, and frozen in LN in the presence of cryoprotectants have been revived (Bajaj 1983 a, b, c) and the results are summarized below:

5.1 Fusion and Freezing of Protoplasts

Pea protoplasts were obtained from the slices of 2-week-old in vitro-grown seedlings, whereas those of wheat and rice were isolated from callus cells raised from pollen-embryo-derived callus. Two drops of protoplasts of wheat and pea or rice and pea (approx. 10^4 protoplasts per ml) were pipetted onto a cover glass in a Petri dish and fused by using 50% solution (final concentration 25%) of polyethylene glycol. After 1 h they were washed and the solution was replaced by 0.5 M mannitol containing 5% dimethyl sulphoxide. The cover glass, along with the protoplasts, was frozen by subjecting it to the vapours of LN for 5 min. After that it was thawed by placing it on a warm sterile glass slide, and the cryoprotectant solution was replaced by a drop of culture medium. The cover glass was placed in the centre of a tight-lid Falcon Petri dish, and incubated in the dark. The fused products were observed periodically. The revival was judged by the mixing of the cytoplasm of the two parents, and by its streaming.

5.2 Culture of Hybrid Protoplasts

The freshly isolated protoplasts immediately after treatment with PEG showed adhesion (Fig. 2 A). The mesophyll-derived protoplasts of pea were more sensitive to PEG than those of wheat and rice obtained from the callus cell suspensions. At 25% PEG, the callus-derived protoplasts were healthy, those obtained from the mesophyll cells started to burst.

In the control (unfrozen) cultures, the hybrid protoplasts showed mixing of the cytoplasm, yellowing and redistribution of the chloroplasts, and an overall increase in size in 5–7 days. The retrieved fused protoplasts (Fig. 2 C), however, lay quiescent for the first 2 weeks in culture. Thereafter changes in their morphology

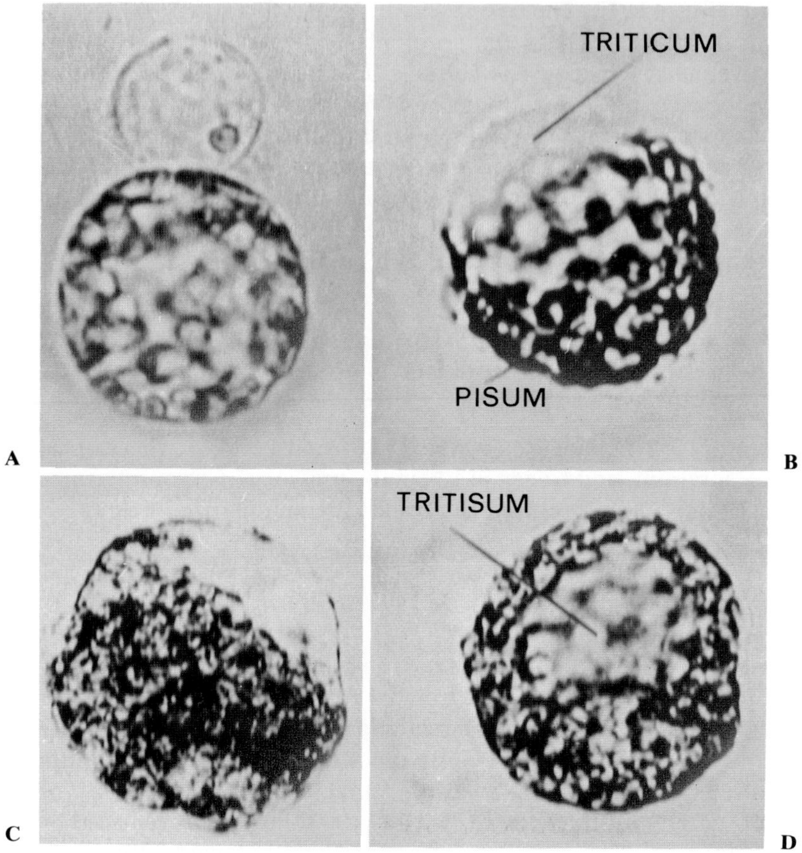

Fig. 2 A–C. Survival of somatic hybrid protoplasts (*Triticum*×*Pisum*) subjected to freezing in liquid nitrogen. **A** Freshly isolated protoplast from cell suspension of wheat (small, colourless) and mesophyll protoplast of pea (large, dark) showing adhesion after treatment with PEG. **B** A fused product at the time of freezing. **C** Same, immediately after thawing at 35 °C. **D** 21 days after culture in a microdrop; note the mixing of the protoplasts and the healthy nature of the retrieved somatic hybrid cell. (Bajaj 1983 a, b)

were observed. There was mixing of the cytoplasm (Fig. 2 D), and visible cyclosis in the cytoplasm. With the passage of time, chloroplasts became scanty and yellowish. The data on their survival are summarized in Table 4. The results in wheat and rice were comparable; however, survival in wheat×pea was slightly higher (12.2–17.9%) as compared to rice×pea (10.8–13.9%). Budding of the protoplasts was a common phenomenon. The heterokaryon showed two nuclei of different sizes, and occasionally fusion of the nuclei was also observed in the somatic hybrid cells. Only in one instance did the retrieved hybrid cell undergo repeated division (Bajaj 1983 a).

Table 4. Survival of somatic hybrid protoplasts of rice × pea, wheat × pea and rice × sorghum frozen in liquid nitrogen in the presence of cryoprotectants (DMSO 5% + mannitol 0.5 M), and thawed at 35 °C. (Bajaj 1983c)

Hybrid protoplasts	Exp. No.	No. of frozen hybrid protoplasts observed	No. showing signs of growth	% Survival
Oryza sativa × *Pisum sativum* (Orysum)	1	37	4	10.8
	2	59	7	11.8
	3	43	6	13.9
Triticum aestivum × *Pisum sativum* (Tritisum)	1	75	11	15.0
	2	41	15	12.2
	3	67	12	17.9
Oryza sativa × *Sorghum bicolor*	1	29	3	10.3
	2	36	5	13.8

6 Summary and Conclusions

For all practical purposes, freezing of protoplasts is similar to that of isolated cells and cell suspensions; however, due to the absence of cell wall they require an appropriate osmotic cum cryoprotectant to protect them from bursting or causing undue damage to the plasmalemma. There are three types of cryoinjury to the protoplasts, i.e. intracellular ice formation, expansion-induced lysis, and loss of osmotic responsiveness. Thorough understanding of various cryobiological aspects of protoplasts will enable less damage and higher survival.

The work conducted so far has demonstrated that the retrieved cultures retain their morphogenetic and biosynthetic potential to regenerate complete plants. Isolated protoplasts have a number of potential uses. They offer a system in which the effect of freezing on plasmalemma can be studied directly in situ. They are also an ideal system for quantitative cryomicroscopic studies, which allows for a direct analysis of intracellular ice formation. Moreover, cryopreservation of protoplasts reduces the number of steps in cloning.

It is at the cellular membranes that biological integrity appears to be insulted by freezing, so it is at the membrane level that biochemical understanding of cyroprotectants must be sought, and in this regard isolated protoplasts are an ideal tool.

References

Bajaj YPS (1976) Regeneration of plants from cell suspensions frozen at −20, −70 and −196 °C. Physiol Plant 37:263−268

Bajaj YPS (1983a) Survival of somatic hybrid protoplasts of wheat × pea and rice × pea subjected to −196 °C. Indian J Exp Biol 21:120−122

Bajaj YPS (1983b) Somatic hybridization and cryopreservation studies on rice × pea and wheat × pea. In: Potrykus I, Harms CT, Hinnen A, Hutter R, King PJ, Shillito RD (eds) Protoplasts 1983, 6th Protoplast Symp, Basel. Birkenhäuser, Basel, pp 248−249

Bajaj YPS (1983c) Cryopreservation of germplam of cereals – Progress and prospects. In: Sakamoto S (ed) Proc 6th Int Wheat Genetics Symp, Kyoto, Kyoto Univ, pp 565–574

Bajaj YPS (1986) In vitro preservation of genetic resources. Int Symp Nuclear Techniques and in vitro culture for Plant Improvement. IAEA, Vienna, pp 43–57

Bajaj YPS (1988a) Cryopreservation, and the retention of biosynthetic potentials in cell cultures of medicinal and alkaloid producing plants, I. In: Bajaj YPS (ed) Biotechnology in agriculture and forestry 4. Medicinal and aromatic plants. Springer, Berlin Heidelberg New York Tokyo, pp 169–187

Bajaj YPS (1988b) Regeneration of plants from frozen (−196 °C) protoplasts of *Atropa belladonna* L., *Datura innoxia* Mill. and *Nicotiana tabacum* L. Indian J Exp Biol 26:289–292

Bajaj YPS, Gosch G, Ottma M, Weber A, Grobler A (1978) Production of polyploid and aneuploid plants from anthers and mesophyll protoplasts of *Atropa belladonna* and *Nicotiana tabacum*. Indian J Exp Biol 16:947–953

Dowgert MF, Steponkus PL (1983) Effect of cold acclimation on intracellular ice formation in isolated protoplasts. Plant Physiol 72:978

Furner IJ, King J, Gamborg OL (1978) Plant regeneration from protoplasts isolated from a predominantly haploid suspension culture of *Datura innoxia* (Mill.). Plant Sci Lett 11:169–176

Gosch G, Bajaj YPS, Reinert J (1975) Isolation, culture, and induction of embryogenesis in protoplasts from cell suspensions of *Atropa belladonna*. Protoplasma 86:405–410

Hauptmann RM, Widholm JM (1982) Cryostorage of cloned amino acid analog-resistant carrot and tobacco suspension cultures. Plant Physiol 70:30–37

Mazur RA, Hartmann JX (1979) Freezing of plant protoplasts. In: Sharp WR, Larson PO, Paddock EF, Raghavan V (eds) Plant cell and tissue culture: Principles and applications. Ohio State University Press, Columbus, p 876

Steponkus PL (1985) Cryobiology of isolated protoplasts: Applications to plant cell cryopreservation. In: Kartha KK (ed) Cryopreservation of plant cells and organs. CRC Press, Boca Raton, pp 49–60

Steponkus PL, Dowgert MF (1981) Phenomenology of intracellular ice nucleation in isolated protoplasts. Plant Physiol 67:S-58

Steponkus P, Wiest SC (1979) Freeze-thaw induced lesions in the plasma membrane, p. 231. In: Lyons JM, Graham DG, Raison JK (eds) Low temperature stress in crop plants: The role of the membrane. Academic Press, New York

Steponkus PL, Dowgert MF, Evans RY, Gordon-Kamm W (1982) Cryobiology of isolated protoplasts. In: Li PH, Sakai A (ed) Plant cell hardiness and freezing stress. Academic Press, New York, p 459

Steponkus PL, Dowgert MF, Ferguson JR, Levin RL (1983) Cryomicroscopy of isolated plant protoplasts. Cryobiology 21:209

Takeuchi M, Matasushima H, Sugawara Y (1980) Long-term freeze-preservation of protoplasts of carrot and Marchantia. Cryo Lett 1:519

Takeuchi M, Matsushima H, Sugawara Y (1982) Totipotency and viability of protoplasts after long-term freeze preservation. pp 797–798. In: Fujiwara A (ed) Plant Tissue Culture 1982. Maruzen, Tokyo

Weber G, Roth EJ, Schweiger H-G (1983) Storage of cell suspensions and protoplasts of *Glycine max* (L.) Merr., *Brassica napus* (L.), *Datura innoxia* (Mill.), and *Daucus carota* (L.) by freezing. Z Pflanzenphysiol 109:29–39

Wiest SC, Steponkus PL (1978) Freeze-thaw injury to isolated spinach protoplasts and its stimulation at above-freezing temperatures. Plant Physiol 62:599

Withers LA (1980) Preservation of germplasm. Int Rev Cytol 11B:101–136

Section II Regeneration of Plants from Protoplasts

II.1 Regeneration of Plants from Rice Protoplasts

J. KYOZUKA[1], K. SHIMAMOTO[1], and H. OGURA[2]

1 Introduction

Rice is one of the three most important crops in the world, next to wheat in total cultivated area, and to corn in production. The cultivated area is 145×10^6 ha and production is 466×10^6 Mt (FAO 1985). It evolved in the humid tropics as a semi-aquatic plant. This plant has unique adaptive nature to hot and humid environments and is grown widely in both tropical and temperate regions, over as broad a range as 50 °N to 40°S. There are two species of cultivated rice, *Oryza sativa* L. and *Oryza glaberrima* Steud; the former is a common rice and the latter is endemic to West Africa. *O. perennis* is considered to be the ancestor of *O. sativa,* and *O. glaberrima* is thought to originate from *O. breviliglata.*

The reproducible plant regeneration from cultured protoplasts is a prerequisite for utilization of genetic manipulation techniques such as somatic hybridization and gene transfer. In the Gramineae, stable transformation has been achieved (Potrykus et al. 1985; Lörz et al. 1985; Fromm et al. 1986); however, transgenic plants have not been obtained due to lack of efficient plant regeneration from protoplasts. Also, protoplast culture is useful for isolation of biochemical and agronomically important mutants (Negrutiu et al. 1984).

Recently, there have been several reports describing the regeneration of plants from rice protoplasts (Table 1). Some reports did not give full details (Fujimura et al. 1985; Coulibaly and Demarly 1986), and also, either low plating efficiencies or low plant regeneration frequencies were noted (Abdullah et al. 1986; Toriyama et al. 1986; Yamada et al. 1986) and the reproducibilities are uncertain. In all these cases, special suspension cultures with fast-growing ability had to be selected to obtain dividing protoplasts. It often takes months to obtain such cultures (Abdullah et al. 1986; Yamada et al. 1986), so that the cultures selected tend to lose their morphogenic ability during the selection. In addition, the number of genotypes that can be used to initiate such cultures is limited (Abdullah et al. 1986; Yamada et al. 1986).

To establish a simple and reproducible rice protoplast culture system, we have developed novel nurse culture methods which allow plant regeneration from protoplasts with high frequency (Kyozuka et al. 1987). With these methods, somatic

[1] Plantech Research Institute, 1000 Kamoshida, Midori-ku, Yokohama 227, Japan
[2] Department of Agricultural Science, Ishikawa College of Agriculture, Nonoichi, Ishikawa 921, Japan

Biotechnology in Agriculture and Forestry, Vol. 8
Plant Protoplasts and Genetic Engineering I (ed. by Y. P. S. Bajaj)

Table 1. Published conditions for protoplast culture of rice[a]

Cultivar	Protoplast source	Enyzmes		Protoplast yield/g fr. wt.	Culture medium	Plating efficiency	Frequency of plant regeneration	Reference
Nihonbare Sasanishiki	Suspension (mature or immature seed callus)	Cellulase RS Macerozyme R 10	4% 1%	1.0×10^7	R 2 inorganic salts B 5 vitamins 2 mg/l 2,4-D 0.4 M sucrose	10%	5 – 20%	Fujimura et al. (1985)
Yamahoushi	Suspension (anther callus)	Cellulase R 10 Macerozyme	4% 0.2%	2.5×10^6	B 5-NH_4^- 1 mg/l 2,4-D 0.2 mg/l kinetin 0.1 mg/l GA_3 20 g/l sucrose	1%	24%	Toriyama et al. (1986)
Taipei 309 Fujisaka 5	Suspension (mature seed callus, immature leaf callus)	Cellulase RS Pectolyase Y 23	1% 0.1%	3.0×10^6	Kao (1977) 0.5 mg/l 2,4-D 10% glucose	0.35 – 1.25%	0 – 20%	Abdullah et al. (1986)
Moroberekan	Immature leaf-derived callus	Cellulase R 10 Macerozyme R 10 Driserase Pectolyase	1% 1% 0.6% 0.5%	–	Kao and Michayluk (1975) 0.1 mg/l BA 2 mg 2,4-D	–	–	Coulibaly and Demarly (1986)
Fujiminori	Suspension (mature seed callus)	Cellulase RS Macerozyme R 10 Pectolyase Y 23 Driselase	2% 2% 0.1% 1%	8.1×10^6	LS Kao vitamins and organic elements 2 mg/l 2,4-D 0.5 M glucose	8.4%	–	Yamada et al. (1986)
Fujisaka 5 Iwaimochi Koshihikari Nipponbare Norin 14	Suspension (mature or immature seed cells)	Cellulase RS MacerozymeR 10	4% 1%	$4 - 5 \times 10^6$	R 2 inorganic salts MS vitamins 2 mg/l 2,4-D 0.4 M sucrose	1 – 10%	5 – 45%	Kyozuka et al. (1987)

[a] Only examples in which plant regeneration took place are shown.

hybrids of rice and a wild millet (Terada et al. 1987) and those of rice and wild *Oryza* species (Hayashi et al. unpubl.) have been obtained. The procedure used in these studies is described below.

2 Isolation of Protoplasts

The media used are given in Table 2. Suspension cultures derived from immature seed calli or mature seed calli were used for protoplast isolation. Cultivars used were Aoisora, Fujisaka 5, Hatsushimo, Iwaimochi, Koshihikari, Nipponbare, Norin 14, Norin 22, Sasanishiki, Tainan 5, Thukinohikari, Chinsura Boro II and Chyokoto. Small calli, which were derived from big and compact calli initially formed on the scutellum, were suitable for initiation of suspension cultures. Suspension cultures were subcultured every 7 days in R2P medium. Within 1 month

Table 2. Composition of culture media

	MSC	R2S	R2P	N6SA	N6A	N6R
Major elements (mg/l)						
KNO_3	1900	4000	4000	2830	2830	2830
NH_4NO_3	1650	–	–	–	–	–
$(NH_4)_2SO_4$	–	335	335	463	463	463
$MgSO_4 \cdot 7H_2O$	370	250	250	185	185	185
$CaCl_2 \cdot 2H_2O$	440	150	150	166	166	166
$NaH_2PO_4 \cdot H_2O$	–	273	273	–	–	–
KH_2PO_4	170	–	–	400	400	400
Iron and minor elements (mg/l)						
Na_2EDTA	37.3	7.5	7.5	37.3	37.3	37.3
$FeSO_4 \cdot 7H_2O$	27.8	5.5	5.5	27.8	27.8	27.8
$MnSO_4 \cdot 4H_2O$	22.3	1.6	1.6	4.4	4.4	4.4
$ZnSO_4 \cdot 7H_2O$	8.6	2.2	2.2	1.5	1.5	1.5
$CuSO_4 \cdot 5H_2O$	0.025	0.125	0.125	–	–	–
$CoCl_2 \cdot 6H_2O$	0.025	–	–	–	–	–
KI	0.83	–	–	0.8	0.8	0.8
H_3BO_3	6.2	3.0	3.0	1.6	1.6	1.6
$NaMoO_4 \cdot 2H_2O$	0.25	0.125	0.125	–	–	–
Organic component (mg/l)						
m-Inositol	100	100	100	–	–	–
Nicotinic acid	0.5	0.5	0.5	0.5	0.5	0.5
Pyridoxine HCl	0.5	0.5	0.5	0.5	0.5	0.5
Thiamine HCl	0.5	0.5	0.5	1.0	1.0	1.0
Glycine	2.0	2.0	2.0	2.0	2.0	2.0
Sucrose (g/l)	20	30	136	60	60	60
2,4-D (mg/l)	2.0	2.0	2.0	2.0	2.0	2.0
Agarose (type I, Sigma) (g/l)	8.0	0	0	2.5	5.0	10.0
pH[a]	5.6	5.6	5.6	5.6	5.6	5.6

[a] pH was adjusted before autoclaving for MSC, R2S, N6SA, N6A and N6R. R2P was filter-sterilized. Basal components of the R2 medium are found in Ohira et al. (1973).

after the initiation of suspension culture, isolation of "good" protoplasts became possible.

Protoplasts were isolated from suspension cultures 3–5 days after subculture, using enzyme solution consisting of 4% Cellulase RS (Kinki Yakult, Japan), 1% Macerozyme R10 (Kinki Yakult, Japan), 0.5% MES and 0.4% mannitol, pH 5.6. Older (6–7-day-old) cultures gave much lower protoplast yield. Usually $4-6\times 10^6$ protoplasts/g fr. wt. cells were obtained irrespective of the genotype and the origin of suspension culture.

3 Culture of Protoplasts

The presence of nurse cells was absolutely required to induce high frequency division of protoplasts. In maize, nurse cultures were used to increase the division rate of cultured cells (Smith et al. 1984) and protoplasts (Ludwig et al. 1985; Imbrie-Milligan and Hodges 1986). In our laboratory, two different methods of nurse culture (mixed nurse method and Millicell method) were used (Fig. 1), and the mixed nurse method was twice as effective as the Millicell method.

The use of agarose-bead-type culture is essential in order to obtain reproducible colony formation. When cultured in liquid medium, most of the protoplasts adhered to the bottom of the dish and failed to divide.

Cultured protoplasts started to divide after 3–4 days in R2P medium (Fig. 2B), and after 7–10 days they formed compact colonies of small cytoplasmically rich cells (Figs. 2C, D, E) with plating efficiencies shown in Table 3. Within 2–3 weeks of culture, numerous colonies became visible and agarose blocks were transferred onto N6SA medium (Fig. 2F).

In order to obtain high plating efficiency, protoplasts had to be nursed for 10 days (Table 4A). Also, species specificity in the capacity to support protoplast division was examined (Table 4B), and only rice cells showed an appreciable effect on induction of division. Another factor examined was protoplast density; the plating efficiency did not vary greatly in the range of $5\times 10^4 - 1\times 10^6$ protoplasts per ml; in addition, a density as low as 1×10^4 per ml gave a 2–3% plating efficiency.

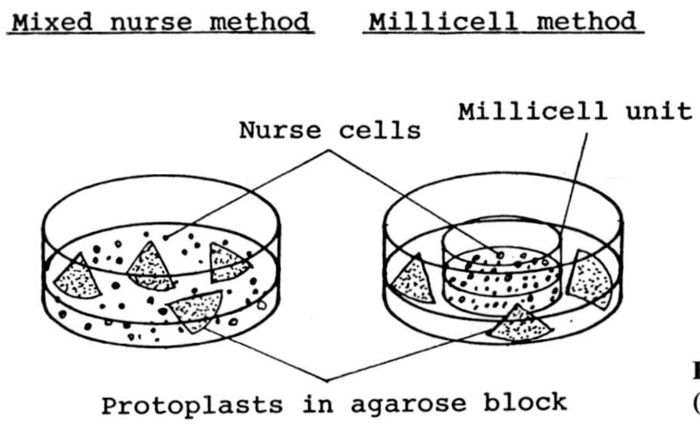

Fig. 1. Nurse culture methods. (Kyozuka et al. 1987)

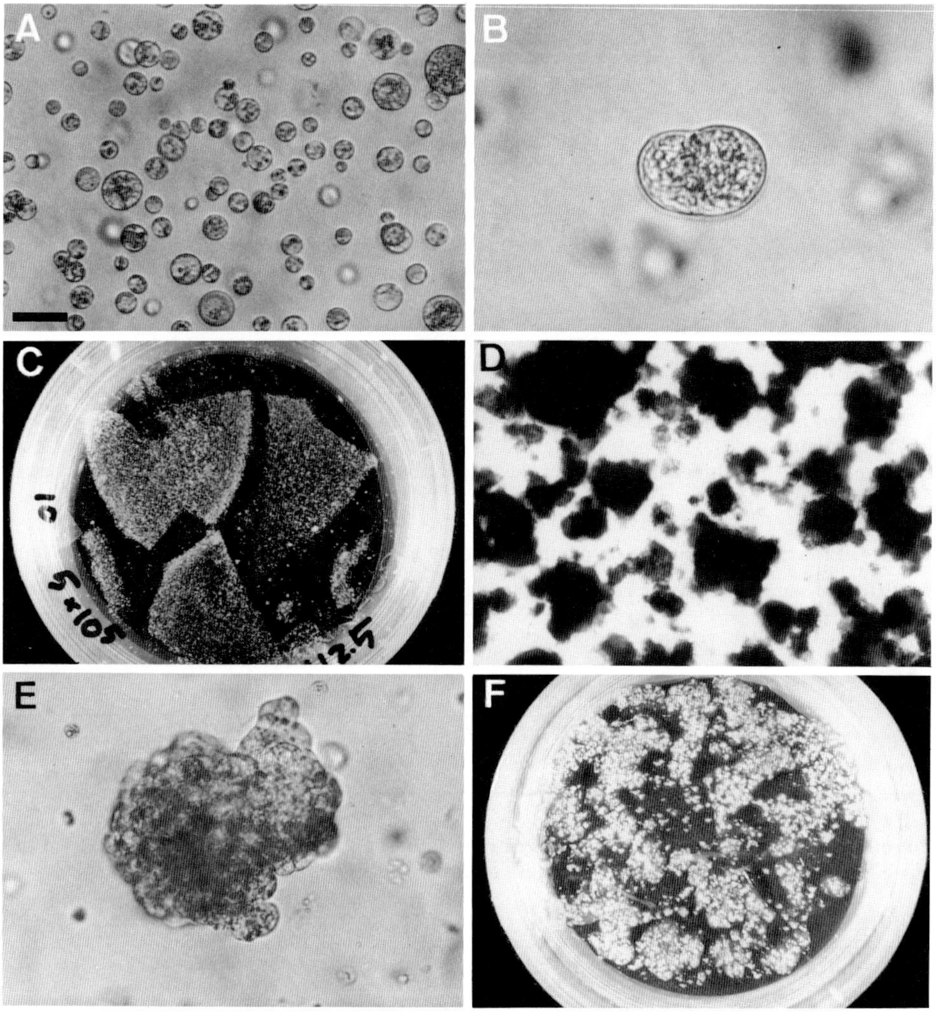

Fig. 2. Colony development from rice protoplasts. **A** Freshly isolated protoplasts of rice. **B** A dividing protoplast in R2P medium. **C** Numerous visible colonies (days 10 in R2P medium). **D** A microscopic view of protoplast-derived colonies in agarose blocks. **E** A compact colony. **F** Growing colonies on the soft agar medium (N6SA) (3 weeks of culture). (Kyozuka et al. 1987) Bar = 20 μm

4 Regeneration of Plants

Shoots as well as roots were regenerated from protoplast-derived callus within 3–8 weeks after the transfer of colonies onto N6R medium (Fig. 3A). Although the frequency of shoot formation varied among cultivars from approximately 10 to 50%, plants were regenerated from all the cultivars tested. Kinetin and zeatin had a stimulating effect at either 2 or 5 mg/l and high concentration (10 mg/l) of cytokinins had a negative effect on shoot regeneration (Table 5).

Table 3. Plating efficiencies of rice protoplasts isolated from various cultivars and cultures[a]. (Kyozuka et al. 1987)

Cultivar	Plating efficiency (%)	
	Immature seed callus	Mature seed callus
Nipponbare	NT[b]	5.2[c]
		5.1
Iwaimochi	1.1	4.5
	2.3	2.7
Norin 14	4.2	NT
	4.8	
Fujisaka 5	3.0	NT
	1.5	

[a] The mixed nurse method was used.
[b] NT: not tested.
[c] Two numbers in each column represent the values obtained from two independent experiments.

Table 4. Effect of nurse culture on plating efficiency of rice protoplasts (Kyozuka et al. 1987)

A. Effect of duration of nurse culture on plating efficiency[a]

No. days	0	3	5	6	7	8	10
	Plating efficiency (%)[b]						
Exp. 1	0	0.1	1.5	1.6	2.6	3.9	4.2
Exp. 2	0	0.7	1.2	2.4	2.2	4.7	4.8

[a] The mixed nurse method was used.
[b] Protoplasts of Norin 14 isolated from suspension cultures of immature callus origin were used and plating efficiency was obtained after 10 days of culture.

B. Effect of species of nurse cells on plating efficiency[a]

Nurse cells	Plating efficiency (%)[b]
Rice Oc	4.2
Rice Norin 14	1.0
Triticum monococcum	0.1
Daucus carota	0
Nicotiana tabacum	0
– Nurse	0

[a] The mixed nurse method was used.
[b] Plating efficiency was obtained after 10 days of culture.

The presence of scutellum-like structures and the observation that the shoot and the root emerged on the same axis but in opposite directions, suggested that plants regenerated through somatic embryogenesis (Fig. 3B, C). Plantlets were transferred to hormone-free N6 medium (Chu et al. 1975) containing 0.8% agarose (type I, Sigma) for further growth (Fig. 3D, E, F).

Fig. 3. Plant regeneration from protoplast-derived calli of rice. **A** Shoot and root formation from protoplast-derived calli (3–4 weeks after the transfer to N6R medium). **B** An embryo-like structure developed from a protoplast-derived callus (Nipponbare) (2–3 weeks after the transfer on N6R medium). **C** An embryo-like structure developed from a protoplast-derived callus (Norin 14). **D** Developing plantlets. **E** Mature rice plants derived from protoplasts. **F** Protoplast-derived plants growing in the field. (Kyozuka et al. 1987)

5 Somaclonal Variation

One hundred and twenty-six plants derived from protoplasts of four rice cultivars were individually cultivated in the submerged paddy field. Most of protoplast-derived rice plants were diploids; however, three tetraploids and one aneuploid (trisomic) were found in Nipponbare, one triploid and four tetraploids in Iwaimochi, one tetraploid in Fujisaka 5 and two tetraploid in Norin 14. Examples

Table 5. Effect of various cytokinins on the frequency of plant regeneration from protoplast-derived calli of rice. (Kyozuka et al. 1987)

Hormone	Conc. (mg/l)	Frequency of plant regeneration (%)[a]	
		Iwaimochi (I)	Iwaimochi (M)
Hormone-free		23.4	25.7
Kinetin	2	18.5	25.7
	5	18.5	17.1
	10	7.4	11.4
BAP[b]	2	3.7	2.9
	5	48.1	8.6
	10	14.8	2.9
Zeatin	2	11.1	11.4
	5	18.5	25.7
	10	18.5	5.7

Suspension cultures of the following origins were used for protoplast isolation: (I) (immature seed callus), (M) mature seed callus).
[a] For each experiment 22 – 54 protoplast-derived calli were tested.
[b] 6-benzylaminopurine.

Table 6. Somaclonal variation observed in protoplast-derived plants of four rice cultivars. (Ogura et al. 1987)

Cultivar	Number of plants	No. of normal plants (%)	Somaclonal variants		
			Chromosomal variants (%)	Low seed fertility variants (%)	Undetermined variants (%)
Nipponbare	73	61 (83.6)	4 (5.5)	8 (11.0)	
Fujisaka	11	9 (81.8)	1 (9.1)	1 (9.1)	
Norin 14	13	11 (84.6)	2 (15.4)		
Iwaimochi	29	21 (72.4)	5 (17.2)	2 (6.9)	1 (3.4)
Total	126	102 (81.0)	12 (9.5)	11 (8.7)	1 (0.8)

of somatic and meiotic chromosomes are shown in Fig. 4. Table 6 shows number and frequency of somaclonal variants of four cultivars. Among these variants, chromosomal variants always accompanied by abnormal phenotypes, e.g. abnormal panicles (Fig. 5b, d) and/or dwarfism. Plants with less than 40% seed fertility were classified as low seed fertility variants. They were all diploids showing no gross chromosomal anomalities and most of them exhibited an almost normal appearance.

Nineteen percent of protoplast-derived plants showed somaclonal variation; in total, however, the frequency is not as high as other results of tissue culture-derived rice plants. Oono (1975) reported that about one-third of anther-derived rice plants showed some anomalies. Oono (1981) also observed that 72% of 1121 regenerated plants from rice seed callus showed some mutated morphological

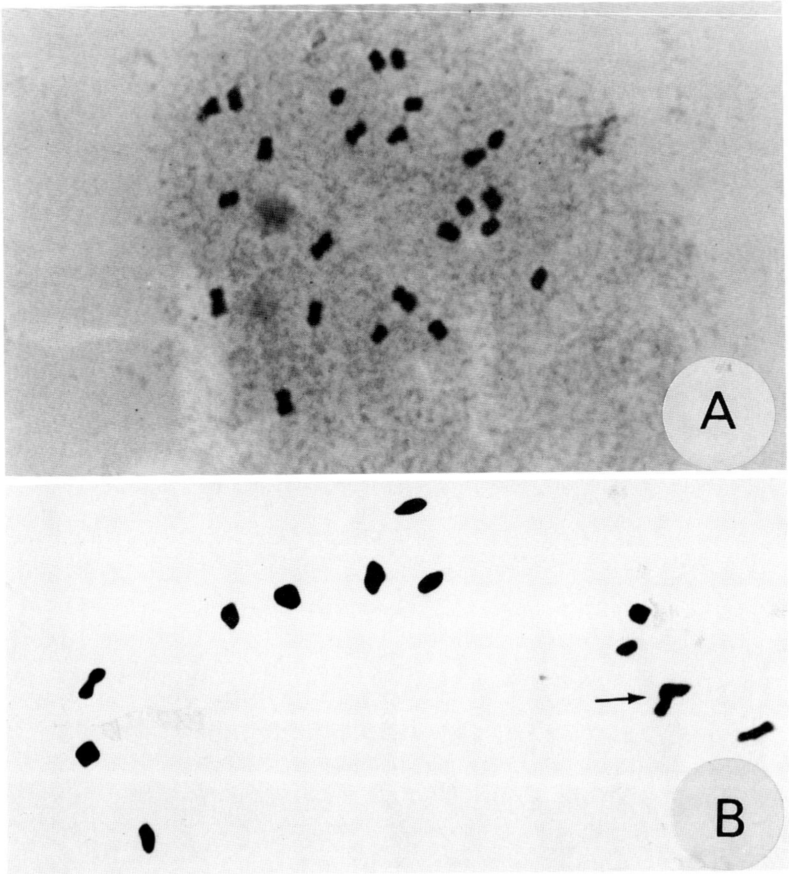

Fig. 4 A, B. Chromosomes of Nipponbare. **A** Metaphase chromosomes of root-tip cell of a protoplast-derived plant, showing normal 2n = 24 chromosome number. **B** Metaphase chromosome configuration of a PMC of a protoplast-derived trisomic plant, showing $1_{III}+11_{II}$. (*Arrow* indicates a trivalent chromosome. (Ogura et al. 1987)

characters. Sun et al. (1983) reported high frequency of somaclonal variation among the plants regenerated from the callus from young panicle or the dehusked seeds of 18 varieties of rice. They observed polyploid or albino plants in a number of lines in the second generations (T_2) of regenerated plants, and found that only 24.4% of the lines were normal in 950 T_2 pedigree lines of four varieties.

The frequency of chromosomal variants was 10% in our study, which does not seem to be high. Nishi and Mitsuoka (1969) reported that among 16 plants obtained through anther culture, one was triploid and seven were pentaploids. Chen and Lin (1976) observed 23 tetraploids among 165 regenerated plants of five japonica varieties of rice. Oono (1981) identified two trisomics, five triploids and one tetraploid in 70 anther-derived rice plants. Likewise, Bajaj and Bidani (1986) reported a wide variability in chromosome number and the production of haploids to polyploids and aneuploids from pollen-, embryo-, and endosperm-derived callus.

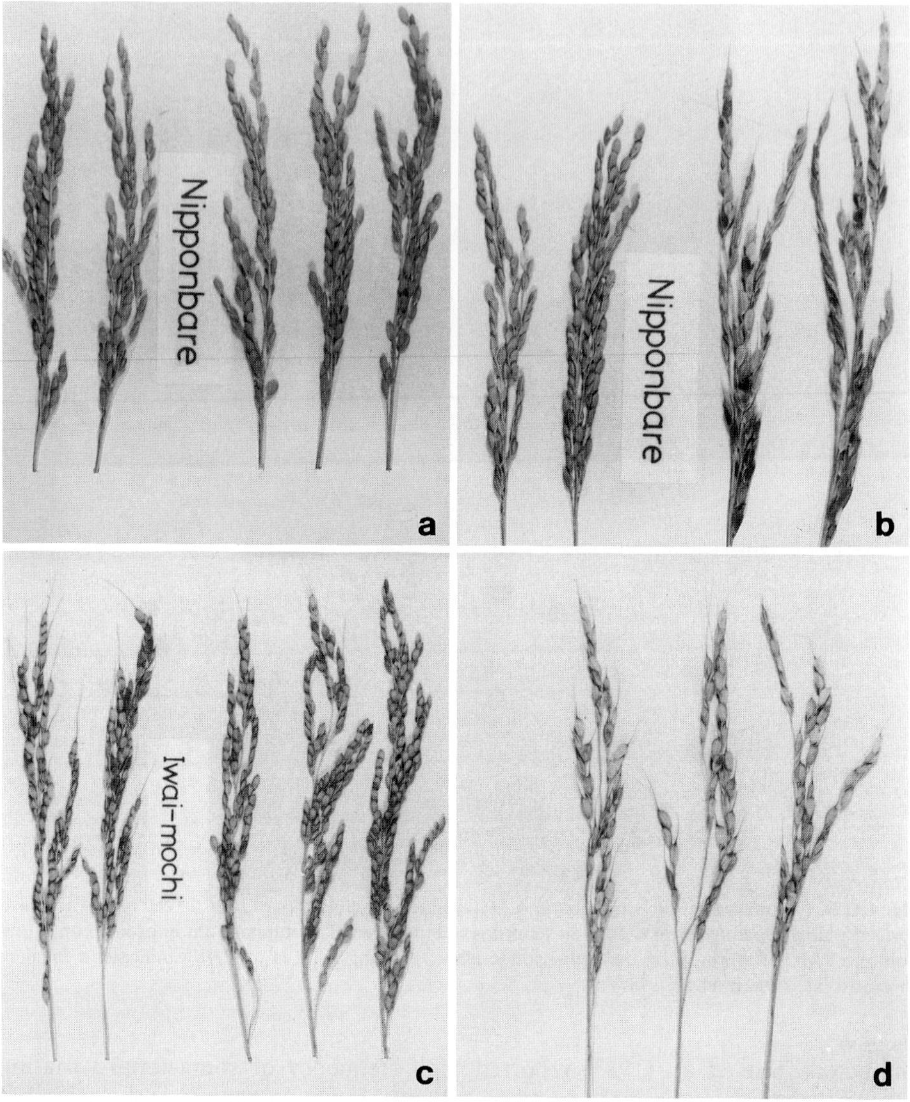

Fig. 5. Panicles of protoplast-derived plants. **a** *Right three* are protoplast derived, and *left two* are control of Nipponbare. **b** *Right two* are protoplast-drived tetraploid, and *left two* are control of Nipponbare. **c** *Right three* are protoplast-derived and *left two* are control of Iwaimochi. **d** Protoplast-derived triploid of Iwaimochi. (Ogura et al. 1987)

6 Agronomic Traits of Protoplast-Derived Plants

Protoplast-derived regenerants and the respective seed-sown control plants cultivated in the paddy field are shown in Fig. 6a. Protoplast-derived plants, except apparent somaclonal variants, were compared as to several agronomic traits

Fig. 6. a Protoplast-derived plants cultivated in the paddy field. **b** Dried specimens of harvested rice plants of Iwaimochi: *Left two* are protoplast-derived plants having increased number of panicles and *right tow* are control. (Ogura et al. 1987)

Table 7. Comparison of agronomic characters between protoplast-derived plants and seed-sown control plants in a cultivar Nipponbare. (Ogura et al. 1987)

	No. of plants	Heading date (range)	Maturation date	Culm length		Panicle length		Panicle no. per m^2	Spikelet no. per panicle	Seed fertility (%)	1000-grain weight (g)	Grain yield (kg/10 a)
				Mean ± S.E.	C.V. (%)	Mean ± S.E.	C.V. (%)					
Protoplast-derived plants	25	Aug. 20 (11 – 23)	Oct. 2	68.2 ± 0.9	6.6	18.0 ± 0.3	7.8	388.5	74.4	91.5	20.2	534.2
Control plants	10	Aug. 19 (15 – 19)	Sept. 29	69.0 ± 1.0	4.3	19.5 ± 0.6	9.7	272.0	78.8	93.1	20.3	405.1

C.V. = Coefficient of variation.

with the respective control plants. Protoplast-derived plants showed more stem number, which resulted in more panicle number than the respective control plants. An example of dried specimens of Iwaimochi is seen in Fig. 6B. Agronomic traits of protoplast-derived plants of four cultivars were (as compared with control); a slightly short or similar culm length, less number of spikelets per panicle, slightly lower seed fertility and similar or lighter weight per 1000 dehusked kernels. Grain yield of protoplast-derived plants was more than that of respective control plants of four cultivars. Table 7 shows a comparison of agronomic characters between protoplast-derived and control plants in Nipponbare. Comparison in other cultivars is described by Ogura et al. (1987). As a whole, protoplast-derived rice plants, excluding somaclonal variants, showed comparable grain yield and less high variations than the seed-sown control. These results indicate the possibility of using protoplasts in rice breeding. Selfed seeds of protoplast-derived plants exhibiting desirable traits are individually cultivated for breeding studies.

7 Conclusions and Prospects

A simple and reproducible plant regeneration system from rice protoplasts has been described. The presence of nurse cells supported division of protoplasts, which do not easily divide otherwise. Accordingly, there is no need to select a special "cell line" to obtain high plating efficiency. In addition, little influence of the genotype was observed. This method has been employed in culture of protoplasts of indica varieties of rice, and preliminary results indicate that, although the response varied among cultivars, some divided with high frequency and plants were regenerated.

Somatic hybridization of rice was possible with the culture system described here (Terada et al. 1987; Hayashi et al. unpubl.).

A preliminary study indicates that wheat (*Triticum aestivum*) protoplasts are able to form colonies by the nurse culture system. This suggests that the nurse culture method described here can be applied to other cereal protoplasts that do not form colonies under the standard culture conditions.

8 Protocol

Media used in the following protocol are listed in Table 2.

A. *Initiation of Suspension Culture*

(1) Dehusk immature seeds or mature seeds using forceps.

(2) Sterilize immature seeds in 70% ethanol for 30 s and in 0.24% sodium hypochlorite for 30 min without shaking. Mature seeds are sterilized in 70% ethanol for 30 s and in 1.2% sodium hypochlorite for 45–60 min with vigorous shaking. Then wash seeds two times in sterilized distilled water.

(3) Place seeds (1/well) into each well of 12-well dish, each containing 3 ml of MS medium (Murashige and Skoog 1962) for callus induction (MSC). Proliferation of the scutellum and the formation of calli are not influenced by the orientation of seed at inoculation.

(4) Calli develop from the scutellum in 3–4 weeks; place them in a 125-ml flask containing 15 ml of suspension medium (R2S). Only secondary calli derived from the primary, compact callus should be used to initiate suspension cultures.
(5) Place suspension cultures on a gyrotary shaker at low speed (80 rpm) and incubate in the light (ca. 3000 lx) at 27 °C for about 4 weeks. They are subcultured every 7 days.

B. Isolation and Culture of Protoplast

(1) Place 3–5-day-old suspension culture in a 10-cm Petri dish and pipette out culture medium. Then add 20 ml of the enzyme solution consisting of 4% Cellulase RS (Kinki Yakult, Japan), 1% Macerozyme R10 (Kinki Yakult, Japan), 0.5% MES and 0.4 M mannitol, pH 5.6.
(2) Incubate the Petri dish in the dark at 30 °C for 3–4 h without shaking.
(3) After the incubation, gently pipette enzyme-protoplast mixture a few times and pour it into a glass funnel with a fine nylon mesh (pore size 20 μm). Add 4× volume of KMC (Harms and Potrykus 1978) solution to the filtered protoplast suspension.
(4) Centrifuge protoplast suspension at 800 rpm for 10 min.
(5) Dispose supernatant and wash protoplasts twice in KMC solution by centrifugation (800 rpm, for 5 min each), then count the number of protoplasts.
(6) Resuspend protoplasts in culture medium (R2P) at twice the desired final cell density.
(7) Mix 1 ml of protoplast suspension with an equal volume of 1.2% agarose (Sea Plaque, FMC Corp., USA) in R2P medium.
(8) Cut solidified agarose into small blocks and transfer to a 6-cm Petri dish containing 5 ml of R2P medium.
(9) Add nurse cells (ca. 50 mg/plate) in the liquid part of the culture (mixed nurse method) or inside a culture plate insert (Millicell HA 30 mm diameter, 0.45 μm pore size, Millipore USA) placed in the centre of the plate (Millicell method).
(10) Culture dishes in the dark at 27 °C with slow shaking (ca. 30 rpm) on the rotary shaker.
(11) After 10 days, transfer agarose blocks to a new plate containing 5 ml of R2P medium and remove nurse cells completely by washing with the culture medium two to three times.
(12) After 2–3-week incubation, transfer agarose blocks containing visible colonies onto N6 soft agar medium (N6SA), and further incubate dishes for 1 week at 27 °C under light.
(13) Pick up microcalli individually using forceps and place them on the same medium with a higher agarose concentration (0.5%) (N6A).
(14) Transfer colonies about 2 mm in diameter onto the regeneration medium (N6R) with or without cytokinin.
(15) Within 3–8-weeks shoots and roots emerge. Transfer shoots (ca. 3 cm height) to a plastic container containing 30 ml of hormone-free N6R medium with 0.8% agarose (Type I, Sigma).
(16) Transfer regenerated plants of 10–15 cm tall to soil in a growth room (27 °C, 3000 lx), then to greenhouse.

References

Abdullah R, Cocking EC, Thompson JA (1986) Efficient plant regeneration from rice protoplasts through somatic embryogenesis. Biotechnology 4:1087–1090

Bajaj YPS, Bidani M (1986) In vitro induction of genetic variability in rice (*Oryza sativa* L.). In: Siddiqui KA, Faruqui AM (eds) New genetical approaches to crop improvement. PIDC Printing, Karachi, pp 63–74

Chen CC, Lin MH (1976) Induction of rice plantlets from anther culture. Bot Bull Acad Sci 17:18–24

Chu CC, Wang CC, Sun CS, Hsu C, Yin KC, Chu CY, Bi FY (1975) Establishment of an efficient medium for anther culture of rice through comparative experiments on the nitrogen sources. Sci Sin 16:659–668

Coulibaly MY, Demarly Y (1986) Regeneration of plantlets from protoplasts of rice, *Oryza sativa* L. Z Pflanzenzücht 96:79–81

FAO (1985) Production yearbook, FAO Rome

Fromm ME, Taylor LP, Walbot V (1986) Stable transformation of maize after gene transfer by electroporation. Nature 319:791–793

Fujimura T, Sakurai M, Akagi H, Negishi T, Hirose A (1985) Regeneration of rice plants from protoplasts. Plant Tissue Cult Lett 2(2):74–75

Harms CT, Potrykus I (1978) Fractionation of plant protoplast types by iso-osmotic density gradient centrifugation. Theor Appl Genet 53:57–63

Imbrie-Milligan CW, Hodges TK (1986) Microcallus formation from maize protoplasts prepared from embryogenic callus. Planta 168:395–401

Kao KN (1977) Chromosomal behaviour in somatic hybrids of soybean-*Nicotiana glauca*. Mol Gen Genet 150:225–230

Kao KN, Michayluk MR (1975) Nutritional requirements for growth of *Vicia hajastana* cells and protoplasts at a very low population density in liquid media. Planta 126:105–110

Kyozuka J, Hayashi Y, Shimamoto K (1987) High frequency plant regeneration from rice protoplasts by novel nurse culture methods. Mol Gen Genet 206:408–413

Lörz H, Baker B, Schell J (1985) Gene transfer to cereal cells mediated by protoplast transformation. Mol Gen Genet 199:178–182

Ludwig SR, Somers DA, Petersen WL, Pohlman RF, Zarowitz MA, Gengenbach BG, Messing J (1985) High frequency callus formation from maize protoplasts. Theor Appl Genet 71:344–350

Murashige T, Skoog F (1962) A revised medium for rapid growth and bioassays with tobacco tissue cultures. Physiol Plant 15:473–497

Negrutiu I, Jacobs M, Caboche M (1984) Advances in somatic cell genetics of higher plants – the protoplast approach in basic studies on mutagenesis and isolation of biochemical mutants. Theor Appl Genet 67:289–304

Nishi T, Mitsuoka S (1969) Occurrence of various ploidy plants from anther and ovary culture of rice plant. Jpn J Genet 44:341–346

Ogura H, Kyozuka J, Hayashi Y, Koba T, Shimamoto K (1987) Field performance and cytology of protoplast-derived rice (*Oryza sativa*); High yield and low degree of variation of four japonica cultivars. Theor Appl Genet 74:670–676

Ohira K, Ojima K, Fujiwara A (1973) Studies on the nutrition of rice cell culture 1. A simple, defined medium for rapid growth in suspension culture. Plant Cell Physiol 14:1113–1121

Oono K (1975) Production of haploid plants of rice (*Oryza sativa*) by anther culture and their use for breeding. Bull Natl Inst Agric Sci D26:139–222

Oono K (1981) *In vitro* method applied to rice. In: Thorpe TA (ed) Plant tissue culture methods and applications in agriculture. Academic Press, New York, pp 273–298

Potrykus I, Saul MW, Petruska J, Paszkowski J, Shillito RD (1985) Direct gene transfer to cells of a gramineaceous monocot. Mol Gen Genet 199:183–188

Smith JA, Green CE, Gengenbach BG (1984) Feeder layer support of low density populations of *Zea mays* L. suspension cell. Plant Sci Lett 36:67–72

Sun ZX, Zhao CZ, Zheng KL, Qi XF, Fu YP (1983) Somaclonal genetics of rice, *Oryza sativa* L. Theor Appl Genet 67:67–73

Terada R, Kyozuka J, Nishibayashi S, Shimamoto K (1987) Plantlet regeneration from somatic hybrids of rice (*Oryza sativa* L.) and barnyard grass (*Echinochloa oryzicola* Vasing). Mol Gen Genet 210:39–43

Toriyama K, Hinata K, Sasaki T (1986) Haploid and dioloid plant regeneration from protoplasts of anther callus in rice. Theor Appl Genet 73:16–19

Yamada Y, Yang ZQ, Tang DT (1986) Plant regeneration from protoplast-derived callus of rice (*Oryza sativa* L.). Plant Cell Rep 5:85–88

II.2 Isolation and Culture of Potato Protoplasts

M. ZUBA and H. BINDING[1]

1 Introduction

1.1 Importance and Distribution of Potato

Potato, *Solanum tuberosum*, is one of the most important crops in the temperate climate zones. It occupies the fifth position among the cultivated plant species with respect to yield and the fourth position as measured by area of cultivation (Barocka and Ross 1985). The present annual production amounts to about 300 million t harvested on about 22 million ha (Geisler 1983). The cultivation is being increased, particularly in developing countries. It tolerates all climate zones, from temperate to tropic, and from humid to arid.

The tuber accumulates high amounts of starch, but also storage proteins are formed. The value of potato as a food crop is due to the combination of richness in calories with content of highly qualified proteins and lipids, as well as of minerals. As an additional field of application, potato has recently become one of the most promising sources for the production of bioalcohol.

The gene centre of tuberous *Solanum* species is located in South America, particularly in the Andes in Peru. Whereas highly variable types of tubers are grown in the native countries, breeding in the industrial countries aims mainly at size of tuber combined with resistance to diseases caused by viruses, bacteria, fungi, nematodes and the beetle *Leptinotarsa decemlineata*.

1.2 Significance and Use of Potato Protoplasts

The whole range of application of protoplast technologies in plant breeding is taken to be valuable in preservation and improvement of potato (Bajaj 1987; Wenzel et al. 1987). The fields of application are: investigations in virus biology, isolation of somaclonal variants, induction of variant clones by mutagenesis, production of somatic hybrids and cybrids, and DNA transfer.

Two aspects will be stressed here because they have been repeatedly discussed with particular reference to potato: namely, subcloning in order to break somaclonal variation, and intraspecific somatic cell fusion. In both proposals it is taken into account that the crop is propagated exclusively by cloning and that

[1] Botanisches Institut, Biologiezentrum, Olshausenstraße 40–60, D-2300 Kiel 1, FRG

Biotechnology in Agriculture and Forestry, Vol. 8
Plant Protoplasts and Genetic Engineering I (ed. by Y. P. S. Bajaj)

the clones are highly heterozygous, particularly increased by the $4x = 2n = 48$ status of somatic cells.

It is evident that isolated protoplasts as single cell systems are particularly appropriate for the isolation of subclones. As somaclonal variants are accumulated in vegetatively propagated plants, single cell cloning is useful in repeated selection of the original, cultivar-specific genotype. The selection of new genotypes within cultivar clones is also a significant aim of single cell cloning (see Jones 1984). Usually applied selection in plant clones is less reliable because somaclonal variant cell lines may be fixed and hidden by their location in certain layers of the apex.

Cross-breeding has been inevitable for establishing, for instance, heterosis and, particularly, for introducing selected characters from wild species into the crop. Due to the high-grade heterozygocity, however, cross-breeding in potato is very time-consuming and requires a high number of individuals. Wenzel et al. (1979; see also 1987) proposed a protocol of potato breeding comprising classical procedures such as crossing and selection with haploid technology and protoplast fusion. The first step is the establishment of homozygous diploid 2x lines by two-step haploidization and autodiploidization of the monohaploids. Selected 2x lines can be used for cross-breeding with wild *Solanum* species, as initiated by Peloquin et al. (1966). Combination of valuable 2x lines by crossing may result in useful combinations and already in heterosis. The 4x status can then be constituted by fusion of two of these heterozygous 2x lines. Fusion brings about the addition of unreduced complements of the 2x lines and, hence, offers the opportuniy to accumulate complete sets of valuable genes and to produce further heterotic increase of the productivity.

Selected valuable 2x lines which arose directly from single haploidization steps may also be combined by fusion to new 4x cultivars. In this case, the advantage of protoplast fusion lies in the possibility of selecting at the gametic stage in a somehow intermittent generation and compiling only those genotypes produced by meiosis which provide advantageous characters.

1.3 Brief Review of the Work Done on Isolated Protoplasts of *Solanum*

Solanum species, as well as clones and cultivars, protoplasts of which were regenerated to plants, are listed in Table 1.

Basically, two different strategies were successful: Shepard and Totten (1977) introduced a protocol including preculture of leaves as protoplast sources and a series of four different culture media which were especially formulated for potato. The technique was successful with the cultivar Russet Burbank and, later, with several other cultivars and a number of dihaploid clones. The protocol of Binding et al. (1978) was based on experience with *Solanum dulcamara* (Binding and Nehls 1977): Tips of plantlets cultured in vitro were used, and this procedure was applied frequently and successfully for *Solanum* species, cultivars and clones. Relatively good homogeneity, as well as significantly high variability of the regenerant populations, were reported.

Table 1. Members of the Genus *Solanum* of which plants have been regenerated from protoplasts

	Selected reference
S. aculeatissimum	Gleddie et al. (1984)
S. aviculare	Gleddie et al. (1985)
S. brevidens	Barsby and Shepard (1983) Nelson et al. (1983)
S. chacoense	Butenko et al. (1977)
S. dulcamara – 2n	Binding and Nehls (1977)
1n	Binding and Mordhorst (1984)
S. etuberosum	Barsby and Shepard (1983)
S. fernandezianum	Barsby and Shepard (1983)
S. gilo	Gleddie et al. (1984)
S. khasianum	Kowalczyk et al. (1983)
S. luteum	Binding et al. (1981)
S. melongena	Saxena et al. (1981); Jia and Potrykus (1981)
S. nigrum	Nehls (1978)
S. pennellii	Haberlach et al. (1985); Hassanpour-Estahbanati and Demarly (1985)
S. phureja	Schuchmann et al. (1980)
S. phureja × *chocoense*	Grun and Chu (1978)
S. pinnatisectum	Sidorov et al. (1984)
S. torvum	Gleddie et al. (1984)
S. tuberosum – Cultivars	
Atlantic	Shepard (1980a)
Bard	Gunn and Shepard (1981)
Bintje	Bokelmann and Roest (1983)
Bison	Shepard (1982)
Desiree	Foulger and Jones (1986)
Dorita	Haberlach et al. (1985)
Early Priekulski	Kuchko (1985)
Gatchinskii	Sidorov et al. (1984)
Granola	Binding et al. (1988a)
Feltwell	Gunn and Shepard (1981)
Fortyfold	Karp et al. (1982)
Foxton	Gunn and Shepard (1981)
Katahdin	Shepard (1982)
King Edward	Foulger and Jones (1986)
Kufri Sindhuri	Bajaj (1986)
Majestic	Creissen and Karp (1985)
Maris Bard	Thomas (1981)
Maris Piper	Gunn and Shepard (1981)
Pentland Crown	Creissen (1984)
Russet Burbank	Shepard and Totten (1977)
Teal	Gunn and Shepard (1981)
Zarevo	Sidorov et al. (1984)
S. tuberosum – 4x Clones	Carlberg et al. (1983) Shahin (1984) Haberlach et al. (1985)
S. tuberosum – 2x Clones	Binding et al. (1978) Wenzel et al. (1979) Shahin (1984) Sidorov et al. (1984) Haberlach et al. (1985) de Vries and Bokelmann (1986) Debnath et al. (1986) Masson et al. (1987) Schilde-Rentschler et al. (1987) Zuba, unpublished
S. uporo	Li and Constabel (1984)
S. viarum	Kowalczyk et al. (1983)
S. xanthocarpum	Saxena et al. (1982)

After the pioneering intergeneric somatic hybridization of potato (×) tomato (Melchers et al. 1978), several interspecific and intraspecific combinations have been investigated (see Table 3). The regeneration of fusants was partly obtained by one of the techniques mentioned above or by following slightly modified protocols.

2 Isolation of Protoplasts

2.1 Plant Material

Protoplasts have been isolated from parts of greenhouse-grown plants, of shoots grown on agar media and of suspension culture cells. Plant material was transferred from greenhouse to aseptic conditions by one of the established procedures of surface sterilization, such as, for instance, by incubation in hypochlorite and 70% ethanol (e.g. Shepard and Totten 1977) or 0.1% $HgCl_2$ with 0.1% sodium-dodecylsulphate (cf. Binding 1974). Leaves from greenhouse were best conditioned when the plants were kept at a 6-h light period.

Explants for starting in-vitro cultures were shoot segments of greenhouse-grown plants including a node with a axillary bud (Bokelmann and Roest 1983). It appeared also useful to take the tips of etiolated stems because decontamination was particularly reliable (Zuba and Binding, unpubl.). Subculture and propagation were done by transfer of tips and shoot segments carrying at least one nearly expanded leaf. Usually, roots are soon formed so that rather plantlet than shoot cultures are established. Propagation by adventitious shoots is possible on media with 6-BA as the only phytohormone or with the addition of 5% liquid coconut endosperm (Zuba and Binding, unpubl.).

Mostly, shoots or plantlets were grown on B5 media (Gamborg et al. 1968) or on MS media (Murashige and Skoog 1962). Whereas hormone-free media were sufficient for obtaining appropriate protoplast donor material (Carlberg et al. 1983; Haberlach et al. 1985; Debnath et al. 1986; Masson et al. 1987), it was reported that cytokinins and GA_3 improved the results. 6-BA was applied at concentrations of 0.05 to 0.5 mg/l (Nelson et al. 1983; Binding and Kollmann 1985; Fish and Karp 1986; Foulger and Jones 1986); media with either 2.5 mg/l GA_3 (Bokelmann and Roest 1983) or with 0.2 mg/l kinetin (Shahin 1984) or with 0.25 mg/l 6-BA and 0.1 mg/l GA_3 (Karp et al. 1982) were also used. The significance of the sucrose concentration in the media for the yields of isolated protoplasts was investigated in the cv. Desiree. Increase from the usual concentration of 2% to 3% resulted in doubling of the protoplast yield (Foulger and Jones 1986).

The cultures in test tubes, Petri dishes or other vessels were kept in tissue culture rooms at air temperatures between 25 and 28 °C and irradiated for 16 h per day under white fluorescent light.

Suspension cultures as protoplast sources were used by Gleddie et al. (1985), de Vries and Bokelmann (1986) and Kuchko (1985). The plant material was grown in modified MS media in darkness.

Usually, light intensities around 3000 lx were applied. Harvesting of shoot tips devoid of expanded leaves was easier when the shoots grew somehow etiolated at about 1000 lx (Zuba and Binding, unpubl.). Masson et al. (1987) proposed to increase the light intensity.

2.2 Preconditioning of the Material

From the beginning of protoplast culture in embryophytes (Binding 1966), it has been repeatedly stressed that the physiological state of the donor cells plays a significant role in successful regeneration. Potato leaves, especially when they were harvested from greenhouse plants, needed to be pretreated in order to avoid bursting of the protoplasts by starch grains, to avoid browning of the protoplasts by phenol oxidation, and to obtain physiological stabilization and conditioning of the cells. Shepard and Totten (1977) proposed an incubation of leaves or leaf segments in conditioning medium for 16 to 24 h (see also Shepard 1980a; Haberlach et al. 1985). Carlberg et al. (1983) suspended leaf segments for 1 h in a solution of 0.3 M sorbitol and 0.05 M $CaCl_2$.

Sufficiently viable protoplasts can be isolated from shoot tips of in vitro cultures grown under the conditions described in Section 2.1. However, increased plating efficencies were obtained by a 1- to 2-d dark period prior to protoplasting. Treatment of shoot cultures by the herbicide SAN 6706 appeared also favourable (Uhrig 1981).

2.3 Isolation of Protoplasts

Potato protoplasts have been isolated without exception by enzymic digestion of the cell wall. The penetration of the agents into leaves from the greenhouse was increased by pretreating the lower epidermis with a soft nylon brush until the colour turned to light green (Shepard and Totten 1977; Shahin 1984). Slicing of the plant material, in particular from in vitro cultures (Binding et al. 1978) was the most common preparation before the enzyme incubation. By a screening experiment, Foulger and Jones (1986) reconfirmed that slicing is the most efficient procedure with respect to the number of isolated protoplasts.

The enzyme activities needed for the digestion of cell walls were established in the incubation solutions by adding one enzyme preparation with pectinase and one with cellulase activities. All products commonly used for protoplasting in higher plants were appropriate; it is thus not necessary to give particular references. The osmotic stabilization was established by mannitol at concentrations of 0.3 to 0.6 M. The vitality of the protoplasts was increased by additional 5 to 10 mM Ca^{2+} ions as $CaCl_2$ or $Ca(NO_3)_2$ (see e.g. Binding et al. 1978; Fish and Karp 1986). The mineral salts of culture media were incorporated into the enzyme solution by Shepard and Totten (1977, see Sect. 7.2.1) and Haberlach et al. (1985). Haberlach et al. (1985) added 0.5 mg/l 6-BA and 2 mg/l NAA to the enzyme solution.

Shepard and Totten (1977) and Haberlach et al. (1985) used 2% polyvinylpyrrolidon (w/v; 10000 Da) to protect the protoplasts from phenolic compounds which are liberated during slicing of the plant material, particularly in case of leaves. The agent was not needed for the preparation of shoot tips from in vitro cultures and from etiolated greenhouse plants (unpubl.).

No significant variations in incubation procedures for protoplasting, or in collection and purification of the isolated protoplasts have been described. Appropriate procedures are outlined in Section 7.

3 Culture of Protoplasts

A number of protocols for the culture of potato protoplasts have been successfully applied. In some cases, regeneration was initiated in liquid culture media, and the plastocytes and cell clusters were embedded into media containing final agar concentrations of 0.1% to 0.2% agar after 2 to 14 d (Binding et al. 1978; Thomas 1981; Bokelmann and Roest 1983; de Vries and Bokelmann 1986; Debnath et al. 1986). A low agar concentration allows pipetting of the twice-concentrated agar medium and mixing it with the cell suspension at room temperature. Freshly isolated protoplasts were frequently embedded into culture media which were gelled by agar (Shepard and Totten 1977; Carlberg et al. 1983; Haberlach et al. 1985), by agarose (Shepard 1980a; Binding et al. 1986, 1987, 1988a), or by alginate (Schilde-Rentschler et al. 1987).

Embedding of protoplasts or cell clusters into solidified media bears two predominant advantages: (1) Locally high cell titres can be prepared which are a good precondition for initiation of mitotic activity (Binding et al. 1978); and (2) changes of culture media are much better tolerated, probably because the substances migrate only by diffusion within the solidified media and not by floatation, as compared to liquid cultures.

It has already been mentioned that high cell densities were beneficial for the initiation of mitoses. The applied densities ranged between 1×10^4 (Thomas 1981) and 5×10^5 (de Vries and Bokelmann 1986). In the cases of particularly high titres, dilution must and may follow a few days after plating (e.g. Puite et al. 1986). However, when densely packed protoplasts were allowed to communicate with high amounts of reserve media in case of layers on bottom agar or solidified droplets in liquid media, reduction of cell densities was not needed for cell proliferation. Feeder cells in liquid media (Schilde-Rentschler et al. 1987) or in agar-containing bottom media (Kuchko 1985; Puite et al. 1986) were used to support early development of the protoclones.

The protoplast cultures were usually kept at 24° to 26 °C and continuous light of 400 to 800 lx.

The composition of the culture media obviously plays a significant role in protoplast regeneration efficiency. However, the inorganic components of various culture media were useful. Osmolarities of usually 0.4 to 0.5 Osm were adjusted by mannitol or sucrose. Appropriate formulations of the inorganic ingredients were marked by comparably low concentrations of nitrogen ions (Shepard and

Totten 1977; Binding et al. 1978). Special attention was repeatedly paid to external phytohormones: one of the cytokinins, 6-BA or zeatin, was combined with NAA and 2,4-D. The concentrations varied largely in the protoplast culture media. It was already stressed by Binding et al. (1978) that a broad range of concentrations was tolerated. Divergent regeneration efficiencies of different genotypes also suggest different optimum amounts of phytohormones for different clones and cultivars. This is, however, not evidenced by prevailing investigations.

It has been already mentioned that phenolic compounds mainly produced by damaged cells interfere with the development of cell clones. Carlberg et al. (1983) proposed the addition of 1% activated charcoal to the culture media. Noxious browning of the cultures was limited to some genotypes as well as to particular experimental conditions. Culturing the protoplasts in solidified droplets surrounded by liquid media seems to be particularly appropriate for reducing damage by phenolic compounds (Binding et al. 1988a).

Turgescence of the regenerated cells was established by gradual or one-step reduction of the osmolarity of the culture medium. Fish and Karp (1986) brought the liquid suspension in contact with agar medium of lower osmolarity already at the 6th day of culture. Cell clusters embedded in agarose lenses survived transfer onto low osmolar agar media from the 9th day of culture on (Binding et al. 1988a). A culture period of 14 d in high osmolar media is needed when the protoclones are grown in soft agar media (Binding et al. 1978). Commonly, the cell colonies were exposed to low osmolar media after around 4 weeks. Mostly, these media contained the nutrient solutions of B5 or MS or derivatives.

The media contained reduced auxin levels as compared to the protoplast culture media. They were either used as an intermittent step of culture of about 4 weeks and shoot organization was obtained at a subsequent culture period (see e.g. Shepard and Totten 1977) or the media directly allowed organized growth (see e.g. Binding et al. 1978).

4 Regeneration of Plants and the Characters of Regenerated Plants

4.1 Regeneration of Plants

The course of plant regeneration from isolated protoplasts is illustrated in Fig. 1. The efficiencies have been calculated in some experimental conditions. They varied largely with the genotypes and the experimental set up. Relatively significant calculations were possible when the protoplast cultures were initiated in liquid media.

The plating efficiencies in terms of the relation of mitotically active clones to the total number of plated protoplasts reached from small values like 4% – 10% (Thomas 1981; Bokelmann and Roest 1983) up to 30% – 60% (Shepard and Totten 1977; Binding et al. 1978; Shahin 1984; Masson et al. 1987). Up to about 50% of microcalli were able to grow to shoot-forming calli. The shoots are easily rooted and transferred to greenhouse conditions without noticeable defects.

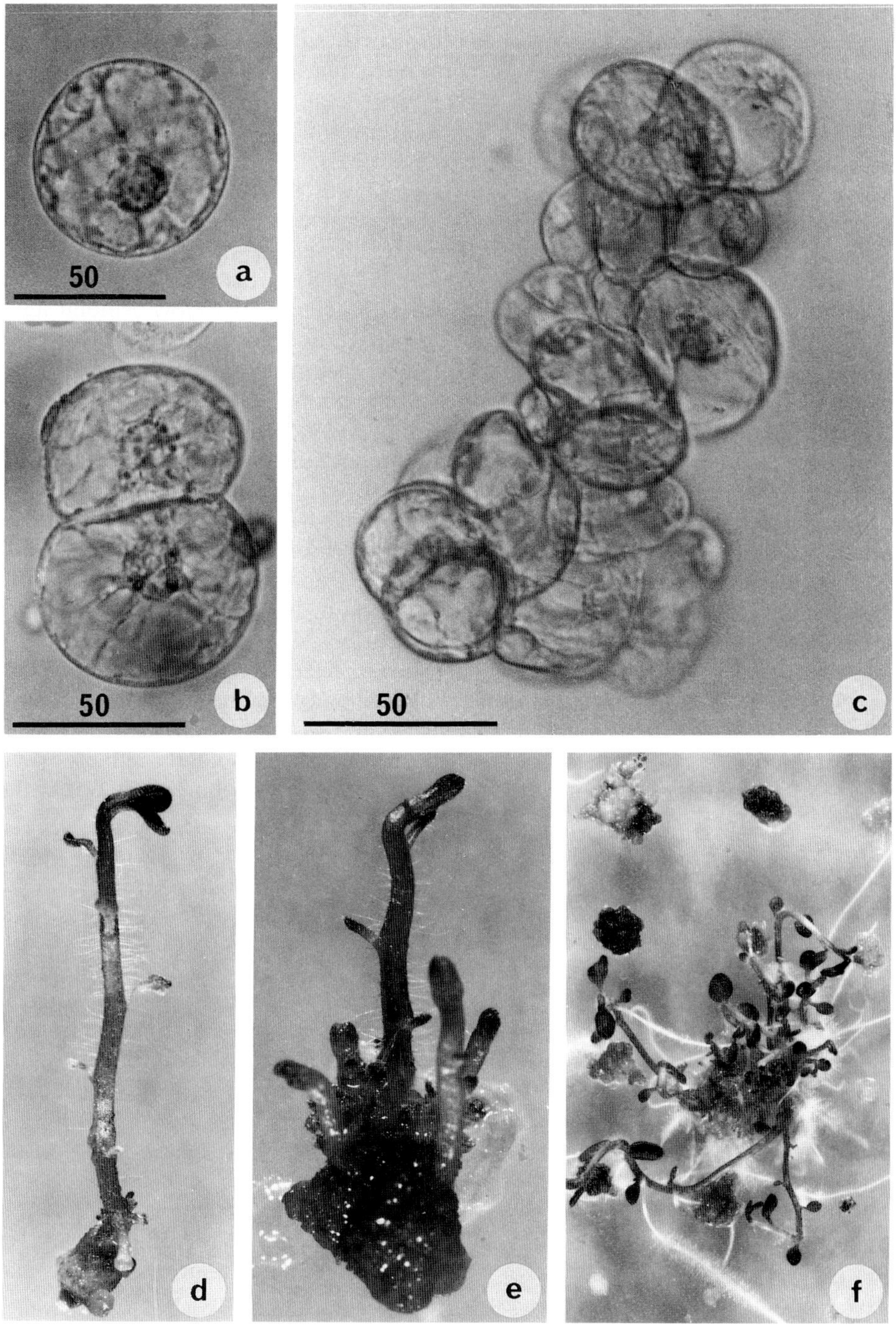

Fig. 1. Protoplast regeneration from $H^2$258. Unless stated otherwise, labelling of bars means μm. **a** Freshly isolated protoplast 410×. **b** First division (3 d) 470×. **c** Callus (11 d) 520×. **d** Callus with single regenerated shoot (34 d) 2.8×. **e** Callus with some regenerated shoots (35 d) 2.8×. **f** Rooted regenerants (42 d) 1.4×

It is nearly impossible to estimate regeneration efficiencies after plating the protoplasts in agarose media because protoclones aggregate and, hence, cannot be identified as individuals. But it became clear that good survival rates are obtained.

Shoot formation of the protoclones of several dihaploid clones and 4x cultivars was obtained during the first subculture period in low osmolar media when the 8p medium (Kao and Michayluk 1975) or derivatives (e.g. medium V-KM; Binding et al. 1978) were used for protoplast culture. Fastest regeneration was achieved when the protoplasts were plated in agarose media at locally high cell densities (streak type culture; Binding et al. 1988a): first shoot tips appeared 24 d after plating of the protoplasts – 14 d after transfer onto low osmolar agar medium. As can be seen from Section 7, Shepard and Totten (1977) introduced a sequence of culture media for the formation of protoclones, increase of calli, initiation of shoot formation and growth of the shoots. Great differences in the capability of shoot regeneration have been found between different genotypes (Binding et al. 1978; Shepard 1980a; Gunn and Shepard 1981; Barsby and Shepard 1983; Haberlach et al. 1985; Debnath et al. 1986; Foulger and Jones 1986; Binding et al. 1988a). It cannot be deduced from the published results if these differences were correlated to different hormone requirements of the investigated clones. Variable hormone contents were applied to different genotypes by different research groups. 6-BA, kinetin, zeatin, NAA, IAA and ABA were used in different combinations and concentrations, but even identical cultivars were treated with different hormone supplies in different laboratories.

Regenerated shoots usually formed roots already on the shooting medium. Media without phytohormones or with an auxin, GA_3 and/or charcoal enhanced root induction. The plantlets survived transfer to soil particularly reliably when the roots measured only about 3 mm and when the soil was thoroughly wet at the transfer (Binding 1985).

4.2 Characters of Regenerated Plants

Protoplast regeneration experiments are being carried out at present with different aims. On the one hand, genetic stability during regeneration is preferred when, for instance, an original genotype should be separated from clonal-variant lines or when a particular genetic constitution should be manipulated by protoplast fusion or transformation. On the other hand, variability in the protoplast stage makes possible the selection of protoclones possessing a new valuable character. However, also in this case, somaclonal stability during growth of the protoclones is demanded. Some results with potato protoplast regenerants gave indications of how the degree of genotypic stability might be influenced.

It became evident – also in investigations with other species – that fast regeneration provides the best guarantee for preservation of the genetic constitution (cf. Binding 1985; Debnath et al. 1986). It is hence not surprising that relatively good homogeneity – at least with regard to leaf shapes (see Fig. 2) – in protoclone populations was obtained with the agarose droplet technique which lead to organized growth already after 24 d of culture (Binding et al. 1988a; see Sect. 7.1).

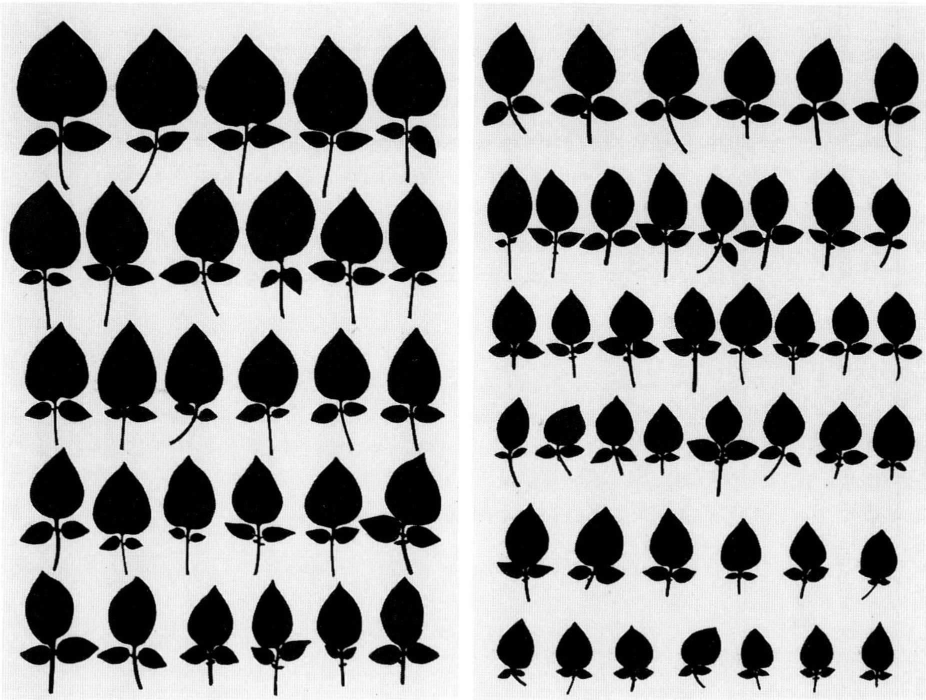

Fig. 2. Leaves from greenhouse-grown regenerants of H^2258 showing fairly good homogeneity of the third pinnate leaf from bottom

Comparatively homogeneous populations were also obtained in earlier experiments in which 3 months were needed for regeneration (Binding et al. 1978; Wenzel et al. 1979). Variation in several characters was described by Thomas et al. (1982; see Table 2), even though procedures were followed which were almost identical to those of the papers mentioned before.

It is remarkable that particular high variability was frequently reported for protoclones obtained by the application of original or modified procedures of Shepard and Totten (1977). This instability was either a result of the series of culture media (see Sect. 7.2), or it depended on particular instabilities of the investigated genotypes. Some influence on the cytogenetic stability of cytokinins was found by Sree Ramulu et al. (1983): zeatin apparently evoked higher variability than 6-BA.

Genetic instability was observed already in early developmental stages of the protoclones. This was established by cytophotometric analysis of the DNA content of the nuclei (Carlberg et al. 1984; Sree Ramulu et al. 1984a). It must be considered when the degree of variation during the protoclone development is estimated that clonal and somaclonal variation may already be present in the protoplast donor material (Fish and Karp 1986; see also Binding 1985). Variability has also been found within single callus populations (e.g. Sree Ramulu et al. 1983). This may commonly be a consequence of mutation during protoplast

Table 2. Assessment of ten morphological characters in Maris Bard, and plants derived from it by protoplast culture. (Thomas et al. 1982)

Shoot numbers from which plants are derived (number of plants scored in brackets)		A Leaf colour	B Leaf gloss	C Variegation	D Terminal leaflet shape	E Lateral leaflet shape	F Fusion of terminal and lateral leaflets	G Overlapping of terminal and lateral leaflets	H Anthocyanin pigment on leaf	I Vigour	J Height
Maris Bard tuber	(1)	2	2	0	2	2	0	3	0	4	3
P1	(1)	3	2	0	1	none	–	–	0	1	1
P2	(5)	V1–4	V1–3	1	V1–2	V2–3	0	V1–3	V0–1	V1–4	V1–3
P23X	(1)	4	3	0	2	2	1	3	0	2	3
P3	(7)	4	3	1	V1–2	V1–2	1	V1–3	0	3	3
P4	(9)	4	2	0	2	2	0	2	0	3	3
P45X	(11)	4	2	0	V2–3	V2–3	0	2	0	3	3
P5	(6)	4	2	1	2	2	0	2	0	3	3
P7	(6)	4	2	0	1	1	2	3	0	3	3
P8	(8)	4	2	0	2	2	0	3	0	3	3
P910	(2)	4	3	0	1	1	2	3	0	2	3
P11	(1)	4	3	0	2	2	0	3	0	3	3
P12	(2)	4	2	0	1	2	2	3	0	2	3
P13	(3)	V1–4	2	0	2	2	2	3	0	2	3
P16	(5)	V1–4	V1–3	1	2	2	0	2	1	V1–3	V2–3
P17	(1)	2	2	0	2	2	2	3	0	3	3
P22	(3)	4	3	0	1	1	3	–	0	3	3
P30	(6)	4	3	2	1	1	2	3	0	3	3
P33	(8)	4	3	0	1	1	0	3	1	3	3
P35	(5)	2	2	0	2	2	2	2	0	4	3
P38	(2)	4	3	0	n. d.	n. d.	0	n. d.	1	2	3
P39	(1)	2	2	0	2	2	2	1	0	3	3
P40	(7)	2	2	0	2	2	0	3	0	4	3
P41	(1)	4	2	0	1	1	0	3	0	2	3

A: 1 pale – 4 dark green; B: 1 matt – 3 heavy gloss; C: 0-none; 1-some plants with; 2-all plants with variegated patches; D & E: 1-round (length over width 1.0 – 1.2); 2-broad (ratio 1.3 – 1.5); 3-narrow (ratio 1.6 – 2); F: 0-no fusion; 1-some plants with some fused terminal and laterals; 2-all plants with some fusion; 3-both laterals fused; G: 1-terminal solitary – 4-terminal large overlap with laterals. H: 0-none; 1-some anthocyanin; I: 1 very slow – 4 vigorous growth; J: 1-prostrate form; 2-low; 3-normal height; V = score varies between different plants regeneration: however, cell or callus grafting cannot be excluded in some cases (see Binding 1985; Binding et al. 1987). Furthermore, temporary malformations are often simply caused by the transfer to greenhouse conditions.

Segregation of somaclonal-variant protoclones is best managed by a repeated cloning via isolated protoplasts under conditions which are, at most, apt to preserve the genetic constitution. This procedure was used, for instance, for the isolation of subclones of fusants *Solanum nigrum* (×) *tuberosum* (Binding et al. 1982).

The most frequently described mutant characters were altered morphology, polyploidy and aneuploidy (see for instance: Shepard and Totten 1977; Binding et al. 1978; Secor and Shepard 1981; Karp et al. 1982; Thomas et al. 1982; Bokelmann and Roest 1983; Sree Ramulu et al. 1983, 1984a, b; Carlberg et al. 1984; Creissen and Karp 1985; Bajaj 1986; Gill et al. 1986; Debnath et al. 1986). Fish and Karp (1986) found that euploid shoots were more readily rooted than aneuploids, which resulted in a selection against the aneuploids when plants were regenerated. The occurrence of chromosome and gene mutation was presumed and taken as the explanation of morphological variation (Shepard 1980a; Sree Ramulu et al. 1984b). A chlorophyll-deficient mutant was reported by Sree Ramulu et al. (1984b). Landsmann and Uhrig (1985) found mutation in a 25S-rRNA gene. Variation in mitochondrial DNA was reported by Kemble and Shepard (1984).

Resistant mutants against potato pathogens are of particular interest. It is evident that such mutants are monogeneic, and hence may be easily overcome by a mutant pathogen organism. Matern et al. (1978; see also Shepard 1980b) found clones resistant to *Alternaria*; this character, however, was lost during vegetative propagation. Resistants against *Phytophthora* at the first phase of infection were selected from protoclones by the use of purified exotoxin of the pathogen (Foroughi-Wehr and Stolle 1985). Schuchmann (1985) was able to select strains with increased tolerance against *Fusarium* exotoxin. Toxin resistance of protoclones was also investigated by Creissen (1984).

5 Fusion of Protoplasts

The hitherto investigated fusion combinations including *Solanum tuberosum* as at least one partner are compiled in Table 3.

Fusion was induced either by the action of calcium ions and high pH (Keller and Melchers 1973), frequently combined with incubation in PEG solutions (Kao and Michayluk 1974; Wallin et al. 1974), or by the application of electric tensions (Zimmermann and Scheurich 1981). Fusion by the action of polyethylene solution, Ca^{2+} and high pH was carried out by Butenko and Kuchko (1980), Binding et al. (1982), Barsby et al. (1984), Austin et al. (1985a) and Debnath and Wenzel (1987). However, PEG was found to be harmful for potato protoplasts (Binding et al. 1982). The action of PEG should be re-examined with a highly purified preparation (e.g. PEG 1500/783641 of Boehringer). Melchers et al. (1978) in-

Table 3. References on somatic hybridization experiments with potato which resulted in plants

Intraspecies fusion within *Solanum tuberosum*
1x (×) 1x: Uijtewaal et al. 1987
2x (×) 2x: Austin et al. 1985b; Debnath and Wenzel 1987; Karlsson et al. 1987; Möllers et al. 1987; Schilde-Rentschler et al. 1987; Zuba unpubl.
Interspecies, intragenus fusion of *Solanum tuberosum*
(×) *S. brevidens*: Barsby et al. 1984; Austin et al. 1985a, b, 1986; Helgeson et al. 1986; Kemble et al. 1986; Ehlenfeld and Helgeson 1987; Fish and Jones 1987; Fish et al. 1987
(×) *S. bulbocastanum*: Helgeson et al. 1987
(×) *S. chacoense*: Butenko and Kuchko 1980; Butenko et al. 1984
(×) *S. nigrum*: Binding et al. 1982; Gressel et al. 1984
(×) *S. phureja*: Puite et al. 1986
Intergenus fusion of *Solanum tuberosum*
(×) *Lycopersicon esculentum*: Melchers et al. 1978; Smillie et al. 1979; Ninnemann and Jüttner 1981
(×) *Lycopersicon pimpinellifolium*: Okamura 1987
(×) *Nicotiana plumbaginifolia*: De Vries et al. 1987
(×) *Nicotiana tabacum*: Sharzhynskaya et al. 1982

cubated the protoplasts in solutions of $CaCl_2$ buffered to pH 10.5 not including PEG treatment.

Recently, a modification of the calcium/high pH technique was developed (Binding et al. 1988b). The protoplasts are embedded into agarose gel droplets containing isoosmolar calcium nitrate, and flooded by calcium nitrate solution adjusted to pH 10 to 10.5 by ammonia. After 20 min of incubation, the fluid is replaced by culture medium. Hybrid as well as cybrid plants were recovered. The electrofusion technique has been successfully adapted to potato hybridization (Puite et al. 1986; Schilde-Rentschler et al. 1987; De Vries and Tempelaar 1987).

The procedures and markers for isolation, identification and analysis of fusion products were largely the same as in systems of other plant species and may be looked up in other surveys (e.g. Binding and Nehls 1982; Schieder 1982; Gleba and Sytnik 1984; Nehls et al. 1986). Whereas enough characters are usually available in interspecific combinations to investigate the populations grown from a fusion experiment (Melchers et al. 1978; Smillie et al. 1979; Ninnemann and Jüttner 1981; Binding et al. 1982; Gressel et al. 1984), much fewer markers are available in intraspecific combinations. Heterokaryons have been enriched by flow cytometric sorting (Puite et al. 1987). An albino was used by Barsby et al. (1984); but such mutants have been rare in potato. Heterosis-like growth of hybrid calli was found to be appropriate for selection in potato (Debnath and Wenzel 1987). Furthermore, isoenzymes, ptDNA and mtDNA restriction fragments were useful for confirmation of the hybrid nature (Butenko and Kuchko 1980; Barsby et al. 1984; Austin et al. 1985a; Kuchko 1985; Kemble et al. 1986; Debnath and Wenzel 1987). The selection by differential phytotoxin response was proposed by Möllers et al. (1987).

Different degrees of fertility were observed in fusion hybrids in which potato was involved: *S. tuberosum* (×) *brevidens* produced functional ovules (Ehlenfeld and Helgeson 1987); hybrid *S. nigrum* (×) *tuberosum* was sterile, whereas a cybrid with uniparental *S. nigrum* nucleus was self-fertile (Binding et al. 1982); all plants of *S. tuberosum* (×) *Lycopersicon esculantum* were sterile (Melchers et al. 1978).

6 Conclusions and Prospects

Various techniques are available for reliable plant regeneration from isolated potato protoplasts. The choice depends on the intended results with respect to genetic stability or clonal variation, the latter, however, is better induced by mutagens. Genotypic differences in regeneration efficiencies which have been frequently found must be overcome.

The most important objective in potato breeding is the establishment of resistance against various diseases. Search for resistant mutants within protoclones is at present an important field of application of protoplast technology in potato; however, these mutants will probably only in particular cases provide a stable breakthrough in the control of a disease; usually, monogenic resistance is easily overcome by mutant pathogens. This is the reason also for moderate pessimism with regard to the transfer of resistance factors by transformation experiments. Best resistance systems are found in wild relatives of potato and are controlled by a number of genes often acting at different stages of the disease process. These complexes are most reliably introduced into the crop by cross-breeding at the 2x level and subsequent fusion in order to re-establish the highly heterozygous, heterotic 4x stage; or by fusion at the 1x level, backcrossing introduction into selected 2x homozygotic lines and fusion-hybridization resulting in a 4x cultivar (cf. Wenzel et al. 1979, 1987). Interspecific fusion at higher ploidy levels of potato was done by Helgeson et al. (1986). Intraspecific protoplast fusion also has some value for heterosis breeding starting from selected 2x clones without prior introduction of foreign genes.

In summary, incorporation of protoplast technologies into potato breeding programmes is significantly indicated for this highly heterozygous, vegetatively propagated crop. The application is particularly promising because powerful procedures for regeneration and fusion are already available.

7 Protocols

Two procedures for plant regeneration from isolated potato protoplasts (Shepard and Totten 1977; Binding et al. 1978) have been widely applied in original or modified forms. The protocols outlined in the following sections are based on these two procedures. The culture media are compiled in Section 7.3. Recently, another highly efficient procedure has been published by Masson et al. (1987).

7.1 Protocol Modified from Binding et al. (1978)

This protocol is presently being used in the authors' laboratory and has been successfully applied to two cultivars and several 2x clones (Binding et al. 1988a).

7.1.1 Conditions for the Isolation of Protoplasts

Plant material: About 3-mm-long tips of shoots of in-vitro cultures. Plantlets are grown on agar media MS or B5 without phytohormones at 16 h d, 2000 lx by white fluorescent light, and 25 °C. Last subculture before harvesting for protoplast isolation is carried out on media with 0.5 µM 6-BA. Shoot tips are cut off for protoplasting. The cultures may be kept in the dark for 1 to 2 d before harvesting. Culture vessels are usually plastic Petri dishes of 9 cm in diameter. The harvested tips are sliced by a razor blade in a droplet of the enzyme solution.

Protoplast isolation: The enzyme solution contains 0.3% Macerozyme R-10, 1% Cellulase Onozuka R-10 (both from Yakult Honsha Co., Higashi-Shinbashi, Minatoku, Tokyo; other preparations with adequate enzyme activities are also useful), 0.5 M mannitol, 5 µM $Ca(NO_3)_2$. Sliced plant material of about 100 mg is suspended in 12 ml enzyme solution either in a 9-cm Petri dish on standing or in a glass vessel on a gyratory shaker at about 30 rpm at 25 °C and 500 lx. Incubation lasts 4 to 15 h. Purification and plating of the protoplasts: The incubation mixture is taken up by a pipette and passed through a steel grid of 40 µM mesh into a centrifuge test tube. The grid used is folded, forming a flat cone of about 3 cm in diameter. Pelleting at 100 g is followed by washing with 0.3 M NaCl, which contains the major salts of medium V-KM at 1/10th concentration. The protoplasts are spun down at 80 g. Both centrifugations last 5 min. After decantation of the supernatant, the protoplasts are suspended in the restfluid, taken up by a Pasteur pipette and placed into a 3-cm plastic Petri dish. The protoplasts are allowed to settle, and most of the supernatant is removed. Agarose-containing culture medium V-KM is added at about the same volume as that of the protoplast-containing droplet; the mixture is taken up and dispensed in form of 10- to 15-µl droplets. After gelling, 1 ml liquid medium V-KM is added, the dishes are sealed with parafilm and placed into a growth chamber. Just 10^3 protoplasts are sufficient for a culture when they are enclosed in one droplet forming a streak of high cell titre. Commonly, about 2×10^4 protoplasts are plated in one dish, distributed to about 10 droplets.

7.1.2 Conditions for the Culture of Protoplasts

The dishes with the protoplasts are kept at 25 °C and illuminated by white fluorescent light of about 800 lx. The liquid medium is exchanged at 2 d intervals. Ten to 14 d after plating are sufficient for the formation of cell colonies which survive the transfer from the high osmolar medium V-KM to agar medium B5. For this purpose, the agarose gel lenses are picked up by forceps and placed onto B5C agar medium in a 5-cm Petri dish.

7.1.3 Conditions for the Regeneration of Plants

The cultures are transferred into a tissue culture room which is kept at about 25 °C and provides a 14-h d by white fluorescent light. An appropriate light intensity is 3000 lx. After 1 week, the clumps of microcalli are flattened to obtain monolayers from the calli. When the protoclones have reached diameters of about 1 mm, they are separated and placed into a 9-cm dish, about 100 into each. Shoot tips appear from the 24th d after protoplast isolation on; shoots elongate on the same medium. Roots are also formed. More efficient root formation is induced on hormone-free media MS or B5. Auxins, GA_3 and activated charcoal were reported to increase shoot formation, but are not needed as concluded from the experiences of the authors. Plantlets measuring at least 1 cm in length reliably survive planting into soil such as is usually taken for growing greenhouse plants. No increased air humidity is necessary when the soil is thoroughly wet at planting.

7.2 Protocol Modified from Shepard and Totten (1977)

Originally, the procedure was applied to regeneration of mesophyll protoplasts from a greenhouse plant. Shahin (1984) outlined a modification, mainly taking advantage of in-vitro cultured plantlets; but leaves and not shoot tips (see Sect. 7.1) are harvested for protoplast isolation. Other modifications are taken from Shepard (1980a).

7.2.1 Conditions for the Isolation of Protoplasts

Plant material: (1) leaf discs from greenhouse-grown plants. Plants are grown at 24 °C, 70%–75% relative humidity and a 6-h photoperiod at 7000 lx. Leaves are cut off and surface-sterilized. They are then treated on the lower side with a soft-nylon brush until their colour turns light green. Leaf discs of not more than 2 cm length of side are preconditioned by incubation in 200 ml liquid medium A, but without sucrose, at 4 °C for 16 to 24 h. (2) leaves of plantlets grown on agar media. Plantlets are grown on MS agar media containing 1.5 μM kinetin at 25 °C and a 16-h photoperiod at 6000 lx light intensity. The plantlets are kept for 2 d in darkness before harvesting of leaves. These are then treated with a nylon brush as indicated above.

Protoplast isolation: The enzyme solution contains 0.1% Macerozyme R-10, 0.5% Cellulase Onozuka R-10, 0.3 M sucrose, 2% PVP, the mineral salts of medium A, and MES buffer, pH adjusted to 5.6. Small slices (about 1 mm^2) of the leaf discs or leaflets are suspended in the enzyme solution, briefly exposed to vacuum of low strength and incubated on a gyratory shaker at 40 rpm for 4 h.

Purification and plating of the protoplasts: The incubation mixture is passed through four layers of cheese-cloth and spun at 350 g for 8 min. The floating protoplasts are washed by additional medium A and a repeated centrifugation. The protoplasts are embedded into CL-agarose-medium at a cell titre of 1×10^4 per ml. The gel is layered onto agar medium R; the dishes are sealed with parafilm. Alternate segments of protoplast suspension and medium R can also be prepared, or reserve medium is not used from the beginning.

7.2.2 Conditions for the Culture of Protoplasts

The cultures are kept at 24 °C and 500 to 800 lx. If reserve agar is not used from the beginning, the suspension gel is layered onto agar media with identical composition of non-solidifying components as the protoplast culture media 10 to 14 d after plating. The dishes are transferred to continuous light of 4000 lx. After 10 d, the microcalli are placed separately onto the same type of media (30 to 40 colonies per dish). In the case when reserve medium is used from the beginning, the protoclones are left on the original medium until they have reached sizes of 2 mm in diameter and are then transferred onto medium C, where they proliferate to 3 to 5 mm sizes within 28 d.

7.2.3 Conditions for the Regeneration of Plants

Protoclones are transferred onto regeneration media about 30 d after the last subculturing indicated in Section 7.2.2. Shepard (1980a) proposed medium D; Shahin used different media for 4x and 2x lines (see Shahin 1984). For rooting and transfer of the plantlets to greenhouse conditions see Section 7.1.3.

7.3 Culture Media

The basic formulations of the culture media are compiled in Table 4. In cases in which molar concentrations are calculated from gram data of publications, rounded values are given. Indication of molarities is convenient particularly for inorganic compounds and osmotic agents because it admits

Table 4. Culture media-basic formulations

	V-KM[a]	B5[b]	C[c]
Major Salts – mM			
NH_4NO_3	3.5		1.3
$(NO_4)_2 \cdot SO_4$		1.5	
KNO_3	14.5	24.8	18.8
$CaCl_2$	5.0	1.0	3.0
$MgSO_4$	4.0	1.0	1.5
KH_2PO_4	0.5		1.0
NaH_2PO_4		1.1	
Fe-EDTA – mM		0.12	
$Na_2 \cdot EDTA$	0.1		0.05
$FeSO_4 \cdot 7H_2O$	0.1		0.05
Na/Fe-EDTA			
Minor Salts – μM			
$MnSO_4$	23.0	59.0	50.0
$MnCl_2$			50.0
H_3BO_3		49.0	16.0
$ZnSO_4$	5.0	7.0	2.5
KJ	1.5	2.6	0.5
Na_2MoO_4	0.4	1.0	0.05
$CuSO_4$	0.06	0.1	0.05
$CoCl_2$	0.04	0.1	
Organic Acids – mM			
Sodium pyruvate	0.18		
Fumaric acid	0.21		
Citric acid	0.30		
Malic acid	0.35		
Additional sugars – mM			
Xylose	1.7		
Ribose	1.7		
Fructose	1.4		
Mannose	1.4		
Rhamnose	1.4		
Cellobiose	0.7		
Vitamins – μM			
Ascorbic acid	1.0		
Nicotinic acid			41.0
Nicotinamide	8.0	8.0	
Cholin chloride	7.2		
Pyridoxine·HCl	5.0	5.0	2.4
Thiamine·HCl	3.0	30.0	1.5
Glycine			27.0
Ca-D-Panthothenate	2.0		
Folic acid	0.9		1.1
Biotin	0.04		0.2
Vitamin A	0.035		
Vitamin D_3	0.025		
Vitamin B_{12}	0.015		
pH-Value	5.8	5.8	5.6 (–5.8)

[a] Binding and Nehls 1977; the medium is composed of inorganic salts of V-47 (Binding 1974) and organic part of the 8p medium (Kao and Michayluk 1975). [b] Gamborg et al. 1968. [c] Shepard and Totten 1977.

Table 5. Complete culture media and their application

	V-KM[a]	V-NaCl[b]	B5[c]	A[d]	Cl[e]	R[f]	C[g]	D[h]
Major salts	V-KM	V-KM × 10^{-1}	B5	10^{-1} × C[i]	4 × C^9	C^9	C	C[i]
Fe-EDTA	V-KM	V-KM × 10^{-1}	B5	5^{-1} × C	C	C	C	C
Minor salts	V-KM	V-KM × 10^{-1}	B5	5^{-1} × C	C	C	C	C
Organic acids	V-KM							
Sucrose (mM)	0.7		58	350	200	100	7	7
Glucose (mM)	380							
Mannitol (mM)	1.4				25	100	300	200
Sorbitol (mM)	1.4				25			
Xylitol (mM)					25			
Additional sugars	V-KM							
Vitamins	V-KM		B5	10^{-1} × C	C	C	C	C
Coconut water (%)	2		(5)					
Casein hydrolys. (%)	25			10	5	10	10	10
Adenine sulphate (μM)							100	100
myo-inositol (mM)	0.56			0.056	25		0.56	0.56
MES-buffer (mM)							5	5
6-BA (μM)	2.5		2.5	2.2	1.8	1.8	2.2	
Zeatin (μM)								2.3
NAA (μM)	5.0			11.0	5.4	5.4	5.4	
IAA (μM)								0.6
2,4-D (μM)	0.5 (0)							
ABA (μM)								0.4 – 0.8
CA_3 (μM)			(0.03 – 30)					0.6
NaCl (mM)		200						
Agar (%)	(0.1)		0.8			0.4	1.0	1.0
Agarose (%)	(1.0)				0.2			
Activ. charcoal (%)	(0.1)					(0.1)		

mM and μM values are partly calculated from mg × l^{-1} values; [a] Medium V-KM is used for protoplast culture (Sect. 7.1.2). Embedding in agarose gel is recommended (Binding et al. 1988a). Uptake of protoclones in soft agar is indicated when protoplast culture was started in liquid medium (Binding et al. 1978). 2,4-D may be omitted to increase genetic stability; [b] V-NaCl is the wash solution for protoplast sedimentation (Sect. 7.1.1); [c] Medium B5 serves for plantlet culture as well as for shoot regeneration (Sect. 7.1.1, 7.1.3). Plantlet propagation is done on agar media without phytohormones. Adventitious shoots are induced in presence of 6-BA. Coconut water (medium B5C; Binding et al. 1981) and GA_3 (Bokelmann and Roest 1983; Debnath et al. 1986) may be added for shoot formation or elongation. MS media (Murashige and Skoog 1962) are equally suited; [d] Medium A is used for enzyme solution and wash solution causing floatation of the protoplasts (Sect. 7.2.1); [e] Medium Cl is taken for protoplast culture (Sect. 7.2.2); [f] Medium R serves as reserve medium for protoplast culture (Sect. 7.2.2). Activated charcoal may be added (Carlberg et al. 1983); [g] Medium C leads to protoclone callus proliferation (Sect. 7.2.2); [h] Medium D admits shoot formation (Sect. 7.2.3). Optimum amount of ABA depends on the genotype (Shepard 1980a). GA_3 may be useful (Shahin 1984; Aviv and Galun pers. commun.); [i] Media A, Cl and R contain no ammonia salt. Ammonium nitrate is present in medium D at 3.3 mM concentration easy comparison of different media. As a consequence, it is advisable to prepare molar stock solutions. Table 5 shows the complete compositions of the culture media with regard to particular purposes. Some modifications which are not mentioned in the preceding parts of Section 7 but were found to be beneficial by particular investigators are indicated in parenthesis. They are probably useful in particular genotypes. Rooting media are not included in the tables because of the minute significance of their composition (see Sect. 7.1.3).

7.4 Application of the Protocols

In the experience of the authors it is recommended to try protoplast regeneration of a given genotype at first following the protocol described in Section 7.1. Prominent advantages of this procedure are: (1) Plantlets grown in vitro are the best conditioned protoplast sources because they are grown under controlled, axenic conditions; (2) shoot apical protoplasts are poor in secondary metabolites which, in higher amounts, would interfere with the viability, and reveal good regeneration; (3) dense package of the protoplasts in agarose gel lenses and rich organic compound of the culture media make possible early reduction of the osmolarity of the culture media; (4) transfer of the protoclones from the protoplast culture media directly onto agar media which allow plantlet regeneration avoids intermittent prolonged callus proliferation. All these factors are supposed to be best guarantees for the preservation of the original genotypes. Modifications of the proposed protocol which are indicated in Section 7.3, as well as the application of some or all steps of the protocol outlined in Section 7.2 may be tested for a given genotype if sufficient success is not obtained with the procedure described in Section 7.1.

References

Austin S, Baer MA, Helgeson JP (1985a) Transfer of resistance to potato leaf roll virus from *Solanum brevidens* into *Solanum tuberosum* by somatic fusion. Plant Sci 39:75–82

Austin S, Baer MA, Ehlenfeldt M, Kazmierczak PJ, Helgeson JP (1985b) Intra-specific fusions in *Solanum tuberosum*. Theor Appl Genet 71:172–175

Austin S, Ehlenfeld MK, Baer MA, Helgeson JP (1986) Somatic hybrids produced by protoplast fusion between *S. tuberosum* and *S. brevidens*: phenotypic variation under field conditions. Theor Appl Genet 71:682–690

Bajaj YPS (1986) Cryopreservation of potato somaclones. In: Semal J (ed) Somaclonal variations and crop improvement. Martinus Nijhoff, Dordrecht, pp 244–250

Bajaj YPS (ed) (1987) Biotechnology in agriculture and forestry 3. Potato. Springer, Berlin Heidelberg New York Tokyo

Barker H, Harrison BD (1982) Infection of potato mesophyll protoplasts with five plant viruses. Plant Cell Rep 1:247–249

Barocka KH, Ross H (1985) Kartoffel. In: Hoffmann W, Mudra A, Plarre W (eds) Lehrbuch der Züchtung landwirtschaftlicher Kulturpflanzen; Band 2: Spezieller Teil. Paul Parey, Berlin, pp 212–287

Barsby TL, Shepard JF (1983) Regeneration of plants from mesophyll protoplasts of *Solanum* species of the *Etuberosa* group. Plant Sci Lett 31:101–105

Barsby TL, Shepard JF, Kemble RJ, Wong R (1984) Somatic hybridization in the genus *Solanum: S. tuberosum* and *S. brevidens*. Plant Cell Rep 3:165–167

Binding H (1966) Regeneration und Verschmelzung nackter Laubmoosprotoplasten. Z Pflanzenphysiol 55:305–321

Binding H (1974) Regeneration von haploiden und diploiden Pflanzen aus Protoplasten von *Petunia hybrida* L. Z. Pflanzenphysiol 74:327–356

Binding H (1985) Regeneration of plants. In: Fowke LC, Constabel F (eds) Plant protoplasts. CRC Press, Boca Raton, Florida, pp 21–37

Binding H, Kollmann R (1985) Regeneration of protoplasts. In: Schäfer-Menuhr A (ed) In Vitro Techniques. Propagation and Long Term Storage, CEC, Nijhoff Junk, Dordrecht Boston Lancaster, pp 93–99

Binding H, Mordhorst G (1984) Haploid *Solanum dulcamara* L.: Shoot culture and plant regeneration from isolated protoplasts. Plant Sci Lett 35:77–79

Binding H, Nehls R (1977) Regeneration of isolated protoplasts to plants in *Solanum dulcamara* L. Z Pflanzenphysiol 85:279–280

Binding H, Nehls R (1982) Techniques of somatic cell hybridization by fusion of protoplasts. In: Shay JW (ed) Techniques in somatic cell genetics. Plenum Press, New York London, pp 471–492

Binding H, Nehls R, Schieder O, Sopory SK, Wenzel G (1978) Regeneration of mesophyll protoplasts isolated from dihaploid clones of *Solanum tuberosum*. Physiol Plant 43:52–54

Binding H, Nehls R, Kock R, Finger J, Mordhorst G (1981) Comparative studies on protoplast regeneration in herbaceous species of the *Dicotyledoneae* class. Z Pflanzenphysiol 101:119–130

Binding H, Jain SM, Finger J, Mordhorst G, Nehls R, Gressel J (1982) Somatic hybridization of an atrazine resistant biotype of *Solanum nigrum* with *Solanum tuberosum*. Part I: clonal variation in morphology and atrazine sensitivity. Theor Appl Genet 63:273–277

Binding H, Görschen E, Jörgensen J, Krumbiegel-Schroeren G, Ling HQ, Mordhorst G, Witt D, Kollmann R (1986) Protoplast regeneration and fusion in agarose lenses. In: Somers DA, Gengenbach BG, Biesboer DD, Hackett WP, Green CE (eds) VI Internat Congr of Plant Tissue and Cell Culture-Abstracts, Univ Minnesota, Minneapolis, p 318

Binding H, Witt D, Monzer J, Mordhorst G, Kollmann R (1987) Plant cell graft chimeras obtained by co-culture of isolated protoplasts. Protoplasma 141:64–73

Binding H, Görschen E, Jörgensen J, Krumbiegel-Schroeren G, Ling HQ, Rudnick J, Sauer A, Zuba M, Mordhorst D (1988a) Protoplast culture in agarose media with particular emphasis to streaky culture lenses. Botanica Acta 101:233–239

Binding H, Zuba M, Rudnick J, Mordhorst G (1988b) Protoplast gel fusion. J Plant Physiol, in press

Bokelmann GS, Roest S (1983) Plant regeneration from protoplasts of potato (*Solanum tuberosum* cv. Bintje). Z Pflanzenphysiol 109:259–263

Butenko RG, Kuchko AA (1980) Somatic hybridization of *Solanum tuberosum* L. and *Solanum chacoense* Bitt. by protoplast fusion. In: Ferenczy L, Farkas GL (eds) Advances in protoplast research. Acad. Kiado, Budapest, pp 293–300

Butenko RG, Kuchko AA, Vitenko VA, Avetisov VA (1977) Production and cultivation of isolated protoplasts from mesophyll of leaves of *Solanum tuberosum* L. and *Solanum chacoense* Bitt. Sov Plant Physiol 24:541–546

Butenko RG, Chromova LM, Sednina GV, Kuchko AA, Burgutin AB, Komarnitzky IK (1984) New forms of potato created by cell and protoplast culture methods. In: Röhlich P, Bácsy E (eds) Tissue culture and RES. Akadémiai Kiadó, Budapest, pp 515–527

Carlberg I, Glimelius K, Eriksson T (1983) Improved culture ability of protoplasts by use of activated charcoal. Plant Cell Rep 2:223–225

Carlberg I, Glimelius K, Eriksson T (1984) Nuclear DNA content during the initiation of callus formation from isolated protoplasts of *Solanum tuberosum* L. Plant Sci Lett 35:225–230

Creissen GP (1984) Protoplast regeneration and selection for toxin resistance in *Solanum*. Ph. D. Thesis, University of London

Creissen GP, Karp A (1985) Karyotypic changes in potato plants regenerated from protoplasts. Plant Cell Tissue Organ Cult 4:171–182

Debnath SC, Wenzel G (1987) Selection of somatic fusion products in potato by hybrid vigour. Potato Res 30:371–380

Debnath SC, Schuchmann R, Wenzel G (1986) Regeneration capacity of potato protoplasts isolated from single cell derived donor plants. Acta Bot Neerl 35:233–241

De Vries SE, Bokelmann GS (1986) Regeneration of callus and plants from cell suspension protoplasts of dihaploid potato (*Solanum tuberosum* L.). J Plant Physiol 122:199–203

De Vries SE, Tempelaar MJ (1987) Electrofusion and analysis of potato somatic hybrids. In: Bajaj YPS (ed) Biotechnology in agriculture and forestry. 3. Potato. Springer, Berlin Heidelberg New York Tokyo, pp 211–222

De Vries SE, Van Everdink WJ, Ferwerda MA, Jacobsen E, Pijnacker LP, Feenstra WJ (1987) Studies on gene mapping in the potato using somatic hybrids of *Solanum tuberosum* + *Nicotiana plumbaginifolia*. In: Puite KJ, Dons JJM, Huizing HJ, Kool AJ, Koornneef M, Krens FA (eds) Current plant science and biotechnology in agriculture, Progress in protoplast research. Kluwer Acad Publ, Dordrecht Boston London, pp 215–218

Ehlenfeldt MK, Helgeson JP (1987) Fertility of somatic hybrids from protoplast fusions of *Solanum brevidens* and *S. tuberosum*. Theor Appl Genet, 73:395–402

Fish N, Jones MGK (1987) Production and field assessment of somatic hybrids between dihaploid *Solanum tuberosum* and *Solanum brevidens*. In: Puite KJ, Dons JJM, Huizing HJ, Kool AJ, Koornneef M, Krens FA (eds) Current plant science and biotechnology in agriculture, Progress in plant protoplast research. Kluwer Acad Publ, Dordrecht Boston London, p 201

Fish N, Karp A (1986) Improvements in regeneration from protoplasts of potato and studies on chromosome stability. 1. The effect of initial culture media. Theor Appl Genet 72:405–412

Fish N, Karp A, Jones MGK (1987) Improved isolation of dihaploid *Solanum tuberosum* protoplasts and the production of somatic hybrids between dihaploid *S. tuberosum* and *S. brevidens*. In Vitro Cell Dev Biol 23:575–580

Foroughi-Wehr B, Stolle K (1985) Resistenzselektion in vitro am Beispiel des Systems Kartoffel/*Phytophthora infestans* (Mont.) de Bary. Nachrichtenbl Dtsch Pflanzenschutzdienst 37: 170–173

Foulger D, Jones MGK (1986) Improved efficiency of genotype-dependant regeneration from protoplasts of important potato cultivars. Plant Cell Rep 5:72–76

Gamborg OL, Miller RA, Ojima K (1968) Nutrient requirements of suspension cultures of soybean root cells. Exp Cell Res 50:151–158

Geisler G (1983) Ertragsphysiologie von Kulturarten des gemäßigten Klimas. Paul Parey, Berlin

Gill BS, Kam-Morgan LNW, Shepard JF (1986) Origin of chromosomal and phenotypic variation in potato protoclones. J Hered 77:13–16

Gleddie S, Keller WA, Setterfield G (1985) Plant regeneration from tissue, cell and protoplast cultures of several wild *Solanum* species. J Plant Physiol 119:405–418

Gressel J, Cohen N, Binding H (1984) Somatic hybridization of an atrazine resistant biotype of *Solanum nigrum* with *Solanum tuberosum*. 2. Segregation of plastomes. Theor Appl Genet 67:131–134

Grun P, Chu LJ (1978) Development of plants from protoplasts of *Solanum* (Solanaceae) Am J Bot 65:538–543

Gunn RE, Shepard JF (1981) Regeneration of plants from mesophyll-derived protoplasts of British potato (*Solanum tuberosum* L.). Plant Sci Lett 22:97–101

Haberlach GT, Cohen BA, Reichert NA, Towill LE, Helgeson JP (1985) Isolation, culture and regeneration of protoplasts from potato and several related *Solanum* species. Plant Sci 39:67–74

Hassanpour-Estahbanati A, Demarly Y (1985) Plant regeneration from protoplasts of *Solanum pennellii*: Effect of photoperiod applied to donor plant. J Plant Physiol 121:171–174

Helgeson JP, Hunt GJ, Haberlach GT, Austin S (1986) Somatic hybrids between *Solanum brevidens* and *Solanum tuberosum*: Expression of a late blight resistance gene and potato leaf roll resistance. Plant Cell Rep 3:212–214

Helgeson JP, Haberlach GT, Pohlman J, Austin S (1987) Somatic fusions of *Solanum species*. In: Puite KJ, Dons JJM, Huizing HJ, Kool AJ, Koornneef M, Krens FA (eds) Current plant science and biotechnology in agriculture, Progress in plant protoplast research. Kluwer Acad Publ, Dordrecht Boston London, pp 205–207

Jia J-F, Potrykus I (1981) Mesophyll protoplasts from *Solanum melongena* var. *depressum* Bailey regenerate to fertile plants. Plant Cell Rep 1:71–72

Jones MGK (1984) Protoplast and somaclonal variation research on potato. In: Innovative methods for propagating potatoes, CIP, Lima, pp 305–318

Kao KN, Michayluk MR (1974) A method for high-frequency intergeneric fusion of plant protoplasts. Planta 15:355–367

Kao KN, Michayluk MR (1975) Nutritional requirements for growth of *Vicia hajastana* cells and protoplasts at a very low population density in liquid media. Planta 126:105–110

Karlsson SB, Eriksson T (1987) Somatic hybridization between anther-derived dihaploid lines of *Solanum tuberosum* L. (potato). In: Puite KJ, Dons JJM, Huizing HJ, Kool AJ, Koornneef M, Krens FA (eds) Current plant science and biotechnology in agriculture, Progress in plant protoplast research. Kluwer Acad Publ, Dordrecht Boston London, pp 197–198

Karp A, Nelson RS, Thomas E, Bright SWJ (1982) Chromosome variation in protoplast-derived potato plants. Theor Appl Genet 63:265–272

Keller WA, Melchers G (1973) The effect of high pH and calcium on tobacco leaf protoplast fusion. Z Naturforsch Sect C Biosci 28:737–741

Kemble RJ, Shepard JF (1984) Cytoplasmic DNA variation in a potato protoclonal population. Theor Appl Genet 69:211–216

Kemble RJ, Barsby TL, Wong RSC, Shepard JF (1986) Mitochondrial DNA rearrangements in somatic hybrids of *Solanum tuberosum* and *Solanum brevidens*. Theor Appl Genet 72:787–793

Kowalczyk TP, Mackenzie IA, Cocking EC (1983) Plant regeneration from organ explants and protoplasts of the medicinal plant *Solanum khasianum* CB Clarke var. *Chatterjeeanum* Sengupta (Syn. *Solanum viarum* Dunal). Z Pflanzenphysiol 111:55–68

Kuchko AA (1985) Interspecies somatic hybridization of wild and cultivated potatoes. In: Butenko RG (ed) Plant Cell Culture. MIR, Moscow, pp 154–174

Landsmann J, Uhrig H (1985) Somaclonal variation in *Solanum tuberosum* detected at the molecular level. Theor Appl Genet 71:500–505

Li GG, Constabel F (1984) Plant regeneration with callus and protoplasts of *Solanum uporo* Dun. J Plant Physiol 117:137–142

Masson J, Lecerf M, Rousselle P, Perennec P, Pelletier G (1987) Plant regeneration from protoplasts of diploid potato derived from crosses of *Solanum tuberosum* with wild *Solanum* species. Plant Sci 53:167–176

Matern U, Strobel G, Shepard J (1978) Reaction to phytotoxins in potato population derived from mesophyll protoplasts. Proc Natl Acad Sci USA 75:4935–4939

Melchers G, Sacristán MD, Holder AA (1978) Somatic hybrid plants of potato and tomato regenerated from fused protoplasts. Carlsberg Res Commun 43:203–218

Möllers C, Deimling S, Wenzel G (1987) Fusion of dihaploid *Solanum tuberosum* protoplasts and trials of hybrid selection with phytotoxins. In: Puite KJ, Dons JJM, Huizing HJ, Kool AJ, Koornneef M, Krens FA (eds) Current plant science and biotechnology in agriculture, Progress in plant protoplast research. Kluwer Acad Publ, Dordrecht Boston London, pp 199–200

Murashige T, Skoog F (1962) A revised medium for rapid growth and bioassays with tobacco tissue culture. Physiol Plant 15:473–497

Nehls R (1978) Isolation and regeneration of protoplasts from *Solanum nigrum* L. Plant Sci Lett 12:183–187

Nehls R, Krummbiegel-Schroeren G, Binding H (1986) Development of protoplast fusion products. In: Reinert J, Binding H (eds) Results and problems in cell differentiation 12, Differentiation of protoplasts and of transformed plant cells. Springer, Berlin Heidelberg New York Tokyo, pp 67–108

Nelson RS, Creissen GP, Bright SWJ (1983) Plant regeneration from protoplasts of *Solanum brevidens*. Plant Sci Lett 30:355–362

Ninnemann H, Jüttner F (1981) Volatile substances from tissue cultures of potato, tomato and their somatic fusion products. Comparison of gas chromatographic patterns for identification of hybrids. Z Pflanzenphysiol 103:95–108

Okamura M (1987) Regeneration and evaluation of somatic hybrid plants between *Solanum tuberosum* and *Lycopersicon pimpinellifolium*. In: Puite KJ, Dons JJM, Huizing HJ, Kool AJ, Koornneef AJ, Krens FA (eds) Current plant science and biotechnology in agriculture, Progress in plant protoplast research. Kluwer Acad Publ, Dordrecht Boston London, pp 213–214

Peloquin SJ, Hougas RW, Gabert AC (1966) Haploidy as a new approach to the cytogenetics and breeding of *Solanum tuberosum*. In: Riley R, Lewis EJ (eds) Chromosome manipulations and plant genetics. Oliver and Boyd, Edinburgh, pp 21–28

Puite KJ, Roest S, Pijnacker LP (1986) Somatic hybrid plants after electrofusion of diploid *Solanum tuberosum* and *S. phureja*. Plant Cell Rep 5:262–265

Puite KJ, Ten Broeke WRR, Schaart JG, Schel JHN (1987) Flow cytometric sorting of potato heterokaryons with inhibition of cell wall synthesis before sorting. In: Puite KJ, Dons JJM, Huizing HJ, Kool AJ, Koornneef M, Krens FA (eds) Current plant science and biotechnology in agriculture, Progress in plant protoplast research. Kluwer Acad Publ, Dordrecht Boston London, pp 209–210

Saxena PK, Gill R, Rashid A, Maheshwari SC (1981) Plantlet formation from isolated protoplasts of *Solanum melongena* L. Protoplasma 106:355–359

Saxena PK, Gill R, Rashid A, Maheshwari SC (1982) Plantlets from mesophyll protoplasts of *Solanum xanthocarpum*. Plant Cell Rep 1:219–220

Schieder O (1982) Somatic hybridization: a new method for plant improvement. In: Vasil I, Scowcroft WR, Frey KJ (eds) Plant improvement and somatic cell genetics. Academic Press, New York, pp 239–253

Schilde-Rentschler L, Boos G, Ninnemann H (1987) Somatic hybridization of diploid potato lines, a tool in potato breeding. In: Puite KJ, Dons JJM, Huizing HJ, Kool AJ, Koornneef M, Krens FA (eds) Current plant science and biotechnology in agriculture, Progress in plant protoplast research. Kluwer Acad Publ, Dordrecht Boston London, pp 195–196

Schuchmann R (1985) In vitro Selektion auf Fusarium Resistenz bei der Kartoffel. Ph. D. Thesis, TU München

Schumann U, Opatrny Z, Koblitz H (1980) Plant recovery from long term callus cultures and from suspension culture derived protoplasts of *Solanum phureja*. Biochem Physiol Pflanz (BBP) 175:670–675

Secor GA, Shepard JF (1981) Variability of protoplast-derived potato clones. Crop Sci 21:102–105

Shahin EA (1984) Isolation, culture and regeneration of potato leaf protoplasts from plants preconditioned in vitro. In: Vasil IK (ed) Cell Culture and Somatic Cell Genetics of Plants. Academic Press, Orlando, San Diego New York London Toronto Montreal Sydney Tokyo, pp 381–390

Sharzhynskaya MV, Cherep NN, Gleba Y (1982) Potato and tobacco hybrid cell lines and plants obtained by cloning individual protoplast fusion products. Sov Cytol Genet 6:42–48

Shepard JF (1980a) Abscissic acid-enhanced shoot initiation in protoplast-derived calli of potato. Plant Sci Lett 18:327–333

Shepard JF (1980b) Mutant selection and plant regeneration from potato mesophyll protoplasts. In: Rubenstein I, Gengenbach B, Phillips RL, Green CE (eds) Genetic improvement of crops: emergent techniques. Univ of Minnesota Press, Minneapolis, pp 185–219

Shepard JF (1982) Cultivar dependent cultural refinements in potato protoplast regeneration. Plant Sci Lett 26:127–132

Shepard JF, Totten RE (1977) Mesophyll cell protoplasts of potato: isolation, proliferation and plant regeneration. Plant Physiol 60:313–316

Sidorov VA, Kuchko AA, Gleba YY (1984) Genetic modification of *Solanum tuberosum* through protoplast culture and fusion. In: Röhlich P, Bácsy E (eds) Tissue Culture and Research Akád Kiadó, Budapest, pp 529–534

Smillie RM, Melchers G, v. Wettstein D (1979) Chilling resistance of somatic hybrids of tomato and potato. Carlsberg Res Commun 44:127–132

Sree Ramulu K, Dijkhuis P, Roest S (1983) Phenotypic variation and ploidy level of plants regenerated from protoplasts of tetraploid potato (*Solanum tuberosum* L. cv. Bintje). Theor Appl Genet 65:329–338

Sree Ramulu K, Dijkhuis P, Roest S, Bokelmann GS, de Groot B (1984a) Early occurrence of genetic instability in protoplast cultures of potato. Plant Sci Lett 36:79–86

Sree Ramulu K, Dijkhuis P, Roest S (1984b) Genetic instability in protoclones of potato (*Solanum tuberosum* L. cv. Bintje): new types of variation after vegetative propagation. Theor Appl Genet 68:515–519

Thomas E (1981) Plant regeneration from shoot-culture derived protoplasts of tetraploid potato (*Solanum tuberosum* cv. Maris Bard). Plant Sci Lett 23:84–88

Thomas E, Bright SWJ, Franklin J, Lancaster VA, Miflin BJ (1982) Variation amongst protoplast-derived potato plants (*Solanum tuberosum* vc. Maris Bard). Theor Appl Genet 62:65–68

Uhrig H (1981) Regeneration of protoplasts of dihaploid potato plants bleached by a herbicide (SAN 6706). Mol Gen Genet 181:403–405

Uijtewaal BA, Suurs LCJM, Jacobsen E (1987) Protoplast fusion of monohaploid (2n = x = 12) potato clones: identification of somatic hybrids using malate dehydrogenase as a biochemical marker. Plant Sci 51:277–284

Wallin A, Glimelius K, Eriksson T (1974) The induction of aggregation and fusion of *Daucus carota* protoplasts by polyethylenglycol. Z Pflanzenphysiol 74:64–80

Wenzel G, Schieder O, Przewozny T, Sopory SK, Melchers G (1979) Comparison of single cell culture derived *Solanum tuberosum* L. plants and a model for their application in breeding programmes. Theor Appl Genet 55:49–55

Wenzel G, Debnath SC, Schuchmann R, Foroughi-Wehr FR (1987) Combined application of classical and unconventional techniques in breeding of disease resistant potatoes. In: Richardson DE, Jellis GJ (eds) The production of new potato varieties: Technological advances. Cambridge University Press, pp 277–288

Zimmermann U, Scheurich P (1981) High frequency fusion of plant protoplasts by electric field. Planta 151:26–32

II.3 Isolation and Culture of *Citrus* Protoplasts

ALIZA VARDI[1] and ESRA GALUN[2]

1 General Account

1.1 Economic Importance of *Citrus* Cultivars

Citrus fruits rank second only to grapes in world fruit production. *Citrus* production was estimated in 1985 at a total of approximately 60 million metric tons (FAO 1985). Oranges comprise the most important *Citrus* product in world trade, followed by grapefruit, mandarins (which have recently increased, especially on the European market) and lemons. In addition to fresh fruits, large quantities of sweet orange, grapefruit and mandarin are processed into juice, concentrates, squashes and other products.

1.2 Distribution and Origin of *Citrus* Cultivars

The cultivars of *Citrus* are grown in tropical and subtropical climates, roughly covering a belt which circumvents the globe approximately between latitudes 40 °N and 40 °S. There are about 90 *Citrus*-producing countries, the most prominent among them being Brazil, USA, Mexico, Spain, Italy and China (FAO 1985).

The commonly cultivated citrus-type species belong to the genus *Citrus*. This genus also contains species of the subgenus *Papeda*, which do not bear edible fruits. In addition to *Citrus*, the subtribe Citrinae includes five other genera: *Eremocitrus, Poncirus, Fortunella, Microcitrus* and *Clymenia*. *Poncirus* is used as a rootstock for *Citrus* cultivars and the kumquat fruits (*Fotunella* spp.) have a limited market (Swingle and Reece 1967). Important *Citrus* cultivars are listed in Table 1.

The subtribe Citrinae ("true *Citrus* fruit trees") is native to a large Asiatic area from the Himalayan foothills in India to North-Central China, extending to Burma, Thailand and Indonesia in the South and to the Philippines, North-East Australia, New Guinea and New Caledonia in the South-East. The centers of origin of the main commerical *Citrus* cultivars were not defined with certainty because of probable hybridization in ancient times; but, using evidence from old

[1] Institute of Horticulture, Agricultural Research Organization, The Volcani Center, Bet Dagan 50250, Israel
[2] Department of Plant Genetics, The Weizmann Institute of Science, Rehovot 76100, Israel

Biotechnology in Agriculture and Forestry, Vol. 8
Plant Protoplasts and Genetic Engineering I (ed. by Y. P. S. Bajaj)

Table 1. Important cultivars of *Citrus* and their affiliation with trivial and true species

Common name	Species	Examples of cultivars
Orange	*Citrus sinensis* (L.) Osbeck	Shamouti, Valencia, Washington Navle, Hamlin, Trovita
Grapefruit	*Citrus paradisi* Macf.	Marsh-seedless, Duncan, Foster, Thompson, Ruby
Mandarin	*Citrus reticulata* Blanco[a]	Dancy, Nagpur, Wilking, Satsuma, Clemantine
Lemon	*Citrus limon* (L.) Burm.	Villafranca, Eureka, Lisbon
Citron	*Citrus medica* L.[a]	Corsican, Diamonte, Earle
Lime	*Citrus aurantifolia* (Christm.) Swing.	Everglade, Mexican
Pummelo	*Citrus grandis* (L.) Osbeck[a]	Kao-Phuang, Kao Panne

[a] True species according to Barrett and Rhodes (1976).

Chinese writings Cooper and Chapot (1977) support the contention that almost all presently known *Citrus* cultivars originated in China. This contention is also in accord with our own recent study on the chloroplast genomes of *Citrus* cultivars (Green et al. 1986).

1.3 Reproduction Mechanisms and their Bearing on Breeding

The reproduction of *Citrus* cultivars is inflicted with mechanisms which complicate genetic studies and breeding. These mechanisms include nucellar embryogeny (Cameron and Frost 1968; Frost and Soost 1968), heterozygosity, self-incompatibility (Soost 1964), sterility (Iwamasa 1966) and an extended juvenile period. While it is relatively easy to breed, by sexual hybridizations, new mandarin types (some of the mandarin types are mono-embryonic), it is rather difficult to improve, by cross-breeding, the polyembryonic cultivars of sweet orange, grapefruit and lemon. Because of these obstacles, *Citrus* breeders lack essential information on the mode of inheritance of desirable traits (Vardi and Spiegel-Roy 1978).

1.4 Relevance of Cell Manipulation to *Citrus* Improvement

There is an urgent need to introduce disease resistance and stress resistance (e.g. salinity, drought) into *Citrus* cultivars and rootstocks. Only a few studies have been devoted to the use of mutagenesis to improve *Citrus* cultivars (e.g. Hensz 1971), and none of these was aimed at improving rootstocks.

Sources for disease- and stress resistance are known to be harboured in *Citrus* relatives, but sexual hybridization between the latter and rootstocks or cultivars is difficult (Barrett 1977). Thus cell-manipulation methodologies, and especially protoplast manipulation, should be of great relevance for *Citrus* breeding. The latter approach will be useful if functional *Citrus* trees can be derived from isolated protoplasts.

1.5 Plant Regeneration from Protoplasts of Fruit Trees

Except for *Citrus* (which will be handled below), the attempts to regenerate fruit trees from isolated protoplasts are still in an initial phase. Protoplast culture has been described in some cases and Ochatt and Caso (1986) reported regeneration of shoots from protoplasts of the wild pear (*Pyrus communis*). The present authors are not aware of any report on functional fruit trees which were derived from protoplasts.

Citrus and Microcitrus Trees derived from Protoplasts. Kochba et al. (1972) described a system by which ovule-derived nucellar callus of *Citrus* could be regenerated into embryos and plants. We utilized this system to isolate protoplasts from Shamouti orange nucellar callus and showed that these protoplasts can be cultured in vitro and regenerated embryos (Vardi et al. 1975) from which plants were derived (Galun et al. 1977; Vardi 1977). As in other protoplast systems, the division of *Citrus* protoplasts was manifested only beyond a certain plating density (ca. 10^5 protoplasts per ml); however, sparse plating could be compensated for by plating over a layer of division-arrested (X-irradiated) protoplasts from either *Citrus* callus or *Nicotiana* leaf-mesophyll (Vardi and Raveh 1976). The protoplast-to-tree system was extended (Vardi et al. 1982) to include, in addition to orange (*C. sinensis*), four other "species" of *Citrus*: grapefruit (*C. paradisi*), lemon (*C. limon*), sour orange (*C. aurantium*) and mandarin (*C. reticulata*). It should be noted that protoplast-derived *Citrus* trees have a remarkable similarity to the plants from which the protoplasts were derived (Figs. 1, 2). Moreover, even fruits

Fig. 1. Protoplast-derived young orange tree (cultivar: nucellar Shamouti); photographed 9 months after protoplast plating

Fig. 2. Flower, floral buds and young fruit on a protoplast-derived Villafranca lemon tree; photographed 6 years after protoplast plating

of such protoplast-derived trees are virtually identical to the fruits of the original trees (see Figs, 3, 4). Recently, the protoplast system was further extended to a relative of *Citrus*: fruit-bearing *Microcitrus* trees were derived from protoplasts at a high efficiency (Vardi et al. 1986).

Focussing on Trovita oranges, Kobayashi et al. (1983) also showed by methodologies which were identical to those used by Vardi and collaborators, that nucellar-callus-derived protoplast can be cultured *in vitro* and ultimately resulted in plants. Kobayashi and co-workers noted the uniformity of protoplast-derived Trovita oranges even though the protoplasts were derived from 5-year-old nucellar callus (Kobayashi et al. 1984), and subsequently amended the procedure to provide a high frequency of regeneration (Kobayashi et al. 1985).

Somatic Hybrids and Cybrids. Ohgawara et al. (1985) fused protoplasts of Trovita orange (*C. sinensis*) with those of *Poncirus trifoliata*. By selecting regenerated plants which had leaves which were identical to those of the sexual hybrid between these species, these workers obtained true somatic hybrids. Vardi et al. (1987) showed recently that when the donor-recipient protoplast-fusion procedure (e.g. Galun and Aviv 1986) is employed with *Citrus* protoplasts it can lead to the production of cybrid trees.

Fig. 3. Protoplast-derived Dancy mandarin tree at the second fruiting season; 7 years after protoplast plating ►

Fig. 4. Mandarin fruits of the cv "Dancy"; *P* fruits from a protoplast-derived tree; *C* fruits from a regular Dancy tree
▼

2 Isolation of Protoplasts

2.1 Plant Material

All the established protoplast-to-tree systems in *Citrus* were based on protoplasts which were isolated from nucellar-callus. Attempts were made to culture protoplasts which were isolated from cotyledons, from seedling leaves and from leaves of mature trees. None of these attempts resulted in sustained cell division. Protoplasts from cotyledons underwent a few (no precise data were provided) cell divisions (Burger and Hackett 1982).

Nucellar callus is obtained by procedures described by Kochba et al. (1972) and Vardi et al. (1982). Briefly, ovules are isolated from young fruits, commonly several weeks after anthesis; the ovules are then cultured in media devoid of growth regulators, but containing malt extract. A few of these ovules start to initiate an embryogenic callus. This callus is then maintained on the same medium; the malt extract can be removed from the medium once the callus is established. This procedure is probably possible in all *Citrus* cultivars which are polyembryonic and in some monoembryonic cultivars.

Before protoplast isolation, the nucellar callus should be subcultured at least twice, at intervals of 2 to 3 weeks, on nutrient medium (MT) of Murashige and Tucker (1969) which lacks growth regulators as described by Vardi et al. (1982) or on MT supplemented with 100 μg/ml malt extract and 20 μg/ml adenine (Kobayashi et al. 1983).

2.2 Enzymatic Maceration

To maintain an appropriate osmotic pressure during maceration and protoplast culture, mannitol or a combination of mannitol and sucrose are added to the medium up to a total of 0.7 M.

The specific commercially available enzymes and the duration of callus exposure to these enzymes strongly affect protoplast yield and division capability. Moreover, the calli from different cultivars differ in their response. The source of pectinase is probably the main factor affecting plating efficiency of *Citrus* protoplasts. Table 2 provides an example. It summarizes an experiment in which calli from five lines were macerated by an enzyme combination which included Cellulase (R-10) and Driselase, and in addition contained either Koch-Light pectinase or the pectinase Macerozyme (R-10). The type of pectinase used did not affect the yield of the protoplasts (data not shown), but strongly influenced the plating efficiency. Thus, plating efficiency of Shamouti protoplasts was higher when Macerozyme rather than Koch-Light pectinase was used, while the opposite was observed with Villafranca protoplasts. The combination of Driselase with either Koch-Light pectinase or Macerozyme was detrimental to Duncan protoplasts. The latter protoplasts did give an adequate plating efficiency when macerated in a combination of Cellulase and Macerozyme. The conclusion from this experiment and from numerous other tests conducted by the authors of this

Table 2. Plating efficiency of protoplasts from five callus-lines of *Citrus* macerated with either of two pectinases. Both maceration fluids contained also 0.2% Cellulase (Onozuka R-10) and 0.1% Driselase (Kyowa Hakko Kohyo Co.); plating efficiency is expressed as percentage of plated protoplasts which produced colonies. (Data from Vardi et al. 1982)

Callus line	Koch-Light pectinase (0.3%)	Macerozyme R-10 (0.3%)
Nucellar Shamouti (orange)	3.78	20.65
Shamouti Landau (orange)	6.25	14.10
Murcott (mandarin)	6.78	10.83
Duncan (grapefruit)[a]	0	0.65
Villafranca (lemon)	12.94	6.10

[a] When macerated in a fluid containing 0.2% Cellulase and 0.3% Macerozyme, the plating efficiency was 10.36%.

review is that the appropriate enyzme combination should be determined for each specific callus line. Moreover, enzyme batches of the same manufacturer may vary considerably in time, thus an amendment of the enzyme content in the maceration fluid is required periodically.

2.3 Callus Age

Although no exact numerical data are available on the impact of callus age on protoplast yield, experience has indicated that the best yields are obtained between 2 and 3 weeks after subculture. Thereafter there is a rapid reduction of protoplast yields. Thus, beyond 4 weeks after subculture nucellar calli of *Citrus* could not be macerated into protoplasts.

3 Protoplast Culture

The sugars and sugar alcohols used as a carbon source and osmotic stabilizers in the protoplast culture medium strongly affect division capability (e.g. Vardi 1981; Vardi et al. 1982). Thus, while fructose is detrimental to orange and sour orange protoplasts, this sugar can serve as a carbon source for mandarin protoplasts. On the other hand, galactose is a good carbon source for orange and sour orange protoplasts but strongly reduces the plating efficiency of mandarin protoplasts (Table 3). Sorbitol, used as an osmotic stabilizer, will reduce the plating efficiency of mandarin protoplasts, but is suitable to serve this purpose with sour orange protoplasts. In practice, because sucrose and mannitol are, respectively, an adequate carbon source and an osmotic stabilizer, for all tested *Citrus* protoplasts our current standard procedure calls for 0.35 M sucrose and 0.35 M mannitol in the protoplast-isolation medium and 0.30 M sucrose and 0.30 M mannitol in the protoplast culture medium. Kobayashi et al. (1985) adopted slightly different sucrose and mannitol molarities for Travita orange protoplast culture (Fig. 5).

Table 3. The effects of sucrose (suc), lactose (lac), glucose (gluc), galactose (gal), and fructose (fruc) on the plating efficiency of *Citrus* protoplasts derived from nucellar calli of several cultivars. All media contained mannitol up to a total sugar molarity of 0.6 M. Data are plating efficiencies: colonies counted 5 weeks after plating as percent of plated protoplasts. (Data from Vardi 1981)

Callus line	10% suc	5% lac	5% gluc	5% gal	5% fruc
Nucellar Shamouti (orange)	14.1	22.7	11.8	15.7	0
Shamouti Landau (orange)	11.6	21.5	7.0	9.7	0
Sour Orange	20.7	31.3	18.7	20.5	0
Villafranca (lemon)	6.2	13.7	4.1	7.5	1.4
Murcott (mandarin)	25.5	6.5	8.5	1.7	5.1

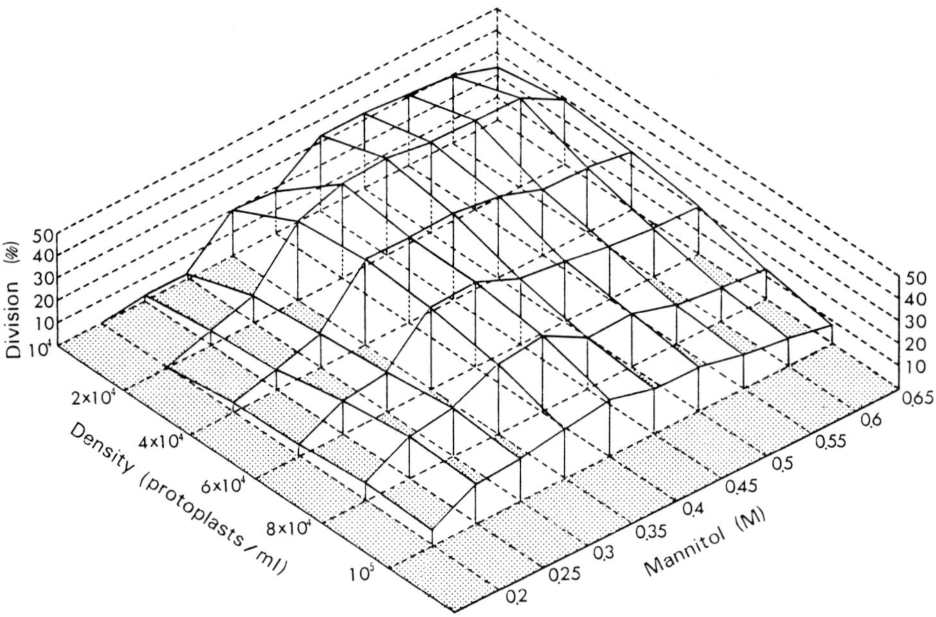

Fig. 5. Effects of protoplast density and mannitol concentrations on cell division of *Citrus* protoplasts after 50 days. (Kobayashi et al. 1985)

As noted above, *Citrus* protoplasts require a minimal plating density for adequate plating efficiency (i.e., 10^5 protoplasts/ml). A much lower plating density can be employed when the protoplasts are plated over a feeder layer of division-arrested (e.g. X- or gamma-irradiated) protoplasts (Vardi et al. 1975; Vardi and Raveh 1976). Table 4 (from Vardi and Raveh 1976) demonstrates that without a feeder layer, orange protoplast plated at a density of 5×10^3 protoplasts/ml or less will not divide, while when plated over a feeder layer of either orange or tobacco protoplasts, division is attained even at a density of 500 protoplasts/ml. Plating at low densities, over a feeder layer, is a convenient procedure after protoplast fusion when the number of viable fusion products is variable.

Table 4. Colony formation by orange protoplasts plated at different densities, with or without feeder layers. Irradiation of orange and tobacco protoplasts was at doses of 15KR and 5KR respectively; feeder densities were 10^5 and 5×10^4 per ml for orange and tobacco, respectively; data are expressed as number of colonies per ml, 6 weeks after plating. (After Vardi and Raveh 1976)

Plating density (protoplasts per ml)	Without feeder layer	With feeder layer		
		Irradiated protoplasts		Non-irradiated tobacco protoplast[a]
		Orange	Tobacco[a]	
10^5	7000	–	–	–
5×10^3	0	149	175	60
10^3	0	82	20	20
5×10^2	0	24	9	12

[a] Tobacco protoplasts were cultured in MT medium, devoid of hormones – thus unable to divide.

4 Embryogenesis

The usual sequence of events after plating of *Citrus* protoplasts is the formation of cell colonies, then development of callus and finally production of embryos from the callus. The length of the callus phase is affected by various factors, among them the source of protoplasts. Commonly, embryos become apparent at 6 or 7 weeks after protoplast plating (Vardi et al. 1982; Kobayashi et al. 1985). Embryogenesis of nucellar-callus was stimulated when sucrose was replaced by certain sugars or sugar alcohols such as galactose, lactose, raffinose, maltose and glycerol (e.g. Kochba and Spiegel-Roy 1973; Button and Kochba 1977; Button 1978; Kochba et al. 1978; Ben-Hayyim and Neumann 1983). The same stimulation occurs in protoplast-derived callus of *Citrus*. We are currently utilizing the embryo-inducing effect of glycerol in our procedure of plant regeneration from *Citrus* protoplasts (Vardi et al. 1987). Thus, when protoplast-derived calli reach a diameter of 1–2 mm, they are transferred to MT medium containing 4% sucrose. This removal from high-osmoticum-containing medium usually facilitates callus growth. Calli reaching a diameter of approximately 5 mm are transferred to MT medium with 2% glycerol as the sole carbon source. Embryos are then produced profusely within a few weeks. These embryos are isolated individually and cultured further in MT containing 4% sucrose and 1.5 g/l malt extract.

5 Regeneration of Plants

The embryos which attain normal shape and have expanded cotyledonary leaves are then transferred individually to test tubes containing MT with 2% sucrose and 0.02–0.05 ppm naphthaleneacetic acid (NAA). The addition of NAA promotes root formation and axis elongation. Thus plantlets are produced in the test tubes.

Plantlets with several leaves are planted in sterilized Jiffy turf pots and are irrigated with a sterilized solution composed of half-strength macro- and microelements of MT. When roots emerge from the turf pots, the young plants are planted in 1-l plastic pots containing a soil-perlite mixture, covered with plastic bags and transferred to a shaded greenhouse. The plants are acclimatized by gradual opening of the plastic bags.

A shorter procedure of regenerating *Citrus* embryos to plants was suggested by Gmitter and Moore (1986). By the latter procedure the medium is supplemented with gibberellic acid (1 mg/l) to stimulate embryo "germination". These authors reported that plant survival was increased when embryos with developed root were transferred directly to soil.

6 Fusion of Protoplast

The capability of isolated *Citrus* protoplasts to divide in culture, to differentiate into embryos and ultimately to produce fruit-bearing trees paved the way for somatic hybridization and cybridization by protoplast fusion. This cell-manipulation approach should be a very useful tool in *Citrus* breeding because of the hurdles in conventional breeding which exist in this genus (see Introduction). Although androgenesis in *Poncirus trifoliata* was reported (e.g. Hidaka et al. 1979), it is limited to very few *Citrus* cultivars and, even where possible, results in diploid rather than in haploid androgenic plants (Vardi unpubl.). Thus, in the absence of haploid protoplasts, somatic hybridization in *Citrus* will result in tetraploid plants. Such plants can serve in further breeding. For example, fusion-derived tetraploids may be crossed with diploids to produce triploid (possibly seedless) cultivars. Ohgawara et al. (1985) fused Trovita orange protoplasts with protoplasts of *Poncirus trifoliata* and obtained somatic-hybrid plants. These hybrids had leaf shapes which were intermediate between leaves of *C. sinensis* and *P. trifoliata*. The analysis of rDNA also indicated that the fusion-derived plants were true hybrids.

Using the donor-recipient protoplast-fusion method (e.g. Galun and Aviv 1983), we recently showed (Vardi et al. 1987) that fusion of gamma-irradiated (donor) protoplasts with iodoacetate-treated (recipient) protoplasts can result in *Citrus* plants which have the nuclear-coded morphology of the recipient but contain novel mitochondrial genomes. The production of such cybrids having the nuclear genome of a given cultivar but alien organelles, has a twofold advantage. First, cybrids may furnish important information on cytoplasmic inheritance which is lacking in *Citrus*. Second, cybridization may have a direct impact on *Citrus* improvement because organelles may contribute important breeding traits, such as resistance to pathogens and to herbicides.

7 Conclusions and Prospects

The protoplast-to-tree system of *Citrus* excels in its efficiency and its methodological convenience. The source of protoplasts, nucellar callus, can be maintained for years without losing its regeneration capability. Furthermore, this callus is efficiently cultured on hormone-free medium and is cytologically stable (retaining its euploidity).

We shall not elaborate the various potential utilizations of the protoplast-system in *Citrus* because they do not differ from utilizations of other protoplast systems. Nevertheless, the following points should be noted. Because of limitations in sexual reproduction in *Citrus*, the protoplast system becomes a useful tool to increase variability and to select lines at the cell and colony levels for specific traits (e.g. stress resistance). Furthermore, novel methods to produce transgenic plants by integrating specific genes into protoplasts should lead in *Citrus* to functional plants. The integration and expression of such genes can be followed at the cell-colony level. Therefore, functional *Citrus* trees harboring specific alien genes should be an achievable goal.

Finally as the protoplast-systems of *Citrus* enable intra- and intergeneric hybridization and cybridization, the road is open to design and produce novel *Citrus* cultivars and rootstocks which cannot be produced by conventional breeding.

8 Protocol

Optimal Conditions/Medium for the Isolation of Protoplasts

(1) Use the nucellar callus subcultured for several times (at 2- to 3-week intervals) on MT medium (Murashige and Tucker 1969).
(2) Place about 0.5 g callus in a 10-cm Petri dish, add 10 ml maceration fluid, maintain overnight at 26 °C. The maceration medium should consist of 0.2% Macerozyme R-10, 0.3% Cellulase R-10 (both from Kinki Yakult Mfg Co. Japan) and 0.1% Driselase (Kyowa Hukko Kogyo Co., Japan), dissolved in a solution composed of half-strength MT macroelements and containing 0.35 M sucrose and 0.35 M mannitol. Adjust the pH of the maceration fluid to 5.7 and filter sterilize (0.2 μm) before use. The enzyme composition may be amended for specific nucellar callus lines.
(3) Isolate protoplasts by sequential filtering through 50 μm and 30 μm nylon screens.
(4) Collect filtrate (which contains the protoplasts) in centrifuge tubes and centrifuge at 100 g for 5 min. Remove the supernatant and add liquid culture medium (MT) containing 0.3 M sucrose and 0.3 M mannitol. Repeat the centrifugation two more times.
(5) Suspend the protoplasts at 2× optimal plating-density (i.e. at approximately 2×10^5 cells per ml) in liquid MT medium.

Optimal Conditions/Medium for Culture of Protoplasts

(1) Add a given volume of melted MT medium (containing 1.2% agar at approximately 43 °C) to an equal volume of suspended protoplasts and plate as 4-ml aliquots in 5-cm Petri dishes.
(2) Incubate the plates in continuous dim light at 26 °C. Cell divisions are expected to start 1 to 2 weeks after plating.

Optimal Condition/Medium for Regeneration of Plants

(1) Isolate colonies which reached a diameter of 1–2 mm and transfer them, for further growth, to Petri dishes containing solidified (1% agar) MT with 4% sucrose.

(2) When calli reach a diameter of approximately 5 mm, transfer part of each callus to the embryo-regeneration medium. The latter is MT medium containing 2% glycerol rather than sucrose. After approximately 6 weeks numerous green embryos should be formed.
(3) Isolate the embryos individually and transfer them to Petri dishes containing solidified (1% agar) MT with 4% sucrose and 1.5 g/l malt extract. Up to this stage the Petri dishes are maintained in dim light at 26 °C.
(4) When embryos reach a size of 5–7 mm, choose those which have a normal morphology and expanded cotyledonary leaves. Transfer these embryos individually to test tubes containing MT with 2% sucrose and 0.05–0.02 mg/l NAA to promote root formation and axis elongation. Maintain the test tubes at relatively high light intensity (white fluorescent tubes, ca. 300 FC).
(5) When several leaves have developed and roots have formed, plant the plantlets in autoclaved Jiffy turf posts (No. 7) and place the turf pots in sterile plastic boxes containing half-strength MT macro- and microelements.
(6) When roots start to emerge from the turf pots, transplant the pots with the rooted plants into plastic containers (ca. 1-l volume) with a soil-perlite mixture. Cover the containers with plastic bags and transfer them to a shaded greenhouse.
(7) After 3–4 weeks start acclimatization by gradually opening the plastic bags in order to lower humidity until the plants can withstand the greenhouse conditions.

Acknowledgement. Our *Citrus* cell-manipulation studies were supported by grant no. I-468-82 from the United States-Israel Binational Agricultural Research and Development Fund (BARD).

Contribution from the Agricultural Research Organization, The Volcani Center, Bet Dagan, Israel, No. 1966E, 1987 series.

References

Barrett HC (1977) Intergeneric hybridization of *Citrus* and other genera in *Citrus* variety improvement. Proc Int Soc Citric 1977, 2:586–589

Barrett HC, Rhodes AM (1976) A numerical taxonomic study of affinity relationships in cultivated *Citrus* and its close relatives. Syst Bot 1:105–136

Ben-Hayyim G, Neumann H (1983) Stimulatory effect of glycerol on growth and somatic embryogenesis in *Citrus* callus cultures. Z Pflanzenphysiol 110:331–337

Burger DW, Hackett WP (1982) The isolation, culture and division of protoplasts from *Citrus* cotyledons. Physiol Plant 56:324–328

Button J (1978) The effect of some carbohydrates on the growth and organization of *Citrus* ovular callus. Z Pflanzenphysiol 88:61–68

Button J, Kochba J (1977) Tissue culture in the *Citrus* industry (refs p 207–248). In: Reinert J, Bajaj YPS (eds) Applied and fundamental aspects of plant cell, tissue, and organ culture. Springer, Berlin Heidelberg New York, pp 70–92

Cameron JW, Frost HB (1968) Genetic breeding and nucellar embryony, Vol II. In: Reuther W, Batchelor LD, Webber HJ (eds) The Citrus industry. University of California Press, Berkeley, pp 325–381

Cooper C, Chapot H (1977) Fruit production with special emphasis on fruit for processing, Vol II. In: Nagy S, Show PE, Veldhuis MK (eds) Citrus science and technology. AVI Publ, Westport, pp 1–27

FAO (1985) Production yearbook. FAO Rome 39:194–196

Frost HB, Soost RK (1968) Seed production: development of gametes and embryos, Vol II. In: Reuther W, Batchelor LD, Webber HJ (eds) The Citrus industry. University of California Press, Berkeley, pp 292–334

Galun E, Aviv D (1983) Cytoplasmic hybridization. In: Evans DA, Sharp WR, Ammirato PV, Yamada Y (eds) Handbook of Plant Cell Culture, Vol I. Macmillan, New York, pp 358–392

Galun E, Aviv D (1986) Organelle transfer. Methods in Enzymology 118:595–611

Galun E, Aviv D, Raveh D, Vardi A, Zelcer A (1977) Protoplasts in studies of cell genetics and morphogenesis. In: Barz W, Reinhard E, Zenk MH (eds) Plant tissue culture and its bio-technological application. Springer, Berlin Heidelberg New York, pp 302–312

Gmitter FG, Moore GA (1986) Plant regeneration from undeveloped ovules and embryogenic calli of *Citrus*: Embryo production, germination and plant survival. Plant Cell Tissue Organ Cult 6:139–147

Green RM, Vardi A, Galun E (1986) The plastome of *Citrus*: Physical map, variation among *Citrus* cultivars and species and comparison with related genera. Theor Appl Genet 72:170–177

Hensz RA (1971) Star Ruby, a new deep-red-fleshed grapefruit variety with distinct tree characteristics. J Rio Grande Val Hortic Soc 25:54–58

Hidaka T, Yamada Y, Shichijo T (1979) In vitro differentiation of haploid plants by anther culture in *Poncirus trifoliata*. Jpn J Breed 29:248–254

Iwamasa M (1966) Study on sterility in genus *Citrus* with special reference to the seedlessness. Bull Hortic Res Stn Jpn (Ser B) 6:1–77

Kobayashi S, Uchimiya H, Ikeda I (1983) Plant regeneration from Trovita orange protoplasts. Jpn J Breed 33:119–122

Kobayashi S, Ikeda I, Nakatani M (1984) Induction of nucellar callus from orange (*Citrus sinensis* Osb.) ovules and uniformity of regenerated plants (Japanese with English abstract). Bull Fruit Tree Res Stn Ser E (Akitsu) 5:32–54

Kobayashi S, Ikeda I, Uchimiya H (1985) Conditions for high frequency embryogenesis from orange (*Citrus sinensis* Osb.) protoplasts. Plant Cell Tissue Organ Cult 4:249–259

Kochba J, Spiegel-Roy P (1973) Effect of culture media on embryoid formation from ovular callus of Shamouti orange (*Citrus sinensis*). Z Pflanzenzücht 69:156–162

Kochba J, Spiegel-Roy P, Saad S, Neumann H (1978) Stimulation of embryogenesis in *Citrus* tissue culture by galactose. Naturwissenschaften 65:261

Kochba J, Spiegel-Roy P, Safran H (1972) Adventive plants from ovules and nucelli in *Citrus*. Planta 106:237–245

Murashige T, Tucker DPH (1969) Growth factor requirement of *Citrus* tissue culture. In: Chapman HD (ed) Proc 1st Int *Citrus* Symp Vol 3, Riverside CA, pp 1155–1161

Ochatt SJ, Caso OH (1986) Shoot regeneration from leaf mesophyll protoplasts of wild pear (*Pyrus communis* var. *Pyraster* L.). J Plant Physiol 122:243–249

Ohgawara D, Kobayashi S, Ohgawara E, Uchimiya H, Ishii S (1985) Somatic hybrid plants obtained by protoplast fusion between *Citrus sinensis* and *Poncirus trifoliata*. Theor Appl Genet 71:1–4

Soost RK (1964) Self-incompatibility in *Citrus grandis* (Linn.) Osbeck. Proc Am Soc Hortic Sci 84:137–140

Swingle WT, Reece PC (1967) The botany of *Citrus* and wild relatives, Vol I. In: Reuther W, Batchelor LD, Webber JH (eds) The Citrus Industry. University of California Press, Berkeley, pp 190–430

Vardi A (1977) Isolation of protoplasts in *Citrus*. Proc Int Soc Citric 1977, 2:575–578

Vardi A (1981) Protoplast-derived plants from different *Citrus* species and cultivars. Proc Int Soc Citric 1981, 1:149–152

Vardi A, Raveh D (1976) Cross feeder experiments between tobacco and orange protoplasts. Z Pflanzenphysiol 78:350–359

Vardi A, Spiegel-Roy P (1978) *Citrus* breeding, taxonomy and the species problem. Proc Int Soc Citric 1978, 1:51–57

Vardi A, Spiegel-Roy P, Galun E (1975) *Citrus* cell culture: Isolation of protoplasts, plating densities, effect of mutagens and regeneration of embryos. Plant Sci Lett 4:231–236

Vardi A, Spiegel-Roy P, Galun E (1982) Plant regeneration from *Citrus* protoplasts: Variability in methodological requirements among cultivars and species. Theor Appl Genet 62:171–176

Vardi A, Hutchinson DJ, Galun E (1986) A protoplast-to-tree system in *Microcitrus* based on protoplasts derived from sustained embryogenic callus. Plant Cell Rep 5:412–414

Vardi A, Breiman A, Galun E (1987) *Citrus* cybrids: Production by donor-recipient protoplast-fusion and verification by mitochondrial-DNA restriction profiles. Theor Appl Genet 75:51–58

II.4 Regeneration of Plants from *Linum* Protoplasts

M.N. BARAKAT[1] and E.C. COCKING[2]

1 Economic Importance and Breeding Objectives in *Linum*

The genus *Linum* of the Linaceae family contains over 200 species, of which *Linum usitatissimum* (flax) is the only one widely grown. This crop is grown either for its fibre (linen flax) or for its seed (oil flax, linseed), or for both (dual purpose flax), depending on the cultivar used and the cultural and climatic conditions. Linen thread is used in gloves, footware, netting and sports gear. Linseed oil has many industrial purposes such as paint, varnishes, oil cloth, the manufacture of linoleum, printer's ink, soaps and patent leather, and recently as a salt-resistant coating for cement surfaces of highways and bridges.

Several of the wild species of *Linum* (flax species) possess many agronomically valuable genes, including resistance to diseases such as flax rust, *Melampsora lini*. These wild species are also a reservoir of genes for other important characteristics including drought resistance, winter hardiness, and an unexplored reservoir of genes for oil quality (Plessers 1966; Yermanos et al. 1966) which would be of great value if they could be transferred to flax (Seetharam 1972).

The basic fundamental problem in attempting such improvements by sexual hybridization is that crosses between wild *Linum* species with nine-chromosome numbers ($2n = 18$) and *Linum usitatissimum* ($2n = 30$) or any other species with 15 chromosomes ($2n = 30$) fail to produce seeds (Gill and Yermanos 1967a, b; Bari and Godward 1970; Seetharam 1972). Unfortunately, characteristics of most interest that are not found in *Linum usitatissimum* are present only in some of the nine-chromosome species (Beard and Comstock 1980).

Since it has been established that somatic cell hybridization overcomes sexual incompatibility barriers to hybridization between some species (Cocking et al. 1981), it was thought realistic to try to utilize somatic hybridization of these sexually incompatible *Linum* species.

[1] Tissue Culture and Plant Breeding Group, Crop Science Department, Faculty of Agriculture, Alexandria University, Alexandria, Egypt
[2] Plant Genetic Manipulation Group, Botany Department, Nottingham University, Nottingham NG7 2RD, U.K.

Biotechnology in Agriculture and Forestry, Vol. 8
Plant Protoplasts and Genetic Engineering I (ed. by Y.P.S. Bajaj)

2 Source Material for Protoplast Isolation

2.1 In Vitro Seedling Material

Seeds of *Linum* species were surface-sterilized in 10–20% Domestos bleach for 30 min and washed in six changes of sterile tap water. Seeds were germinated on agar solidified (0.8% w/v; Sigma) Murashige and Skoog (1962) based medium (MS) containing 3% sucrose, but lacking growth regulators and incubated in the dark. Incubation was in the light for the production of in vitro-grown shoots and when green cotyledons were used for the isolation of protoplasts (2000 lx, daylight fluorescent tubes) at 23±2 °C.

2.2 Cell Suspension Culture

Callus was induced on aseptic seedling cotyledon explants using agar solidified (0.8% w/v) MS medium containing 3% sucrose, 2.0 mg/l NAA and 0.5 mg/l BAP (MSP-1). Cultures were maintained at 23±2 °C in continuous fluorescent light (2000 lx). Cell suspensions were initiated by transfer of 2 g fr. wt. to 50–75 ml of MSP-l liquid medium in 250-ml Erlenmeyer flasks, with agitation on a rotary shaker (120 rpm, 2000 lx, 23±2 °C). Cell suspensions were subcultured at intervals of 7 days to the same fresh medium.

3 Protoplast Isolation

3.1 Protocol for Isolation

Aseptic shoots of *Linum* spp. were excised after 3 weeks; young cotyledons and hypocotyls were excised after 7 days. Shoots and cotyledons were cut into approximately 1 mm-wide strips; however, hypocotyls and roots (3 days) were cut longitudinally. Tissues were plasmolyzed in CPW solution containing 13% w/v mannitol (Frearson et al. 1973) for 2 h prior to incubation in enzymes. Cells of suspension cultures 3–4 days after subculturing were allowed to settle and incubated in suitable enzyme mixture. The enzyme mixtures were E_1 2% w/v Rhozyme HP 150 (Rohm and Hass Ltd., Philadelphia, USA), 4% w/v Meicelase (Meiji Seika Kaisha Ltd., Tokyo, Japan) and 0.03% w/v Macerozyme (Yakult Biochemicals Ltd., Nichinomiya, Japan) in CPW salts solution with 13% w/v mannitol (CPW 13M) pH 5.8; E_2 2% w/v Rhozyme HP150, 2% w/v Meicelase P and 0.03% w/v Macerozyme R-10 in CPW 13M and E_3 2% w/v Cellulase R-10 (Kinki Yakult Manufacturing Co. Ltd., Japan) and 0.05% w/v Macerozyme R-10 in CPW 13M. Enzyme incubation was for 16 h at 25 °C on a rotary shaker (30–40 rpm). Protoplasts were released by gentle squeezing and filtered through a nylon sieve (64 µm pore size). Etiolated cotyledon protoplasts were pelleted by centrifugation at 100 *g* for 10 min and washed by centrifugation using three changes of CPW

13 M. Root, green cotyledons, cell suspension and shoot protoplasts were pelleted by centrifugation at 100 *g* for 5 min and then purified by flotation in CPW 21% w/v sucrose (100 *g* for 10 min). Hypocotyl protoplasts were pelleted by centrifugation at 100 *g*, 5 min; in this step most of the protoplasts were highly vacuolated and were floating on the surface of the enzyme solution. Protoplasts that had pelleted were purified by flotation in CPW 21% w/v sucrose (100× g for 10 min). Floating protoplasts were washed twice with CPW 13% w/v mannitol, resuspended in the culture medium, and counted.

3.2 Yield of Protoplast

It was possible to release protoplasts from all the sources of plant material investigated as well as the nine cultivars of *Linum usitatissimum*. Yields ranged from 1.0×10^6/g fr. wt. (cell suspension culture, Table 1) to 25.7×10^6 fr. wt. (green cotyledon, Table 2).

Protoplasts were also readily produced from a range of seedling tissues from cell suspension cultures of six wild *Linum* species. Highest yield of protoplasts was from cotyledons, with significantly lower numbers of protoplasts from hypocotyl and cell suspensions (Tables 3, 4). Yields were comparable from both roots and shoots (Table 3).

The protoplast population derived from all sources consisted largely of unvacuolated protoplasts, except in the case of hypocotyl protoplasts, which consisted of unvacuolated and vacuolated protoplasts.

The yield of cotyledon protoplasts from wild species was significantly lower compared with the release of protoplasts from cotyledons of *Linum usitatissimum* cultivars. This could be due to the small size of the cotyledons of the wild species, resulting in more damage during the cutting of the cotyledon for protoplast isolation. The age of seedlings was critical for high yield. The yield was 6.5×10^6 protoplasts/g fresh weight of etiolated cotyledon (seedling 4 days old), while the yield was 6.7×10^6 and 3.9×10^6 protoplasts/g fresh weight of etiolated and green cotyledon (10 days old); highest yields were obtained 7 days after germination (Table 1). Green cotyledon protoplasts may be a suitable alternative to leaf mesophyll protoplasts or shoot culture protoplasts for somatic hybridization assessments. It has been suggested that protoplasts isolated from etiolated cotyledons could be used, instead of cell suspension protoplasts, for fusion with green cotyledons or leaf mesophyll protoplasts.

Several factors are known to influence the release of root protoplasts; dark germination of seeds increases the yield of *Avena* (Ruesink and Thimann 1966) and *Phaseolus* (Xu et al. 1981). The stage of germination is also important. *Pisum* cortical root explants have been found very suitable for protoplast release, 48 to 56 h after germination; increased starch content of the tissues after longer periods of germination reduced protoplast yield (Crowder et al. 1979). These factors were taken into consideration in this investigation with *Linum* seeds. Root protoplasts could also provide an alternative to cell suspension culture protoplasts as a source of colourless protoplasts for fusion with green cotyledon protoplasts or leaf mesophyll protoplasts (Lu et al. 1982). Seedling root protoplasts have previously

Table 1. Isolation, yield and plating efficiencies, of *Linum usitatissimum* cv. Precederia protoplasts. (Barakat 1985)

Source of protoplasts	Age (days)	Enzyme	Yield[c] (10^6)/g fr. wt.	Plating efficiency							
				Dark condition				Light condition			
				MB_5P	KP8	KM8P:KM8	MSP-1	MB_5P	KP8	KM8P:KM8	MSP-1
Root	3[a]	E_1	4.4	24.8	32.6	33.2	12.4	26.5	38.4	38.4	16.9
Hypocotyl	7[a]	E_1	2.5	22.9	29.6	23.7	18.2	25.6	30.2	31.5	26.0
Etiolated cotyledon	7[a]	E_1	21.0	20.2	27.2	22.5	25.6	26.7	32.4	35.0	27.6
Green cotyledon	7[a]	E_1	15.7	30.8	32.5	35.3	24.7	30.4	37.1	41.4	28.1
Shoots	14[a]	E_3	7.9	25.9	34.1	31.5	28.1	36.6	44.0	41.4	38.2
Cell suspension	3[b]	E_1	1.0	23.5	29.9	33.7	26.4	35.4	42.5	45.3	32.4

[a] From seed germination.
[b] After subculture.
[c] Average at least three isolations.

Table 2. Protoplast isolation, yield and plating efficiency of root and green cotyledon protoplasts for *Linum usitatissimum* varieties in KP8 and KM8P:KM8 media (diffuse light). (Barakat 1985)

Variety	Explant	Age[a] (days)	Enzyme	Yield[b] (10^6)/g fr. wt.	Plating efficiency	
					KP8	KM8P:KM8
Natasha	Root	3	E_1	3.3	45.0	49.5
	Cotyledon	7	E_1	14.3	47.6	48.3
H-1	Root	3	E_2	3.4	48.1	48.2
	Cotyledon	7	E_2	16.0	37.9	42.4
Giza-4	Root	3	E_1	3.9	50.4	49.2
	Cotyledon	7	E_1	15.8	48.3	49.7
Entry 58	Root	3	E_2	4.4	46.7	48.7
	Cotyledon	7	E_2	15.8	47.7	47.0
Entry 115	Root	3	E_1	4.4	49.0	47.6
	Cotyledon	7	E_1	14.6	48.0	49.5
Bison	Root	3	E_2	4.5	46.5	50.1
	Cotyledon	7	E_2	25.6	39.5	46.6
McGregor	Root	3	E_2	5.4	47.8	48.3
	Cotyledon	7	E_2	25.7	46.4	45.9
Culbert	Root	3	E_2	5.1	46.1	49.7
	Cotyledon	7	E_2	25.0	43.8	46.2

[a] From seed germination.
[b] Average of three experiments.

Table 3. Protoplast isolation, yield and plating efficiency of some wild species of *Linum*. (Barakat and Cocking 1985)

Species/source of protoplast	Enzyme mixture	Protoplast yield (10^6)/g fr. wt.	Plating efficiency	
			K8P	KM8P:KM8
Linum alpinum				
Root	E_2	6.8	48.0	48.5
Hypocotyl	E_2	1.1	42.0	42.0
L. narbonense				
Root	E_2	4.3	37.1	39.1
Hypocotyl	E_2	0.7	39.9	39.9
Shoots	E_3	4.3	43.0	44.3
L. grandiflorum				
Root	E_2	5.7	48.6	46.0
Hypocotyl	E_2	1.0	40.3	46.2
Shoots	E_3	7.2	43.0	44.5
L. strictum				
Root	E_2	6	45.6	41.3
Hypocotyl	E_2	1.2	39.6	38.5
Shoots	E_3	5.1	43.5	44.3
L. lewissii				
Shoots	E_3	2.3[a]	47.7[a]	51.7[a]

[a] Data from one experiment.

Table 4. Isolation, yield and plating efficiency of protoplasts from cotyledons and cell suspension cultures of some wild species of *Linum*. (Barakat and Cocking 1985)

Species/source of protoplasts	Enzyme	Yield (10^6)/g fr. wt.	Plating efficiency			
			MB_5P	KP8	KM8P:KM8	MSP-1
Linum aplinum						
Etiolated cotyledon	E_1	10.7	35.2	39.7	45.2	40.7
Green cotyledon	E_1	7.5	34.6	46.5	42.1	33.6
Cell suspension	E_1	0.8	34.1	39.8	38.2	34.0
L. narbonense						
Etiolated cotyledon	E_1	12.4	35.9	45.3	45.8	34.9
Green cotyledon	E_1	11.5	36.2	42.4	46.5	32.6
Cell suspension	E_2	2.3	32.4	40.6	51.4	33.4
L. grandiflorum						
Etiolated cotyledon	E_1	15.4	27.6	42.1	44.5	31.0
Green cotyledon	E_1	7.4	32.2	44.3	41.0	32.4
Cell suspension	E_1	1.3	30.9	36.2	36.9	23.1
L. strictum						
Etiolated cotyledon	E_1	11.1	25.2	39.4	42.9	35.1
Green cotyledon	E_1	8.6	30.8	44.1	47.4	31.5
Cell suspension	E_1	1.9	34.2	48.4	47.8	36.7
L. altaicum						
Green cotyledon	E_2	5.3	–	19.4	24.3	–
Cell suspension	E_2	1.5	–	30.1	33.3	–
L. lewissii						
Green cotyledon	E_2	3.6	–	39.8	40.6	–
Cell suspension	E_2	1.2	23.4	37.4	39.4	38.3

been used in fusion studies (Power et al. 1970). Root protoplasts may also have some advantages over cell suspension culture protoplasts, since root protoplasts may be more stable in their chromosome number than cell suspension culture protoplasts. Moreover, the yield of root protoplast was higher than cell suspension protoplasts.

Hypocotyl tissues have been proposed as sources for the production of stable protoplasts (Eriksson et al. 1978), since these protoplasts often divide more readily are more resistant to physical shock than leaf protoplasts. Furthermore, advantages with this type of plant material are that minimal space is needed for growth under sterile conditions, little time is required for growth and the hypocotyls give high yields of protoplasts (Glimelius 1984). Hypocotyl protoplasts can be used instead of cell suspension protoplasts for fusion with mesophyll protoplasts. Chuong et al. (1985) fused hypocotyl protoplasts with mesophyll protoplasts of *Brassica*.

4 Protoplast Culture

Protoplasts of *Linum usitatissimum* cv. Precederia were cultured in different culture media, and several cultural conditions were compared. This information was utilized as a guide for assessments of other cultivars of *L. usitatissimum* and the wild *Linum* species. The protoplasts were cultured in:

a) 4 ml liquid medium
b) 1 ml liquid medium over 4 ml agar solidified medium
c) 1 ml liquid medium over 4 ml agar solidified medium with protoplasts plated onto a sterile filter paper laid on the agar surface (Santos et al. 1980).

The following media were used:

i. MB_5P (Gamborg and Shyluk 1976). This was diluted with MB_5, the same components as in MB_5P except for hormones and sugar (0.1 mg/l zeatin, 0.2 mg/l NAA, 0.1 mg/l, 2,4-D, 20 g/l sucrose and 10 g/l glucose) in the proportion of 2:1, 1:1 and 1:2 v/v.
ii. K8P (Kao 1977) containing 0.2 mg/l 2,4-D, 0.5 mg/l zeatin and 1 mg/l NAA. This was diluted with K8 (Kao 1977) containing 0.1 mg/l 2,4-D, 0.2 mg/l zeatin, 1 mg/l NASA, 2% (w/v) sucrose and 1% (w/v) glucose, in the proportion of 2:1, 1:1 and 1:2 v/v.
iii. KM8P:KM8 (2:1, v:v, Kao and Michayluk 1975). This was diluted in the proportion of 1:1, 1:2 and 0.1 v/v of KM8P:KM8.
iv. MSP-1 (Murashige and Skoog 1962 medium modified by adding 2 mg/l NAA and 0.5 mg/l BAP) with 9% w/v mannitol. This was diluted with MSP-1 containing 6% w/v mannitol, 3% w/v mannitol and finally MSP-1 with no mannitol.

The plating density of protoplasts was adjusted to 0.25, 0.5 and 1×10^5/ml. The osmotic concentration of the culture medium was lowered regularly at weekly intervals. The osmoticum was lowered by the addition of small volumes (5 drop or 0.5–1 ml) of liquid medium of lower osmotic pressure to the liquid medium or liquid over agar. Another procedure was also used for dilution; agar blocks containing cell colonies following 2 weeks of culture were transferred to fresh agar-solidified medium with lowered osmoticum. In the case of the filter paper culture method, dilution was facilitated by transfer of the filter paper to the surface of fresh agar solidified medium with lowered osmoticum. The plating efficiency (number of dividing protoplasts expressed as a percentage of the total protoplast population) was estimated after 7 days from three experiments, each experiment was replicated three times.

First divisions occurred within 48 h in the case of cotyledon, hypocotyl and cell suspension protoplasts and within 48–72 h in those from roots and shoots. Division was best at 2.5×10^4/ml for root protoplasts and at 5×10^4/ml for all other sources of plant material investigated. High plating efficiencies were obtained, for all sources of protoplasts, in either K8P (Kao 1977) or in KM8P:KM8 (Kao and Michayluk 1975). However, the plating efficiency of green cotyledon protoplasts of *L. altaicum* was lower than other species (Tables 1, 2, 3 and 4). The present investigation highlights the usefulness of the complex media, originally developed by Kao and Michayluk (1975) and Kao (1977) to stimulate protoplast division at low densities, and subsequently used for culturing seedling protoplasts of a number of genera (Berry et al. 1982; Lu et al. 1982, 1983a, 1983b; Ahuja et al. 1983; Glimelius 1984). Although the best plating efficiency was obtained using Kao and Michayluk medium, the growth of colonies was more rapid in MB_5P liquid over agar (Fig. 1). The finding that protoplast division is improved when in diffuse light (Table 1) is in agreement with the earlier results of Gamborg and Shyluk (1976).

The procedure for lowering the osmotic concentration of the media, which prevented browning in liquid media as the colonies developed, was a critical factor for sustained colony growth of *Linum* species. Previously a filter paper technique was used in the culture of *Medicago sativa* protoplasts to reduce browning (Santos et al. 1980) and also for *Pteridium aquilinum* protoplasts (Partenan 1981), but this has since been rendered unnecessary by using cotyledon tissue protoplasts of *Medicago sativa* (Lu et al. 1982). In the present study, cotyledon protoplasts of *Linum usitatissimum* cv. Precederia on filter paper continued to divide and gave rise to large colonies (Fig. 2). Colony formation from root and hypocotyl protoplasts of *L. usitatissimum* cv. Precederia using filter paper support was rarely achieved. The role of a filter paper substrate in the development of cell colonies is not clear, but the reduction of protoplast browning may be due to absorption of enzymes remaining from protoplast isolation and improvement of gaseous exchanges (Santos et al. 1980). In previous studies it was observed that *Linum* hypocotyl protoplasts divided once or twice, but then deteriorated after about 10 days of culture. When, however, the dishes were gently shaken after the initial cell divisions, the droplets mixed and the cells became attached to the bottom surface of the Petri dish and continued to grow (Gamborg and Shyluk 1976). In the present investigation, sustained development of colonies without significant browning was achieved either by culturing protoplasts in liquid media and, after 7 days, washing colonies free from the original medium and transferring them to diluted liquid over agar, or by culturing protoplasts immediately on liquid over agar. The best-sustained colony growth occurred following the addition of small volumes (5–20 drops) of diluted medium after 1 week, in the case of cotyledon and cell suspension protoplasts which were cultured on liquid over agar. However, in the case of the culture of protoplasts isolated from seedling roots, hypocotyl and in vitro-grown shoots, browning was minimized by transferring agar blocks containing cell colonies after 2 weeks of culture to fresh agar solidified medium with lower osmoticum. For several major crops, browning during cell culture still remains a problem, for example sunflower and cotton (Barakat unpubl).

5 Plant Regeneration from Protoplast-Derived Callus

Hypocotyl protoplast-derived callus of *Linum usitatissimum* were reported to regenerated only roots (Gamborg and Shyluk 1976), and later plants from protoplast-derived callus of several cultivars of *L. usitatissimum*, and the wild *Linum* species, *L. strictum* and *L. lewissii* were obtained (Barakat and Cocking 1983, 1985).

Colonies derived from different sources of protoplasts were transferred, after 2–4 weeks, to regeneration media. The regeneration media used were MSP-1,

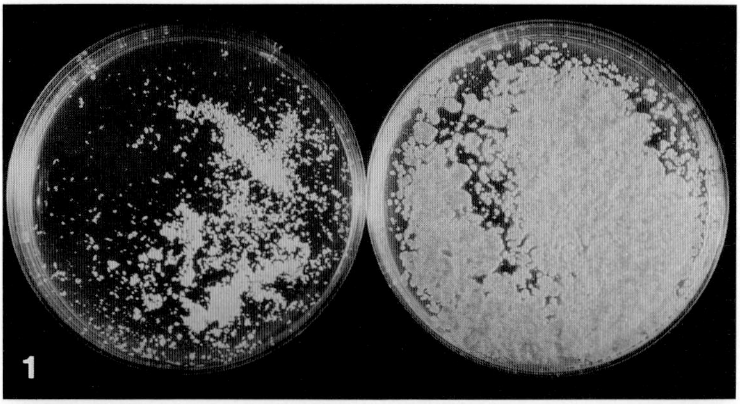

1

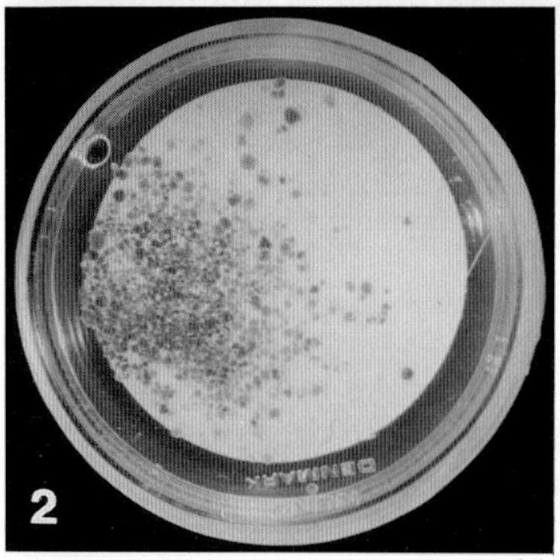

2

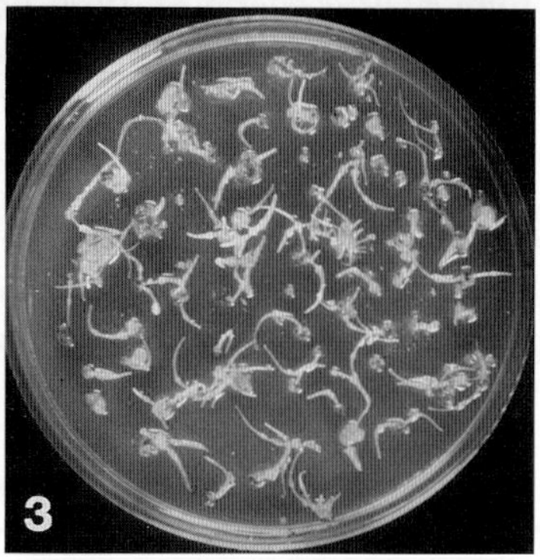

3

5

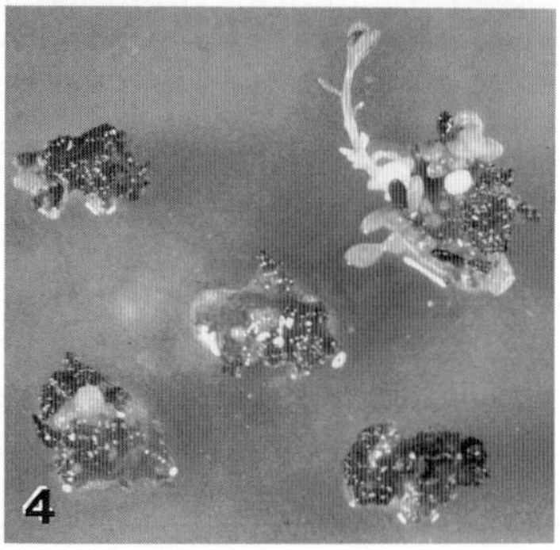

4

6

MSP-3, MSD-3, MSD-4, MSZ, MSF, B_5-1 and B_5-2, as previously described (Barakat and Cocking 1983, 1985).

Results from this investigation show that plant regeneration is possible from protoplasts of *L. usitatissimum*, isolated from root tissues of the cultivars Precederia, Giza-4 and Entry-115, and also isolated from cotyledon tissues of the cultivars Precederia and H-1. Modified MS agar medium containing BAP (1.13 mg/l) and NAA (0.02 mg/l) was effective for shoot induction from root and cotyledon protoplast-derived callus of several cultivars of *L. usitatissimum*. The modified MS medium containing BAP (0.5 mg/l) and NAA (0.1 mg/l) was effective only for shoot induction from root protoplast-derived callus of the Precederia cultivar.

Shoot regeneration of *L. strictum* from protoplast-derived callus occurred only at low frequency (1%). However, the fact that regeneration was possible confirms the totipotency of seedling protoplasts of *Linum* species utilizing the MS agar medium containing BAP (1.13 mg/l) and NAA (0.1 mg/l). High frequency (70%) shoot regeneration was obtained in the case of *L. lewissii* from either in vitro-grown shoot protoplasts or from cell suspension protoplasts. To facilitate shoot regeneration from in vitro-grown shoot protoplasts, colonies were transferred to MSD-4 (MS containing 0.5 mg/l NAA and 0.5 mg/l BAP) and after three subcultures to fresh MSD-4 medium, at 6–8 week intervals, high frequency of shoots was obtained. To obtain high frequency of shoot regeneration from cell suspension protoplasts of *L. lewissii*, either colonies or calli which had been produced on solidified MSP-1 were transferred to MSP-1 liquid medium, and after three subcultures to the same fresh medium of MSP-1 at 6–8-week intervals. Embryogenic structures (Fig. 3) were transferred to BGS solidified medium (Power et al. 1984), and shoot regeneration was obtained (Fig. 4). The success of shake cultures may be related to the more efficient washing away or diluting out of endogenously produced auxins or other substances, that may otherwise suppress regeneration (Chen and Galston 1967). MSD-4 has possibly been found suitable for shoot regeneration from root protoplasts of brassicas (Xu et al. 1982), and also suitable for shoot regeneration from callus of solanaceous species (Power et al. 1984). Root formation was induced when shoot tips (with three to four leaves)

Fig. 1. Cell colonies derived from cotyledon protoplasts of *Linum usitatissimum* cultured on solid medium. **a** Culture in KM8P:KM8 medium. **b** Culture in MB_5P medium. (Barakat and Cocking 1983)

Fig. 2. Cell colonies derived from cotyledon protoplasts after 15 days of culture on solid medium plated onto a sterile filter paper laid on the surface of K8P agar medium. (Barakat 1985)

Fig. 3. Embryogenic structures (derived from cotyledon protoplasts of *L. lewissii*) after two subcultures in MSP-1 liquid medium. (Barakat and Cocking 1985)

Fig. 4. Shoot development from *L. lewissii* embryogenic cultures on BGS solidified medium. (Barakat and Cocking 1985)

Fig. 5. Whole-plant regeneration from *Linum usitatissimum* established in the greenhouse at flowering stage. Age 20 weeks

Fig. 6. Whole-plant regeneration from *L. lewissii* in vitro-grown shoot protoplast-derived callus under greenhouse conditions. Age 24 weeks were transferred to MS-based agar without phytohormones. Plants were established under greenhouse conditions (Figs. 5, 6).

6 Protoplast Fusion

It was possible to fuse several combinations of protoplasts using various fusion protocols, and to identify fusion procedures with suitable combinations of protoplasts for producing a maximum number of heterokaryons. Protoplasts were fused both chemically and electrically. The chemical fusion, high pH/Ca^{2+}, polyethylene glycol (PEG), PEG-high pH/Ca^{2+} and PEG using the small-scale method were as previously described (Barakat 1985). The small-scale protocol, as a chemical fusion procedure, gave the maximum number of heterokaryons when green cotyledon protoplasts of *L. usitatissimum* cv. Precederia were fused with cell suspension protoplasts of *L. strictum* (up to 4.5%), and also when green cotyledon protoplasts of *L. strictum* were fused with root protoplasts of *L. usitatissimum* (up to 5.5%).

Green cotyledon protoplasts of *L. usitatissimum* were fused electrically with cell suspension protoplasts of *L. lewissii* using a DC pulse (200–400 V) for approximately 2 µs. The optimum frequency of heterokaryons was 7.2% following fusion by 300 V DC pulse for 2 µs. Heterokaryons and *Linum* parental protoplasts divided (P.E. = 38.3%) when they were cultured together in K8P liquid medium.

7 Conclusions

Overall these studies on isolationculture and plantregeneration from protoplasts of wild species and cultivated varieties of *Linum* now provide a firm baseline for a range of genetic manipulations. The opportunity now exists to undertake a programme of somatic hybridization to facilitate the introgression of desirable genes from wild *Linum* species into cultivated varieties.

8 Protocol

(1) Optimal Conditions for the Isolation of *Linum* Protoplasts. Protoplasts are isolated from hypocotyls of seedlings after 7 days' germination. After cutting longitudinally, hypocotyl segments are plasmolysed in CPW solution containing 13% w/v mannitol and then incubated in 2% w/v Rhozyme HP 150, 4% w/v Meicelase and 0.03% w/v Macerozyme CPW 13M pH 5.8 for 16 h at 25 °C on a rotary shaker (30–40 rpm). Protoplasts are released by gentle squeezing and filtering through a nylon sieve (64 µm pore size) and pelleted by centrifugation at 100 *g* for 5 min. Protoplasts are purified by flotation in CPW 21% w/v sucrose (100 *g* for 10 min).

(2) Optimal Conditions for the Culture of *Linum* Protoplasts. The optimal plating density for the hypocotyl protoplasts is 5×10^4/ml. Protoplasts are cultured in KM8P:KM8 liquid medium in diffuse light. After 7 days colonies formed are washed free from the original KM8P:KM8 medium and then transferred to diluted KM8P:KM8 liquid medium on top of agar solidified KM8P:KM8 medium.

(3) Optimal Conditions for the Regeneration of Plants from Protoplast-Derived Colonies. After culture in diluted KM8P:KM8 liquid medium on top of agar solidified KM8P:KM8 medium for 3 weeks colonies are transferred to MSP-1, MSP-3, MSD-4, MSZ, MSF, B_5-1, B_5-2 and BGS. Different species of *Linum* respond differently and a screening for shoot regeneration or embryoid formation on these media provides the basis for the establishment of plant regeneration from the protoplast-derived colonies.

References

Ahuja PS, Lu DY, Davey MR, Cocking EC (1983) An assessment of the cultural capabilities of *Trifolium repens* L. (White Clover) and *Onobrychis viciifolia* Scop. (Sainfoin) mesophyll protoplasts. Plant Cell Rep 2:269–272

Barakat MN (1985) Protoplast and tissue culture studies for somatic hybridisation in the genus *Linum*. Ph.D. Thesis. Nottingham University

Barakat MN, Cocking EC (1983) Plant regeneration from protoplast-derived tissues of *Linum usitatissimum* L. (Flax.). Plant Cell Rep 2:314–317

Barakat MN, Cocking EC (1985) An assessment of the cultural capabilities of protoplasts of some wild species of *Linum*. Plant Cell Rep 4:164–167

Bari G, Godward MBE (1970) Interspecific crosses in *Linum*. Euphytica 19:443–446

Beard BH, Comstock VE (1980) Flax. In: Hybridisation of crop plants. American Society of Agronomy – Crop Science Society of America, pp 357–366

Berry SF, Lu DY, Pental D, Cocking EC (1982) Regeneration of plants from protoplasts of *Lactuca sativa* L. Z Pflanzenphysiol 108:31–38

Chen HR, Galston AW (1967) Growth and development of *Pelargonium* pith cells in vitro. Initiation of organised development. Physiol Plant 20:533–539

Chuong PV, Pauls KP, Beversdorf WD (1985) A simple culture method for *Brassica* hypocotyl protoplasts. Plant Cell Rep 4:4–6

Cocking EC, Davey MR, Pental D, Power JB (1981) Aspects of plant genetic manipulation. Nature 293:265–270

Crowder AI, Landgreen CR, Torrey JG (1979) Cultivar differences in starch content and protoplast yields from root cortical explants of *Pisum sativum*. Physiol Plant 46:85–88

Eriksson T, Bonnett HT, Glimelius K, Wallin A (1978) Protoplast isolation, cultivation and development. In: Thorpe T (ed) Frontiers of plant tissue culture. Univ Calgary Press, Canada, pp 131–139

Frearson EM, Power JB, Cocking EC (1973) The isolation, culture and regeneration of *Petunia* leaf protoplasts. Dev Biol 33:130–137

Gamborg OL, Shyluk JP (1976) Tissue culture, protoplasts and morphogenesis in flax. Bot Gaz 137:301–306

Gill KS, Yermanos DM (1967a) Cytogenetic studies on the genus *Linum*. I. Hybrids among taxa with 15 as the haploid chromosome number. Crop Sci 7:623–627

Gill KS, Yermanos DM (1967b) Cytogenetic studies on the genus *Linum*. I. Hybrids among taxa with nine as the haploid chromosome number. Crop Sci 7:627–631

Glimelius K (1984) High growth rate and regeneration capacity of hypocotyl protoplasts in some Brassicaceae. Physiol Plant 61:38–44

Kao KN (1977) Chromosomal behaviour in somatic hybrids of soybean – *Nicotiana glauca*. Mol Gen Genet 150:225–230

Kao KN, Michayluk MR (1975) Nutritional requirements for growth of *Vicia hajastana* cells and protoplasts at very low population density in liquid media. Planta 126:105–110

Lu DY, Pental D, Cocking EC (1982) Plant regeneration from seedling cotyledon protoplasts. Z Pflanzenphysiol 107:59–63

Lu DY, Cooper-Bland S, Pental D, Cocking EC, Davey MR (1983a) Isolation and sustained division of protoplasts from cotyledons of seedlings and immature seeds of *Glycine max* L. Z Pflanzenphysiol 111:389–394

Lu DY, Davey MR, Cocking EC (1983b) A comparison of the cultural behaviour of protoplasts from leaves, cotyledons and roots of *Medicago sativa* L. Plant Sci Lett 31:87–99

Murashige T, Skoog F (1962) A revised medium for rapid growth and bioassays with tobacco tissue cultures. Physiol Plant 15:473–497

Partenman CR (1981) Filter paper as a support and carrier for plant protoplast cultures. In vitro 17:77–80

Plessers AG (1966) The variation in fatty acid composition of the seed of *Linum* species. Can J Genet Cytol 8:328–335

Power JB, Cummins SE, Cocking EC (1970) Fusion of isolated plant protoplasts. Nature 225:1016–1018

Power JB, Chapman JV, Wilson D (1984) Laboratory manual, plant tissue culture. Dept. Botany, Univ Nottingham, Nottingham, UK

Ruesink AW, Thimann UV (1966) Protoplast preparations from higher plants. Science 154:280–281

Santos AVP dos, Outka DE, Cocking EC, Davey MR (1980) Organogenesis and somatic embryogenesis in tissues derived from leaf protoplasts and leaf explants of *Medicago sativa.* Z Pflanzenphysiol 99:261–270

Seetharam A (1972) Interspecific hybridisation in *Linum*. Euphytica 21:489–495

Xu ZH, Davey MR, Cocking EC (1981) Isolation and sustained division of *Phaseolus aureus* (Mung Bean) root protoplasts. Z Pflanzenphysiol 104:289–298

Xu ZH, Davey MR, Cocking EC (1982) Plant regeneration from root protoplasts of *Brassica*. Plant Sci Lett 24:117–121

Yermanos DM, Beard BH, Gill KS, Anderson MP (1966) Fatty acid composition of seed oil of wild species of *Linum*. Agron J 58:30–32

II.5 Isolation and Culture of *Glycine* Protoplasts

C. A. NEWELL[1] and H. T. LUU[2]

1 Introduction

1.1 Importance and Distribution

The soybean, *Glycine max* (L.) Merr., first emerged as a domesticated plant in the eastern half of north China around the eleventh century B. C., according to historical, linguistic and geographical evidence (Hymowitz 1970). By the first century A. D., it had probably spread through central and south China and into Korea; from there it expanded into countries throughout the far east, and many land races were developed (Hymowitz and Newell 1980). The crop's rise to a position of worldwide economic importance has taken place relatively recently, and only within the last 50 years have plant breeders made concerted efforts to develop improved cultivars, which are adapted to the USA, by artifical hybridization. Approximately 51 million hectares of soybean are currently grown worldwide in 44 countries, with the USA (25 million ha) and Brazil (9 million ha) being the largest producers. In the USA alone, this adds up to a crop value of 11 billion dollars Anonymous 1986).

The soybean is responsible directly or indirectly for a wide range of useful products. In the Orient, the traditional uses still predominate, with human nutrition foremost in importance. Human foods include fermented soy products such as soy sauce, natto, miso, and tempeh; and non-fermented products such as soybean sprouts, soy milk, and tofu. Western countries, on the other hand, have developed the soybean primarily as a source of oil and high-protein meal. Soybean oil is used in countless modern-day products such as margarine, salad oils, emulsifers, paints, resins, plastics and lubricants, to name but a few. The meal is used mainly as a source of animal feed for production of milk, poultry, beef and pork (Mounts et al. 1987).

The soybean industry in the USA is based on germplasm introduced primarily from Japan, Korea, and the People's Republic of China. Considering the value of the crop, it is perhaps surprising that the high-yielding varieties developed in the USA are based on an extremely narrow germplasm range. Although many introduced cultivars have been used in breeding programs across the USA and Canada, fewer than 20 introductions have given rise to the major cultivars grown throughout the USA (Bernard 1983).

[1] Monsanto Company, 700 Chesterfield Village Parkway, St. Louis, MO 63198, USA
[2] The Upjohn Company, 301 Henrietta Street, Kalamazoo, MI 49001, USA

Biotechnology in Agriculture and Forestry, Vol. 8
Plant Protoplasts and Genetic Engineering I (ed. by Y. P. S. Bajaj)

The genus *Glycine*, to which the soybean belongs, currently comprises 14 species. *Glycine soja* Sieb. & Zucc., the wild progenitor of the soybean, is native to China, Japan, Korea, Taiwan, and the USSR; like the soybean, it is annual and diploid (2n = 40) and can be readily crossed with the soybean. The remaining 12 species are perennial and indigenous to Australia. *Glycine arenaria* Tind., *G. canescens* F.J. Herm., *G. clandestina* Wendl., *G. curvata* Tind., *G. cyrtoloba* Tind., *G. falcata* Benth., *G. latifolia* (Benth.) Newell & Hymowitz, *G. latrobeana* (Meissn.) Benth., and *G. microphylla* (Benth.) Tind. are diploid (2n = 40). *Glycine tabacina* (Labill.) Benth. and *G. tomentella* Hayata encompass diploid and tetraploid (2n = 80) races, as well as aneuploids (2n = 38, 78) in *G. tomentella*; both species extend northward from Australia to South China and eastward to several Pacific islands. The perennial species can be crossed with the soybean only with difficulty. Embryo rescue is required to obtain F_1 hybrids, and the resulting plants are sterile. Doubling the chromosome number with colchicine has had limited success in restoring fertility, and so far no attempts at backcrossing have been successful (Hymowitz and Singh 1984; Newell et al. 1987).

1.2 Significance and Use of Protoplasts

Protoplasts can be readily isolated from soybean cell suspensions and plant organs and have been used fairly extensively for fusion and physiological studies. It is possible to regenerate callus from protoplasts, but there are no reports in the literature to date that describe reproducible and efficient plant regeneration from soybean protoplasts. Thus protoplast technology in soybean has limited usefulness at present, and its full potential will only be realized when the obstacle of plant regeneration has been overcome.

Protoplast regeneration in soybeans would allow exploitation of developing techniques such as microinjection, electroporation and electrofusion as methods of gene transfer. Currently, in vitro gene transfer into soybeans must make use of explants that show capacity for regeneration, and *Agrobacterium tumefaciens* or *A. rhizogenes* are the vectors of choice. However, the types of soybean explant that will regenerate are restricted in number (Kerns et al. 1986) and furthermore, soybeans are variable in their susceptibility to *Agrobacterium* (Byrne et al. 1987). These factors may make electroporation or microinjection of single cells or protoplasts viable alternatives to gene transfer by *Agrobacterium* once protoplast regeneration is achieved.

Wild perennial *Glycine* species represent a reservoir of variation that is not present in the soybean germplasm, and which is already known to include some genes for disease resistance that are potentially of agronomic importance. Given the difficulties attendant upon interspecies crosses in *Glycine*, protoplast fusion may facilitate hybridization between the soybean and wild perennials that may not otherwise be possible even with embryo rescue. Protoplast fusion followed by regeneration would have the added advantage of producing plants combining somatic parental genomes, and the extra step of colchicine doubling would be obviated. In addition, it would be possible to take advantage of the procedure being developed in other crop systems whereby selected cytoplasmic organelles are combined with nuclei from separate donors, or nuclei are combined with chromosomes or chromosome fragments. The combination of cytoplasm from wild species and soybean nuclei could facilitate development of a cytoplasmic male sterile system in soybeans; addition of chromosomes or chromosome fragments followed by recombination might circumvent the sterility problems associated with interspecific hybridization. Thus, protoplast technology is viewed as a means of accessing variation in wild germplasm that has so far proven intractable with conventional methods.

1.3 Review of Work on Protoplast Isolation

Protoplasts have been successfully isolated and cultured from almost every part of the soybean plant (Table 1). Early work described protoplast isolation from cell suspensions using an enzyme mixture containing Cellulase and Pectinase (Kao et al. 1970, 1971), and root nodules (Davey et al. 1973) using cellulase, macerozyme and hemicellulase. Protoplasts from cell suspensions reformed cell walls within 24 h and up to 50% of protoplasts divided within 3–4 days to give small colonies.

Plant tissue proved somewhat more recalcitrant to protoplast isolation, however, and it was not until 1980 that Zieg and Outka reported that immature soybean pods (less than 1×4 cm) could be treated with Pectinase and Cellulase to give protoplasts with a yield of 4.6×10^6 protoplasts/g of starting tissue. Other plant parts that have since yielded protoplasts include root tips from 3-day-germinated seeds (1.2×10^6/g) (Xu et al. 1982), cotyledons from immature 5–7-mm seeds (up to 0.7×10^6/pair of cotyledons) and 4–7-day-old seedlings (0.34×10^6/pair of cotyledons) (Lu et al. 1983), mechanically isolated primary leaf mesophyll cells (Schwenk et al. 1981) and first trifoliolate leaf mesophyll cells (Rees et al. 1985), expanded leaves and shoots from newly established sterile shoot cultures (Oelck et al. 1983), leaflets of 10–12-day-old seedlings germinated in culture (Gamborg et al. 1983; Tricoli et al. 1986), fully expanded trifoliate leaves from 3-week-old plants (Tricoli et al. 1986), and immature and mature leaves from 5-week-old plants (Lin 1983). Protoplasts could also be isolated with good efficiency (2×10^6/ml cells) from callus derived from cotyledons, shoots and roots if the callus cultures were 3–4 months old, and were placed in liquid medium for at least 2 days before isolation (Bojsen and Wyndaele 1985). Low yields were obtained from newly initiated callus and callus that had not been cultured in liquid medium prior to protoplast isolation.

Pectolyase is commonly added to the enzyme mixture for isolation of leaf mesophyll protoplasts (Schwenk et al. 1981; Gamborg et al. 1983; Lin 1983; Oelck et al. 1983; Franceschi et al. 1984; Rees et al. 1985). The conversion of mechanically isolated, first trifoliolate leaf mesophyll cells to protoplasts was found to be dependent upon plant age (Rees et al. 1985), as the highest rates of conversion were obtained for plants up to 24 days old, but declined thereafter. Lu et al. (1983) found that while enzyme mixtures containing pectolyase would release protoplasts from seedling cotyledons of all five varieties tested, an enzyme mixture lacking pectolyase would not; but the latter did release protoplasts from immature seed cotyledons, indicating a difference in the composition of mature cotyledonary cell

Table 1. Soybean tissue from which protoplasts have been successfully isolated and cultured

Origin of protoplasts	Enzyme mixture	Culture medium	Growth response	Reference
Cell suspension	Cellulase and Pectinase in B5 + 0.1 M sucrose, 0.3 M sorbitol, 0.005 M citric acid	Conditioned B5 + 0.13 M sorbitol, 0.05 M sucrose, 0.001 M $CaH_4(PO_4)_2 \cdot 2H_2O$	Cell clusters	Kao et al. (1970)
Immature pods from greenhouse-grown plants	Macerozyme (3%), Cellulase R-10 (5%) in 0.6 M mannitol + 8 mM $CaCl_2$, or Macerozyme (3%) in 0.6 M mannitol for 2 h followed by Cellulase R-10 (5%) in 0.6 M mannitol + 8 mM $CaCl_2$	Kao (1977) medium	Callus, roots	Zieg and Outka (1980)
1 cm root tips from germinating seeds in culture	Rhozyme (2%), Meicelase (4%), Macerozyme (0.3%) in 13% mannitol + KH_2PO_4 (27.2 mg/l), KNO_3 (101 mg/l), $CaCl_2 \cdot 2H_2O$ (1.48 g/l), $MgSO_4 \cdot 7H_2O$ (240 mg/l), KI (0.16 mg/l), $CuSO_4 \cdot 5H_2O$ (0.025 mg/l), ampicillin (400 mg/l), gentamycin (10 mg/l), tetracycline (10 mg/l)	Modified B5 + chlorophenoxyacetic acid (1 mg/l), BA (1 mg/l), sucrose (3%), mannitol (7%), or: Modified B5 + glucose (1%), sorbitol (3.5%) mannitol (3.5%), casein hydrolysate (200 mg/l) + 2,4-D (2 mg/l), NAA (0.5 mg/l), BA (0.5 mg/l), or 2,4-D (0.25 mg/l), NAA (0.5 mg/l), BA (0.1 mg/l); or: KM8P : KM8, 2 : 1 (Kao and Michayluk 1975)	Callus, roots	Xu et al. (1982)
First and second leaves of 10–12-day-old seedlings germinated in culture	Pectolyase Y-23 (0.1%), Cellulase R-10 (0.2%) in 8% mannitol + $CaCl_2 \cdot 2H_2O$ (0.12%)	SL salts (Collins and Phillips 1982), B5 vitamins (Gamborg 1975), calcium and multivitamins (Gamborg et al. 1979), casamino acids (125 mg/l), L-glutamine (100 mg/l), sodium pyruvate (5 mg/l), sodium citrate (10 mg/l), ribose (125 mg/l), sorbitol (27.3 g/l), sucrose (150 mg/l), m-inositol (150 mg/l), glucose (36 g/l), 2,4-D (0.22 mg/l), picloram (0.6 mg/l), BA (0.1 mg/l)	Culture cell suspension	Gamborg et al. (1983)

Hypocotyls of dark-grown 7-day-old *Glycine canescens* seedlings	Macerozyme R-10 (0.5%), Cellulase R-10 (5%) in 1:3 dilution of R-medium major salts (Shepard 1980), 0.3 M sucrose, 5 mM MES	R-medium major salts (Shepard 1980), Cl-medium osmoticum (Shepard 1980), thiamine-HCl (0.5 mg/l), glycine (2 mg/l), nicotinic acid (5.0 mg/l), pyridoxine-HCl (0.5 mg/l), folic acid (0.5 mg/l), biotin (0.05 mg/l), ammonium succinate (0.65 mM), CL-medium iron and minor elements (Shepard 1980) in 1:1 dilution, NAA (0.1 or 1.0 mg/l) or 2,4-D (1.0 mg/l), BA (0.4 mg/l), agarose (0.45%)	Plants	Newell and Luu (1985)
4–7-day-old seedling cotyledons	Pectolyase Y-23 (0.1%), Onozuka R-10 (0.2%) in 0.12% $CaCl_2 \cdot 2H_2O$ + 8% mannitol , or: Pectolyase Y-23 (0.1%), Onozuka R-10 (0.2%), Rhozyme (1%) in CPW salts (Frearson et al. 1973) + 9% mannitol, or: Pectolyase Y-23 (0.1%), Onozuka RS (0.2%) in CPW + 9% mannitol, or: Pectolyase Y-23 (0.1%), Onozuka RS (0.2%), Rhozyme (1%) in CPW + 9% mannitol	Kao (1977) medium	Callus	Lu et al. (1983)
5–7 mm immature seed cotyledons	Meicelase (4%), Rhozyme (2%), Macerozyme R-10 (0.3%) in CPW + 9% mannitol	Kao (1977) medium	Callus	
Expanded leaves, shoots from 4–6-week-old seedlings germinated in culture	Cellulase (1%), Macerozyme (0.2%), Pectolyase (0.05%) in 0.6 M mannitol	B5 medium + mannitol (730 mOsm), 2,4-D, NAA, BA	Callus, roots	Oelck et al. (1983)

Table 1 (Continued)

Origin of protoplasts	Enzyme mixture	Culture medium	Growth response	Reference
Cell suspension derived from crown gall tumor	Cellulase R-10 (2%), Hemicellulase (1%), Pectinase (1%), Driselase (2%), Sorbitol (10 mg/l) in Miller medium (Miller 1968) lacking hormones	K8P without hormones (Kao and Michayluk 1975)	Callus	Pederson et al. (1983)
Cell suspension	Cellulase R-10 (1%), Pectolyase Y-23 (0.2%), Driselase (2%) in 0.7 mM KH_2PO_4 + 6 mM $CaCl_2 \cdot 2H_2O$ + 0.15 M sorbitol + 0.15 M mannitol + 0.1 M glucose + 3 mM MES	SL salts (Collins and Phillips 1982), B5 vitamins (Gamborg 1975), calcium and multivitamins (Gamborg et al. 1968), casamino acids (125 mg/l), sodium citrate (10 mg/l), L-glutamate (100 mg/l), sodium pyruvate (5 mg/l), ribose (125 mg/l), sorbitol (27.3 g/l), sucrose (0.15 g/l), m-inositol (0.15 g/l), glucose (36 g/l), 2,4-D (0.22 mg/l), picloram (0.06 mg/l), BA (0.1 mg/l)	Callus	Chowdhury and Widholm (1985)
10-day-old seedling primary leaves from seeds germinated in culture or trifoliolate leaves from 3-week-old greenhouse-grown plants	Cellulase CEL (2%) or Cellulase CELF (2.65%), Pectolyase Y-23 (0.5%) in 0.4 M mannitol, 10 mM $CaCl_2$, 1% BSA, 5 mM MES	Protoplasts encapsulated in alginate beads + medium MPC + 10% actively growing soybean cell suspension	Callus	Tricoli et al. (1986)

walls compared with those of a developing seed. Mesophyll protoplasts from seedling primary leaves or young trifoliolate leaves have been successfully cultured following encapsulation in alginate beads (Tricoli et al. 1986). The protoplasts were induced to form small colonies by culturing the beads in an actively growing soybean cell suspension which acted as a nutrient source. Eighteen days after encapsulation, the alginate matrix could be dissolved to release viable callus colonies. An advantage of this method is that protoplasts can be embedded at low cell densities and yet will still produce colonies, in contrast to the usual culture conditions where high plating densities are required.

With respect to organized plant tissue, protoplast isolation and subsequent culture has been most successful when immature or rapidly expanding plant organs have been used. Soybean protoplasts generally show wall regeneration within 24 h, the first cell divisions in 48 h, and small colony formation in 7 days. Division efficiencies (number of dividing protoplasts expressed as a percentage of the total population) of up to 50% have been obtained for root and seedling tissue (Lu et al. 1983; Xu et al. 1982). Protoplast-derived callus from pod (Zieg and Outka 1980), root (Xu et al. 1982), and mesophyll tissue (Oelck et al. 1983) has differentiated to produce roots. Small green nodular areas have been observed in callus from immature seed cotyledons (Lu et al. 1983) but to date no shoot regeneration from protoplast-derived callus has been reported for soybean.

Owing to the absence of a protoplast to plant regeneration procedure for the soybean, attention has focussed on related wild species in the hope that they would be more amenable to manipulation in culture. Protoplasts isolated from cell suspensions of *G. tabacina* and *G. soja* were cultured to produce new suspension cultures, which formed embryo-like structures (Gamborg et al. 1983). Newell and Luu (1985) reported the regeneration of plants from *G. canescens* seedling hypocotyl protoplasts with a low but reproducible frequency; this procedure for protoplast isolation and plant regeneration is described below. Regeneration of plants from *G. canescens* suspension culture protoplasts was also reported recently by Myers et al. (1986), as well as from seedling cotyledon protoplasts of *G. clandestina* and *G. canescens* (Hammatt et al. 1987).

2 Isolation of Protoplasts

2.1 Plant Material

Seed of *G. canescens* Plant Introduction (PI) 399478 was sterilized by washing for 5 min in absolute ethanol, 10 min in 20% Clorox solution with a drop of Tween 20, and rinsing three times with sterile distilled water. Seeds were scarified under a flow hood by nicking the side opposite the hilum with a scalpel, and were germinated in Magenta GA-7 (Magenta Corporation, Chicago, USA) boxes on MS medium (Murashige and Skoog 1962) with 0.6% agar, pH 5.6, in the dark at 24 °C. After 7 days the hypocotyls were approximately 6–7 cm long. The hypocotyls were cut into 2–5-mm sections and placed into a side-arm Erlenmeyer flask with a filter-sterilized "soak" medium composed of major salts, minor salts, and or-

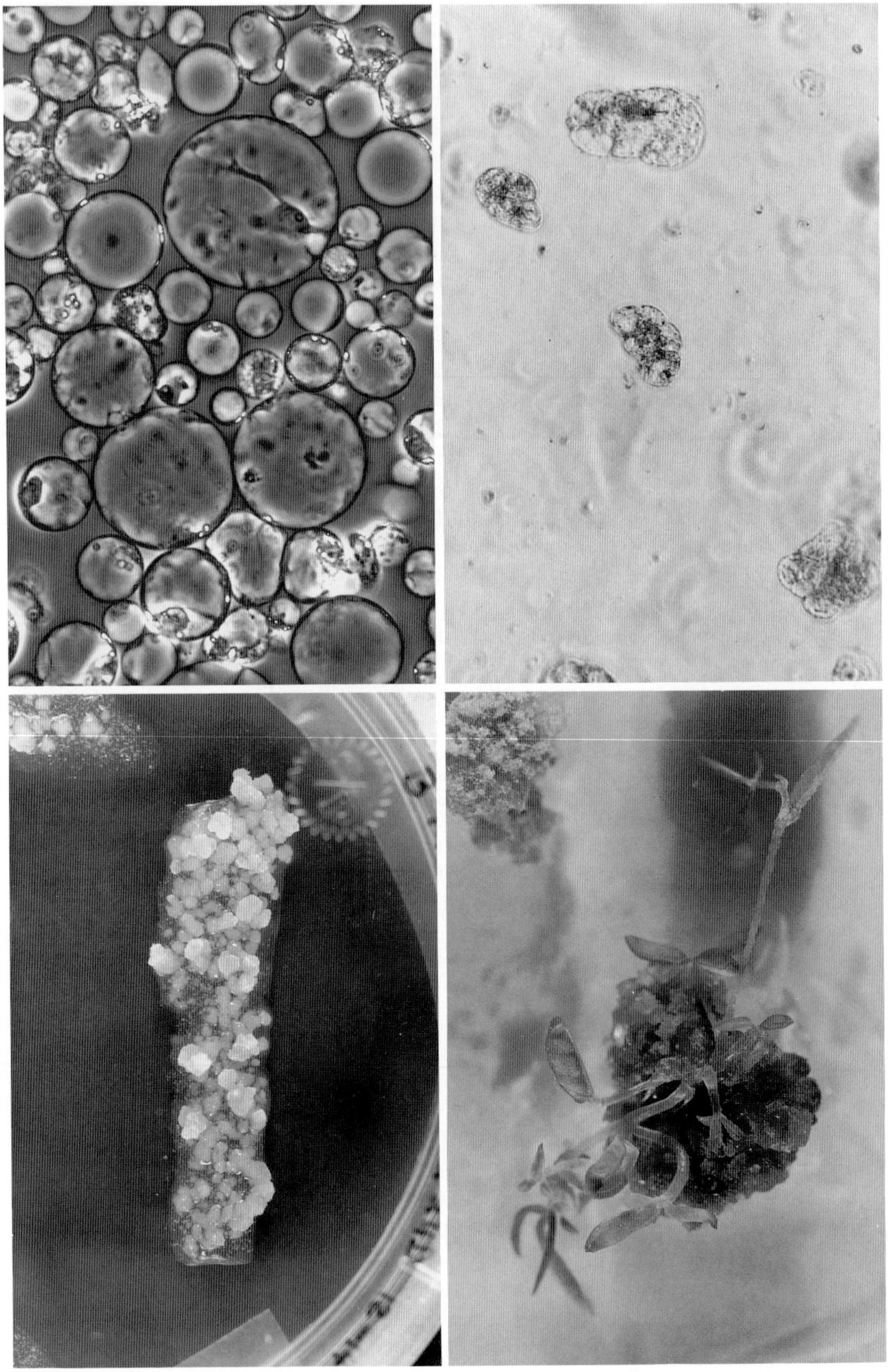

A B C D

ganic addenda as in the R-medium of Shepard (1980), 0.2 M sucrose, 0.1 M mannitol, 1 mg/l naphthalenacetic acid (NAA), and 0.4 mg/l 6-benzylaminopurine (BA), at pH 5.6. Tissue was immersed overnight (24 h) in a refrigerator at 4 °C.

2.2 Enzymes

The preliminary "soak" medium was removed with a Pasteur pipette and replaced with an enzyme solution containing 0.5% Macerozyme R-10 and 5% Cellulase Onozuka R-10 (Yakult Pharmaceutical Industry, Japan) dissolved in one quarter strength R-medium major salts, 0.3 M sucrose, and 5 mM 2(N-morpholino)ethane sulfonic acid (MES) at pH 5.6. Approximately 10 ml of enzyme solution was added for each gram of plant tissue. The tissue was infused under vacuum for 15 min, then incubated for 5 h on a slowly rotating (ca. 50 rpm) platform in a water bath at 28 °C. The digested hypocotyl suspension was filtered through two layers of cheese cloth into 20-ml narrow-necked ampules and centrifuged at 1000 rpm for 10 min. Undamaged protoplasts floated to the surface and formed a band which was removed with a Pasteur pipette. Protoplasts were resuspended in a "rinse" solution of major salts of R-medium and 0.3 M sucrose at pH 5.6, and re-centrifuged for 10 min. Floating protoplasts were again removed and resuspended in a small amount of rinse solution ready for plating. Counts of protoplast number/ml were obtained with a haemocytometer. The resulting hypocotyl protoplasts were colorless and variable in size (Fig. 1); the yield averaged 1.4×10^6/g of tissue over several isolations.

3 Culture of Protoplasts

The protoplasts were plated at a density of 10^5/ml in a nutrient medium composed of R-medium major salts, CL-medium (Shepard 1980) osmoticum and organic addenda, with the exception that casein was replaced by 0.65 mM ammonium succinate, half-strength iron and minor elements of the CL-medium, 0.1 or 1 mg/l NAA or 1 mg/l 2,4-dichlorophenoxyacetic acid (2,4-D), and 0.4 mg/l BA at pH 5.6. This medium was solidified with 0.45% (w/v) low temperature gelling agarose (Type VII: Sigma Chemical Company). The protoplasts were placed into two opposite sections of a quadrant Petri plate, which had slits along the base of the divisions between the sections to allow diffusion between plating and reservoir medium (Shepard 1980). The reservoir medium contained the same salts, organic addenda and growth regulators as the plating medium except that 1.3 mM ammonium succinate and 0.5 mM ammonium chloride replaced casein. The osmoticum was 0.1 M sucrose and 0.1 M mannitol, while 0.6% washed agar replaced the

Fig. 1A–D. A Freshly isolated hypocotyl protoplasts of *Glycine canescens* (×410). **B** Protoplast-derived colonies from hypocotyls of *G. canescens* approximately 14 days after isolation (×165). **C** Protoplast-derived colonies from hypocotyls of *G. canescens* approximately 6 weeks after isolation (×3.3). **D** Regeneration of shoots on protoplast-derived callus of *G. canescens* 18 weeks after isolation agarose. The reservoir medium was poured into two opposite sections of quadrant Petri-plates and allowed to set before the plating medium containing protoplasts was added to the remaining quadrants. The ammonium source was varied at the plating stage by using either casein or ammonium succinate in the reservoir medium. Both plating and reservoir media originally contained casein hydrolysate (Shepard 1980). However, better colony growth was obtained when ammonium succinate replaced 50 mg/l casein in the plating medium, so that 19 mM NO_3 and 0.65 mM NH_4 were present. Similarly, when reservoir medium containing either 100 mg/l casein or 1.3 mM ammonium succinate +0.5 mM ammonium chloride was tested, it was found that colonies formed earlier and developed faster when the reservoir medium contained succinate. In the presence of medium containing casein, the colonies grew slowly and often did not develop beyond the initial stages. All shoots which were obtained eventually originated from experiments utilizing ammonium succinate. Culture plates were kept initially in a low-light growth cabinet (ca. 6 μE m^2 s, cool white fluorescent tubes) at 26 °C. Cell divisions were observed as early as 3 days after plating in some isolations, but the time varied from 3–14 days. The survival rate averaged 32% following plating, and on average, 16% of the surviving protoplasts developed into small colonies (Fig. 1B).

4 Regeneration of Plants

Approximately 14–21 days after plating, the agarose appeared opaque due to development of many small colonies. The agarose was then layered onto medium C of Shepard (1980) with half-strength iron and minor elements and 1.65 g/l ammonium nitrate instead of ammonium chloride. Growth regulators comprised 0.5 mg/l BA and 0.1 mg/l NAA. The nitrogen source was varied by using either 19 mM NO_3 + 2 mM NH_4 as in Medium-C (Shepard 1980) or 40 mM NO_3 + 20 mM NH_4 as in MS medium. The latter supported better colony growth and was responsible for all shoots that were subsequently obtained. Cultures were exposed to low light (ca. 50 μE m^2 s, cool white fluorescent tubes, 17-h photoperiod) at 26 °C. After 4 weeks the colonies formed a dense layer of growth in the agarose (Fig. 1C), and at this time they were large enough to be transferred individually to medium of the same composition. Once individual calli reached 1–1.5 mm in diameter they showed a survival rate of 80–100% upon transfer to fresh medium. After a further 4 weeks of growth the calli were placed on regeneration medium consisting of MS salts, MS organic constituents with 1 mg/l thiamine-hydrochloride, 3% sucrose, and combinations of BA (0.5, 1, 2, 5 mg/l) with NAA or indole-3-acetic acid (IAA) (0, 0.1, 0.5 mg/l). Approximately 6600 calli from four separate isolations were transferred to the 20 MS combinations. These were selected on the basis that similar media promoted shoot regeneration from seedling hypocotyl sections of *G. canescens* (Kameya and Widholm 1981).

Shoots formed approximately 4 weeks after transfer to regeneration medium (Fig. 1D). Six shoots were obtained from the four experiments, and shoot induction occurred on MS + 2 mg/l BA + 0.1 mg/l NAA or IAA, MS + 0.5 mg/l BA

Fig. 2. Plant of *G. canescens* regenerated from hypocotyl protoplast-derived callus

+0.1 mg/l NAA, and MS+1 mg/l BA+0.5 mg/l NAA. One shoot per 40–80 calli was produced on each successful combination. Shoots were not formed if NAA was replaced by 2,4-D throughout the procedure; if the MS salts were replaced with B5 (Gamborg et al. 1968) media components at the shoot regeneration stage; and if BA was replaced by kinetin in the final stages. Small calli which were transferred directly to the MS shoot regeneration medium without a second 4-week period on Medium-C turned black and died. Alteration of the levels of medium components such as ammonium, nitrate, sucrose, and mannitol was tried in an attempt to increase the efficiency of shoot induction, but these changes did not result in shoot formation. Decreasing the NAA concentration from 1 mg/l to 0.1 mg/l in the reservoir medium at plating led to a denser growth of colonies in the agarose; one shoot developed after the second transfer to Medium-C, but did not continue to grow. Regenerated shoots appeared to originate via organogenesis rather than embryogenesis, as there was no indication of somatic embryo development. The purified protoplast preparations did not appear to contain noticeable amounts of contaminating cell debris, thus the resulting calli were judged to be protoplast rather than cell origin. Although small colonies could be distinguished in the agarose, with such a density of plating it was not possible to ascertain at later stages whether colonies were single-cell in origin or whether they may have arisen through coalition of small groups of cells. The shoots usually branched to give a small clump in the presence of BA. The shoot clump could be removed from the callus and further multiplied by transferal to B5 medium containing BA (0.2 mg/l) and indole-3-butyric acid (IBA) (0.05 mg/l). Shoots 2–3 cm in length were rooted in B5 medium containing 0.5 mg/l IAA. Shoots which could not be rooted readily were grafted onto 2-week-old soybean seedlings and grown in the greenhouse before cuttings were taken and rooted with commercial rooting powder (Fig. 2). The resulting plants were morphologically similar to *G. canescens* and set seed normally.

5 Fusion of Protoplasts

Protoplasts from soybean cell suspension cultures have been successfully fused with leaf mesophyll protoplasts from a number of species (Table 2), using polyethylene glycol (PEG) as the fusing agent. Heteroplasmic fusion products were identified by the combination of dense cytoplasm, amyloplasts, and undifferentiated colorless-plastids from cultured soybean cells and green chloroplasts from mesophyll protoplasts. The green heterokaryons typically became oval in shape with the onset of wall formation, like the parental cultured cells, while leaf mesophyll cells often remained spherical and did not divide upon subsequent culture (Kao and Michayluk 1974; Kao et al. 1974; Kartha et al. 1974; Constabel et al. 1976; Fowke et al. 1977). The percentage of the total protoplast population that was present as heterokaryons varied considerably, depending upon the species and culture conditions, the highest recording being for *Nicotiana glauca* (39%) 24 h after fusion treatment (Kao 1977). First divisions of the heterokaryocytes were observed within approximately 3 days (Kao and Michayluk 1974; Kartha et al. 1974; Constabel et al. 1976), and after 10–12 days of culture, small colonies developed (Fowke et al. 1976, 1977; Kao 1977; Kao and Michayluk 1974; Kartha et al. 1974).

Differential staining of nuclei in heterokaryons was accomplished with carbol fuchsin (Constabel et al. 1975a; Kao 1977; Kao et al. 1974) and aided investiga-

Table 2. Intergeneric protoplast fusion (between cell suspension derived protoplasts of soybean and leaf-derived protoplasts of other genera mentioned below) products capable of cell division

Protoplast origin	References
Brassica napus	Kartha et al. (1974)
Caragana arborescens	Constabel et al. (1975b)
Colchicum autumnale	Constabel et al. (1976)
Hordeum vulgare	Kao and Michayluk (1974); Kao et al. (1974)
Medicago sativa	Constabel et al. (1975b)
Melilotus officinalis	Constabel et al. (1975b); Fowke et al. (1976)
Nicotiana glauca	Contabel et al. (1976); Kao (1977); Wetter and Kao (1980)
N. langsdorfii	Constabel et al. (1976)
N. rustica	Constabel et al. (1976)
N. tabacum	Chien et al. (1982); Constabel et al. (1976)
Oryza sativa	Niizeki et al. (1986)
Pisum sativum	Constabel et al. (1975a, b); Constabel et al. (1976); Fowke et al. (1977); Kao et al. (1974)
Vicia hajastana	Kao et al. (1974)
Zea mays	Kao et al. (1974)

tion of nuclear behavior. Fusion products between soybean and *Melilotus* (Fowke et al. 1976) and soybean and *Pisum* (Fowke et al. 1977) were found to lose chloroplasts with each successive division of the fusion product; chloroplasts appeared to deteriorate and did not divide to keep pace with cell divisions. Heterokaryons combining one or more nuclei from the parental species typically showed synchronized, or nearly so, nuclear divisions (Kao et al. 1974). Nuclear fusion usually occurred during the first mitotic division after protoplast fusion (Kao et al. 1974; Constabel et al. 1975a), although evidence for nuclear fusion prior to mitosis was occasionally seen (Constabel et al. 1975a; Kao 1977). Ultrastructural studies suggested that the narrow nuclear envelope bridges seen in heterokaryons were involved in the nuclear fusion process (Fowke et al. 1977).

Long-term studies of chromosome behavior in somatic hybrids have been carried out with soybean-*Nicotiana glauca* (Kao 1977) and soybean-*Nicotiana tabacum* (Chien et al. 1982) fusions. In soybean-*N. glauca* heterokaryocytes, chromosomal behavior of soybean and *N. glauca* was not synchronous in the first few cell generations. Chromosomal abnormalities such as anaphase bridges, long chromosomes, and chromosome fragments were found in the cell population approximately 1 month after fusion. The number of *N. glauca* chromosomes in the hybrids decreased with time in culture; remaining *N. glauca* chromosomes showed altered morphology and standard chromosomes were observed rarely, while no obvious changes were seen in the soybean chromosomes. The continued presence of *N. glauca* chromosomes in three out of five hybrid cell lines after 8 months of culture was confirmed by gel electrophoresis analysis of alcohol dehydrogenase and aspartate aminotransferase isoenzyme patterns: two hybrid lines contained the hybrid isoenzyme pattern for aspartate aminotransferase, and a third carried both parental bands for alcohol dehydrogenase (Wetter 1977). After 3 years of culture, a few *N. glauca* chromosomes with modified structure were observed in soybean-*N. glauca* cell lines. Two "back-fusions" of these cell lines with leaf mesophyll protoplasts of *N. glauca* were carried out approximately 27 and 34 months after the initial fusion (Wetter and Kao 1980). This led to cell lines with a higher number of *N. glauca* chromosomes and fewer chromosomal abnormalities than the hybrid parent. Many *N. glauca* chromosomes were still present after 6 months of culturing the back-fusion derivatives, suggesting greater stability of the *N. glauca* chromosomes. The back-fusion technique was thus proposed as a technique that might aid in stabilizing somatic hybrids.

Heterokaryocytes of soybean and *Nicotiana tabacum* were also found to eliminate *Nicotiana* chromosomes during the early cell divisions (Chien et al. 1982). Of a total of 21 hybrid cell lines selected for long-term culture, after 6–7 months five cell lines had retained more than half the *N. tabacum* chromosomes, five cell lines had all but eliminated the *N. tabacum* chromosomes, and the remainder were intermediate. Again there were no obvious changes in the soybean chromosomes, while abnormal *N. tabacum* chromosomes were observed. Somatic hybrids of soybean-*N. tabacum* thus appeared to be more stable than soybean-*N. glauca* hybrids.

Hybrid callus resulting from fusion of rice and soybean protoplasts was described recently by Niizeki et al. (1986). Sorting-out of chloroplasts following fusion could be inferred from changes in callus color coupled with sensitivity to streptomycin (see Chap. III.10, this Vol.).

Spontaneous fusion of protoplasts during isolation was observed in soybeans (Miller et al. 1971) when soybean cell suspension cultures were used as a protoplast source. Up to 25% of protoplasts were found to be multinucleate; these were mostly dinucleate, but up to six nuclei were frequently observed. After 24 h of culture, both mono- and di-nucleates divided. The majority of multinucleates were synchronized with respect to nuclear division, and about 10% of fused protoplasts showed evidence of nuclear fusion prior to nuclear division. Ultrastructural studies by means of light and electron microscopy (Fowke et al. 1975) confirmed phragmoplast and cell plate formation, but did not establish whether all nuclei were finally separated by cell plates with subsequent formation of a multicellular aggregate.

Identification of fusion products has usually been facilitated by fusing colorless suspension cells with plastid-bearing leaf cells, so that heterokaryons can be distinguished by their characteristic morphology. Galbraith and Galbraith (1979) report the successful labelling of cell suspensions with fluorescent probes; suspension culture cells could be labelled with either fluorescein isothiocyanate or rhodamine B prior to protoplast isolation without adverse effects on protoplast viability. Fusion products could be readily identified by the presence of both fluorescent probes.

Immunological methods have been applied to protoplast fusion by Hartmann et al. (1973). Protoplasts from cell suspension cultures of *Vicia hajastana* and soybean could be tightly agglutinated either separately or in combination by immune sera prepared in rabbits; protoplasts remained viable and divided after antibody treatment. Although *Vicia* and soybean nuclei could be distinguished by differential staining, there was no clear evidence for the formation of heterokaryons using this method.

6 Miscellaneous Studies

Soybean protoplasts have been used for diverse investigations including cell wall synthesis, ultrastructural, physiological, and pathological studies.

Fowke et al. (1974) compared cell division between cultured cells and their protoplast derivatives, and found the process to be similar in both. The only consistent difference was the more rapid formation of the phragmoplast and development of a thicker cell plate in protoplasts, indicating that cross wall formation during cytokinesis was slightly modified. The distribution of carbohydrate residues on the plasma membrane was studied by labelling soybean protoplasts with concanavalin-A (Williamson et al. 1976), which then bound haemocyanin. Electron microscope visualization of the haemocyanin distribution after different treatments indicated that carbohydrate binding sites were evenly distributed but mobile within the plane of the membrane. Cell wall formation by protoplasts has been examined by Hanke and Northcote (1974) and Klein et al. (1981). Hanke and Northcote reported that ^{14}C glucose was incorporated into protein, starch, and cellulose, with a small amount incorporated into an acidic pectin, during the first 20 h of wall regeneration. Klein et al. (1981), however, reported the synthesis of a broad spectrum of polysaccharide polymers during the first 3 h of wall regeneration. Using ^{14}C-labelled glucose as a tracer, they also established the unequivocal synthesis of new cellulose during the first few hours of wall regeneration following removal from the enzyme digestion mixture. After 1 h of culture in wall regeneration medium, fibrils could be visualized by electron microscopy on the surface of the plasma membrane.

Chu and Lark (1976) examined cell cycle parameters and DNA synthesis in cell suspensions and protoplasts as a prerequisite for isolating conditional lethal mutants that could be used in genetic studies. Cress et al. (1978) have studied DNA replication in cultured cells and protoplasts. Conversion of cells to protoplasts prior to DNA isolation was found to be beneficial, as it avoided problems with DNA shear and nuclease digestion during isolation of DNA replication intermediates. DNA binding experiments by Kool and Pelcher (1978) showed that binding of ^{14}C-labelled *Salmonella* DNA or ^{3}H-labelled soybean DNA to cultured soybean cells was markedly affected by incubation conditions. Removal of the cell wall to form protoplasts resulted in a tenfold decrease in DNA that was bound in a comparable amount of cells. The DNA was found not be associated with the protoplasts but with cell wall contaminants, and it was concluded that these accounted for the majority of DNA binding. Uptake of exogenous DNA by protoplasts was investigated by Cress (1982). Soybean cell suspension cultures were synchronized with 5-fluorodeoxyuridine, so that cells accumulated in the G_1/S boundary; protoplasts were isolated at varying times after FUdR removal and assayed for plasmid DNA uptake capacity. Protoplasts isolated from S phase cells showed a 2.5–4.5 higher DNA uptake over a 1-h incubation period compared with G_1 or G_2 phase protoplasts.

Soybean leaf mesophyll protoplasts have been used for physiological studies. Franceschi et al. (1984) developed a procedure for isolation and purification of protoplasts from two soybean leaf cell types, mesophyll cells and paraveinal mesophyll cells (PVM), the latter specialized for nitrogen metabolism and storage. Differences in buoyant density, microscope examination and immunochemical assay of PVM glycoprotein indicated that the preparations were pure.

Protoplasts isolated from developing soybean cotyledons have been subjects for analyses of the mechanisms of sugar transport (Lin 1985a, b; Lin et al. 1984; Schmitt et al. 1984). Protoplasts were considered advantageous, since the kinetics of uptake could be studied without involving bulk diffusion and tissue penetration; the cell membrane was readily accessible for challenging with various sugars, sugar analogs, membrane-modifying reagents and binding probes to facilitate carrier identification and isolation. Sucrose uptake by protoplasts from cotyledons at the stage of rapid seed growth resulted from three different mechanisms: a saturable, carrier-mediated, energy-dependent mechanism; a carrier-mediated but non-saturable, or linear, mechanism; and simple diffusion (Lin 1985a; Lin et al. 1984; Schmitt et al. 1984). Further examination of the linear, non-diffusive sucrose mechanism indicated that a proton-sucrose co-transport was involved in sucrose uptake (Lin 1985b).

Uptake of exogenous macromolecules into protoplasts requires that they be permeabilized, and various agents which induce permeabilization have been tested in soybean protoplasts derived from suspension cultures (Cutler and Saleem

1987; Saleem and Cutler 1986). Hypotonic shock and electroporation were both found to be effective in allowing uptake of the fluorescent dye, calcein, although the latter induced ten times more uptake than hypotonic shock treatment. Protoplasts which surviced the treatment were still able to undergo cell division, although at a lower frequency than untreated protoplasts.

Protoplasts isolated from photoautotrophic soybean cell cultures grown in the absence of sucrose and in a 5% carbon dioxide atmosphere under continuous light were compared with protoplasts from heterotrophic cultures (Chowdhury and Widholm 1985). Protoplasts from the photoautotrophic cultures retained the characteristics of the parental culture from which they were derived, and showed greater sensitivity to atrazine, a photosynthetic inhibitor, than heterotrophic protoplasts and cells which were less dependent on photosynthesis.

Parameters for the successful incubation of soybean cell suspension protoplasts with southern bean mosaic virus and cow pea mosaic virus have been examined by Jarvis and Murakishi (1980). Cell suspension-derived protoplasts were considered advantageous in that cell suspensions represented a reliable source of cells for which the growth conditions could be precisely defined and regulated, and sterility could be easily maintained. Soybean protoplasts allowed synchronous infection and replication under sterile conditions without antibiotics and with a high degree of reproducibility, and were considered valuable subjects for extended studies of virus-cell interactions during cell wall regeneration and callus formation.

Successful induction of crown galls on soybean by *Agrobacterium tumefaciens* was achieved by Pederson et al. in 1983. Suspension cultures were derived from isolated crown galls and used as a source of protoplasts; the transformed protoplasts were able to regenerate cell walls, divide, and produce callus in the absence of an exogenous supply of hormones, while untransformed control protoplasts could not sustain cell division in hormone-free medium.

7 Conclusions and Prospects

The field of protoplast culture and fusion has progressed rapidly in recent years, and new techniques to manipulate protoplasts in culture continue to be developed. Although it is evident from published literature that substantial use has been made of soybean protoplasts for a variety of experimental purposes, protoplast studies have not as yet made an impact on the agronomic aspects of soybeans. The potential that protoplast technology holds for soybeans will only be fully realized when a reproducible and efficient regeneration system from protoplasts becomes available. Protoplast to plant regeneration procedures developed successfully for wild *Glycine* species are still in their infancy and have not so far been directly transferable to the cultivated soybean. Some possible ways in which these techniques may benefit crop improvement programs include the following: (1) Insertion of DNA coding for specific desired genes via techniques such as electroporation or electrofusion. (2) Selection for mutations in large populations of single cells or protoplasts that render cells resistant to metabolites such as plant pathogen toxins, thereby potentially increasing genetic variability. (3) Utilization of haploid cell cultures for induction and detection of mutants. (4) Incorporation of selected cytoplasmic organelles from wild related species with soybean nuclei to produce cybrids, a technique which holds promise as a method for developing a cytoplasmic male sterility system in soybean. (5) Insertion of part of a genome from a related species carrying agronomically useful traits not present in soybean germplasm, as a means of exploiting the variation present in wild germplasm.

8 Protocol

(1) Surface sterilize seeds 5 min in absolute ethanol, 10 min in 20% Clorox bleach, then wash three times in sterile water. Germinate seeds 7 days on agar-solidified MS medium at 26 °C in the dark.

(2) Excise hypocotyls and cut into 2–5 mm sections. Immerse overnight at 4 °C in osmoticum containing 0.2 M sucrose, 0.1 M mannitol, R-medium salts and organic addenda (Shepard 1980), 1 mg/l NAA + 0.4 mg/l BA. Digest hypocotyl tissue for 5 h at 28 °C on a rotating platform (approximately 50 rpm) in an enzyme mixture composed of 0.5% Macerozyme R-10, 5% Cellulase Onozuka R-10, dissolved in 1/4 strength R-medium major salts, 0.3 M sucrose, 5 mM MES at pH 5.6.

(3) Filter the digest through two layers of cheesecloth into 20-ml narrow-necked ampoules and centrifuge at 1000 rpm for 10 min. Resuspend floating layer of protoplasts in a rinse solution of R-medium major salts and 0.3 M sucrose, pH 5.6. Recentrifuge 10 min at 1000 rpm. Remove floating protoplasts, resuspend in a few ml rinse solution.

(4) Dilute protoplasts to a density of 10^5/ml in plating medium; R-medium (Shepard 1980) major salts, CL-medium (Shepard 1980) osmoticum and organic addenda containing 0.65 mM ammonium succinate instead of casein, 1/2 strength iron and minor elements of CL-medium, 1 mg/l NAA, 0.4 mg/l BA, solidified with 0.45% low temperature gelling agarose, at pH 5.6. Reservoir medium contains the same salts, organic addenda and growth regulators as the plating medium except that casein is replaced by 1.3 mM ammonium succinate + 0.5 mM ammonium chloride, 0.1 M sucrose and 0.1 M mannitol as osmoticum, and 0.6% washed agar instead of agarose.

(5) Place 10 ml of reservoir medium into each of two opposite sectors of a quadrant Petri plate that has slits cut along the base of the dividing sections (Shepard 1980) and allow to solidify. Pipette 4 ml of plating medium containing protoplasts into the two remaining opposite sectors of the Petri plate. Maintain culture plates at 26 °C in low light (6 μE m^2 s).

(6) After 14–21 days when small colonies are visible in the agarose, remove agarose with a spatula and layer onto Medium-C (Shepard 1980) with 1/2 strength iron and minor elements and 1.65 g/l ammonium nitrate instead of ammonium chloride, 0.5 mg/l BA and 0.1 mg/l NAA. Maintain cultures in low light (50 μE m^2 s, 17-h photoperiod) at 26 °C.

(7) After 4 weeks transfer individual colonies to fresh medium of the same composition. After a further 4 weeks transfer calli to regeneration medium: MS + BA (0.5, 1, 2, mg/l) + NAA (0.1, 0.5 mg/l) or IAA (0.1 mg/l). Shoots which form on this medium can be transferred to B5 + BA (0.2 mg/l) + IBA (0.05 mg/l) to multiply.

(8) Once shoots reach 2–3 cm in length they can be rooted on B5 medium + IAA (0.5 mg/l). When the root system is well developed, the plantlets can be transferred to vermiculite in a growth chamber before being transplanted to the greenhouse.

References

Anonymous (1986) Soya bluebook. American Society of Agronomy, St. Louis, Missouri, USA

Bernard RL (1983) Soybean germplasm, breeding, and genetic activities in the United States. In: Irwin BJ, Sinclair JB, Wang J (eds) Soybean research in China and the United States. Proc of the First China/USA Soybean Symp and Working Group Meeting, INTSOY Series Number 25, Univ Illinois at Urbana-Champaign, Urbana, USA, pp 19–25

Bojsen KM, Wyndaele R (1985) Influence of lowered temperature storage on growth and protoplast isolation of soybean callus. J Plant Physiol 118:95–103

Byrne MC, McDonnell RE, Wright MS, Carnes MG (1987) Strain and cultivar specificity in the *Agrobacterium*-soybean interaction. Plant Cell Tissue Organ Cult 8:3–15

Chien YC, Kao KN, Wetter LR (1982) Chromosomal and isozyme studies of *Nicotiana tabacum-Glycine max* hybrid cell lines. Theor Appl Genet 62:301–304

Chowhury VK, Widholm JM (1985) Callus production from photoautotrophic soybean cell culture protoplasts. Plant Cell Rep 4:289–292

Chu Y, Lark KG (1976) Cell-cycle parameters of soybean (*Glycine max* L.) cells growing in suspension culture: suitability of the system for genetic studies. Planta 132:259–268

Collins GB, Phillips GC (1982) In vitro tissue culture and plant regeneration in *Trifolium pratense* L. In: Earle ED, Demarly Y (eds) Variability in plants regenerated from tissue culture, Praeger Scientific, NY, pp 22–34

Constabel F, Dudits D, Gamborg OL, Kao KN (1975a) Nuclear fusion in intergeneric heterokaryons. A note. Can J Bot 53:2092–2095

Constabel F, Kirkpatrick JW, Kao KN, Kartha KK (1975b) The effect of canavanine on the growth of cells from suspension cultures and on intergeneric heterokaryocytes of canavanine-sensitive and tolerant plants. Biochem Physiol Pflanz 168:319–325

Constabel F, Weber G, Kirkpatrick JW, Pahl K (1976) Cell division of intergeneric protoplast fusion products. Z. Pflanzenphysiol 79:1–7

Cress DE (1982) Uptake of plasmid DNA by protoplasts from synchronized soybean cell suspension cultures. Z Pflanzenphysiol 105:467–470

Cress DE, Jackson PJ, Kadouri A, Chu YE, Lark KG (1978) DNA relication in soybean protoplasts and suspension-cultured cells: comparison of exponential and fluorodeoxyuridine synchronized cultures. Planta 143:241–253

Cutler AJ, Saleem M (1987) Permeabilizing soybean protoplasts to macromolecules using electroporation and hypotonic shock. Plant Physiol 83:24–28.

Davey MR, Cocking EC, Bush E (1973) Isolation of legume root nodule protoplasts. Nature 244:460–461

Fowke LC, Bech-Hansen CW, Constabel F, Gamborg OL (1974) A comparative study on the ultrastructure of cultured cells and protoplasts of soybean during cell division. Protoplasma 81:189–203

Fowke LC, Bech-Hansen CW, Gamborg OL, Constabel F (1975) Electronmicroscope observations of mitosis and cytokinesis in multinucleate protoplasts of soybean. J Cell Sci 18:491–507

Fowke LC, Rennie PJ, Kirkpatrick JW, Constabel F (1976) Ultrastructure of fusion products from soybean cell culture and sweet clover leaf protoplasts. Planta 130:39–45

Fowke LC, Constabel F, Gamborg OL (1977) Fine structure of fusion products from soybean cell culture and pea leaf protoplasts. Planta 135:257–266

Franceschi VR, Ku MSB, Wittenbach VA (1984) Isolation of mesophyll and paraveinal mesophyll protoplasts from soybean leaves. Plant Sci Lett 36:181–186

Frearson EM, Power JB, Cocking EC (1973) The isolation culture and regeneration of *Petunia* leaf protoplasts. Dev Biol 33:130–137

Galbraith DW, Galbraith JEC (1979) A method for identification of fusion of plant protoplasts derived from soybean tissue cultures. Z Pflanzenphysiol 93:149–158

Gamborg OL (1975) Callus and cell culture. In: Gamborg OL, Wetter LR (eds) Plant tissue culture methods, Nat Res Council, Saskatoon, Canada, pp 1–10

Gamborg OL, Miller RA, Ojima K (1968) Nutrient requirements of suspension cultures of soybean root cells. Exp Cell Res 50:151–158

Gamborg OL, Shyluk JP, Fowke LC, Wetter LR, Evans DA (1979) Plant regeneration from protoplasts and cell cultures of *Nicotiana tabacum* sulfur mutant (Su/Su). Z Pflanzenphysiol 95:255–264

Gamborg OL, Davis BP, Stahlhut RW (1983) Cell division and differentiation in protoplasts from cell cultures of *Glycine* species and leaf tissue of soybean. Plant Cell Rep 2:213–215

Hammatt N, Kim H-I, Davey MR, Nelson RS, Cocking EC (1987) Plant regeneration from cotyledon protoplasts of *Glycine canescens* and *G. clandestina.* Plant Sci 48:129–135

Hanke DE, Northcote DH (1974) Cell wall formation by soybean callus protoplasts. J Cell Sci 14:29–50

Hartmann JX, Kao KN, Gamborg OL, Miller RA (1973) Immunological methods for the agglutination of protoplasts from cell suspension cultures of different genera. Planta 112:45–56

Hymowitz T (1970) On the domestication of the soybean. Econ Bot 24:408–421

Hynowitz T, Newell CA (1980) Taxonomy, speciation, domestication, dissemination, germplasm resources and variation in the genus *Glycine*. In: Summerfield RJ, Bunting AH (eds) Advances in legume science. Royal Botanic Gardens, Kew, England, pp 251–264

Hymowitz T, Singh RJ (1984) A soybean × *Glycine tomentella* hybrid: progress and problems. Soybean Genet Newslett 11:90

Jarvis NP, Murakishi HH (1980) Infection of protoplasts from soybean cell culture with southern bean mosaic and cowpea mosaic viruses. J Gen Virol 48:365–376

Kameya T, Widholm J (1981) Plant regeneration from hypocotyl sections of *Glycine* species. Plant Sci Lett 21:289–294

Kao KN (1977) Chromosomal behaviour in somatic hybrids of soybean-*Nicotiana glauca*. Mol Gen Genet 150:225–230

Kao KN, Michayluk MR (1974) A method for high-frequency intergeneric fusion of plant protoplasts. Planta 115:355–367

Kao KN, Michayluk MR (1975) Nutritional requirements for growth of *Vicia hajastana* cells and protoplasts at a very low population density in liquid media. Planta 126:105–110

Kao KN, Keller WA, Miller RA (1970) Cell division in newly formed cells from protoplasts of soybean. Exp Cell Res 62:338–340

Kao KN, Gamborg OL, Miller RA (1971) Cell divisions in cells regenerated from protoplasts of soybean and *Haplopappus gracilis*. Nature New Biol 232:124

Kao KN, Constabel F, Michayluk MR, Gamborg OL (1974) Plant protoplast fusion and growth of intergeneric hybrid cells. Planta 120:215–227

Kartha KK, Gamborg OL, Constabel F, Kao KN (1974) Fusion of rapeseed and soybean protoplasts and subsequent division of heterokaryocytes. Can J Bot 52:2435–2436

Kerns HR, Barwale UB, Meyer MM, Widholm JM (1986) Correlation of cotyledonary node shoot proliferation and somatic embryoid development in suspension cultures of soybean (*Glycine max* L. Merr.). Plant Cell Rep 5:140–143

Klein AS, Montezinos D, Delmer DP (1981) Cellulose and 1,3-glucan synthesis during the early stages of wall regeneration in soybean protoplasts. Planta 152:105–114

Kool AJ, Pelcher LE (1978) Conditions that affect binding of DNA by cultured plant cells and protoplasts: an autoradiographic analysis. Protoplasma 97:71–84

Lin W (1983) Isolation of mesophyll protoplasts from mature leaves of soybeans. Plant Physiol 73:1067–1069

Lin W (1985a) Energetics of sucrose transport into protoplasts from developing soybean cotyledons. Plant Physiol 78:41–45

Lin W (1985b) Linear sucrose transport in protoplasts from developing soybean cotyledons. Plant Physiol 78:649–651

Lin W, Schmitt MR, Hitz WD, Giaquinta RT (1984) Sugar transport into protoplasts isolated from developing soybean cotyledons. I. Protoplast isolation and general characteristics of sugar transport. Plant Physiol 75:936–940

Lu DY, Cooper-Bland S, Pental D, Cocking EC, Davey MR (1983) Isolation and sustained division of protoplasts from cotyledons of seedlings and immature seeds of *Glycine max* L. Z. Pflanzenphysiol 111:389–394

Miller CO (1968) Naturally-occurring cytokinins. In: Wightman F, Setterfield G (eds) Biochemistry and physiology of plant growth substances. Runge, Ottawa, Canada, pp 33–45

Miller RA, Gamborg OL, Keller WA, Kao KN (1971) Fusion and division of nuclei in multinucleated soybean protoplasts. Can J Genet Cytol 13:347–353

Mounts TL, Wolf WJ, Martinez WH (1987) Processing and utilization. In: Wilcox JR (ed) Soybeans: improvement, production, and uses. American Society of Agronomy No. 16, 2nd edn. Madison, Wisconsin, pp 819–866

Murashige T, Skoog F (1962) A revised medium for rapid growth and bioassays with tobacco tissue cultures. Physiol Plant 15:473–497

Myers JR, Kysely W, Lazzeri PA, Hildebrand DF, Collins GB (1986) Protoplast isolation and culture of *Glycine* species with plant regeneration of *G. canescens*. In: Somers DA, Gengenbach BG, Biesboer DD, Hackett WP, Green CE (eds) Abst. VI Int Cong Plant Tissue and Cell Culture, Aug. 1986. Univ Minnesota Press, Minneapolis, Minnesota, USA, p 269

Newell CA, Luu HT (1985) Protoplast culture and plant regeneration in *Glycine canescens* F. J. Herm. Plant Cell Tissue Organ Cult 4:145–149

Newell CA, Delannay X, Edge ME (1987) Interspecific hybrids between the soybean *Glycine max* (L.) Merr. and wild perennial relatives. J Hered 78:301–306

Niizeki M, Tanaka M, Saito K (1986) Response of somatic hybrid calli between rice and soybean and parent calli to streptomycin. In: Somers DA, Gengenbach BG, Biesboer DD, Hackett WP, Green CE (eds) Abstracts. VI Int Cong Plant Tissue and Cell Culture, Univ of Minnesota Press, Minneapolis, Minnesota, USA, p 82

Oelck MM, Rao PS, Ozias-Akins P, Schieder O (1983) Protoplast regeneration of some legume species. In: Potrykus I, Harms CT, Hinnen A, Hutter R, King PJ, Shillito RD (eds) Protoplasts 1983. Poster Proc, Sixth Int Protoplast Symp, Basel, Switzerland. Experientia Suppl. 45 pp 50–51

Pederson HC, Christiansen J, Wyndaele R (1983) Induction and in vitro culture of soybean crown gall tumors. Plant Cell Rep 2:201–204

Rees RT, Cobb AH, Pallett KE (1985) A developmental study of *Glycine max* cell and protoplast isolation in relation to leaf age and photosynthetic competence. Physiol Plant 65:251–256

Saleem M, Cutler AJ (1986) Preparation and characterization of chemically and osmotically permeabilized soybean (*Glycine max* (L.) Merr.) protoplasts. J. Plant Physiol 124:11–21

Schmitt MR, Hitz WD, Lin W, Giaquinta RT (1984) Sugar transport into protoplasts isolated from developing soybean cotyledons: II. Sucrose transport kinetics, selectivity, and modeling studies. Plant Physiol 75:941–946

Schwenk FW, Pearson CA, Roth MR (1981) Soybean mesophyll protoplasts. Plant Sci Lett 23:153–155

Shepard JF (1980) Mutant selection and plant regeneration from potato mesophyll protoplasts. In: Rubenstein I, Gengenbach BG, Phillips RL, Green CE (eds) Genetic improvement of crops, Emergent techniques. University of Minnesota Press, Minneapolis, Minnesota, USA, pp 185–219

Tricoli DM, Hein MB, Carnes MG (1986) Culture of soybean mesophyll protoplasts in alginate beads. Plant Cell Rep 5:334–337

Wetter LR (1977) Isoenzyme patterns in soybean-*Nicotiana* somatic hybrid cell lines. Mol Gen Genet 150:231–235

Wetter LR, Kao KN (1980) Chromosome and isoenzyme studies on cells derived from protoplast fusion of *Nicotiana glauca* with *Glycine max-Nicotiana glauca* cell hybrids. Theor Appl Genet 57:273–276

Williamson FA, Fowke LC, Constabel FC (1976) Labelling of concanavalin-A sites on the plasma membrane of soybean protoplasts. Protoplasma 89:305–316

Xu ZH, Davey MR, Cocking EC (1982) Callus formation from root protoplasts of *Glycine max* (soybean). Plant Sci Lett 24:111–115

Zieg RG, Outka DE (1980) The isolation, culture and callus formation of soybean pod protoplasts. Plant Sci Lett 18:105–114

II.6 Regeneration of Plants from Cabbage (*Brassica oleracea* var. *capitata*) Protoplasts

Y. YAMASHITA[1] and K. SHIMAMOTO

1 Introduction

Cultivated cabbage (*Brassica oleracea* var. *capitata* L.) belongs to 2n = 18 *Brassica* and is thought to originate from West European wild species of *B. oleracea*. It has a long history as a vegetable crop and there is evidence that the Romans cultivated head cabbages (Snogerup 1980). Cabbage is one of the most important vegetables and ranks second next to tomato in cultivated areas as well as in production in the world (FAO 1982). The total cultivated area for cabbage is 1686×10^3 ha and tomato 2483×10^3 ha and production is $36{,}225 \times 1000$ Mt for cabbage and $53{,}892 \times 1000$ Mt for tomato. The distribution and production of cabbage in the world is given in Table 1 (see also Leike 1988).

Protoplast studies of cabbage started rather recently. It has been observed that *Brassica* mesophyll protoplasts tend to aggregate and release brown particles into the medium. In order to overcome this problem, tissues other than leaf have been sought for protoplast sources. Cotyledons (Liu et al. 1982) and seedling roots (Xu et al. 1982) were successfully used for protoplast isolation, and in both studies, protoplast division and shoot formation were described. More recently, improved procedures for the culture of cabbage mesophyll protoplasts have been reported. Bidney et al. (1983) applied the X-plate system (Shepard 1980) for cabbage protoplasts and were able to regenerate whole plants. In addition to the use of the X-plate system, precise growth conditions of mother plants and preconditioning of leaf materials were essential for successful regeneration of whole plants. An agarose-bead type culture (Shillito et al. 1983) was introduced for cabbage (Yamashita and Shimamoto 1986) and whole plants were regenerated from mesophyll protoplasts of green houseplants. Conditions for the formation of subprotoplasts in cabbage have been also reported (Vatsya and Bhaskaran 1981).

Successful somatic hybridization between cabbage and two different varieties of *B. campestris* have been reported (Schenck and Röbbelen 1982). The selection of hybrids in this study is based on the observation that protoplasts from both parents were not able to divide and form colonies under the conditions used. Recent progress in protoplast manipulation of cabbage should stimulate further studies on protoplast fusion and gene transfer in the near future.

[1] Plantech Research Institute, 1000 Kamoshida, Midori-ku, Yokohama 227, Japan

Biotechnology in Agriculture and Forestry, Vol. 8
Plant Protoplasts and Genetic Engineering I (ed. by Y. P. S. Bajaj)

Table 1. The distribution and production of cabbage in the world. (FAO 1982)

Country or continent	Area		Production	
	10^3 ha	%	10^3 Mt	%
Asia	804	47.7	15587	43.0
China	433	25.7	6025	16.6
Japan	105	6.2	3750	10.4
India	79	4.7	478	1.3
Korea	79	4.7	3975	11.0
others	108	6.4	1359	3.8
Europe	317	18.8	7767	21.4
Poland	59	3.5	1220	3.4
Yugoslavia	48	2.8	781	2.2
others	210	12.5	5766	15.9
USSR	412	24.4	9600	26.5
USA	73	4.3	1507	4.2
Other	80	4.7	1764	4.9
Total	1686	100.0	36225	100.0

Table 2. Enzyme solutions for the isolation of cabbage protoplasts

	Mesophyll	Hypocotyl
Cellulase Onozuka R10	0.5%	0
Cellulase Onozuka RS	0	1.0%
Pectolyase Y23	0.004%	0.01%
Potassium dextran sulphate	0.1%	0.1%
Mannitol	0.5 M	0
Sucrose	0	0.4 M

2 Isolation of Protoplasts

Leaf. Young leaves collected from 30–50-day-old plants grown in a greenhouse or growth chamber are suitable for protoplast isolation. Plants are grown in 25-cm pots and watered daily. Plants older than the above materials give similar protoplast yield, but their viability decreases considerably. Enzyme solutions used for protoplast isolation are described in Table 2.

Hypocotyl. Hypocotyls from 3–5-day-old seedlings are used. Vigour of seedlings influences the yield of protoplasts and normally $1-1.5\times10^4$ protoplasts per seedling are obtained. Another factor affecting protoplast yield is the size of the hypocotyl segments incubated with the enzyme solution. Hypocotyls should be cut into segments ca. 1 mm in length, which facilitates entry of enzymes into the inner cells of the hypocotyls.

Purification of Protoplasts. The protoplasts isolated from leaf and hypocotyl segments were further purified. For purification of mesophyll protoplasts, a low speed of centrifugation generally separates intact protoplasts from broken protoplasts and cell debris. For hypocotyl protoplasts, a flotation method is used and it routinely gives pure populations of intact protoplasts.

3 Culture of Protoplast (Figs. 1–9)

Mesophyll Protoplasts. Use of agarose-bead type culture is essential in order to induce sustained division of protoplasts. When cultured in liquid medium, mesophyll protoplasts often aggregate, quickly become brown and die. Medium components greatly influence the frequency of first division. Particularly, a high concentration of ammonium ion in the medium decreases the division frequency (Table 3). Although the presence of ammonium ion is inhibitory for the first division of the protoplasts, it is required to obtain further cell divisions. Therefore, in order to recover colonies, use of two media in sequence is needed. The media are changed when most of the viable protoplasts have entered the division phase. The timing of medium change is critical to reproducibly obtain high colony-forming efficiency. After growth in bead-type culture for 4 weeks, pieces of agarose containing growing cell colonies are transferred onto solid C-1 medium for further growth. Several cultivars were examined for their capacities to form colonies and no appreciable differences were observed.

Hypocotyl Protoplasts. Compared to mesophyll protoplasts, hypocotyl protoplasts are easier to handle and are able to divide in a broader range of media. K3 or KM8p is suitable for the induction of first division which starts on day 2, as well as for sustained division. Hypocotyl protoplasts can be cultured in liquid medium as well as in agarose beads, while the latter gives a slightly slower growth rate. Hypocotyl protoplasts develop loose cell clusters in liquid culture, but in agarose beads semi-compact colonies are formed. Two- to three-week-old colonies can be transferred onto solid K3-1 medium for further growth.

Genotypic differences in colony-forming frequency are tested by using more than 20 cultivars, and we observed no significant differences among them. Generally 20–30% of protoplasts divided in liquid medium.

Table 3. Effects of different media on protoplast division

Media	Division frequency (%)		Reference
	Expt. 1	Expt. 2	
1. MS	7.6	12.8	Murashige and Skoog (1962)
2. MS-NH_4NO_3	20.1	16.2	Murashige and Skoog (1962)
3. K3	8.5	13.1	Nagy and Maliga (1976)
4. KM8P	0.03	NT	Kao and Michayluk (1975)
5. NN67	0.0	NT	Nitsch and Nitsch (1967)

NT = not tested

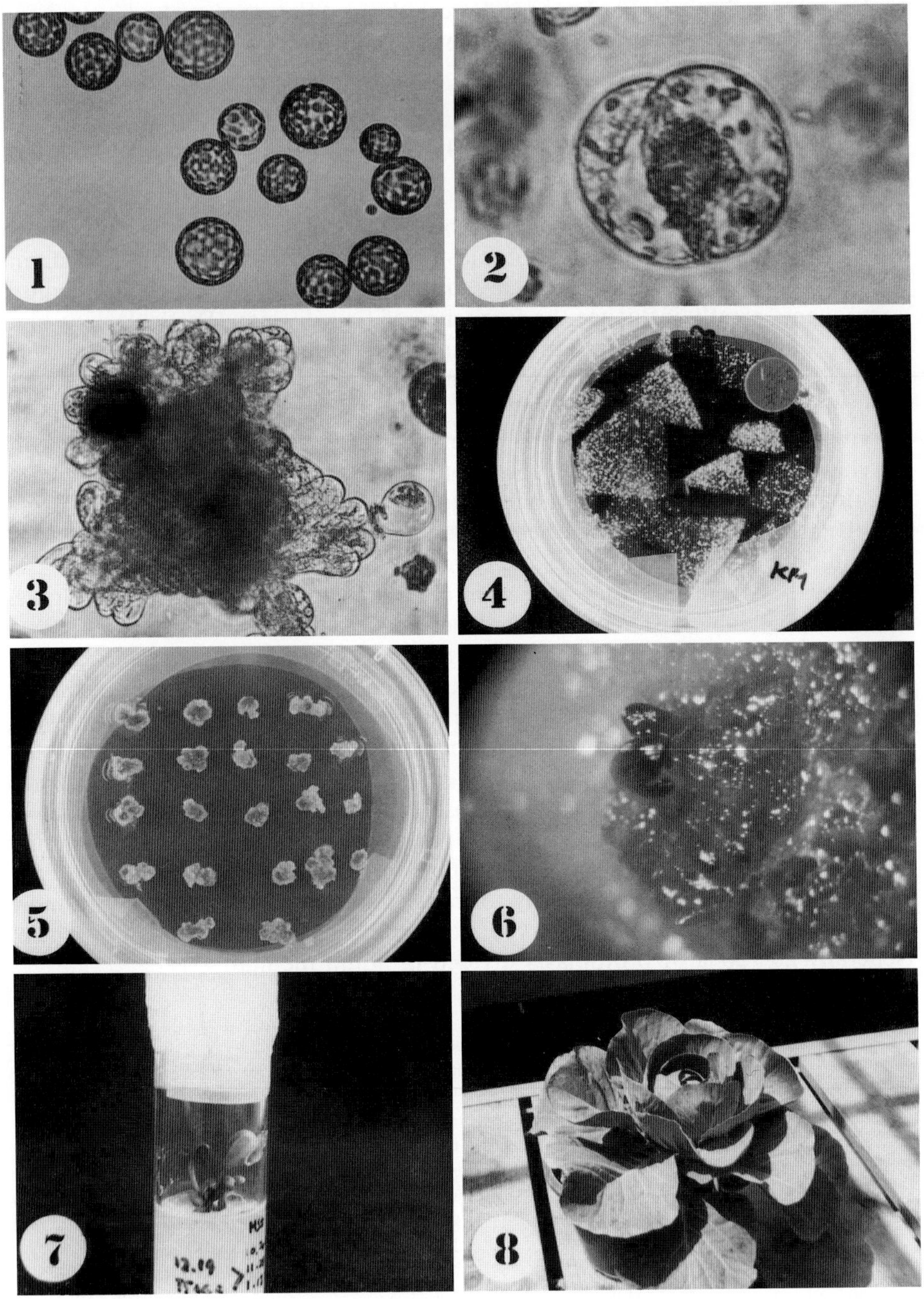
1
2
3
4
5
6
7
8

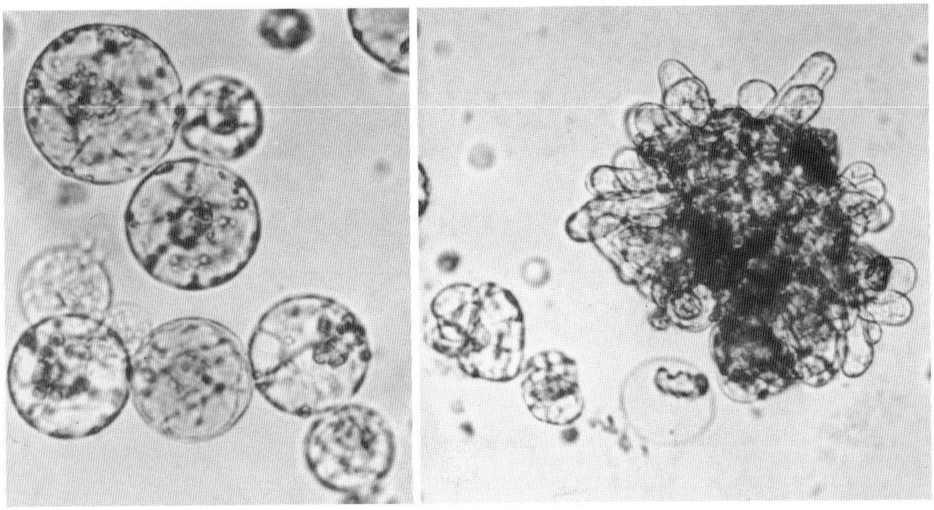

Fig. 9. Culture of cabbage hypocotyl protoplasts. **A** Freshly isolated protoplasts. **B** Colonies derived from protoplasts (2 weeks in culture)

4 Regeneration of Plants

Mesophyll Protoplasts. Small calli (ca. 1 mm) on solid C-1 medium (0.3 M mannitol) become green in colour and can be individually picked 2 weeks later. The size of colonies at the time of transfer is an important factor determining shoot formation in the later stage. When calli are too small, they are not able to survive the transfer. Only healthy calli with proper size (>2 mm) are able to form shoot primordia within 2 weeks. Kinds and concentrations of plant hormones in the medium exert a marked effect on efficiency of plant regeneration (Table 4). Proper osmotic pressure (0.2–0.3 M mannitol) is also required for shoot formation. Factors influencing shoot formation are summarized in Table 5. When shoots are visible, calli should be transferred to T5 medium for shoot elongation and

Fig. 1. Freshly isolated mesophyll protoplasts of cabbage

Fig. 2. First division (5 days in MS-N medium)

Fig. 3. Protoplast-derived colonies in KM-1 medium (3 weeks in culture)

Fig. 4. Protoplast-derived colonies in KM-1 medium (4 weeks in culture)

Fig. 5. Protoplast-derived calli on C-1 medium (6 weeks in culture)

Fig. 6. Leaf primordia appeared on C-1 medium (8 weeks after protoplast isolation)

Fig. 7. A regenerated plantlet (11 weeks after protoplast isolation)

Fig. 8. Protoplast-derived plant grown in greenhouse

Table 4. Effects of different cytokinins on shoot formation

	Shoot formation (%)
1. 0.03 mg/l GA_3 + 1.0 mg/l BAP	6
2. 0.03 mg/l GA_3 + 2.0 mg/l BAP	12
3. 0.03 mg/l GA_3 + 1.0 mg/l kin	2
4. 0.03 mg/l GA_3 + 2.0 mg/l kin	10
5. 0.03 mg/l GA_3 + 1.0 mg/l 2iP	10
6. 0.03 mg/l GA_3 + 2.0 mg/l 2iP	2

GA_3 = Gibberellic acid
BAP = 6-Benzylaminopurine
kin = Kinetin
2iP = 2-Isopentenylpurine

Table 5. Factors influencing shoot formation

1. Basal medium	C-1
2. GA_3	0.03 – 0.1 mg/l
3. Auxin	beneficial with GA_3
4. Cytokinin	1.0 – 2.0 mg/l
5. Sucrose	0.25%
6. Osmoticum	0.2 – 0.3 M mannitol

rooting. On T5 medium the majority of shoots develop roots and there is little problem of rooting.

Hypocotyl Protoplasts. In contrast to the C-1 medium described above, K3-2 regeneration medium stimulates callus growth as well as shoot formation. Typically pale green calli grow well on this medium and shoots appear on proliferating calli. The frequency of shoot formation varies among cultivars and the most responsive ones form shoots from 10 – 15% of calli on the regeneration medium. Occasionally roots are also formed. On K3-2 medium shoots elongate quickly and when they touch the lids of the dishes they are transferred to T5 medium for further elongation and rooting. Shoots grow well on T5 medium, but occasionally root formation becomes a problem. Whether or not actively growing roots are obtained appears to be determined by the quality of the shooting calli. Acclimation of plants can be done in the same way as for plantlets derived from mesophyll protoplasts.

5 Characterization of Protoplast-Derived Plants

A total of 25 regenerated plants were grown to maturity and no obvious phenotypic changes were observed. Chromosome analysis of 21 plants revealed that 14 plants were diploid and 7 plants were tetraploid (Figs. 10, 11). Other than ploidy change we observed no karyotypic changes. Twelve plants were induced to flower and all of them set the normal number of seeds (Fig. 12). Progeny of those

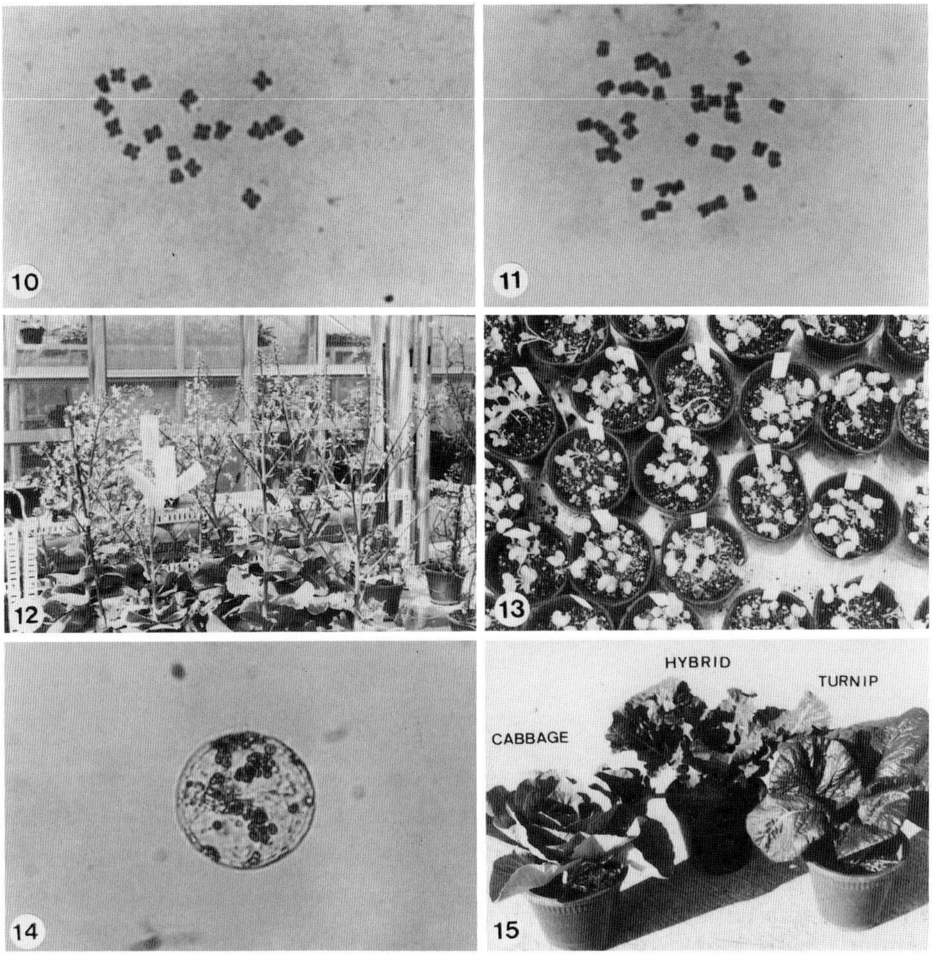

Fig. 10. Chromosomes (2n = 18) of a protoplast-derived cabbage plant

Fig. 11. Chromosomes (2n = 36) of a protoplast-derived cabbage plant showing ploidy change

Fig. 12. Protoplast-derived cabbage plants showing normal flowering

Fig. 13. Test of progeny from protoplast-derived plants

Fig. 14. A hybrid cell between cabbage hypocotyl protoplast and mesophyll protoplast of *B. campestris* var. *rapa* (2 h after fusion treatment)

Fig. 15. A somatic hybrid (*center*) between cabbage and *B. campestris* var. *rapa*

Table 6. Chromosome number of somatic hybrids[a] (Terada et al. 1987)

B. oleracea	*B. campestris*	Plant no.	Chrom. no.
cv. Shutoku	var. *rapa*	F 1	56
	cv. Tokyo	F 2	56
	Komatsuna	F 3	57
		No. 1	56
cv. Shutoku	var. *rapa*	416-4	56, 36
	cv. Misugi	416-5	49, 33
	Komatsuna	516-1	38
		516-2	38
cv. Shutoku	var. *pekinensis* cv. Michihli	412-11	38
cv. Shutoku	Bird's rape	725-3	56

[a] Hybrid nature of all the plants was identified by isozyme analysis and plant morphology.

plants was examined for germinability and seedling characteristics, and it was found that they were not different from control (Fig. 13). No morphological abnormalities in protoplast-derived plant were observed and the only genetic change we have seen is a ploidy change.

6 Protoplast Fusion

For protoplast fusion, hypocotyl protoplasts were found to be excellent material. By using a fusion method of Menczel and Wolfe (1984), we are able to reproducibly obtain a high fusion frequency. When cabbage hypocotyl protoplasts are fused with mesophyll protoplasts of turnip (*B. campestris* var. *rapa*) routinely more than 10% fusion frequency was obtained. Fused cells contain chloroplasts from turnip and show active cytoplasmic streaming and they are easily identified under the microscope (Fig. 14). By following the protocol for culture of hypocotyl protoplasts (Protocol 2) 22 hybrids between cabbage and three different varieties of *B. campestris* (Fig. 15) were obtained (Terada et al. 1987), and ten plants were analyzed for their chromosome numbers (Table 6). Some of the hybrids flowered, among which plant F3 set normal seeds.

7 Conclusion

Both mesophyll protoplasts (Protocol 1) and hypocotyl protoplasts (Protocol 2) of cabbage can be manipulated to regenerate plants routinely. When two protoplast sources are compared, variability in colony-forming efficiency and regeneration frequency is larger for mesophyll protoplasts than for hypocotyl protoplasts. Also hypocotyl protoplasts are more suitable for fusion and transformation studies because of their relative resistance to polyethylene glycol treatment. One disadvantage of hypocotyl protoplasts is that a large number of seeds are needed (100–200 seeds per experiment). For the limited amounts of seed materials leaves are the only tissue that provide enough numbers of dividing protoplasts.

No gross phenotypic abnormalities have been observed among protoplast-derived plants which have been regenerated according to the protocols described here. Ploidy change appears to be the only genetic change that has been noticed. The protocols described here can be employed for various studies making use of protoplasts, such as somatic hybridization, transformation and virus infection.

8 Protocol

Protocol 1 (Mesophyll protoplasts)

Media used in the following protocol are listed in Table 7.

(1) Collect leaves (1st to 5th) from 30–50-day-old plants grown in either a greenhouse or growth chamber. Wash them well in commercial detergent and water successively.

(2) Sterilize leaves in 70% ethanol for 30 s and in 3% sodium hypochloride for 20 min with gentle shaking. Then wash in sterile distilled water three times.

(3) Remove water from leaf surface by placing leaves on sterile filter paper.

(4) Remove lower epidermis with fine forceps, cut out midribs and cut leaves into 5×5 mm pieces. Place them in enzyme solution (20 ml) in 10-cm Petri dish. Approximately 1 g of leaf material is placed into one Petri dish.

(5) Incubate the Petri dish in the dark at 30 °C for 2 h with gentle shaking (30 strokes per min).

(6) After incubation, gently pipette enzyme-protoplast mixture a few times and pour it into a glass funnel fitted with four layers of cheesecloth.

(7) Centrifuge filtered protoplast suspension at 100 *g* for 3 min. This generally pellets intact protoplasts, while broken protoplasts and cell debris remain in the supernatant.

(8) Wash protoplasts twice in 0.5 M mannitol by centrifugation.

(9) Count number of protoplasts by using haematocytometer.

(10) Wash once in culture medium (MS-N) and resuspend protoplasts in the medium (1×10^5 protoplasts/ml).

(11) Place 1.25 ml of protoplast suspension into a 6-cm Petri dish, add 1.25 ml of molten MS-N medium (1.2% Sigma Type I agarose, 40 °C) and mix well quickly.

(12) Cut solidified agarose into pieces (1×1 cm) and transfer them into a 6-cm Petri dish containing 5 ml of liquid medium.

(13) Culture dishes in dark at 25 °C with slow shaking (ca. 30 rpm) on the rotary shaker.

(14) Seven to 10 days later (most of the protoplasts have divided) remove MS-N medium and change to KM-1 medium.

(15) After 3-week incubation, transfer agarose blocks containing visible colonies onto solid C-1 medium (0.3 M mannitol) and spread colonies on the surface of the medium with forceps.

(16) Incubate dishes for 2 weeks at 25 °C under light (ca. 3000 lx).

(17) Pick calli (1 mm) individually and transfer onto fresh C-1 medium (0.2 M mannitol).

(18) Shoot primordia emerge within 2 weeks, transfer those calli possessing shoot primordia onto T5 medium for shoot elongation and rooting.

(19) When shoots reach ca. 5 cm, transfer plantlets into glass container (10×12 cm) containing vermiculite and hyponex.

(20) When plants touch the lid of the container, remove lid, cover the container with a plastic bag and secure the plastic bags with rubber bands.

(21) Two to four weeks later cut edges of the plastic bag to stimulate air exchange.

(22) Transfer plants into pots and grow further in the growth chamber or greenhouse.

Table 7. Culture media for mesophyll protoplasts

Component	MS-N	KM-1	C-1	T5
Major salts (mg/l)				
NH_4NO_3		600	1650	
KNO_3	1900	1900	2400	1200
$CaCl_2\ 2H_2O$	440	600	440	220
$MgSO_4\ 7H_2O$	370	300	370	185
KCl		300		
KH_2PO_4	170	170	170	170
Iron and minor elements (mg/l)				
Na_2 EDTA	37.3	37.3	18.6	9.3
$FeSO_4\ 7H_2O$	27.8	27.8	13.9	7.0
H_3BO_3	6.2	3	3.1	3.1
$MnCl_2\ 4H_2O$			6.3	6.3
$MnSO_4\ 4H_2O$	22.3	10		
$ZnSO_4\ 7H_2O$	8.6	2	4.3	4.3
KI	0.83	0.75	0.42	0.42
$Na_2MoO_4\ 2H_2O$	0.25	0.25	0.13	0.13
$CuSO_4\ 5H_2O$	0.025	0.025	0.013	0.013
$CoCl_2\ 6H_2O$	0.025	0.025		
$CoSO_4\ 7H_2O$			0.015	0.015
Organic components (mg/l)				
Thiamine HCl	0.5	1.0	0.5	0.75
Glycine	2.0		2.0	1.0
Nicotinic acid	0.5	1.0	5.0	2.5
Pyridoxine HCl	0.5	1.0	0.5	0.25
Folic acid		0.4	0.5	0.25
Biotin		0.01	0.05	0.025
p-Aminobenzoic acid		0.02		
Choline chloride		1.0		
Riboflavin		0.2		
Ascorbic acid		2.0		
Vitamin A		0.01		
Vitamin D_3		0.01		
Vitamin B_{12}		0.02		
Casein hydrolysate		50	400	
myo-Inositol	100	100		
Adenine sulfate			40	20
Sodium pyruvate		20		
Citric acid		40		
Malic acid		40		
Fumaric acid		40		
Sucrose (%)	1.0	1.0	0.25	3
Glucose (%)	0.5	0.5		
Mannitol (M)	0.5	0.5	0.3/0.2	
MES (mM)			5	
Hormones (mg/l)				
2,4-D	1.0	1.0		
NAA	0.5	0.5		0.05
BAP	0.5	0.5	2.0	
GA_3			0.03	0.03
Agarose (%)	Type VII		Type I	
	1.0		0.75	0.75
pH	5.8	5.8	5.8	5.8

Table 8. Culture media for hypocotyl protoplasts

Component	KM-2	K3-1	K3-2
Major salts (mg/l)			
NH_4NO_3		250	
KNO_3	= KM-1	2500	
$(NH_4)_2SO_4$		134	
$CaCl_2\ 2H_2O$		900	
$MgSO_4\ 7H_2O$		250	
KCl			
KH_2PO_4			
Iron and minor elements (mg/l)			
Na_2 EDTA		37.3	
$FeSO_4\ 7H_2O$		27.8	
H_3BO_3		3	
$MnSO_4\ 4H_2O$		22.3	
$ZnSO_4\ 7H_2O$	= KM-1	2	
KI		0.75	
$Na_2MoO_4\ 2H_2O$		0.25	
$CuSO_4\ 5H_2O$		0.025	
$CoCl_2\ 6H_2O$		0.025	
Organic components (mg/l)			
Thiamine HCl		10	
Glycine			
Nicotinic acid		1	
Pyridoxine HCl	= KM-1	1	
Folic acid			
Biotin			
p-Aminobenzoic acid			
Choline chloride			
Riboflavin			
Ascorbic acid			
Vitamin A			
Vitamin D_3			
Vitamin B_{12}			
Casein hydrolysate		400	
myo-Inositol		100	
Adenine sulphate		40	
Sodium pyruvate			
Citric acid			
Malic acid			
Fumaric acid			
Sugars			
Sucrose		0.2 (M)	1 (%)
Glucose	0.4 (M)		
Fructose	250 (mg/l)		
Ribose	250 (mg/l)		
Xylose	250 (mg/l)		
Mannose	250 (mg/l)		
Rhamnose	250 (mg/l)		
Cellobiose	250 (mg/l)		
Sorbitol	250 (mg/l)		
Mannitol	250 (mg/l)		
MES (mM)		5	5
Hormones (mg/l)			
2,4-D	1.0	0.1	

Table 8 (Continued)

Component	KM-2	K3-1	K3-2
NAA	0.5		
BAP	0.5	1.0	1.0
GA_3			0.03
		Type I	
Agarose (%)		0.25	0.75
pH	5.8	5.8	5.8

2,4-D = 2,4-dichlorophenoxyacetic acid
NAA = 1-Naphthaleneacetic acid
MES = 2-N-Morpholino-ethane sulfonic acid
Reference C-1 T5: Bidney et al. (1983)

Protocol 2 (Hypocotyl protoplast)

A protocol described below is a modification of a published procedure (Glimelius 1984) originally devised for *B. napus* protoplasts. Media used in the following protocol are listed in Table 8.
(1) Surface-sterilize seeds in 70% ethanol for 30 s in 4% sodium hypochlorite for 30 min with vigorous shaking, and wash in distilled water three times.
(2) Place sterilized seeds (20–30 seeds/10 cm Petri dish) on germination medium (1/10 conc. MS, 0.2% sucrose, 0.8% Bactoagar) and incubate at 25 °C for 4–5 days.
(3) Remove cotyledons and roots and transversely cut hypocotyls into 0.5–1 mm segments.
(4) Incubate hypocotyl segments in plasmolysis buffer (0.3 M sorbitol, 0.05 M $CaCl_2$) for 1 h at room temperature.
(5) Replace plasmolysis buffer with enzyme solution (40–50 hypocotyls/20 ml enzyme solution/10 cm Petri dish), vacuum for 15 min and incubate for 2 h at 30 °C with gentle shaking.
(6) After incubation, gently pipette enzyme-protoplast mixture several times to release protoplasts from hypocotyl segments.
(7) Filter enzyme solution with 133-μm nylon mesh and add 1/3 volume of 16% sucrose solution.
(8) Distribute protoplast suspension into test tubes (ca. 10 ml/tube) and layer 1 ml of W5 solution (125 mM $CaCl_2$, 155 mM NaCl, 5 mM KCl, 5 mM glucose) on top of the suspension.
(9) Centrifuge for 8 min at 100 *g* and transfer protoplasts banded at interphase with Pasteur pipette to another test tube.
(10) Wash protoplasts in W5 solution twice by centrifugation (100 *g*, 3 min) and count number of protoplasts.
(11) Suspend protoplasts in KW-2 medium (10^5 protoplasts/ml) and culture in 6 cm Petri dish (3 ml/dish) under dim light at 25 °C.
(12) Seven to ten days later add 3 ml of fresh KW-2 medium into the dish and culture further for 2 weeks.
(13) Pipette ca. 1 ml of protoplast suspension, drop onto solid K3-1 medium and spread colonies well on the medium.
(14) Incubate for 2 weeks in light (ca. 3000 lx) at 25 °C.
(15) Pick individual colonies and transfer onto K3-2 medium for shoot induction. Shoots usually appear within a month after the transfer. Non-shooting calli are transferred to fresh K3-2 medium for shoot induction.
(16) Transfer calli with shoots onto T5 medium.
(17) The method for acclimation of regenerated plants follows protocol 1.

References

Bidney DL, Shepard JF, Kaleikau E (1983) Regeneration of plants from mesophyll protoplasts of *Brassica oleracea*. Protoplasma 117:89–92

FAO (1982) Production Yearbook, FAO Rome, Italy

Glimelius K (1984) High growth rate and regeneration capacity of hypocotyl protoplasts in some Brassicaceae. Physiol Plant 61:38–44

Kao KN, Michayluk MR (1975) Nutritional requirements for growth of *Vicia hajastana* cells and protoplasts at a very low population density in liquid media. Planta 126:105–110

Leike H (1988) Cabbage (*Brassica oleracea* var *capitata* L.). In: Bajaj YPS (ed) Biotechnology in agriculture and forestry 6. Crops II. Springer, Berlin Heidelberg New York Tokyo, pp 220–225

Lui DY, Pental D, Cocking EC (1982) Plant regeneration from seedling cotyledon protoplasts. Z Pflanzenphysiol 107:59–63

Menczel L, Wolfe K (1984) High frequency of fusion induced in freely suspended protoplast mixtures by polyethylene glycol and dimethylsulfoxide at high pH. Plant Cell Rep 3:196–198

Murashige T, Skoog F (1962) A revised medium for rapid growth and bioassays with tobbaco tissue cultures. Physiol Plant 15:473–497

Nagy JI, Maliga P (1976) Callus induction and plant regeneration from mesophyll protoplasts of *Nicotiana sylvestris*. Z Pflanzenphysiol 78:453–455

Nitsch C, Nitsch JP (1967) The induction of flowering in vitro in stem segment of *Plambago indica* L. I. The production of vegetative buds. Planta 72:355–370

Schenck HR, Röbbelen G (1982) Somatic hybrids by fusion of protoplasts from *Brassica oleracea* and *B. campestris*. Z Pflanzenzücht 89:278–288

Shepard JF (1980) Mutant selection and plant regeneration from potato mesophyll protoplasts. In: Rubenstein I, Gengenbach B, Phillips RL, Green CE (eds) Genetic improvement of crops, emergent techniques. University of Minnesota Press, Minneapolis, pp 185–219

Shillito RD, Paszkowski I, Potlykus I (1983) Agarose plating and bead type culture technique enable and stimulate development of protoplast-derived colonies in a number of plant species. Plant Cell Rep 2:244–247

Snogerup S (1980) The wild forms of the *Brassica oleracea* Group (2n = 18) and their possible relations to the cultivated ones. In: Tsunoda S, Hinata K, Gomez-Campo C (eds) *Brassica* crop and wild allies. Japan Sci Soc Press, Tokyo, pp 121–132

Terada R, Yamashita Y, Nishibayashi S, Shimamoto K (1987) Somatic hybrids between *Brassica oleracea* and *B. campestris*: selection by the use of iodoacetamide inactivation and regeneration ability. Theor Appl Genet 73:379–384

Vatsya B, Bhaskaran S (1981) Production of subprotoplasts in *Brassica oleracea* var *capitata* – a function of osmolarity of the media. Plant Sci Lett 23:277–282

Xu ZH, Davey MR, Cocking EC (1982) Plant regeneration from root protoplasts of *Brassica*. Plant Sci Lett 24:117–121

Yamashita Y, Shimamoto K (1984) Plant regeneration from mesophyll protoplasts of *Brassica oleracea* (cabbage). Proceeding of 1st Intn Symp Somatic Cell Genet in Crops. Beijing, China, Oct 1984

II.7 Regeneration of Plants from Chicory (*Cichorium intybus* L.) Protoplasts

Y. CHUPEAU [1]

1 Introduction

Cichorium intybus is a member of the Compositae and has attractive blue flowers. Chicory is the "traditional" western European winter salad, but as a result of recent technical progress which has led to reliable ways of forcing roots to give firm and clean chicons, chicory is now cultivated all over the world. Since most countries are only starting chicory production, one can predict an increasing demand for this very tasty and easy to use salad. Chicory breeding will also become more active with the identification of new objectives, which the application of novel in vitro techniques, including the use of isolated protoplasts, may help to achieve (see Schoofs and de Langhe 1988).

The very first mention of the use of the sedative properties of chicory comes from Ancient Egypt (Smith 1976). Later, in western Europe chicory was cultivated for its salad leaves, especially if grown in the shade or earthed up (Gibault 1912).

1.1 Importance and Distribution of Chicory

Starting in about the 18th century, renewed interest and proper breeding made chicory a salad which came to be more and more cultivated. Botanists separate chicory into two species (which are in fact cross-compatible):

- *Cichorium endivia* L.: (endive in English) usually with curled leaves (scarole in French);
- *Cichorium intybus* L.: chicory, cultivated for its green red or mixed coloured salad leaves. Some varieties are also cultivated for their sweet roots, which when roasted are used with or instead of coffee (Gibault 1912). The "Witloof chicory" (endive in French) is probably derived from a "coffee chicory" type and has been especially bred for its capacity to develop big white buds in darkness, the "chicons", from the lifted roots.

The modern history of the "endive" began in the Botanical Garden of Brussels in the middle of the 19th century with the discovery of the Witloof cultivar

[1] Laboratoire de Biologie Cellulaire I.N.R.A. – 78000 Versailles, France

Biotechnology in Agriculture and Forestry, Vol. 8
Plant Protoplasts and Genetic Engineering I (ed. by Y. P. S. Bajaj)

(Gibault 1912). Then cultivation was restricted to Belgium, with the following schedule:

- mid-May, field sowing leading to about 250000 plants per hectare;
- November, thinning out of leaves and lifting of the sweet tap roots, which are then stored in the cold;
- forcing roots during winter in the dark and at 15 °C; in controlled conditions buds develop at the crown, and about 20 days later are detached from the root and sold as salad.

For seed production roots are field-replanted in March and apical bud development in natural lengthening days leads to a flowering stem; after allogamic fertilization seeds are produced about August.

Nowadays breeding is concerned with shape, colour and taste of chicons and also with production characteristics such as homogeneity and precocity of maturation of roots. Recent developments give sharp firm chicons, well adapted to forcing in controlled non-soil (hydroponic) conditions with defined mineral nutrition in flow solutions (Lesaint et al. 1978). These clean and controlled forcing processes give a very fine salad and have permitted the endive to spread from Europe to Canada, USA, Australia, Chile, Mexico, etc. where hundreds of hectares are devoted to chicory. However, Belgium, France and the Netherlands are still the main producers, France being the first producing country, with more than 15000 ha for root production, leading to about 200000 tons of chicons annually.

1.2 Significance of the Use of Protoplasts

Protoplasts offer two experimental advantages. Firstly, as a large and homogeneous population of independent cells, protoplasts make direct biochemical selection possible. Secondly, cell fusion, due to absence of cell wall provides new ways for broadening genetic variability, not only for nuclear genes by somatic hybridization (Pelletier and Chupeau 1984) but also for cytoplasmic genes. Following the pioneering work with *Nicotiana* protoplasts (Belliard and Pelletier 1978), numerous examples have shown that mitochondrial DNA is recombined after somatic fusion (Pelletier et al. 1985). The reported chloroplastic DNA recombination (Medgyesi et al. 1985) indicates that a better understanding of the process could make it as effective and general as mitochondrial DNA recombination.

In the case of chicory, these novel breeding tools could be applied to select for herbicide resistance and also to search for resistance to various fungi. Fungi can cause severe losses in production during the forcing process (*Sclerotinia, Phytophthora, Phoma*), and especially since toxins have been detected in some parasites such as cytochalasine with *Phoma exigua* (Bousquet and Barbier 1972).

At the moment, interest in cytoplasm manipulations and especially male sterility is not very great for chicory, since allogamy allows easy natural cross-pollination; thus permitting hybrid production after genealogical selection of lines by forced auto-selfing of selected plants (Bannerot and De Coninck 1970). However, self-incompatibility could cause severe barriers in specific breeding programmes for hybrid seed production; such barriers could be aleviated by in vitro multiplication of parent plants by somatic embryogenesis in tissue culture (Heirweigh et al. 1985) or by rapid en masse regeneration of protoplasts of selected parent plants.

In any case, protoplast manipulations and plant regeneration are going to be increasingly needed for the complete utilization of the various potential applications of direct transformation of plant cells (Potrykus et al. 1985).

1.3 State of the Art of Plant Regeneration from Chicory Protoplasts

Chicory is not in major production in most countries, so little work has been done with its protoplasts. An early report (Wagner and Siegelman 1975) was devoted only to the isolation of intact vacuoles. Then three reports (Binding et al. 1981; Crepy et al. 1982; Saksi et al. 1986) revealed that in spite of very good plating efficiencies of 60 to 80% over the first week of culture, difficulties in preventing browning of small cell colonies were the main limitation to plant regeneration. This repeated observation contrasts with the already well-known facility to regenerate plants from in-vitro-cultivated explants from all parts of the adult chicory plant, and especially from the storage root tissue. Furthermore, chicory tissue cultivated in vitro has been extensively used for its capacity to directly differentiate flower buds under certain conditions (for a review see Margara 1977). It is probably worth mentioning reports of "somatic embryogenesis" for *Cichorium* cells cultivated in suspension. *Cichorium endivia* has been regenerated through somatic embryogenesis (Vasil and Hildebrandt 1966), and this capacity has recently been extended to *Cichorium intybus* cv. Carolus (Heirwegh et al. 1985). In both cases illustrations do not unambiguously demonstrate classical embryo formation; however, these reports favour a special, not yet completely understood, type of regeneration for chicory.

Information obtained from work with primary calli, or liquid transplantation of primary calli is not directly applicable for plant regeneration from small cell colonies arising from protoplast development; however, the indications are that the very specific step of plant regeneration, whether via bud or embryo formation, is not a real problem for chicory.

2 Isolation of Protoplasts

2.1 Plant Material

Although adult storage root has been the favoured material for most in-vitro flowering studies, it generally gives very poor protoplast yields. Classically, young leaves are preferred, starting from the apex (Binding et al. 1981) or 21-day-old axenic leaflet (Saksi et al. 1986) or young greenhouse-grown leaflets (Crepy et al. 1982). Use of axenic material propagated in vitro provides various advantages: the inherent sterility of the material grown in vitro avoids the use of sterilant, which may affect viability of the protoplasts, the efficiency of which is sometimes dubious. In our opinion, the major advantage of axenic in vitro-grown leaves lies not only in the control of the medium, but also in the control of the physical parameters such as light, relative humidity and temperature, which may reproducibly provide optimal conditions for a high yield of viable protoplasts.

For greenhouse-grown plants it was established that low light and regular watering were essential for reproducible protoplast yields from young leaves (Crepy et al. 1982). By using controlled growth room conditions, we found (Lenee et al. 1986) that conditions for growing chicory, giving not only the best protoplast yields, but also best development in culture, were the following:

- day: 60 $\mu E\,s^{-1}\,m^{-2}$, 8 h at 25 °C.
- night: 16 h at 17 °C.
- relative humidity 90%.

Very good disinfection is obtained with Bayrochlor contained in tablets, which on complete dissolution liberates hypochlorous acid (one tablet: 1.5 g active chlorine). This sterilant is also very effective on a wide variety of seeds and tissues.

2.2 Factors Influencing Yield

The techniques for protoplast isolation have already been reviewed (see Chap. I, this Vol.; Davey 1983) and therefore the principles and procedures as they apply to chicory will be discussed here.

A very critical factor to ensure not only reproducible yield but also the viability of protoplasts is the osmotic pressure used. It is clear that the osmotic potential of protoplast isolation and culture media has to be at the lowest value which will maintain the physical integrity of the protoplasts; it also has to be reduced gradually for culture, as culture medium evaporates more or less rapidly, thus enhancing the actual concentrations of the various components of the culture medium, and especially that of the osmotic stabilizers. This emphasizes both the necessity for maintaining a high relative humidity around culture vessels and the requirements for a good control of the osmotic state of the donor plant or plantlets. Thus donor plants have to be regularly and profusely watered and grown in the lowest light intensity compatible with normal growth for the particular species. Under these conditions, maceration and culture media of about 700 mosmol are generally suitable. In the case of chicory leaf protoplasts (Crepy et al. 1982), the yield and viability (plating efficiency) were about the same for protoplasts liberated in 0.5 M mannitol or 330 mM KCl + 14 mM $CaCl_2$.

2.3 Enzymes

Besides the advantage of working with favourable osmotic pressures, avoiding light and water stress to donor plants offers another very important advantage:

the ease of cell wall degradation even by low enzyme concentrations. In the case of *Nicotiana* (Chupeau et al. 1974), these plant growth conditions allow protoplast isolation by a mixture of three different commercial crude enzyme preparations, each at low concentration:

- as pectinase: Macerozyme R 10 0.02%,
- as cellulase: Onozuka R 10 0.1%,
  and Driselase 0.05%

Such a mixture is effective for chicory protoplast isolation (Crepy et al. 1982; Saksi et al. 1986) and we also found it generally active for a wide diversity of tissues and species (Chupeau unpubl).

Maceration has to take place between 20 and 23 °C, thus leading to slow plasmolysis and slow concomitant cell wall degradation. In these conditions liberation of protoplasts is effected in 15–16 h, which with overnight macerations renders protoplasts available in the morning. In addition, maceration is done in complete culture medium to maintain, as far as possible, a physiological environment for cell and protoplasts.

Attempts to isolate protoplasts from different tissues (root, stem, bract, callus) resulted in low yields of protoplasts, which ultimately did not develop in culture. In fact, work has been concentrated on leaf protoplasts. In our conditions 2×10^6 protoplasts are routinely prepared out of 1 g fresh weight of leaf tissue.

3 Culture of Protoplasts

The physical parameters of the culture conditions are generally difficult to compare and have often not been properly assessed. However, long days (16–18 h) provided by artificial light are frequently favoured (Saksi et al. 1986; Crepy et al. 1982). Our culture conditions are the following:

- long days: 16 h at 28 °C (incident energy from fluorescent cool white tubes of 50 to 100 $\mu E\, s^{-1}\, m^{-2}$ depending on the age of tubes,
- short nights: 8 h at 24 °C,
- relative humidity controlled to be around 80%.

Controlled relative humidity prevents a too rapid evaporation of culture media. Routinely, protoplasts are kept 2 days in complete darkness before being subjected to the above climate (Enzmann-Becker 1973). One of the important parameters for successful protoplast culture is the initial plating density (Fig. 1). Like other Compositae protoplasts (Lenee and Chupeau 1986), chicory protoplasts exhibit a rather peculiar behaviour. Although the first two or three divisions take place for various initial densities between 10^4 to 10^5 protoplasts per ml, above 10^5 protoplasts per ml cell colonies very rapidly became necrotic (Crepy et al. 1982). This probably indicates that the medium formulation based on MS (Murashige and Shoog 1962) is not really suitable for chicory protoplasts.

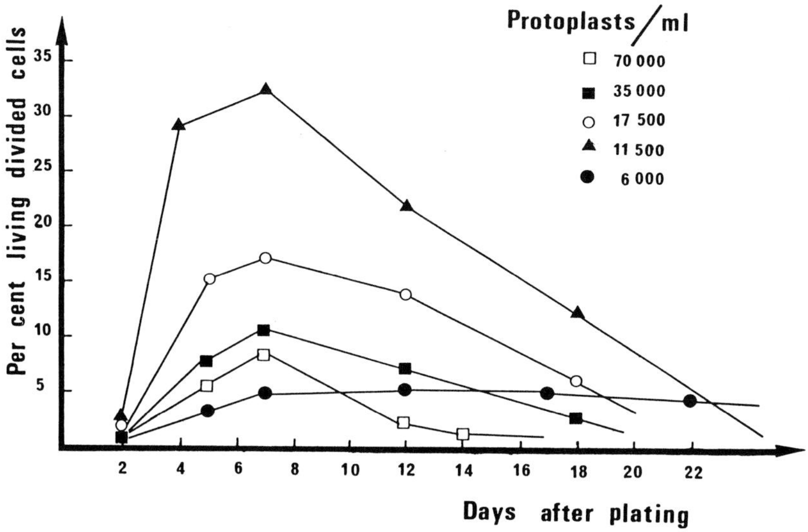

Fig. 1. Effect of plating density on the division and survival of plated protoplasts of chicory cv. Zoom in T_0 culture medium

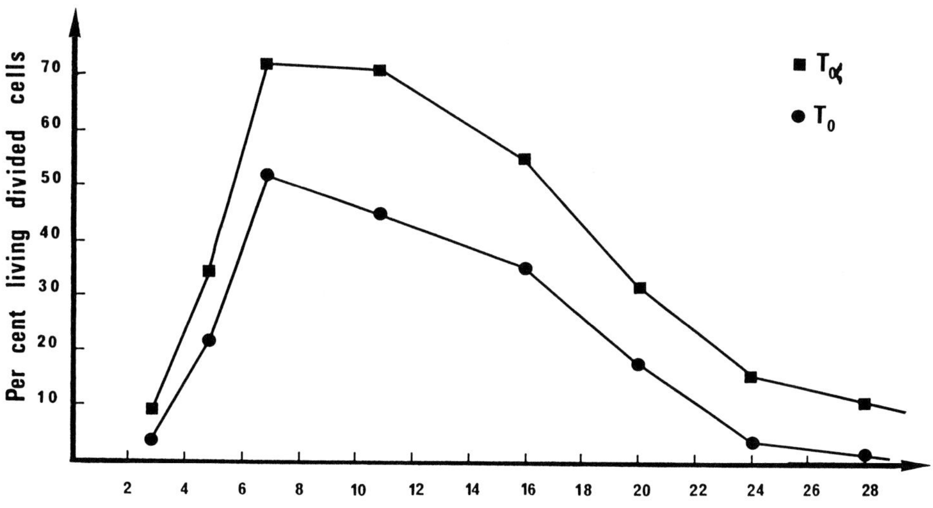

Fig. 2. Compared effect of T_0 and T_α on division and survival of protoplasts of chicory cv. Red Witloof

A more careful survey involving variations in levels of individual salts revealed that this early necrosis was essentially provoked by NH_4NO_3 ions in the culture medium, and that better colony survival could be achieved with 10 mM glutamine as the sole nitrogen source (Fig. 2) (Crepy et al. 1982). For protoplasts of axenic leaflets of cv. Zoom cultivated at initial densities of 2×10^4 protoplasts per ml, it

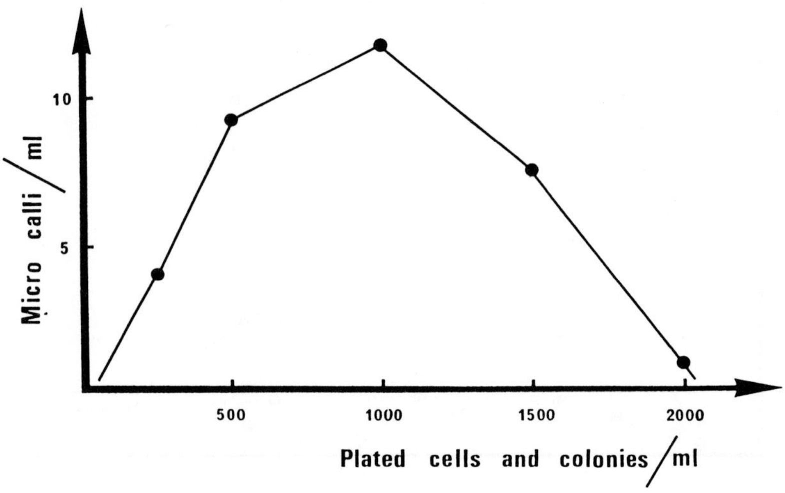

Fig. 3. Effect of cell density on the yield of microcalli of chicory cv. Zoom upon transfer from T_0 to T_{a8} medium

was shown that lowering the sucrose concentration from 20 to 5 g per liter in a glutamine medium led to more than 50% colony formation over the first week of culture and further improved cell colony survival (Saksi et al. 1986). However, cell colonies have to be subcultured to prevent complete cell necrosis. Dilution and subculture towards a low auxin regeneration medium can be achieved by serial subcultures (Saksi et al. 1986) or directly (Crepy et al. 1982). Despite these improvements, actual colony formation rates are still rather low, in the range of 1% and even less for some cultivars (Fig. 3). Poor rates of callus formation by protoplasts in culture appear to be the main limiting step in plant regeneration.

4 Regeneration of Plants

Once cell colonies have reached the stage of microcalli, they are easy to transplant individually to regeneration medium (Table 1). Traditionally, bud regeneration occurs on semi-solid medium and, as glutamine is inhibitory, with NH_4NO_3 as nitrogen source. Protoplast-derived calli transferred to media with such growth regulator balances as IAA 0.5 mg/l + BA 1 mg/l (Crepy et al. 1982) or NAA 0.05 mg/l + BA 0.5 mg/l (Saksi et al. 1986) have proved effective for bud induction about 1 month after transplanting for 20 to 45% of the calli. Buds are easily rooted on medium devoid of growth substances. Transfer to the greenhouse is achieved after the usual precaution of acclimatization to prevent desiccation of

Fig. 4. Development of protoplasts of *Cichorium intybus* cv. Red Witloof. **A** Isolated protoplasts in liquid T_0. **B** Beginning of necrosis in dividing cells 8 days after liquid culture in T_0. **C** Rescued cell colony after dilution. **D** Bud development 2 months after transplanting small calli on R_4 regeneration medium. **E** Rooted plantlets before greenhouse transplantation (Crepy et al. 1982)

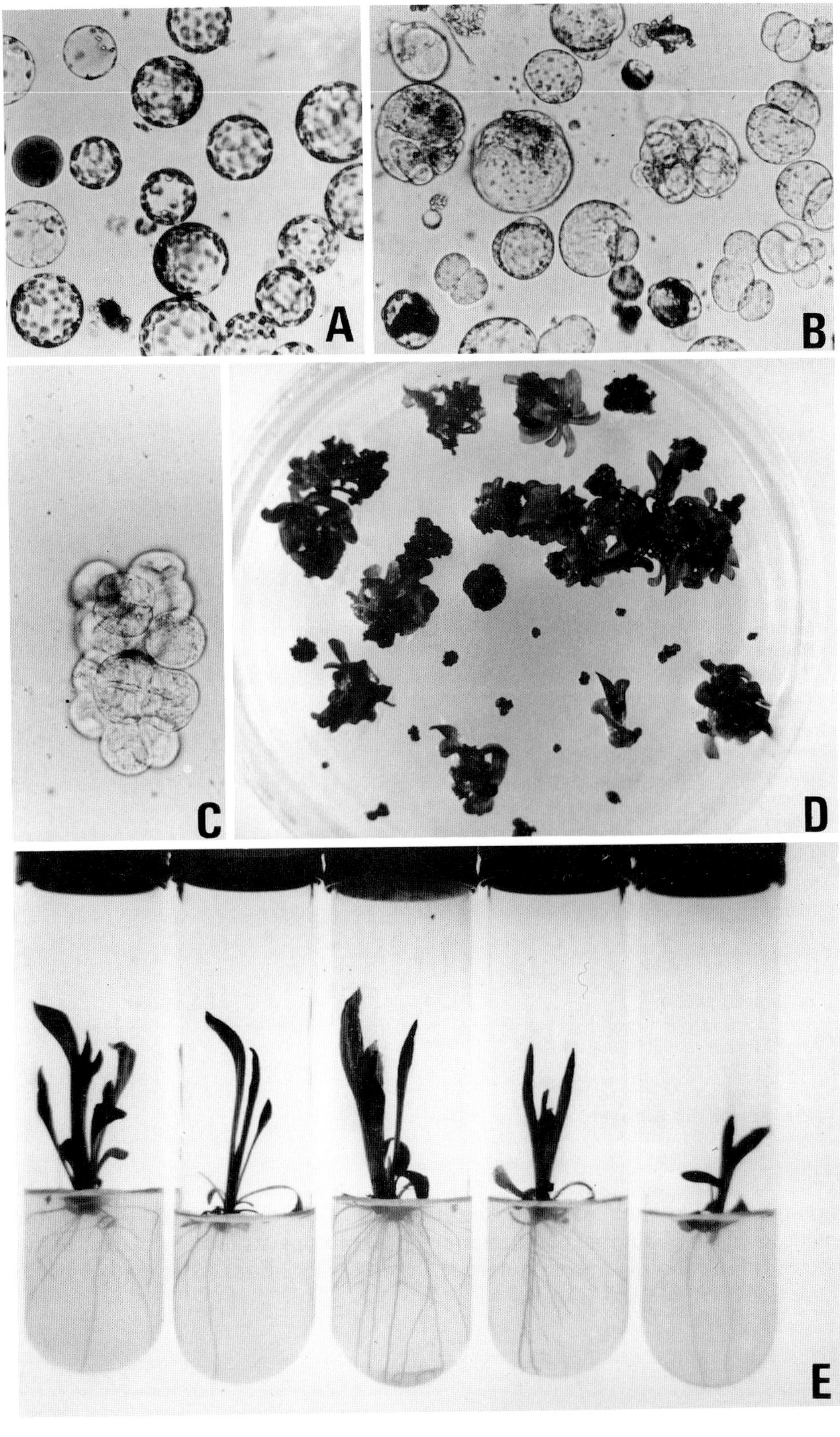
A
B
C
D
E

Table 1. Culture media for chicory (mg/l)

	T_0 Protoplast isolation	T_a Protoplast culture	T_{a8} Cell colony culture	R_4 Bud regeneration	Rooting
Macroelements[a]:					
KCl	–	750	750	–	–
KNO_3	950	–	–	1900	1900
NH_4NO_3	825	–	–	1650	1650
$CaCl_2 \cdot 2H_2O$	220	440	440	440	440
$MgSO_4 \cdot 7H_2O$	185	123	123	370	370
KH_2PO_4	85	68	68	170	170
Microelements[b]:					
$FeSO_4 \cdot 7H_2O$	27.85				
Na_2EDTA	37.25				
$ZnSO_4$	1				
H_3BO_3	1				
$MnSO_4 \cdot 4H_2O$	0.1	id	id	id	id
$CuSO_4 \cdot 5H_2O$	0.03				
$AlCl_3$	0.03				
$NiCl_2 \cdot 6H_2O$	0.03				
KI	0.01				
Vitamins[c]:					
Inositol	100				
Pantenathenate Ca	1				
Biotin	0.01	id	id	id	id
Niacine	1				
Pyridoxine	1				
Thiamine	1				
Growth regulators					
Naphthalene acetic acid	3	–	0.1	–	–
Indolylacetic acid	–	1	–	0.5	–
6-benzyladenine	1	–	1	1	–
Zeatin	–	1	–	–	–
Glutamine	–	750	750	–	–
Sucrose	20000	20000	20000	20000	10000
Mannitol	80000	80000	80000	60000	–
Tween 80	8	8	8	–	–
Bromocresol purple[d]	8	8	8	8	8
MES	700	700	700	–	–
Bacto agar (Difco)	–	–	–	6000	6000

[a] After Murashige and Skoog 1962.
[b] After Heller 1953.
[c] From Morel and Wetmore 1951.
[d] After Roscoe and Bell 1981.

non-hardened in vitro plantlets. The efficiency of the overall process at the moment stands at about one plant for every 1000 protoplasts initially plated. Plants grown in greenhouse appeared normal; however, deviations from mother plants have not been checked (Fig. 4).

5 Conclusions and Prospects

In spite of a rather low plant recovery rate, the process of protoplast isolation and the early stages of culture are not at fault, as indicated by the fairly high plating efficiencies achieved over the first week of culture. On the other hand, plant regeneration by bud formation of cell colonies arising from protoplast development is not difficult to achieve. Rather surprisingly, the usual salt formulations for in vitro culture are not effective for small cell colonies. Yet, numerous variations of practically all of the culture medium components have not given the clue. The beneficial effect of glutamine as the only nitrogen source tends to indicate specific problems for chicory cells in nitrogen assimilation. However, it could also be linked to problems of energy use, since part of the necrosis could be due to a too high level of sugar in the culture medium. Clearly, more basic physiology and metabolism studies have to be conducted on chicory protoplasts. Such basic studies are indispensible if protoplasts are to be a tool in chicory breeding, as the low rate of plant recovery (10^{-3}) seems incompatible with in vitro selection of such rare events as biochemical mutants, transformants or somatic hybrids.

6 Protocol

(1) Use axenic leaflets or growth cabinet-grown leaves. Sterilize non-axenic leaves by half an hour immersion in a Bayrochlor solution (1 liter H_2O + 1 tablet Bayrochlor + 10 ml liquid soap) and rinse carefully by three changes in sterile water before drying leaves on sterile filter paper.
(2) Take off lower epidermis; if difficult, make strips (0.5 mm) with a scalpel before putting about 500 mg fresh weight of leaves in 10 ml of the maceration medium already in Petri dishes.
(3) Macerate peeled or leaf strips either in the complete medium (Table 1) or in a simple saline medium (KCl 2.5% + $CaCl_2 \cdot 2H_2O$ 0.2%). In both cases add enzymes at final concentrations of: Macerozyme 0.02%, Onozuka R10 0.1%, Driselase 0.05% and incubate in the dark at 23 °C. Digestion can be completed by gentle swirling next morning.
(4) Pipette protoplasts and remove undigested tissues by filtration through a 100-μm mesh filter. Wash protoplasts by pelleting and resuspending three times in the saline medium (three centrifugations, less than 10 min and no more rapidly than 100 *g* each time). Resuspend the last pellet in the saline medium to obtain a protoplast titre of about 10^6/ml.
(5) Dilute directly in liquid culture medium (see Table 1) already poured in Petri dishes to 2×10^4; keep dishes 2 days in the dark at 28 °C, before subjecting to an artificial climate such as: 16 h light (50 μE m^2 s) at 28 °C 8 h night at 25 °C.
(6) Between days 7 and 10 transfer small cell colonies by diluting to about 10^3 colonies/ml and 1 month later transfer calli individually to agar solidified regeneration medium in Petri dishes. About 1 month later transfer buds to the rooting medium.

Acknowledgements. Thanks are due to L. Crepy and B. de Coninck for providing information, and to P.K. Evans for reviewing this manuscript. Part of the work was supported by the Biomolecular Engineering Program of the Commission of European Communities-contract no. GBI-6-071-F.

References

Bannerot H, De Coninck B (1971) L'utilisation des hybrides F 1. Nouvelle méthode d'amélioration de la chicorée de Bruxelles. Eucarpia meeting "Chicory", Gembloux (1970), pp 99–118

Belliard G, Pelletier G (1978) Hybridation cytoplasmique par fusion de protoplastes chez *Nicotiana tabacum*. Physiol Veg 16:441–448

Binding H, Nehls R, Kock R, Finger J (1981) Comparative studies on protoplast regeneration in herbaceous species of the Dicotyledoneae class. Z. Pflanzenphysiol 101:119–130

Bousquet JF, Barbier M (1972) Sur l'activité phytotoxique de trois souches de *Phoma exigua* et la présence de cytochalasine B (ou phomine) dans leur milieu de culture. Phytopathol Z 75:365–367

Chupeau Y, Bourgin JP, Missionier C, Dorion N, Morel G (1974) Préparation et culture de protoplastes de divers *Nicotiana*. C R Acad Sci Ser D Paris 278:1565–1568

Crepy L, Chupeau MC, Chupeau Y (1982) The isolation and culture of leaf protoplasts of *Cichorium intybus* and their regeneration into plants. Z Pflanzenphysiol 107:123–131

Davey MR (1983) Recent development in culture and regeneration of plant protoplasts. In: Potrykus I, Harms GT, Hinneau A, Hutter R, King PJ, Shillito RD (eds) International Protoplast Symposium Basel. Lectures Proc Protoplast 1983. Experientia Suppl. 46. Birkhäuser Basel, pp 19–29

Enzmann-Becker G (1973) Plating efficiency of protoplasts of tobacco in different light conditions. Z Naturforsch 28:470–471

Gibault G (1912) Chicorée endive. In: Gibault G (ed) Histoire des Légumes. Librairie Horticole, Paris, pp 107–120

Heirwegh KMG, Banerjee N, Van Nervm K, de Langhe E (1985) Somatic embryogenesis and plant regeneration in *Cichorium intybus* L. (Witloof, Compositae). Plant Cell Rep 4:108–111

Heller R (1953) Recherches sur la nutrition minérale des tissus cultivés in vitro. Ann Sci Nat Bot Biol Sci Paris, pp 213–217

Lenee P, Chupeau Y (1986) Isolation and culture of sunflower protoplasts (*Helianthus annuus* L.) factors influencing the viability of cell colonies derived from protoplasts. Plant Sci 43:69–75

Lenee P, Chupeau MC, Chupeau Y (1986) Regeneration of plants from Compositae protoplasts. In: Magnien E, De Nettancourt D (eds) Genetic engineering of plant and microorganisms important for agriculture. CEC final meeting. Wageningen 1985, pp 75–76

Lesaint C, de Coninck B, Grandjean M, Hatte M (1978) l'utilisation de solutions nutritives pour le forçage de l'endive (*Cichorium intybus*) en hydroponique et son efficience. CR Acad Agric Fr, pp 797–803

Margara J (1977) Schéma du développement en culture in vitro de *Cichorium intybus* L., exemple type de plante bisannuelle de jour long à besoin de vernalisation. Bull Soc Bot Fr 124:491–501

Medgyesy P, Fejes E, Maliga P (1985) Interspecific chloroplast recombination in a *Nicotiana* somatic hybrid. Proc Natl Acad Sci 82:6960–6964

Morel G, Wetmore RH (1951) Fern callus tissue culture. Am J Bot 38:141–143

Murashige T, Skoog F (1962) A revised medium for rapid growth and bio-assays with tobacco tissue cultures. Physiol Plant 15:473–497

Pelletier G, Chupeau Y (1984) Plant protoplast fusion and somatic plant cell genetics. Physiol Veg 22:377–399

Pelletier G, Vedel F, Belliard G (1985) Cybrids in genetics and breeding. Hereditas Suppl 3:49–56

Potrykus I, Paszkowski J, Saul M, Kruger-Lebus S, Muller T, Schoder R, Negrutiu I, Kunzler P, Shillito R (1985) Direct gene transfer to protoplasts: an efficient and generally applicable method for stable alterations of plant genomes. In: Freeling M (ed) Ucla Symposia on Molecular and Cellular Biology, New Series "Plant Genetics". AR Liss, New York, pp 181–199

Roscoe DH, Bell MG (1981) Use of a pH indicator in protoplast culture medium. Plant Sci Lett 21:275–279

Saksi N, Dubois J, Millecamps JL, Vasseur J (1986) Régénération de plantes de chicorée Witloof cv. Zoom à partir de protoplastes: influence de la nutrition glucidique et azotée. CR Acad Sci Paris Ser III 302:165–170

Schoofs J, de Langhe E (1988). Chicory (*Cichorium intybus* L.) In: Bajaj YPS (ed) Biotechnology in agriculture and forestry 6. Crops II. Springer, Berlin Heidelberg New York Tokyo, pp 294–321

Smith PM (1976) Chicory and Endive. In: Simmonds NW (ed) Evolution of crop plants. Longman, London, pp 304–305

Vasil IK, Hildebrandt AC (1966) Variation of morphogenetic behavior in plant tissue cultures. I. *Cichorium endivia*. Ann J. Bot 53:860–869

Wagner GJ, Siegelman HW (1975) Large-scale isolation of intact vacuoles and isolation of chloroplasts from protoplasts of mature plant tissues. Sciences 190:1298–1299

II.8 Regeneration of Plants from Protoplasts of Lettuce and its Wild Species

S. ENOMOTO and K. OHYAMA[1]

1 General Account

Cultivated species of lettuce originated in inland areas of Asia Minor in the Middle and Near East, and its wild (sibling) species are distributed world-wide. Lettuce, including many cultivated species, belongs to the family Compositae and is classified into five types: head lettuce, leaf bunching lettuce, cos or romaine lettuce, cos type and stem lettuce. Lettuce has 2n = 18 chromosomes, except for some artificially bred tetraploid varieties. Breeding of a disease-resistant variety is one of the most important aims to be achieved with regard to lettuce cultivation. Typical diseases in lettuce include microbial diseases such as *Sclerotinia* rot, drop, grey mould, bacterial soft rot and root rot, viral diseases such as Mosaic Virus (LMV, CMV) and Lettuce Big Vein and microplasma diseases such as Yellows. On the other hand, the existence of several wild species of lettuce resistant to Turnip Mosaic Virus (TMV), CMV, downy mildow pathogen, etc., has been reported (Netzer et al. 1976; Prrovdenti et al. 1980).

Breeding of disease-resistant varieties of lettuce has been performed mainly by crossing. However, recent advances in biotechnology have made new breeding techniques possible (see Bajaj 1986). Therefore, development of these new techniques based on cell selection, cell fusion, direct gene transformation, etc. is expected for the future breeding of disease-resistant lettuce varieties. Breeding lettuce to obtain a disease-resistant variety seems to be possible by transforming a lettuce with a disease-resistant genes in a wild species, since wild lettuce species possess disease-resistant genes which do not exist in cultivated species, and culture systems for plant regeneration from protoplasts of wild and cultivated species of lettuce have already been established. These techniques appear especially useful and effective for interspecific hybridization between allied species which are incompatible, or between remote species.

In order to apply such cell fusion and gene transformation techniques, it is necessary to establish systems for the isolation and culturing of protoplasts and plant regeneration from callus. Plant regeneration from callus has been reported by Alconero (1983) using cell suspension of cultivated and wild lettuce species. Plant regenerations from protoplasts, on the other hand, have been reported using cultivated lettuces (Berry et al. 1982; Engler and Grogan 1983; Matsumoto and

[1] Department of Cell Biology, National Institute of Agrobiological Resources, Tsukuba, Ibaraki 305 Japan

Biotechnology in Agriculture and Forestry, Vol. 8
Plant Protoplasts and Genetic Engineering I (ed. by Y. P. S. Bajaj)

Table 1. Protoplast culture of *Lactuca* species

Species/cv	Explant	Enzyme solution	Basal medium	Growth regulator	Growth response	Reference
Avon defiance	Leaf in vitro shoot root cotyledon	2% Rhozyme 2% Meicerase 0.3% Macerozyme CPW	Kp MS	BAP IAA	Plant	Berry et al. (1982)
Capitata cv. Climax	Leaf	0.3% Macerase 0.3% Wallase PVP·MES	MS	2.4-D BAP Zeatin	Plant	Engler and Grogan (1983)
Sinano green Summer jewel Calmar MR Stem lettuce	Leaf	1% Macerozyme 0.8% Cellurase-RS	MS	2.4-D BAP IBA GA_3	Plant	Matsumoto and Ono (1985)
Kaiser	Cotyledon	0.5% Cellurase-RS 0.02% Pectoriase-Y23	MS	NAA BAP	Plant	Suzuki et al. (1986)
Kalmar MR *L. serriola*	Leaf	0.4% Macerozyme 0.5% Cellurase-RS	MS	IAA BAP	Plant	Matsumoto (1987)
L. saligna	Leaf	1.5% Meicelase 0.5% Macerozyme	MS Kp	NAA BAP	Plant	Brown et al. (1987)
Cos lettuce	Cotyledon	0.5% Cellurase-RS 0.02% Pectroase-Y23 CPW	MS		Plant	Murayama et al. (1987)

BAP: 6-benzylaminopurine; 2,4-D: 2.4 dichlorophenoxy acetic acid; IAA: indol acetic acid; IBA: indol-3-butyricacid; NAA: naphthaleneacetic acid; GA_3: gibberellic acid; MES: 2(N-morpholino)ethane sulphonic acid; PVP: polyvinylpyrrolidon; CPW: CPW solution (Frearson et al. 1973).

Ono 1985; Suzuki et al. 1986, Ohyama et al. 1986, and a wild species (Brown et al. 1987). In addition, Engler and Grogan (1984) have reported genetic mutations in a plant regenerated from protoplasts. Recently, remarkable progress in such culture techniques has been made, including breeding a hybrid of cultivated and wild species of lettuce, as reported by Matsumoto (1987). The studies on protoplast culture are summarized in Table 1, and the culture system based on our work is described here.

2 Isolation of Protoplasts

The seeds of *Lactuca sativa* L. cv. Kaiser were surface-sterilized with 70% ethanol for 3 s, placed into 1% of sodium hypochlorite solution for 15 min and then washed three times with sterilized water. They were sown aseptically on a hormone-free MS medium (Murashige and Skoog 1962) supplemented with 2%

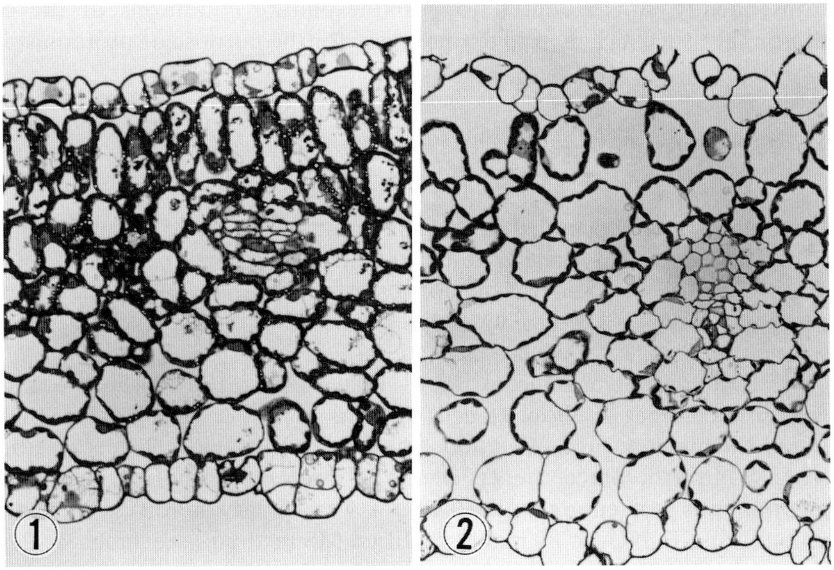

Fig. 1, 2. Light microscopy of lettuce mesophyll tissue (after 5 days of seedling) **1** Illumination 150 lx 24 h (×125), **2** Illumination 5000 lx 24 h (×125)

sucrose. A capped vial (4×10 cm) was used as the culture vessel and aseptic shoot cultures were maintained at 23 °C under 3000 lx white fluorescent lamps at photoperiods of 16 h.

Cotyledons taken from 72 to 96-h shoots were cut into sections, submerged in an enzyme solution (pH 5.8) containing 0.5 M mannitol, 0.5% cellulase Onozuka RS (Yakult Biochemicals Co., Ltd., Nishinomiya, Japan) and 0.02% pectolyase Y 23 (Seishin Pharmaceutical Co., Ltd. Japan) and incubated statically at 25 °C for 16–18 h in the dark. The enzyme mixture were filtered through a layer of Miracloth, and released protoplasts were sedimented by centrifugation at 100 *g* for 2 min and isolated by washing them three times with 0.5 M mannitol solution. The yield of protoplasts under these conditions was about 7×10^5 per 100 mg fresh weight of leaf material. Various explants i.e. cotyledon and mature leaves, root, hypocotyl, callus, etc. can be used for the purpose of protoplast isolation. Among them, the cotyledon is characterized by the stable production of aseptic material within a short period. The cotyledon, however, causes a low yield of protoplasts when young and also shows inefficient cell division during culturing when it is old. Because of such an age-dependent inefficiency of cell division, variation of growth stages of shoot affected by light and temperature conditions can be considered. The state of cytoplasm varied depending on the light condition during shoot culture (Figs. 1, 2). When mesophyll protoplasts obtained under 5000 lx fluorescent lamps were cultured, browning of the culture medium occurred at an early stage of the culturing along with the browning of protoplasts themselves. This result coincided with that reported by Webb et al. (1984), that the ability of adventitions bud formation decreased as the cotyledon became old.

Consequently, it is essential to examine culture conditions of the cotyledon thoroughly, such as light and temperature, for the purpose of protoplast isolation.

3 Protoplast Culture

Protoplasts were suspended in various liquid MS media (Table 2) supplemented with 0.5 M mannitol, 0.5 mg/l 6-benzyladenine (BAP) and 0.1 mg/l naphthalene acetic acid (NAA), and 2.5 ml portion of the suspension was transferred in a plastic Petri dish (6 cm in diameter). After sealing the dishes with Parafilm, they were cultured at 25 °C in the dark.

Cell wall regeneration occurred 2–3 days after culturing (Fig. 4) and each cell grew into colonies of about 10 cells after 10 days of culturing. After that, browning of the medium proceeded and, at the same time, colonies became brown and stopped growth. Since the efficiency of protoplast division in the basal MS medium was not enough, the medium was slightly modified by reducing the amount of NH_4NO_3 to 80 mg/l (modified MS medium, see Table 3). In addition, since aggregation of protoplasts was observed at an early stage of culturing, the

Table 2. Composition (mg/l) of the different media used for protoplasts culture

	A	B	C	D
NH_4NO_3	80	80	80	80
KNO_3	1900	1900	1900	1900
KH_2PO_4	170	170	170	170
$CaCl_2\ 2H_2O$	440	440	440	440
$MgSO_4 7H_2O$	370	370	370	370
$H_3BO\ 4H_2O$	6.2	6.2	6.2	6.2
$MnSO_4\ 4H_2O$	22.3	22.3	22.3	22.3
$ZnSO_4\ 4H_2O$	8.6	8.6	8.6	8.6
KI	0.83	0.83	0.83	0.83
$Na_2MoO_4\ 2H_2O$	0.25	0.25	0.25	0.25
$CuSO_4\ 5H_2O$	0.025	0.025	0.025	0.025
$CoCl_2\ 6H_2O$	0.025	0.025	0.025	0.025
$FeSO_4\ 7H_2O$	27.8	27.8	27.8	27.8
Na_2-EDTA	37.3	37.3	37.3	37.3
Inositol	100.0	100.0	100.0	100.0
Nicotinic acid	0.5	0.5	0.5	0.5
Pyridoxin-HCl	0.5	0.5	0.5	0.5
Thiamine-HCl	0.5	0.5	0.5	0.5
Glycine	2.0	2.0	2.0	2.0
Sucrose	10000.0	10000.0	10000.0	10000.0
D-Mannitol	0.5 (Mol)	6.0 (%)	3.0 (%)	0 (%)
NAA	0.1	0.5	0.5	0.5
BA	0.5	0.5	0.5	0.5
Agarose Type-2 (%)	0.1	0	0	0
pH	5.8	5.8	5.8	5.8

NAA: Naphthalene acetic acid; BA: 6 Benzyl aminopurine; Agarose: Sigma Type II. Medium; B.C.D: Concentration 7, 12, 17 days after the start of protoplast culture.

Table 3. Effect of NH_4NO_3 on cell division and colony formation from protoplasts

NH_4NO_3 (mg/l)	Cell division[a]	Colony formation[a]
1640	+ +	+
240	+ + +	+ +
80	+ + +	+ + +
0	+ +	+ +

[a] Visual estimation; + = Poor; + + = Good; + + + = Very good.

Table 4. Effect of time of addition and mannitol concentration in fresh medium

Exp. No.	Medium A (days)	Medium B (days)	Medium C (days)	Colony formation[a]
1	7[b]	12	17	+ + +
2	7	14	21	+
3	7, 12	17, 22	27, 32	+
4	7, 12, 17	–	–	+

The media contained modified MS salts, 1% sucrose, NAA, BA and 6% (Medium A), 3% (Medium B) or 0% (Medium C) mannitol.
[a] As in Table 2; [b] After the start of protoplast culture.

effect of agarose (Sigma type II) was evaluated and an aggregation-preventing effect was obtained by the addition of 0.1% agarose to the medium. The optimum density of protoplasts for cell division and colony formation was found to be $2-5\times10^4$/ml. The cell division efficiency decreased at lower density, and at a density of more than 10^5/ml, browning of the medium proceeded quickly, which made it difficult to continue the culturing. Such a deterioration of the culture conditions was improved by the intermittent addition of a fresh medium at 7, 12 and 17 days after culturing, which suppressed browning of the cells and medium (Table 4). The modified MS medium was added with 1% sucrose and certain amounts of NAA and BAP and used as the supplementary fresh medium, by stepwise reducing the amount of mannitol to 6%, 3% and 0%. The browning was suppressed by the supplement of the fresh medium. Callus formation was obtained by transferring the colonies, which had been grown 0.3–1 mm in diameter, onto the modified MS medium, which was further supplemented with 0.1 mg/l NAA, 0.5 mg/l BAP, 1% sucrose, 3% mannitol and 0.7% agar (Figs. 5–7).

A problem in lettuce protoplast culture is that browning of medium occurs with the development of colonies, which renders the continuation of the culturing difficult. Various attempts at anti-browning have been reported. Murayama et al. (1987) reported that the browning could be suppressed by adding agarose-entrapped active carbon after protoplast division. According to this report, 2% (w/v) activated charcoal powder is added to a medium and solidified with 1% agarose and 0.5 ml of the entrapped activated charcoal is added to a 6-cm Petri dish. Addition of activated charcoal at an early stage of culturing, however, inhibited pro-

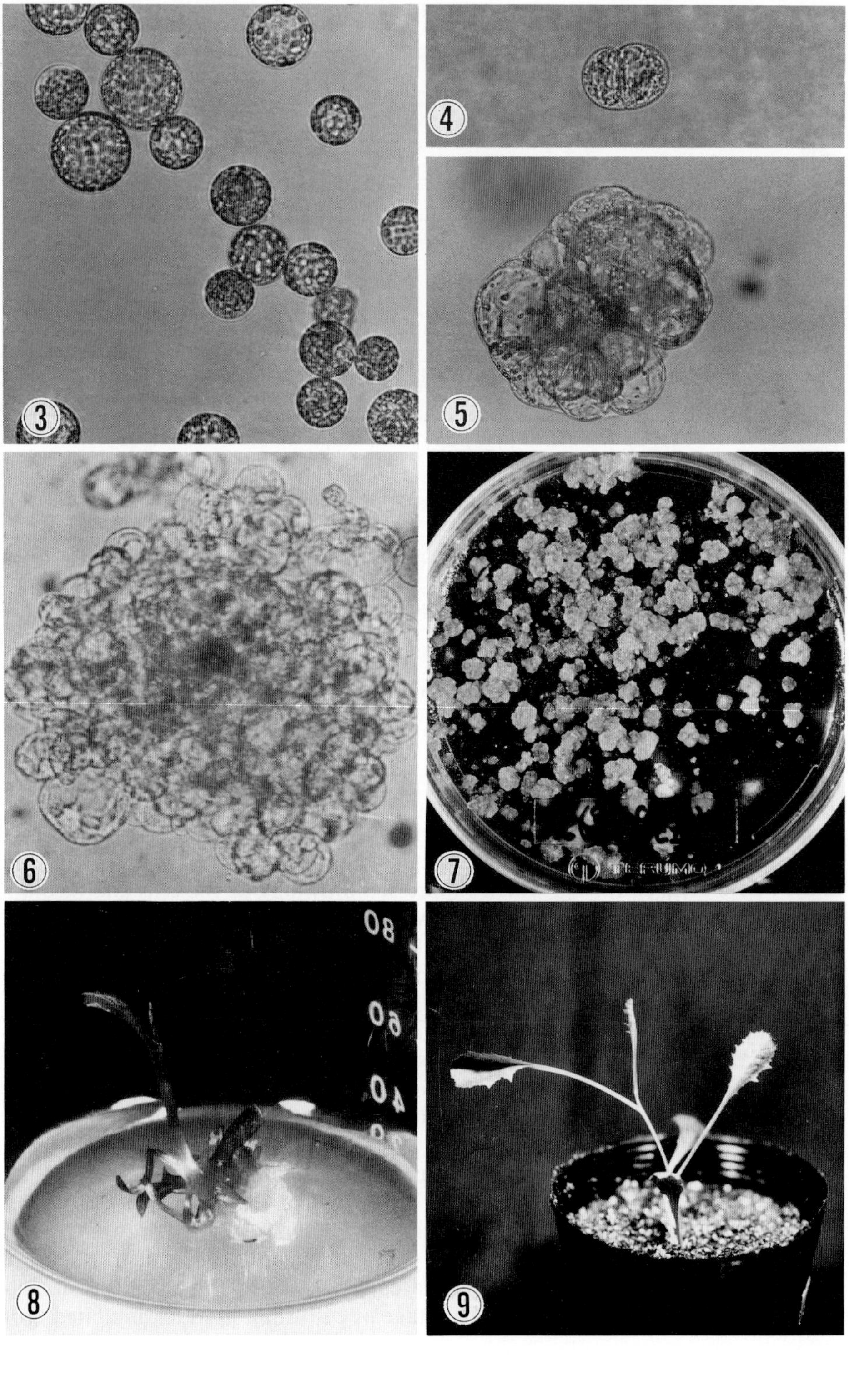
3
4
5
6
7
8
9

toplast division and thus had an inferior effect. By the agarose bead method, protoplasts were entrapped in the modified MS medium supplemented with 0.8% agarose and the solid was cut into small pieces and suspended in a liquid medium. Mannitol-eliminated modified MS medium was the best liquid medium.

4 Plant Regeneration

When calli which have been grown into 2–3 mm on the agar medium were transferred onto the modified MS medium containing 0.5 mg/l NAA and 1 mg/l BAP, half of them showed induction of adventitious buds (Figs. 8, 9). Complete plants were obtained by culturing these adventitious buds on MS medium containing 0.1 mg/l NAA and BAP in order to grow their stems and leaves and then transferring them onto a hormone-free rooting medium which was prepared by adjusting the MS contents to one-tenth. Difficult rooting was sometimes improved by the addition of 0.5–1 mg/l IBA. After acclimatization and potting, plants with well-developed roots grew normally, blossomed and bore seeds. It took 140–150 days to obtain the plant from protoplasts.

5 Protoplast Fusion

Cell fusion between lettuce and a Philadelphia fleabane (*Erigeron philadelphicus* L.) was attempted, aiming at applying the result to a selection method of fused cells, since a Paraquat (herbicide)-resistant biotype was found in *E. philadelphicus* (Watanabe et al. 1982) and both plants belong to the Compositae. Protoplasts of *E. philadelphicus* were isolated by cutting in vitro-grown leaves, shaking them (25 °C, 70 rpm) for 60 min in an enzyme solution containing 0.75% Macerozyme, 2% Cellulase YC and 0.5 M mannitol, and filtering the enzymes through a layer of Miracloth, followed by the same treatment as that for the lettuce protoplast isolation. There was a tendency to decreased activity when the time of treatment was extended. Identification of fused cells is easy in the case of these two plants because of the morphological differences of chloroplasts in their protoplasts. When protoplast fusions by PEG and dextran methods were compared, the former showed superior performance, and a rate of 14% protoplast fusion was

Fig. 3. Freshly isolated protoplasts

Fig. 4. First division after 2–3 days of protoplast culture

Fig. 5. Colony formation after 7 days of protoplast culture (Medium A)

Fig. 6. Colony formation after 12 days of protoplast culture (Medium B)

Fig. 7. Callus formation after 60 days of protoplast culture (MS medium + BA 0.5 g/l + NAA 0.5 mg/l)

Fig. 8. Shoot regeneration from callus after 90 days of protoplast culture (MS medium + BA 1 mg/l + NAA 0.1 mg/l)

Fig. 9. Whole plant obtained from protoplast, after 120 days of protoplast culture obtained using 50% concentration of PEG (1540 Da). When the fused protoplasts were cultured using the medium for lettuce protoplasts, their divisions were induced, but they shrank at the colony stage. Examination of their culture conditions is now in progress. Under the state of protoplasts, Paraquat-resistant individuals of *E. philadelphicus* survived for 3 days with no symptoms in the presence of 0–25 mg/l of the drug, while sensitive ones were affected by 0.025 mg/l of the drug and died at a concentration of 25 mg/l.

6 Conclusions and Prospects

It has been reported that mutations occur in regenerated plants from lettuce protoplasts (Engler and Grogan 1984). This finding could be applied to the increment of mutation frequency in cultured cells and selection of individuals with new characteristics.

In addition to the presence of several viral disease-resistant wild species in sibling species of lettuce, as mentioned already, many other Compositae species can be regarded as useful genetic resources. For example, it is quite possible to establish a cell selection method using a herbicide-resistant species of a Philadelphia fleabane or to breed a lettuce transformed by a new character through cell fusion with other herbicide-resistant species.

Many studies have been reported on the differentiation and growth of organs from lettuce tissue (see Alconero 1988). Growing point cultures and adventitious bud inductions have been reported (Marcia et al. 1967; Koevary et al. 1978; Webb et al. 1984; Michael 1977; Enomoto et al. 1985; Oka et al. 1986). Sasaki (1974, 1975, 1979a, b, 1982) has minutely studied and reported on induction conditions of adventitious bud from hypocotyl tissue. These achievements can be applied to the growth of an individual which has a useful genetic character.

Achievements in the establishment of protoplast culture system and redifferentiation of fused cells have been reported using cultivated and wild species of lettuce. Judging from these technical advances, it seems that production of new breeding materials by manipulation of lettuce cells is quite possible.

7 Protocol

(1) Seed sterilization and germination (cv. Kaiser seed)
- a) 70% Ethyl alcohol; 2–3 s
- b) 1% Na hypochlorite; 10 min
- c) Sterile tap water rinse three times
- d) Seedling medium; MS + 2% sucrose + 0.2% gellangam, pH 5.8 (3000 lx 16 h/light)

(2) Protoplast isolation (4-day seedlings)
- a) Enzyme solution; 0.5% cellulase RS, 0.02% pectinase, 0.5 M Mannitol, pH 5.8
- b) Dark, overnight
- c) Filtration. Centrifugation 100 g, 2 min, 0.5 M mannitol rinse three times

(3) Protoplast culture

a) Modified MS+1% sucrose+0.1% agarose+0.5 mg/l BAP−0.1 mg/l NAA, pH 5.8
b) Protoplast density, 2×10^4/ml
c) Dark conditions, 25 °C
d) Medium addition 7, 12, 17 days after culture
e) Callus formation. MS+2% sucrose+0.2% agar+0.1 mg/l NAA−0.5 mg/l BAP

(4) Plant regeneration. MS+2% sucrose+0.2% agar+0.1 mg/l NAA−0.5 mg/l BAP

(5) Rooting MS/2+2% sucrose+0.7% agar

Days required: Colony formation, 14−17 days.
Callus formation, 30 days.
Plant regeneration, 90 days.

References

Alconero R (1983) Regeneration of plants from cell suspension of *Lactuca saligna*, *Lactuca sativa*, and *Lactuca serriola*. Hortic Sci 18:303−307

Alconero R (1988) Lettuce (*Lactuca sativa* L.). In: Bajaj YPS (ed) Biotechnology in agriculture and forestry 6. Crops II. Springer, Berlin Heidelberg New York Tokyo, pp 351−369

Bajaj YPS (ed) (1986) Biotechnology in agriculture and forestry 2. Crops I. Springer, Berlin Heidelberg New York Tokyo

Berry SF, Lu DY, Pental D, Cocking EC (1982) Regeneration of plants from protoplasts of *Lactuca sativa* L. Z Pflanzenphysiol 108:31−38

Brown C, Lucas JA, Power JB (1987) Plant regeneration from protoplasts of a wild Lettuce species (*Lactuca saligna* L.). Plant Cell Rep 6:180−182

Engler DE, Grogan RG (1983) Isolation, culture and regeneration of lettuce leaf mesophyll protoplasts. Plant Sci Lett 28(2):223−229

Engler DE, Grogan RG (1984) Variation in lettuce plants regenerated from protoplasts. J Hered 75:426−430

Enomoto S, Oka S, Ohyama K (1985) Structural studies in shoot formation from lettuce cotyledons culture in vitro. Jpn Assoc plant tissue culture 9th Symp (Kobe, Japan), pp 98−99

Frearson EM, Power JB, Cocking EC (1973) The isolation, culture and regeneration of *Petunia* leaf protoplasts. Dev Biol 33:130−137

Koevary K, Rappaport L, Morris LL (1978) Tissue culture propagation of head lettuce Hortic Sci 13:39−41

Marcia R, Doerschug C, Miller O (1967) Chemical control of adventitious organ formation in *Lactuca sativa* explants. Am J Bot 54:410−413

Matsumoto E, Ono K (1985) Effects of some factors on plant regeneration from lettuce protoplast. Jpn J Breed 35 (Suppl. 2) 20−21

Matsumoto E (1987) Production of somatic hybrids between *Lactuca sativa* and *L. serriola* by cell fusion. Jpn J Breed 35 (Suppl. 1) 134−135

Michael Samish (1977) Phytochrome-regulated organogenesis in lettuce tissue culture. Nature 270:49−50

Murashige T, Skoog F (1962) A revised medium for rapid growth and bioassays with tobacco tissue cultures. Physiol Plant 15:473−497

Murayama T, Enomoto S, Ohyama K (1987) Progress on mesophyll protoplasts culture of lettuce. Jpn Assoc Plant Tissue Culture 10th Symp (Sendai, Japan), p 212

Netzer D, Globerson D, Sacks J (1976) *Lactuca saligna* L. a new source of resistance to downy mildow (*Bremia lactucae* Reg.). Hortic Sci 11:612−613

Ohyama K, Enomoto S, Suzuki H (1986) Protoplast regeneration system of lettuce (*Lactuca sativa* L.). In: Somers DA, Gengenbach GB, Biesboer DD, Hackett WP, Green CE (eds) Proc VI Int Congr plant cell and tissue culture, Minnesota, p 339

Oka S, Enomoto S, Ohyama K (1986) Some morphological aspects in shoot formation from lettuce cotyledons cultured in vitro. In: Somers DA, Gengenbach GB, Biesboer DD, Hackett WP, Green CE (eds) Proc VI Int Congr plant cell and tissue culture, Minnesota, p 339

Prrovdeti R, Robinson RW, Shail JW (1980) A source of resistance to a strain of cucumber mosaic virus in *Lactuca saligna* L. Hortic Sci 15:528–529

Sásaki H (1974) Physiological and morphological studies on development of vegetable crops: organ formation of lettuce tissue cultured in vitro. J Hokkaido Univ Educ (Sect II B) 25:17–27

Sasaki H (1975) Physiological and morphological studies on development of vegetable crops III. Adventitious bud formation of callus tissues derived from lettuce hypocotyl. J Jpn Soc Hortic 44:138–143

Sasaki H (1979a) Physiological and morphological studies on development of vegetable crops IV. Effect of various media on the adventitious bud formation of lettuce hypocotyl tissue culture in vitro. J Jpn Soc Hortic 47:479–484

Sasaki H (1979b) Physiological and morphological studies on development of vegetable crops V. Action of 2,4-D and kinetin on the adventitious bud formation of lettuce hypocotyl tissue cultured in vitro. J Jpn Soc Hortic 48:61–66

Sasaki H (1982) Effect of temperature and light on adventitious bud formation of lettuce hypocotyl tissue cultured in vitro. J Jpn Soc Hortic 51:187–194

Shimizu S (1977) The encyclopedia of vegetables and horticulture. Yokendou (Tokyo, Japan) (in Japanese), pp 1262–1277

Suzuki Y, Enomoto S, Ohyama K (1986) Regeneration from leaf mesophyll protoplasts of *Lactuca sativa* L. Plant Tissue Cult 3(2):83–85

Watanabe Y, Honma T, Ito K, Miyahara M (1982) Paraquat resistance in *Erigeron philadelphicus* L. Weed Res (Japan) 27:49–54

Webb DT, Torres LD (1984) Interaction of growth regulators, explants age, and culture environment controlling organogenesis from lettuce cotyledon in vitro. Can J Bot 62:586–590

II.9 Regeneration of Plants from Protoplasts of Butterbur (*Petasites japonicus* Fr. Schmidt)

KAZUNORI YABE[1], TAKESHI NISHIO[2] and KENJI TAKAYANAGI[2]

1 Introduction

1.1 Importance and Distribution of Japanese Butterbur

Japanese butterbur (Fuki) is a perennial composite and distributed throughout Japan, Korea and China. Several species related to this plant, e.g. *Petasites fragrans* Presl, are distributed in Europe (Shibata and Shimizu 1978). In Japan, this plants is widely distributed from Hokkaido to Kyushu, and has been cultivated since the tenth century for its succulent petioles as vegetables (Fig. 1). The flower stalk of the butterbur (Fukino-tou) is highly valued as one of the seasonal foods which heralds the arrival of early spring.

In Japan, the area under cultivation of Japanese butterbur is 1034 ha, and annual production amounts to 33000 tons (1984). Butterbur requires plenty of soil moisture, and loam is the most suitable soil texture for this crop. Optimum temperature of growth for cv. Aichi-wase-buki ranges between 10 and 23 °C. In Tokai District (annual mean temperature 15 °C), sprouting of scape begins in January to February, and petioles grow well in the middle of May. However, because of heat and dryness, growth of petioles is inhibited in summer (July to September); they grow in fall again.

Distribution of the subterranean stem of butterbur is 10–15 cm in depth, branching and elongation of the subterranean stem continue from spring through fall, but stop growing during winter. It is possible to harvest the petioles of butterbur from April to May in the open field. In forcing culture in the greenhouse, the subterranean stem clones are stored at 2 °C for 30–40 days during summer. In September, the clones are planted in the greenhouse covered with plastic film, and the petioles can be harvested from November to April. The yields are 60–80 t/ha. As in recent years the virus diseases of butterbur (Tochihara and Tamura 1976) have increased on account of the continuous cropping and vegetative propagation of the rootstock, to counteract the damage, butterbur plants free from cucumber mosaic virus (CMV) and butterbur mosaic virus (BuMV) were obtained by tissue culture (Morishita et al. 1980; Morishita 1986; Yabe et al. 1986b).

[1] Aichi Agric. Res. Centr., Aichi 480-11, Japan
[2] Natl. Res. Inst. Veg. Ornam. Pl. and Tea, Mie 514-23, Japan

Biotechnology in Agriculture and Forestry, Vol. 8
Plant Protoplasts and Genetic Engineering I (ed. by Y. P. S. Bajaj)

Fig. 1. Japanese butterbur cultivation in a plastic film greenhouse

1.2 Significance and Utilization of Protoplast Culture for Butterbur

The major cultivars of Japanese butterbur, i.e. Aichi-Wase-buki, Mizu-buki and Akita-Oo-buki, were selected from wild-grown butterbur (Imazu 1961). There were some distinctions in characteristics and isozyme patterns among these cultivars (Morishita 1987).

Although the butterbur is a dioecious plant, male plants are seldom found in cultivated varieties (Imazu and Fujishita 1961). The ploidy of both Aichi-Wase-buki and Mizu-buki is triploid ($2n = 87$, $x = 29$) and they are usually propagated vegetatively by rhizomes (Imazu and Fujishita 1962a, b). As these facts explain the difficulties in applying cross-breeding to butterbur, it was necessary to apply protopast, cell and tissue culture in butterbur breeding.

2 Isolation of Protoplasts

Aichi-Wase-buki was selected as experimental material, because it is widely grown in the Aichi Prefecture and the other butterbur-growing districts. The rhizomes were used for dissecting the shoot tips, surface-sterilized with 70% ethanol for 60 s and sodium hypochlorite solution (0.5% available chlorine) for 10 min, and rinsed twice with sterile water. The shoot tips were placed on agar media in a plastic box (plant box, 6×6×9.5 cm). Young leaves, 3–5 cm in length, of aseptic plants were cut into small pieces (1 mm in width) and incubated in the enzyme solution; 0.1% Pectolyase Y-23, 0.5% Cellulase Onozuka RS, 0.5% Driselase and

Fig. 2. Plant regeneration from mesophyll protoplasts of butterbur. **A** Isolated protoplasts from mesophyll cells (×160). **B** First cell division (×160). **C** Cell cluster (×160). **D** Cell colony (×80). **E** Shoot regeneration from calluses. **F** Plants regenerated in a plant box

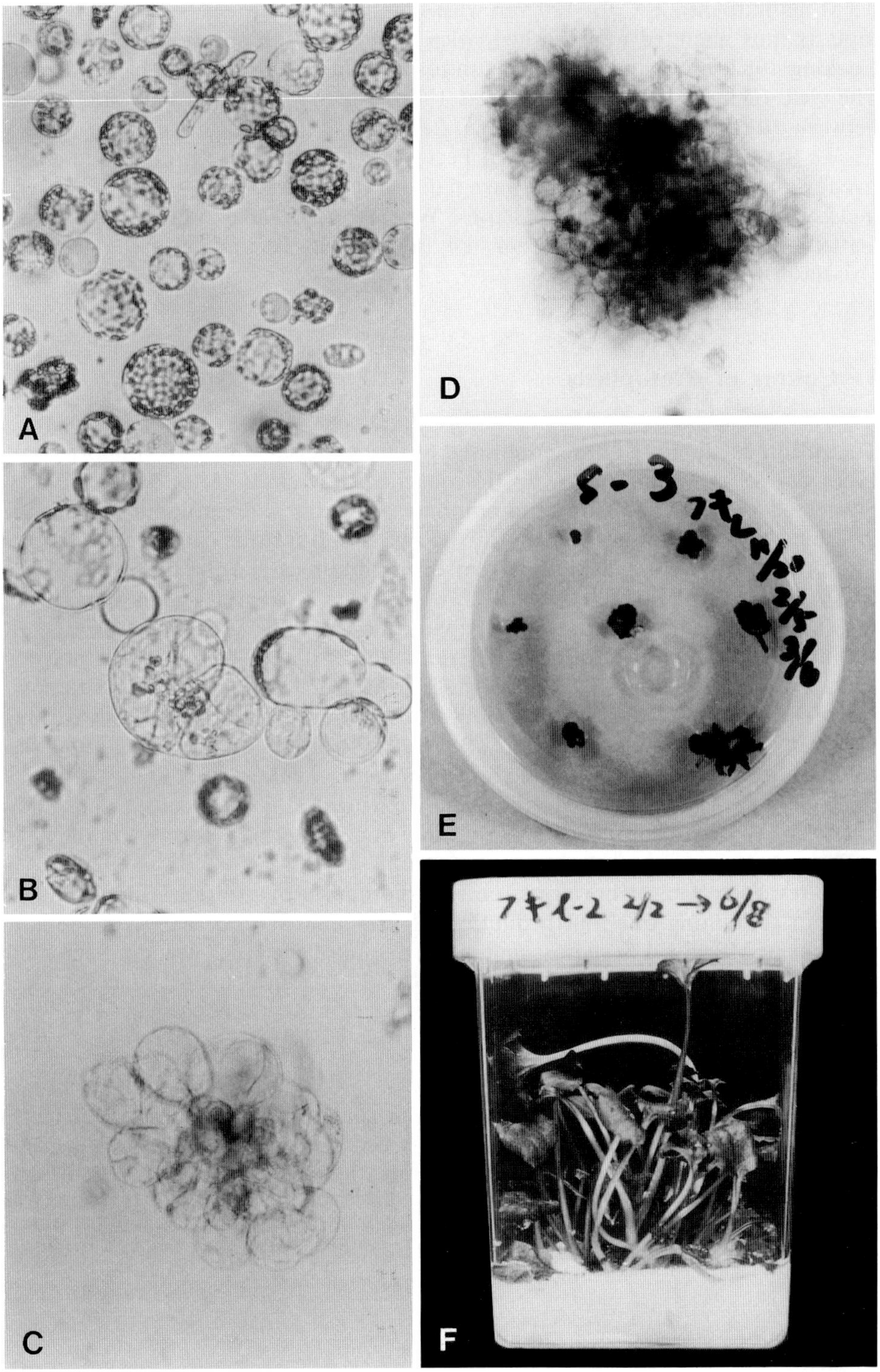

0.5 M mannitol at 27 °C for 120 min with shaking at 80 strokes per min. The protoplasts thus liberated were washed twice with CPW salt solution (Morgan and Cocking 1982) containing 0.5 M mannitol by centrifugation at 100 *g* for 2 min. The yield of mesophyll protoplast was 1.4×10^5 per g fresh wt, their size varied between 30 to 50 µm in diameter (Fig. 2A).

We first tried enzyme mixtures of Pectolyase and Cellulase, which were effective for taro, garlic, scallions and wakegi (Oosawa and Takayanagi 1984), but were ineffective for mature butterbur leaves. We then used axenic plants, and young leaflets were treated with an enzyme mixture of Pectolyase, Cellulase and Driselase.

3 Culture of Protoplasts

The culture media tested were composed of inorganic salts of B5 (Gamborg et al. 1968) or MS and vitamins of 8p (Kao and Michayluk 1975) or MS, and supplemented with various kinds and amounts of auxins and cytokinins, as shown in Table 1. The protoplast density of the initial culture was ca. 1×10^5 cells/ml. The protoplasts were cultured in 6-cm disposable plastic Petri dishes at 25 °C, under dim light.

The media of one-fourth MS salts with one-half MS vitamins supplemented with hormones induced protoplast divison within 7 days (Fig. 2b), while no protoplast division was observed in the media of B5 or MS salts with 8p vitamins. In the medium supplemented with 1 mg/l NAA, and 0.5 mg/l BA, colony formation was observed (Table 1, Fig. 2C, D). After one or one and a half months, the colonies formed were transferred onto an agar medium containing 1/4 MS minerals, 1/2 MS vitamins, 1 mg/l NAA, 0.5 mg/l BA, 1% sucrose and 0.8% agar. With 3-month culture, calli of 2–5 mm in diameter were formed on the agar medium.

To test the effect of hormones on the resumption of protoplast division, the medium containing one-fourth dilution of inorganic salts of MS and one-half vitamins of MS with 1% sucrose and 0.5 M mannitol at pH 5.7 was supplemented with various concentrations of NAA (0.2, 0.5, 1, 2, 5 and 10 mg/l) and BA (0.2,

Table 1. Medium for Japanese butterbur protoplast culture (0.5 M mannitol, pH 5.7). (Yabe et al. 1986a)

Minerals	Vitamins	Sucrose %	Glucose %	NAA mg/l	2,4-D mg/l	BA mg/l	Zeatin mg/l	Protoplast division[a]
B5	8p	1	0.5	0.5	0.5		0.5	No division
1/4 MS	1/2 MS	1		1		0.5		Colony formation
1/4 MS	1/2 MS	1			1	0.5		1st division
1/4 MS	8p	1	1	1	0.25		0.5	No division
1/4 MS	8p	1	1	1	1		0.5	No division

[a] Cultured for 43 days.

Table 2. Effect of NAA and BA concentration on protoplast division[a]. (Yabe et al. 1986a)

BA (mg/l)	NAA (mg/l)					
	0.2	0.5	1	2	5	10
0.2	−	−	−	□	+	+
0.5	−	−	+	□	+	−
1	−	−	−	+	−	−
2	−	−	−	+	−	−

[a] −: no division; +: 2∼5 cells; ○: 6∼10 cells. □: 11∼15 cells.
Cultured for 22 days in the medium containing 1/4 MS minerals, 1/2 MS vitamins, 1% sucrose, 0.5 M mannitol, pH 5.7 and various concentrations of NAA and BA.

Table 3. Effect of NH_4NO_3 and other MS minerals concentration on protoplast division[a]. (Yabe et al. 1986a)

Dilution of MS salts[b]	NH_4NO_3 (mg/l)					
	0	200	400	800	1600	3200
1/1	−	−	−	−	−	−
1/2	+	○	○	○	+	−
1/4	□	×	□	○	+	−
1/8	○	□	○	+	+	+

[a] −: no division; +: 2∼5 cells; ○: 6∼10 cells, □: 11∼15 cells, ×: 16∼20 cells, Cultured for 22 days.
[b] MS basic medium with various concentrations of NH_4NO_3, 1/2 vitamins of MS, 1 mg/l NAA, 0.5 mg/l BA, 1% sucrose, 0.5 M mannitol, pH 5.7.

0.5, 1 and 5 mg/l). The results in Table 2 show that the combination of 2 mg/l NAA, and 0.2−0.5 mg/l BA was the most favourable.

In protoplast culture, it has been said that the concentration of ammonium salt in the MS is too high for the protoplast to survive (Kao et al. 1973; Meyer 1974; Uchimiya and Murashige 1976). To clarify the effects of ammonium salt and MS salts on protoplast division, modified MS medium (diluted NH_4NO_3 concentration) was applied. Preparing modified MS media with various concentrations of ammonium nitrate (0, 200, 400, 800, 1600 and 3200 mg/l) containing the same amount of vitamins (1/2 of MS), 1 mg/l NAA, 0.5 mg/l BA, 1% sucrose and 0.5 M mannitol, the effect of dilution of the media on protoplast division was investigated (Table 3). The most preferable was a one-fourth dilution of MS salts with 200 mg/l NH_4NO_3, equivalent to 1/33 NH_4NO_3 concentration (50 mg/l) of the original MS medium (1650 mg/l NH_4NO_3).

For protoplast culture, the authors had success using the MS medium as a basic one, which has frequently been used for shoot tip culture (Matsubara and Mitsunobe 1978) and organ culture (Morishita et al. 1980; Yabe et al. 1986b).

Concentration of auxin and cytokinin supplemented to the basic medium was 2 mg/l NAA, and 0.2−0.5 mg/l BA, respectively, and satisfactory results were obtained for protoplast division. This hormone balance was the same as that for callus formation from leaf blade in butterbur (Yabe et al. 1986b).

Osmotic regulation during protoplast culture was successful with 0.5 M mannitol at initial culture and transferance to mannitol-free media after one and a half months of culture.

4 Regeneration of Plants

When calli of 2–5 mm in diameter were transplanted on an agar medium, it was found that shoot regeneration occurred about 1 month after transplanting the calli onto MS medium containing 3% sucrose, 1 mg/l BA and 0.8% agar (Fig. 2E). Figure 2F shows one of the regenerated plants from mesophyll protoplasts of butterbur at 6 month of culture.

The effect of hormone balance on shoot regeneration from calli was investigated in various combinations of NAA and BA. The shoot regeneration rate was compared with media containing various concentrations of NAA and BA. The higher the concentration of BA with a small amount of NAA in the media, the higher the rate of shoot regeneration obtained (Table 4). The media containing 0.02 mg/l NAA and 4 mg/l BA gave the best results.

Table 4. The effect of hormone balance on the shoot regeneration from calluses. (Yabe et al. 1986a)

Medium[a] no.	NAA mg/l	BA mg/l	Number of calli transplanted	Number of shoot regenerated calli	Percentage of regenerated calli
S-1	0	1	64	0	0
S-2	0	2	28	4	14.3
S-3	0	4	28	4	14.3
S-4	0.02	1	35	2	5.7
S-5	0.02	2	28	1	3.6
S-6	0.02	4	28	9	32.1
S-7	0.2	1	55	2	3.6
S-8	0.2	2	28	5	17.9
S-9	0.2	4	28	7	25.0

[a] Basic medium: MS salts, MS vitamins, 3% sucrose, 0.8% agar.

5 Fusion of Protoplasts

The major cultivar Aichi-Wase-buki has some defects in yield and quality. To counteract these, somatic hybridization by protoplast fusion has been tried between Aichi-Wase-buki and Mizu buki or Akita-Oo-buki, the latter two cultivars having good characters, high yields as well as high quality. Protoplast fusion was induced by polyethelene glycol (PEG), and high pH and high Ca condition (Fig. 3), but somatic hybrid plants have not yet been obtained.

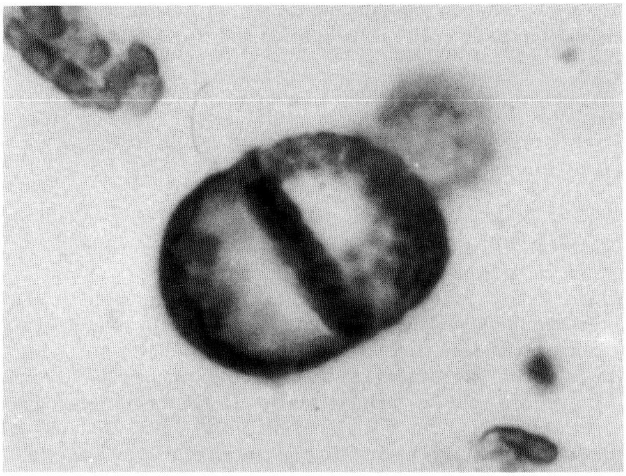

Fig. 3. Protoplast fusion of butterbur. *Left* Aichi-Wase-buki; *right* Mizu-buki. Fused in 33% PEG 4000 containing 10 mM $CaCl_2 \cdot 2H_2O$, 0.7 mM KH_2PO_4, and 0.2 M glucose at pH 5.8

6 Conclusion and Prospects

Achievement of protoplast culture and plant regeneration is a prerequisite for protoplast fusion, induction and selection of useful mutants, and gene transfer into plants. In particular, as butterbur is a vegetatively propagated plant, advantage may be taken of effective induction and selection of useful characters through protoplast culture. It is not necessary to consider the seed fertility of the somatic hybrids obtained. Therefore, the study presented here will contribute to biotechnological work in vegetatively propagated plants.

7 Protocol

(1) Optimal Conditions/Medium for the Isolation of Protoplasts

Mesophyll protoplasts of Japanese butterbur grown in axenic conditions can be prepared in the enzyme solution containing 0.1% Pectolyase Y-23, 0.5% Cellulase Onozuka RS, 0.5% Driselase and 0.5 M mannitol at 27 °C for 120 min with shaking at the rate of 80 strokes per min.

(2) Optimal Conditions/Medium for the Culture of Protoplasts

Initial culture of protoplasts carried out in the medium of one fourth modified MS salts (diluted ammonium nitrate, 200 mg/l NH_4NO_3) with one half MS vitamins supplemented with 1 mg/l NAA and 0.5 mg/l BA. Protoplast density of initial culture is ca. 1×10^5 cells/ml. The culture is kept at 25 °C under dim light. Optimum hormone balance for protoplast division is 2 mg/l NAA and 0.2–0.5 mg/l BA. Callus formation from colonies is induced on agar medium containing 1/4 MS salts, 1/2 MS vitamins, 1 mg/l NAA, 0.5 mg/l BA, 1% sucrose and 0.8% agar.

(3) Optimal Conditions/Medium for the Regeneration of Plants

Shoot regeneration from calli is obtained on MS medium containing 3% sucrose, 1 mg/l BA and 0.8% agar. Optimum hormone balance of shoot regeneration from calluses is 0.02 mg/l NAA and 4 mg/l BA.

Acknowledgments. The authors express sincere thanks to Dr. S. Miyajima, Laboratory of Biotechnology, Division of Agrobiological Resources, Aichi-ken Agricultural Research Center, for his useful suggestions during the preparation of the manuscript. This work was partially supported by a special research grant on Cell Fusion and Nuclear Transfer from the Min. Agri. Forest. and Fish., Japan.

References

Gamborg OL, Miller RA, Ojima K (1968) Nutrient requirements of suspension cultures of soybean root cells. Expt Cell Res 50:151–158

Imazu T (1961) Morphological, ecological and cytological studies on cultivated and wild butterburs, *Petasites japonicus* Maxim. I. On the morphology of leaf, sprouting and flowering habits, and germination of seed. Jpn Soc Hortic Sci 30:233–240

Imazu T, Fujishita N (1961) Morphological, ecological and cytological studies on cultivated and wild butterburs, *Petasites japnicus* Maxim. II. On the morphology of flower, sex reatio and secondary sex characters. J Jpn Soc Hortic Sci 30:291–298

Imazu T, Fujishita N (1962a) Morphological, ecological and cytological studies on cultivated and wild butterburs, *Petasites japonicus* Maxim. III. On the pollen and seed fertilities. J Jpn Soc Hortic Sci 31:23–29

Imazu T, Fujishita N (1962b) Morphological, ecological and cytological studies on cultivated and wild butterburs, *Petasites japonicus* Maxim. IV. On the chromsome numbers. J Jpn Soc Hortic Sci 31:293–302

Kao KN, Michayluk MR (1975) Nutritional requirements for growth of *Vicia hajastana* cells and protoplasts at very low protoplast density in liquid media. Planta 126:105–110

Kao KN, Gamborg OL, Michayluk MR, Keller WA, Miller RA (1973) The effect of sugar and inorganic salts on cell regeneration and sustained division in plant protoplasts. Colloq Int Cent Natl Res Sci 212:207–213

Matsubara S, Mitsunobe T (1978) Meristem culture of japanese butternbur. Abstr Jpn Soc Hortic Sci, Spring Meeting 1978, pp 208–209

Meyer Y (1974) Isolation and culture of tobacco mesophyll protoplasts using a saline medium. Protoplasma 81:363–372

Morgan NA, Cocking EC (1982) Plant regeneration from protoplasts of *Lycopersicon esculentum* Mill. Z Pflanzenphysiol 106:97–104

Morishita M (1986) Butterbur (*Petasites japonicus* Miq.). In: Bajaj YPS (ed) Biotechnology in agriculture and forestry, Vol 2, Crops 1. Springer, Berlin Heidelberg New York, pp 421–428

Morishita M (1987) Characteristics and peroxidase isozyme variation of butterbur (*Petasites japonicus* Miq.) in Japan. Rep Soc Crop Breed, Kinki 32:9–16

Morishita M, Kagi T, Yamada K (1980) Mass production of virus-free butterbur plants (*Petasites japonicus* Maxim.) from callus derived from pith tissue of the spath and petiole in vitro. Bull Osaka Agric Res Cent 17:1–6

Murashige T, Skoog F (1962) A revised medium for rapid growth and bioassays with tobacco tissue cultures. Physiol Plant 15:473–497

Oosawa K, Takayanagi K (1984) Protoplast approach in vegetable breeding I. Isolation and culture conditions of protoplasts from mesophyll cells of vegetable crops. Bull Veg Ornamental Crops Res Stn A-12:9–28

Shibata H, Shimizu S (1978) Chemovars of Japanese butterburs (*Petasites japonicus* Maxim.). Bull Fac Agric Shinushu Univ 15:195–207

Tochihara H, Tamura M (1976) Viruses in Japanese butterbur (*Petasites japonicus* Miq.). Ann Phytopathol Soc Jpn 42:533–539

Uchimiya H, Murashige T (1976) Influence of the nutrient medium on the recovery of dividing cell from tobacco protoplasts. Plant Physiol 57:424–429

Yabe K, Nishio T, Takayanagi K (1986a) Plant regeneration from mesophyll protoplasts of Japanese butterbur (*Petasites japonicus* Fr. Schmidt). Jpn J Breed 36:131–137

Yabe K, Sakurai Y, Iida T, Washida S (1986b) Production of Japanese butterbur (*Petasites japonicus* Fr. Schmidt) virus free plants by leaf and petiole tissue culture. Res Bull Aichi Agric Res Ctr 18:102–109

II.10 Isolation and Culture of Orchid Protoplasts

SABURO YASUGI[1]

1 Introduction

1.1 Importance and Distribution of *Dendrobium*

The Orchidaceae are one of the largest families of flowering plants and the most highly evolved family among the monocotyledones. In the Schlechter system, it is divided into 88 subtribes, over 660 genera, and at least 25000 species. Additionally over 25000 hybrids have been produced by crossing wild and cultivated forms, according to the book *Orchids* (Shuttlework et al. 1970).

Dendrobium, which is one of the largest genera of over 1500 species in orchids, is found throughout the Asian tropics and subtropics including the Himalayas to Thailand, Indochina, Taiwan and China. *Dendrobium nobile*, used in the present investigation, is a very popular deciduous plant, with many horticultural varieties which are often used in hybridization (Thomas 1981).

According to the book *The Orchids: Scientific Studies* (see Luning 1974) *Dendrobium nobile* contains some alkaloids such as dendrobine (Suzuki and Keimatsu 1932), nobiline (Yamamura and Hirata 1964), nobibonine (Onaka et al. 1965), n-methyldendrobinium (Inubushi 1967), n-isopentenyldendrobinium (Inubushi and Sasaki et al. 1964), dendroxine and 6-hydroxydendroxine (Hedman and Leander 1972). Dendrobine was isolated from the drug "Chin Shih Hu" made of dried stems of *D. nobile*, which was the first successful attempt to isolate an alkaloid from an orchid. This drug had long been used in China as a tonic and antipyretic. Therefore, *Dendrobium nobile* is an economically important orchid as well as *Dendrobium phalaenopsis*.

In Japan in 1985, the total agricultural production was 12000 billion yen (Statistical Association of Agriculture and Forestry 1986), of which the gross production of the flower industry was 414.4 billion including 157.6 billion yen in cut flower products and 61.2 billion yen in pot plant products. The gross production of orchids was 4.5 billion yen, including 3.3 billion yen in cut flower products and 1.2 billion yen in pot plant products. Regarding the cut flower industry, the gross amounts of import and export were 5.3 billion yen and 71.8 million yen, respectively. In orchids, the gross amounts of import and export were 1036 million yen and 6.3 million yen, respectively (Table 1).

[1] Department of Horticulture, University of Minamikyushu, Hibarigaoka, Takanabe-cho, Miyazaki Prefecture, Japan 884

Biotechnology in Agriculture and Forestry, Vol. 8
Plant Protoplasts and Genetic Engineering I (ed. by Y. P. S. Bajaj)

Table 1. Production of the flower industry in Japan in 1986

Production of all flowering plants		Import	Export
Gross	414.4 bil.	5.3 bil.	71.8 mil.
Cut flower	157.6 bil.		
Pot flower	61.2 bil.		
Production of orchids		**Import**	**Export**
Gross	4.5 bil.	1036 mil.	6.3 mil.
Cut flower	3.3 bil.		
Pot flower	1.2 bil.		

The conventional methods of multiplication in *Dendrobium nobile* are division, stem-cutting and air-laying using keikis, and the non-conventional methods are aseptic seed culture and mericlone.

1.2 Significance and Use of Protoplasts

Dendrobium nobile has the advantage of growing and becoming popular in the temperate regions, including Japan, because it can generally survive the relatively low temperature. Although there are many intergeneric and interspecific hybrids in orchids, *Dendrobium* has difficulty in producing an intergeneric hybrid or even an intersectional hybrid using a conventional sexual method of hybridization. Hence the protoplast fusion technique may be the means of obtaining intergeneric hybrids of *Dendrobium*. Table 2 gives a brief review of the work done on orchid protoplasts.

2 Isolation of Protoplasts

2.1 Plant Material and Pretreatment

Dendrobium Yukidaruma "Queen" grown in a greenhouse of Minamikyushu University, was used as a source of material. The plants were previously maintained under low intensity illumination (500 lx) for more than 1 week as a pretreatment. The exposure of plants to short photoperiods of dim light was especially beneficial for obtaining consistently high yields of viable potato protoplasts (Shepard and Totten 1977). An expanded leaf (about 15×2 cm) on 2-year-old pseudobulb was excised from a mother plant, and was surface-sterilized by a 0.5% sodium hypochlorite solution with a few drops of Tween 20 for 10 min and 70% (v/v) ethanol for 1 min, followed by rinsing three times in sterile distilled water. The upper epidermis of the leaf was gently scratched with a scalpel to remove it successfully. Then the leaf was cut into small sections (about 2×2 cm) with a scalpel in a sterile Petri dish (100×15 mm).

Table 2. Orchid species in which protoplasts have been isolated

Plant species	Origin of protoplasts	Enzyme mixture	Culture medium	Growth response	Reference
Renantanda	Protocorm	Cellulase (2%)	Knudson C	Some	Teo and
Rosalind Cheok		Maserozyme (1%)	medium	clusters	Neumann
Cattleya	Seedling-	Pectinase (0.5%)	(19) +		(1978a)
Plalaenopsis	leaves	in 0.7 M	2×10^{-6} M 2,4-D		
Dendrobium	Shoots	sorbitol	10^{-6} M kinetin		
Paphiopedilum	Shoots				
Angraecum	Green	Cellulysin (2%)			Price and
gyrimae	sepals	in 0.5 M			Earle
Brassia		sorbitol			(1984)
brachiata	Root tip	Cellulysin (2%)			
Cattleya	Protocorms	Driselase (0.5%)			
aurantiaca	Leaves	Macerozyme (1%)			
	Roots	in 0.2 M			
		mannitol			
Dendrobium	Yellow	Cellulysin (2%)			
Beach Girl ×	green				
D. schulleri	petals				
D. Beach Girl ×	Purple petals				
D. Takami Kodama	Flower buds				
Maxillaria	Leaves				
tenuifolia					
Odontonia	Leaves				
Debutante					

Table 2 (continued)

Plant species	Origin of protoplasts	Enzyme mixture	Culture medium	Growth response	Reference
Paphiopedilum villosum	Petals	Cellulysin (2%)			Price and Earle (1984)
Vanilla planifolia	Leaf				
Dendrobium Yukidaruma Queen	Leaf	Cellulase (1%) Macerozyme (0.2%) Pectolyase (0.01%) in 0.6 M mannitol			Yasugi et al. (1985)
D. aggregtum	Leaves				
Cattleya skinneri alba	Leaves				
Epidendrum radicans	Leaves				
Cymbidium Miracle × Alexandri Golden Hill	Leaves				
Phalaenopsis amabilis Java	Leaves				
Paphiopedilum insigne	Leaves				
Calanthe discolor	Leaves				
Dendrobium Yukidaruma Queen	Leaves		Mod. MS + NAA (2 mg/l) + BA (0.5 mg/l)	embyoids	Yasugi (1986)

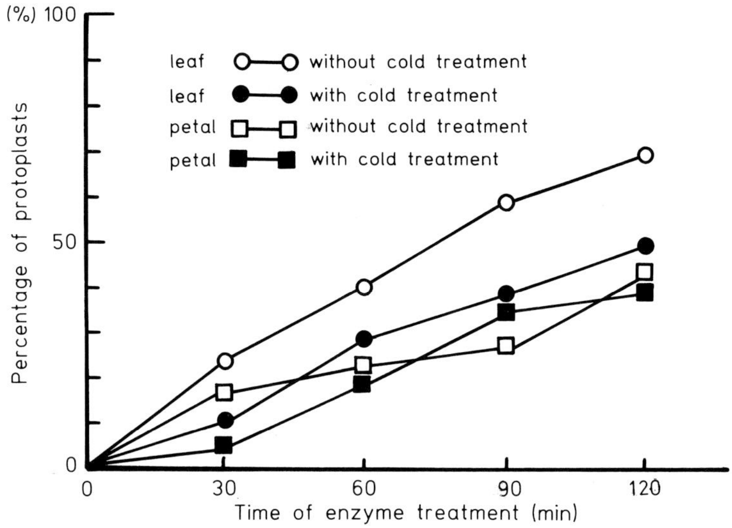

Fig. 1. Effects of cold (5 °C) treatment prior to enzyme treatment on percentage of protoplasts per isolated cells

The age and position of leaves on the mother plant and their environmental conditions influence the yield and viability of isolated protoplasts (Constabel 1975). In general, a dark green leaf or a yellowish senescing leaf yield unstable protoplasts, which degenerate during enzyme treatment or subsequent culture (Nagata and Takebe 1971). Mesophyll cells may vary in the osmotic property and the digestibility of their walls according to season and environmental conditions. The conditioning of leaf tissue under low temperature in a liquid medium prior to protoplast isolation generally enhanced the stability of protoplasts before and after plating (Shepard and Totten 1977). However, in this study, keeping leaves and petals under low temperature prior to isolation of protoplasts rather decreased the percentages of protoplasts in total cells after enzyme treatment (Fig. 1).

2.2 Protoplast Isolation

After leaving in vacuum for 2 min, leaf pieces of about 1 g fresh weight were incubated in 10-ml filter-sterilized enzyme solution in 50-ml Erlenmeyer flask (Pyrex, Japan) with a gentle shaking of 40 strokes per min, amplitude 2.2 cm (Plus Shaker EP-1, Taiyo, Co., Japan). The basic enzyme solution contained 1% Cellulase Onozuka R-10 (Yakult Biochemicals Ltd., Japan), 0.2% Macerozyme R-10 (Yakult Biochemicals Ltd., Japan), 0.01% Pectolyase Y-23 (Seisin Pharmaceuticals Ltd., Japan), 5 mM MES (Sigma Chemical Co., USA) buffered at pH = 5.8, 10 mM $CaCl_2 \cdot 2H_2O$, 0.6 M D-mannitol, and 0.5% potassium dextran-sulphate.

After 3 h incubation, the protoplast suspension was filtered through a 60-µm-mesh stainless steel sieve, then it was centrifuged at 150 g for 5 min. The supernatant was removed with a Pasteur pipette and the protoplasts were resuspended in a 0.6 M mannitol solution. This procedure was repeated three times to remove enzymes and cell debris. The final protoplast suspension (about 1 ml) was carefully layered over 2 ml of 20% sucrose solution, and it was centrifuged at 50 g for 3 min. During centrifugation, viable protoplasts floated to the meniscus of the 20% sucrose solution. The floating protoplasts were then gently pipetted out from the surface with a wide-mouthed pipette (10 cm^3) in order to avoid damage.

2.3 Factors Influening Protoplast Yield

2.3.1 Individual Variability of Plant Material. In general, the protoplast yield is influenced to a large extent by the leaf-to-leaf and plant-to-plant variability. To solve the problem, material plants have to be maintained under precise and constant environment as much as possible prior to isolation of protoplasts. In the present study, all plants were 2-year-old clones propagated by the stem-cutting

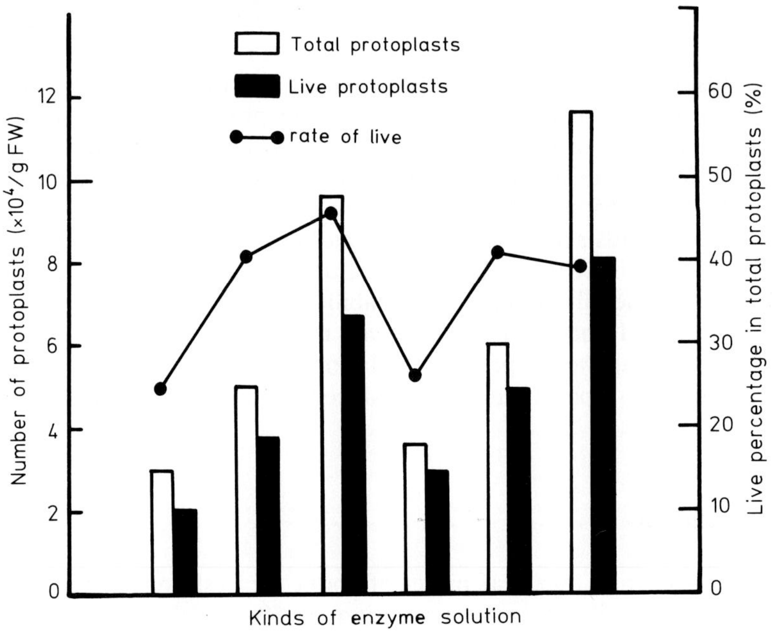

	1	2	3	4	5	6
Driselase	+			+	+	
Pectolyase	+		+	+		+
Cellulase		+	+		+	+
Macerozyme		+		+	+	+

+ indicates that enzyme is contained in enzyme solutions from 1–6

Fig. 2. Effects of kind of enzyme solution on number of isolated total and live protoplasts after enzyme treatment for 2 h method and kept under low intensity illumination (500 lx) for more than 1 week. Unfortunately, considerable variability among replicates has still been observed.

2.3.2 Kinds of Enzyme Solution. Six different kinds of enzyme solutions were used for 2 h in order to isolate mesophyll protoplasts. The yield, quality and plating efficiency of the protoplasts could be improved by prior desalting and purification of the enzyme (Shekhawat and Galston 1983). In this study all enzyme solutions had been centrifuged at 6000 g for 10 min (55 P-72 Ultracentrifugator – Hitachi Factory Co., Japan) and only supernatant was used in order to increase the yield of viable protoplasts.

The best enzyme solution to isolate *Dendrobium* protoplasts in this study included 0.5% Cellulase Onozuka R-10, 0.1% Pectolyase Y-23 and 0.1% Macerozyme R-10 as shown in Fig. 2. Driselase (Kyowa Hakko Kogyo Co. Ltd., Japan) seemed to be less effective than expected.

For isolation of orchid protoplasts, Teo and Neumann (1978a, b) used an enzyme solution containing 2% Cellulase, 1% Macerozyme and 0.5% Pectinase, and Price and Earle (1984) used either 2% Cellulysin or a mixture of 2% Cellulysin, 0.5% Driselase and 1% Macerozyme.

2.3.3 Period of Enzyme Treatment. The yield of protoplasts increased correlating with increase of period of enzyme treatment as shown in Fig. 3. After 5 h incubation in the basic enzyme solution consisting of 1% Cellulase Onozuka R-10, 0.2% Macerozyme R-10 and 0.01% Pectolyase Y-23, the yield of protoplasts achieved was 5.4×10^5/g fresh wt. Some orchids were incubated in the enzyme solution for 1 to 4 days (Price and Earle 1984); however, the enzyme solution is generally deleterious for protoplasts. Therefore, we should find the shortest period of enzyme treatment to obtain high yields of viable and vital protoplasts with a high capacity for proliferation.

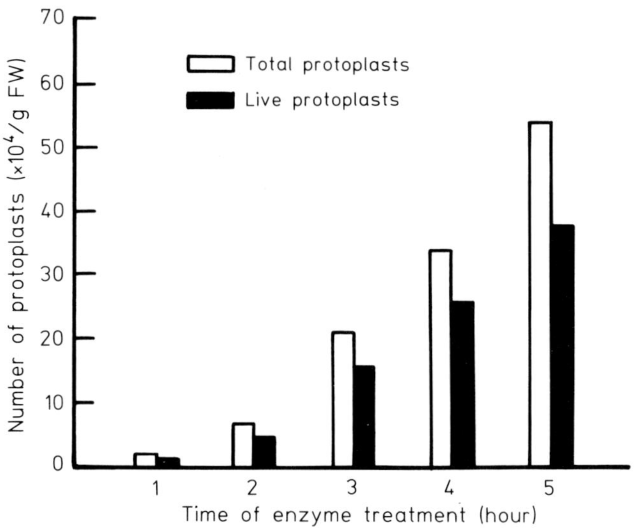

Fig. 3. The number of isolated total and live protoplasts during enzyme treatment for 5 h

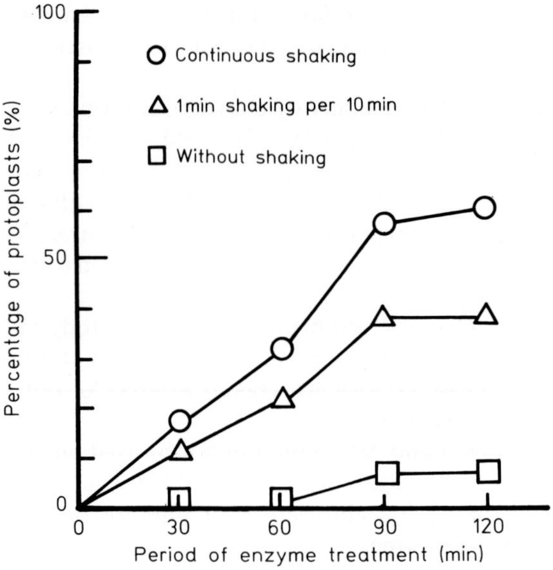

Fig. 4. Effect of shaking on isolation of protoplast during enzyme treatment

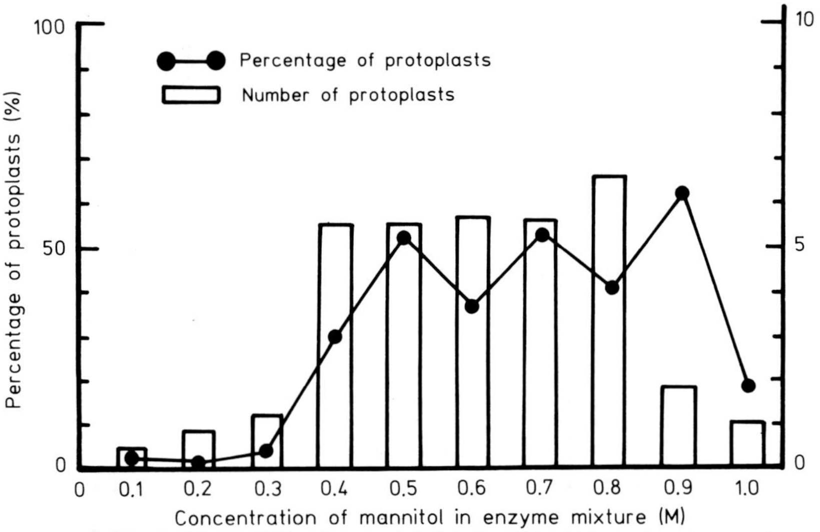

Fig. 5. Effects of the concentrations of mannitol in enzyme solution on percentage of protoplasts in total isolated cells and the number of isolated protoplasts

2.3.4 Effect of shaking on the Yield of Protoplasts. Continuous but gentle shaking of the flask containing leaf pieces in enzyme solution is effective to gain a high yield of protoplasts, as shown in Fig. 4.

2.3.5 Concentration of Osmoticum. When the concentration of mannitol was in the range of 0.4–0.8 M more than 5×10^5/g fresh wt. obtained after 6 h incubation in the basic enzyme solution (Fig. 5). The proper mannitol concentration would be slightly high in osmotic pressure.

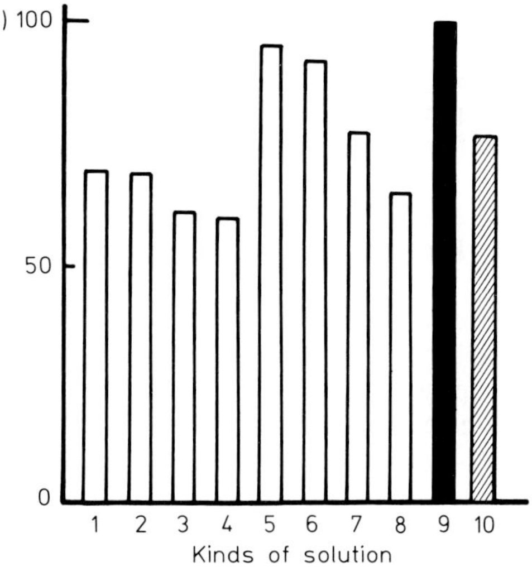

Fig. 6. Effects of kinds of rinse solution on the number of isolated protoplasts. 1. 0.6 M mannitol; 2. 20% sucrose; 3. 25% sucrose; 4. 30% sucrose; 5. CPW solution; 6. A solution + 0.6 M mannitol; 7. 20% sucrose + 0.5% dextransulphate; 8. 20% sucrose and settle for 10 min; 9. original enzyme solution (without rinse); 10. average in eight different rinse solutions

2.3.6 Kind of Rinse Solution. Seven different kinds of rinse solution were examined (Fig. 6). A solution, which was used in this study as the basic culture medium, containing 1/10 strength inorganic salts of Murashige and Skoog's medium (1962) and the organic nutrients of Kao and Michayluk (1974). When CPW solution (Frearson et al. 1973) or a solution with 0.6 M mannitol were used as rinse solution, a high yield of protoplasts was obtained.

2.3.7 Centrifugation. Speed and frequency of centrifugation during rinse of protoplasts also influence the final yield. A high speed destroys the relatively soft protoplasts. In this study, 150 g (KA-1000, Kubota Co., Japan) was applied as a speed. Although we were not able to avoid completely the loss of any protoplasts during rinsing, we had to remove the deleterious enzyme solution as much as possible by means of a centrifugation.

2.3.8 $CaCl_2$ and Dextransulphate. During isolation, protoplasts seem to have the benefit from a higher level of divalent cations like Ca^{2+} than is normally present in an ordinary medium (Constabel and Kao 1974). Dextransulphate in the enzyme solution helps to isolate the majority of mesophyll cells in an intact state (Takebe et al. 1968). In this study, 10 mM $CaCl_2 \cdot 2H_2O$ and 0.5% potassium dextransulphate were added to the enzyme solution.

3 Culture of Protoplasts

3.1 Density and Viability of Protoplasts

Before the isolated protoplasts are placed in culture medium, it is necessary to examine their viability and to adjust their plating density. Viability of protoplasts was determined by staining them with Evans Blue or with fluorescein diacetate using the methods of Larkin (1976). In addition, spherical shape, viable colour and cytoplasmic streaming of protoplasts were confirmed. The density of isolated protoplasts in a given sample was determined by using a Fuchs-Rosenthal modified hemocytometer with a field depth of 0.2 mm.

3.2 Procedure of Plating

The isolated and purified protoplasts were suspended in the liquid medium A at a cell population density of about $5.5-8.2\times10^4$/ml. The protoplasts were resuspended in a liquid medium A supplemented with 2 mg/l NAA (naphthalene acetic acid) and 0.5 mg/l BA (benzyladenine), were poured on the solid medium B containing 1.5% Bacto agar (Difco, USA) in addition to medium A. Culturing in a liquid medium has the advantage of being able to change the composition of the medium by adding a solution or substituting a new medium.

The Petri dishes were sealed with Parafilm M to prevent desiccation and contamination. For orchid protoplast cultures, Knudson C medium (Teo and Neumann 1978a), modified B5 nutrient medium (Price and Earle 1984) and medium A (in this study) were used. Medium A contained the 1/10 strength inorganic salts of MS medium and the organic nutrients of Kao and Michayluk's medium.

3.3 Culture Condition

Petri dishes, containing protoplasts in a liquid medium A on solid agar medium B, were placed in a culture chamber (NK 200RD, Nihon Ikakiki Co., Japan) at 22 °C under continuous illumination at 500 lx from white fluorescent lamps.

The sequence of development of an embryoid from a single isolated protoplast of *Dendrobium* cv. Yukidaruma Queen is shown in Fig. 7. The first division was observed after 5–7 days. Within 2 weeks, protoplasts continuously divided and formed the cell clusters or colonies. When the colonies were transferred and cultured on the agar medium C containing 1 mg/l NAA, 0.25 mg/l 2,4-D (2,4-dichlorophenoxyacetic acid) and 1 mg/l BA under 3000 lx for 2 months, some colonies developed into green embryoids.

Observation after 3 weeks showed 3.2% plating efficiency, 18.64% cell division rate, 55.4% cell viability and 6.2×10^5/ml cell density. After 3 months, 9.6% colonies formed embryoids.

Fig. 7. Sequence of development of an embryoid of *Dendrobium* Yukidaruma Queen from a single isolated protoplast: **A** A single isolated mesophyll protoplast. **B** A viable protoplast after staining with fluorescein diacetate. **C, D, E** 7-day-old dividing protoplasts. Bars in **A, B, C, D, E** = 50 μm. **F, G** Development of colonies from 1-month protoplast culture. **H** Formation of green embryoids on agar medium in 3-month-old culturee. Bars in **F, G, H** = 100 μm

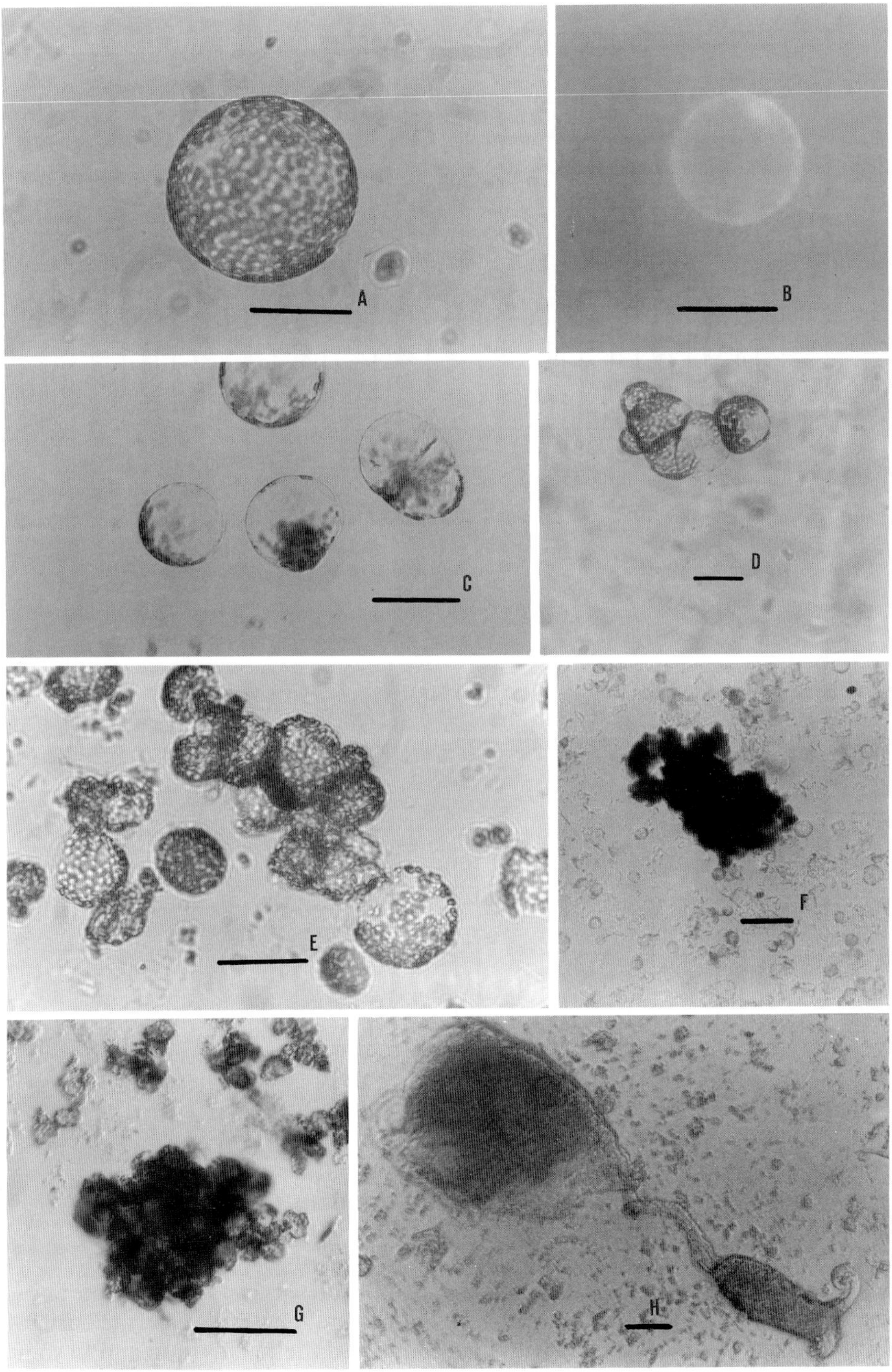
A
B
C
D
E
F
G
H

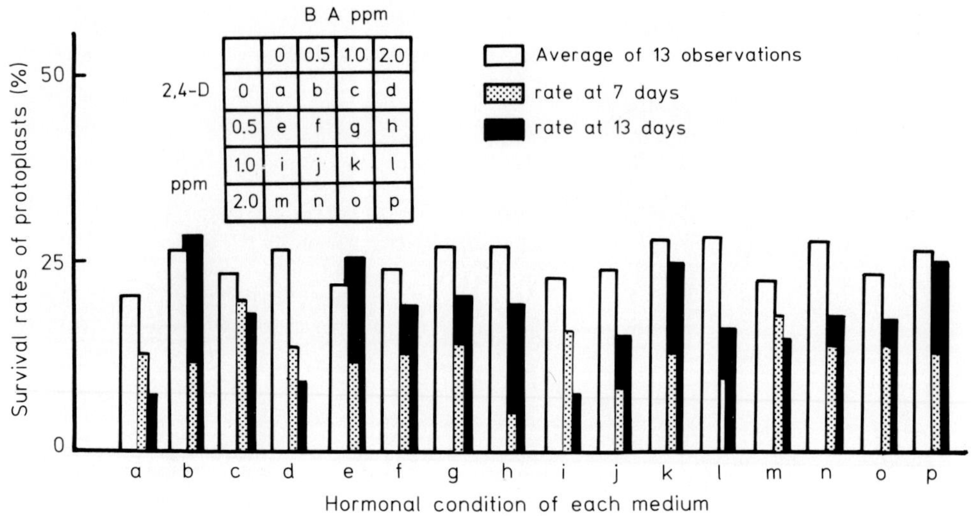

Fig. 8. Effect of growth regulators (combinations of 2,4-D and BA) on survival rates of protoplasts during culture for 2 weeks

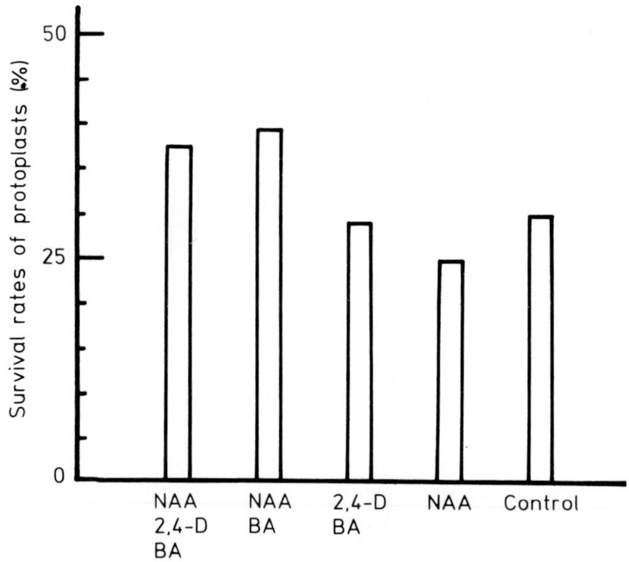

Fig. 9. Effect of growth regulators (NAA, 2,4-D and BA) on survival rates of protoplasts during culture for 2 weeks

3.4 Factors Influencing Survival Rate of Protoplasts During the Initial Culture

1. Growth Regulators. Effects of growth regulators on survival rates of protoplasts are shown in Figs. 8 and 9. Growth regulators were not critical for the survival of protoplasts during initial culture. They promoted the cell division and to some extent increased the number of surviving protoplasts.

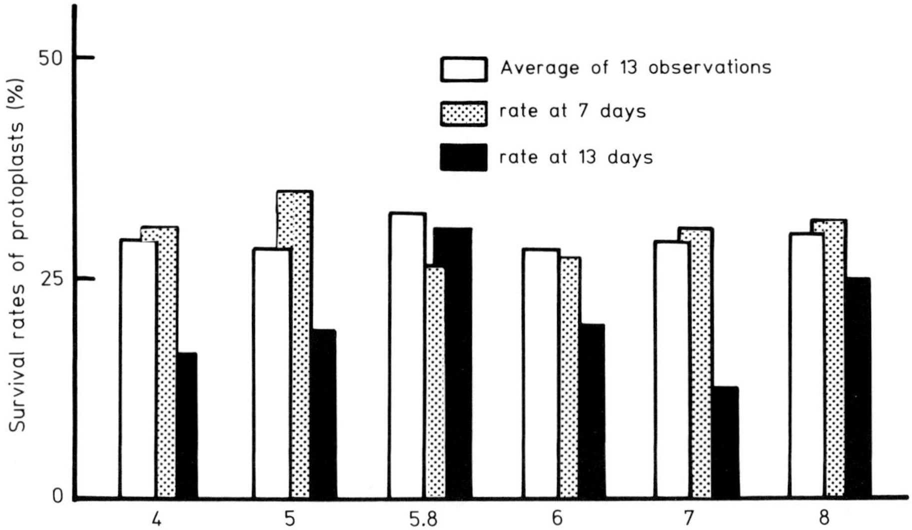

Fig. 10. Effect of pH of medium on survival rates of protoplasts during culture for 2 weeks

2. pH of Medium. Protoplasts were cultured in a liquid medium A with six different levels of pH under 500 lx at 20 °C for 2 weeks. Survival rates of protoplasts are shown in Fig. 10. When the pH was 5.8, the highest survival rate was 32.8% as the average of 13 observations for 2 weeks. It seems that pH is not a critical factor for protoplasts to survive culture when it was in the range of 4–8.

3. Light Intensity. Protoplasts were cultured under three different levels of intensity of light at 20 °C, pH = 5.8 for 2 weeks. The best survival rate was observed under dark condition (Fig. 11). Therefore, light seems not to be a requisite for protoplasts to survive for initial culturing. However, colony formation from the plated protoplasts is influenced by the light intensity during culturing. A light intensity lower than 800 lx gave poor plating efficiency, whereas a light intensity higher than 5000 lx does not improve colony formation (Nagata and Takebe 1971). In the present study, the light intensity was controlled at 500 lx for the first 2 weeks, and then increased to 3000 lx.

4. Temperature. Protoplasts were cultured in a liquid medium A (pH = 5.8) under dark condition for 2 weeks. The best survival rate was observed when the protoplasts were cultured at 5 °C (Fig. 12). This means that low temperture may offer good conditions for protoplast survival.

5. Coconut Milk (CM). The addition of coconut milk does not seem to be necessary for protoplasts to survive during initial culturing (Fig. 13).

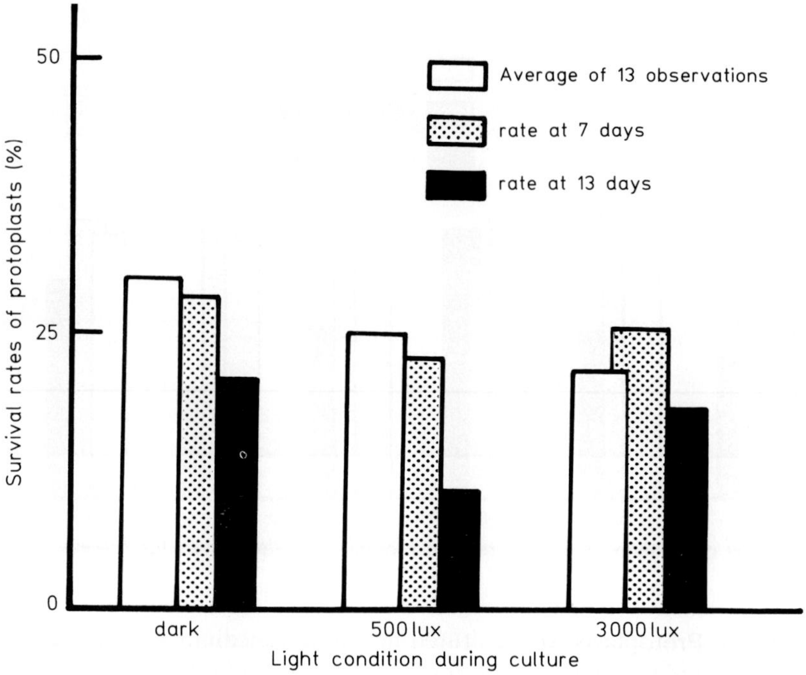

Fig. 11. Effect of light intensity on survival rates of protoplasts during culture for 2 weeks

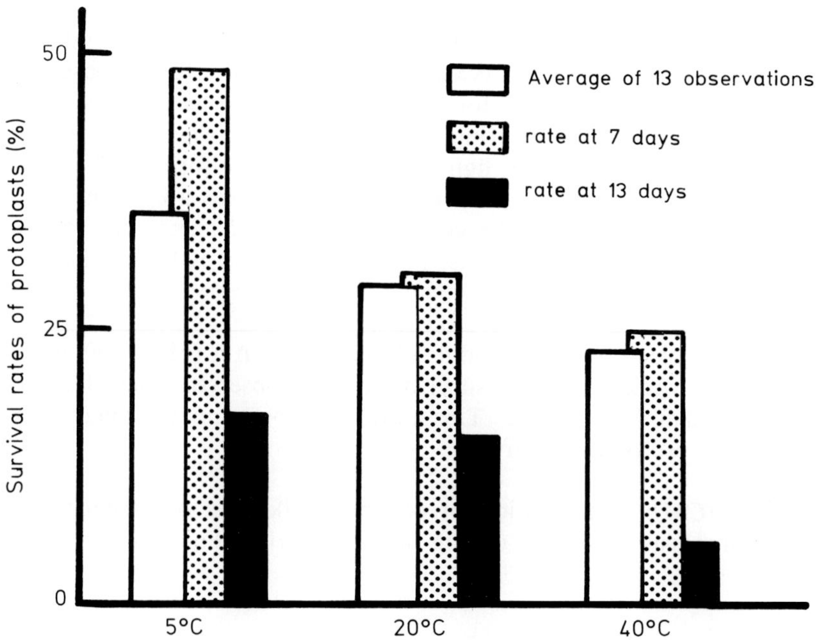

Fig. 12. Effect of temperature on survival rates of protoplasts during culture for 2 weeks

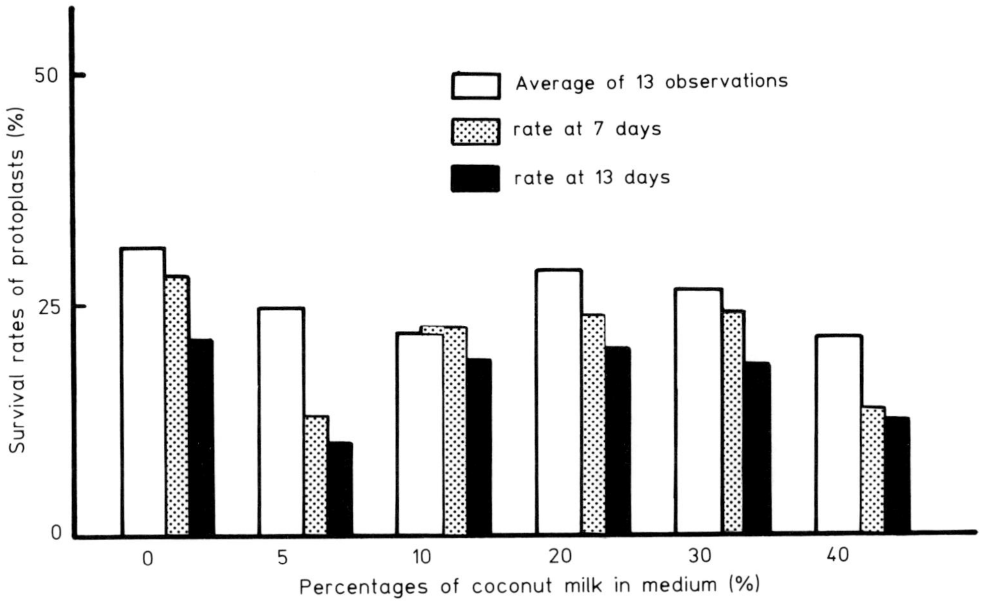

Fig. 13. Effect of coconut milk on survival rates of protoplasts during culture for 2 weeks

4 Fusion of Protoplasts

Isolated protoplasts of *Cattleya skinneri* alba, *Epidendrum radicans*, *Cymbidium* Miracle × Alexandri "Golden Hill", *Phalaenopsis amabilis* "Java", *Paphiopedilum insigne* and *Dendrobium* Malones "Hope" were used as the sources of protoplast fusion.

Two drops of about 200 µl containing protoplasts about 1×10^5/ml of two different orchids were placed in the centre of 6-cm Petri dish (Pyres Iwakigarasu Co., Japan), and four microdrops (each 50 µl) of 33% (w/v) PEG 6000 solution with 0.6 M mannitol and 10 mM $CaCl_2 \cdot 2H_2O$ were gently added to the protoplast suspension. The protoplasts were then allowed to settle for 15 min at room temperature, followed by adding a 40-µl rinse solution containing 0.6 M mannitol and 50 mM/l $CaCl_27 \cdot 2H_2O$. At 15-min intervals, 1 ml rinse solution was added to the protoplast suspension twice. After adding a 4-ml rinse solution, the total solution was centrifuged at 150 *g* for 2 min and washed with a rinse solution twice. Finally, protoplasts were resuspended to 1 ml of liquid medium A, plated in 6-cm Petri dish, and cultured at 20 °C under 500 lx.

Figure 14 shows the various fusion products from different kinds of protoplasts isolated from some orchid genera, i.e. *Cymbidium*, *Phalaenopsis*, *Cattleya*, *Dendrobium* and *Paphiopedilum*.

Table 3 shows the percentages of cell division and plating efficiencies of different kinds of protoplasts cultured for 3 months. The possible hybrid cell clusters or colonies were difficult to identify.

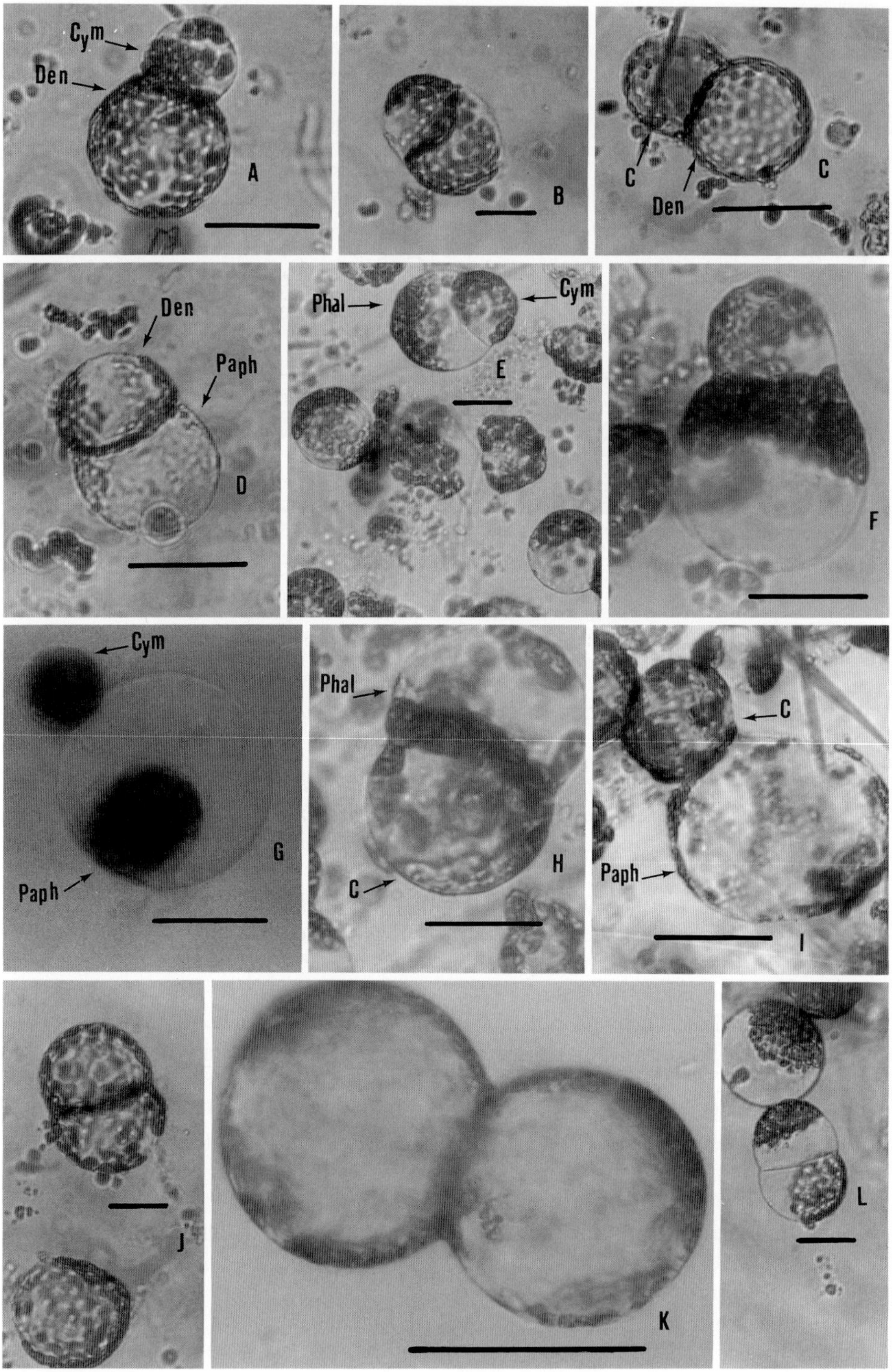
Cym
Den
A
B
C
Den
C
Den
Paph
D
Phal
Cym
E
F
Cym
Paph
G
Phal
C
H
C
Paph
I
J
K
L

Fig. 14A–L.

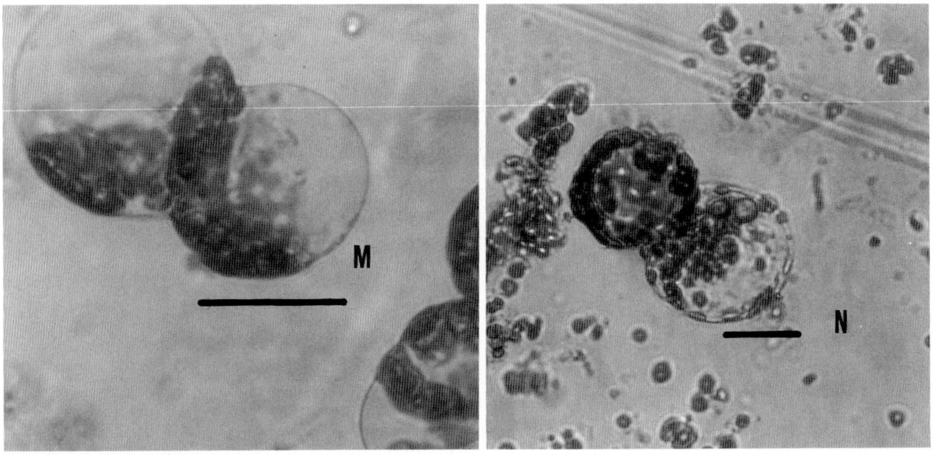

Fig. 14. Various fusions of different kinds of protoplast: **A, B** *Dendrobium* × *Cymbidium*.**C** *Dendrobium* × *Cattleya* (Anthocyanoplast - reddish purple). **D** *Dendrobium* × *Paphiopedilum*. **E, F** *Cymbidium* × *Phalaenopsis*. **G** *Cymbidium* × *Paphiopedilum*. **H** *Cattleya* × *Phalaenopsis*. **I** *Cattleya* × *Paphiopedilum*. **J, K** *Dendrobium* × *Dendrobium*. **L, M** *Cymdidium* × *Cymbidium*. **N** *Cattleya* × *Cattleya* (Anthocyanoplast). Bar = 50 μm in all plates

Table 3. Rates of cell division and plating efficiency of some orchid protoplasts cultured for 3 months

Source of protoplasts	Rate of cell division (%)	Plating efficiency (%)
Cattleya	14.5	1.7
Paphiopedilum	13.8	1.7
Epidendrum	8.3	1.8
Dendrobium Malones Hope	7.4	0.9
Dendrobium × *Epidendrum*	19.4	0.9
Dendrobium × *Cattleya*	10.6	1.2
Dendrobium × *Paphiopedilum*	9.8	2.2
Epidendrum × *Paphiopedilum*	31.0	0.8

5 Conclusions and Prospects

The present investigation describes the protocol to isolate protoplasts of *Dendrobium* and to culture them by multi-stepped media, and subsequent embryoid formation. It would appear desirable to use the enzyme solution containing Cellulase Onozuka R-10, Pectolyase Y-23 and Macerozyme R-10 in order to isolate viable protoplasts from orchid plants. The proper plating density of protoplasts is about $5.5-8.2\times10^4$/ml. For initial protoplast culture, medium A is suitable, which contains the 1/10 strength inorganic salts of MS medium and the organic nutrients of KM medium. Medium A is supplemented with 2 mg/l NAA and 0.5 mg/l BA. When the colonies are formed, they are transferred to Medium C containing 1 mg/l NAA, 0.25 mg/l 2,4-D and 1 mg/l BA. For protoplast survival in the initial culture, relatively low temperature and low light intensity are preferable.

The present investigation also describes the protocol to hybridize the protoplasts isolated form orchid plants. The protoplast suspension (about 200 μl) obtained from two different orchid plants was placed in the centre of a Petri dish, and four microdrops (each 50 μl) of 33% (w/v) PEG (6000 molecular weight) with 0.6 M mannitol and 10 mM $CaCl_2 \cdot 2H_2O$ were gently added in order to induce protoplast fusion.

References

Constabel F (1975) Isolation and culture of plant protoplasts, pp. 11–16. In: Gamborg OL, Wetter LR (eds) Plant tissue culture methods. National Research Council of Canada, Prairie Regional Laboratory, Saskatoon

Constabel F, Kao KN (1974) Agglutination and fusion of plant protoplasts by polyethylene glycol. Can J Bot 52:1603–1606

Frearson EM, Power JB, Cocking EC (1973) The isolation, culture and regeneration of *Petunia* leaf protoplasts. Dev Biol 35:130–137

Hedman K, Leander K (1972) Acta Chem Scand 26:3177

Inobushi Y, Sasaki Y, Tsuda Y, Yasui B, Konita T, Matsumoto J, Katarao E, Nakano J (1964) Tetrahedron 20:2007

Kao KN, Michayluk MR (1974) A method for high frequency intergeneric fusion of plant protoplasts. Planta 115:355–367

Larkin PJ (1976) Purification and viability determination of plant protoplasts. Planta 128:213–216

Luning B (1974) Alkaloids of the Orchidaceae. In: Withner CL (ed) The orchids scientific studies. Wiley-Interscience, New York, pp 352–360

Murashige T, Skoog, F (1962) A revised medium for rapid growth and bioassays with tobacco tissue culture. Physiol Plant 15:473–497

Nagata T, Takebe I (1971) Plating of isolated tobacco mesophyll protoplasts on agar medium. Planta 99:12–20

Onaka T, Kamata S, Maeda T, Kawazoe Y, Natsume M, Okamoto T, Uchimaru F, Shimizu M (1965) Chem Pharm Bull 13:745

Pais MS, Barroso J, Fevereiro P, Oliveira M (1983) Intergeneric fusion of terrestrial orchid protoplasts induced by different fusion promoting agents. In: Protoplasts 1983, Poster Proc. Birkhäuser, Basel

Power JB, Berry SF, Cocking EC, Evans PK, Frearson EM (1976) Isolatin, culture and regeneration of leaf protoplasts in the genus *Petunia*. Plant Sci Lett 7:51–55

Price GR, Earle ED (1984) Sources of orchid protoplasts for fusion experiments. Am Orch Soc Bull 53(10):1035–1043

Shekhawart NS, Galston AW (1983) Mesophyll protoplasts of fenugreek (*Trigonella foenumgraecum*): Isolation, culture and shoot regeneration. Plant Cell Rep 2:119–121

Shepard JF (1982) Regeneration of potato leaf cell protoplast. In: Earle ED, Demarly Y (eds) Variability in plants regenerated from tissue culture. Praeger, New York, pp 47–57

Shepard JF, Totten RE (1977) Mesophyll cell protoplasts of potato. Isolation, proliferation and plant regeneration. Plant Physiol 60:313–316

Shuttlework FS, Zim HS, Dillon GW (1970) Orchids. Western Publishing, Wisconsin, US, pp 4–6

Statistical Association of Agriculture and Forestry (1986) The state of affairs of import and export of floricultural products. J Shinkakijyoho (3):36–39

Suzuki H, Keimatsu I (1932) J Pharm Soc Japan 52:996 (Ger Abstr 162), 1049 (Ger Abstr 183)

Takebe I, Otsuki Y, Aoki S (1968) Isolation of tobacco mesophyll cells in intact and active state. Plant Cell Physiol 9:115–124

Teo CKH, Neumann KH (1978a) The culture of protoplasts isolated from *Renantanda* Rosalind Cheok. Orchid Rev 86:156–158

Teo CKH, Neumann KH (1978b) The isolation and hybridization of protoplasts from orchids. Orchid Rev 86:186–189

Thomas HE (1981) The New York Botanical Garden Illustrated Encyclopedia of Horticulture. (3):1034–1039. Garland, New York

Widholm J (1972) The use of fluorescein diacetate and phenosafranine for determining viability of cultured plant cells. Stain Technol 47:189–194

Yamamura S, Hirata Y (1964) Nippon Kagaku Zasshi 85:377

Yasugi S (1986) Embryoids formation from mesophyll protoplasts of *Dendrobium* (Orchidaceae). Abstr Am J Bot 73(5):688

Yasugi S, Kugimiya M, Katsura N (1986) Isolation of mesophyll protoplasts in orchids. Bull Fac Hortic Minamikyushu University 16:91–99

II.11 Regeneration of Plants from Protoplasts of *Salpiglossis*

CHARLENE J. BOYES[1] and KENNETH C. SINK[2]

1 Introduction

Salpiglossis species are all members of the mainly Neotropical tribe Cestreae of the Solanaceae (D'Arcy 1978). Four *Salpiglossis* species are known: *S. erecta* in Mexico and Texas; *S. sinuata* and *S. spinescens* in Chile; *S. anomala* in Argentina (D'Arcy pers commun). *Salpiglossis sinuata* L. is the only species in cultivation. It comprises a minor portion of the various plant species grown by the bedding plant industry. *Salpiglossis* is grown as an annual outdoors primarily for the showy flowers which come in an array of colours. The species has received sparse scientific study in terms of breeding, biochemistry, genetics and physiology.

The significance of *Salpiglossis* protoplasts is in their potential for cell fusions with other solanaceous bedding plant species such as *Browallia, Nicotiana alata, Petunia hybrida* and other *Petunia* species. Various attempts at intergeneric sexual crossings between *Salpiglossis sinuata* and the three other genera have not succeeded. Thus, cell fusion is a potential means of gene transfer, with the inital interest focused on gene(s) controlling flower pigments and floral morphology.

Tissue culture studies on *Salpiglossis* have been limited, but they include successful reports on the initiation and maintenance of callus and regeneration of plants, that should enable further work on somatic cell genetics. Hughes et al. (1973) first reported tissue culture of *S. sinuata* with the induction of callus on anthers placed on Nitsch's (1969) medium containing 1 mg/l kinetin and 5 mg/l indoleacetic acid (IAA). Depending on stage of development at the time of culture, 27 to 46% of the anthers formed callus. None of the calli, however, originated from pollen. Shoots occurred after 5 months culture of the somatic-derived callus on Nitsch medium and those rooted in Murashige and Skoog (1962) medium (MS) containing 0.04 mg/l kinetin and 10 mg/l IAA. Lee et al. (1977) placed young leaf discs of *S. sinuata* on MS medium to devise an in vitro propagation method. Optimum shoot production occurred at 0.1 mg/l naphthalene acetic acid (NAA) and 2 mg/l kinetin in 15–20 days. The shoots rooted at 60% efficiency in MS medium and NAA was found to be inhibitory to rooting. Boyes and Sink (1981a) reported the results of morphogenetic studies on leaf explants and derived callus cultures to the auxins 2,4-dichlorophenoxyacetic acid (2,4-D), IAA or NAA and the cytokinins 6-benzylaminopurine (6-BAP), 6 (γ-γ dimethylallyl

[1] Plant Genetics Incorporation, 1930 Fifth Street, Davis, CA 95616, USA
[2] Department of Horticulture, Michigan State University, East Lansing, MI 48824, USA

Biotechnology in Agriculture and Forestry, Vol. 8
Plant Protoplasts and Genetic Engineering I (ed. by Y.P.S. Bajaj)

amino) purine (2ip) or kinetin singly or NAA+6-BAP. Callus that varied in friability was obtained on MS+2,4-D at various concentrations. More uniform, dividing friable callus was obtained on Uchimiya and Murashige (1974) medium (UM). Second subcultures of callus growing on these media were tested for subsequent shoot regeneration. Shoots occurred within a 6-week period and growth was optimum on MS+0.1 or 1 mg/l 2ip. As in other tissue culture studies on *Salpiglossis*, rooting was the most difficult stage in obtaining whole plants. It was found that low levels of 2,4-D (0.001 mg/l) in MS medium gave 75% rooting of regenerated shoots and subsequent transfer to soil and survival in the greenhouse was 85–90% successful.

2 Isolation of Protoplasts

2.1 Plant Material and Callus Cultures

Protoplasts were found to be isolated most efficiently from callus derived initially from leaf explants taken from young seedlings of *S. sinuata* (Boyes and Sink 1981 a). Seedlings were established from seed germinated on a soil-less planting medium at 22±2 °C under 16-h photoperiod (74–84 $\mu E/m^2/s$). After approximately 3 weeks' growth, the seedlings were transplanted to individual cell packs and grown for an additional 5 to 9 weeks in the greenhouse at 22–25 °C. Natural day-light was supplemented for 16 h each day with 550 $\mu E/m^2/s$ cool white fluorescent lighting. Seedlings were feritilized three times weekly with 200 ppm N prepared from 20N/8.6P/16.6K commerical fertilizer.

Calli were initiated from leaf explants removed from fully expanded leaves from the lower half of 8- to 12-week-old seedlings. Excised leaves were surface-sterilized in sodium hypochlorite (5–7%) for 30 min and rinsed six times with sterile distilled water. Leaf pieces, (1.0×0.5 cm), excluding the midrib and margins, were excised from the leaves and the lower epidermal layer was peeled off with jewelers' forceps. Leaf explants were cultured, with the exposed surface down, on UM medium, which was used for both callus induction and maintenance. Leaf explants were cultured in Petri dishes under a 16-h photoperiod at 74–84 $\mu mE/m^2/s$ at 22±2 °C. After 2 to 4 weeks, friable, creamy yellow callus formed around the periphery of the cultured leaf pieces. Within 6 weeks, the calli could be removed from the explants and subcultured. Pieces of calli slightly less than 1 cm^2 were subcultured every 3 to 4 weeks on UM medium.

2.2 Protoplast Isolation and Purification

Protoplasts were isolated from calli 7 to 10 days after subculture. One or two grams of callus were pressed through a 350-µm sieve into a Petri dish containing cell/protoplast wash (CPW) solution (Frearson et al. 1973) amended with 0.44 M mannitol. Cells and small cell aggregates that passed through the sieve were allowed to stand in the CPW+mannitol solution for one half hour for pre-plasmolysis. Pre-plasmolysis also seemed to deter excessive enlargement of the protoplasts during initial culture. After pre-plasmolysis, cells were pelleted by centrifugation for 2 min at 80 *g*. The supernatant was removed and replaced with 10 ml of an enzyme solution composed of CPW salts, 0.44 M mannitol, 2% Cellulase R-10, 0.75% Pectinase (Sigma), and 1% Driselase. The cells were resuspended in the enzyme mixture and incubated for 5 h at 27 °C on a gyratory shaker at 50 rpm in diffuse light (11 $\mu E/m^2/s$). After 5 h incubation, the enzyme solution was centrifuged at 80 *g* for 2 min to pellet the protoplasts and debris. The enzyme solution was removed and replaced with CPW salts amended with 0.44 M sucrose and the cells were resuspended. This slurry was centrifuged at 100 *g* for 10 min; the meniscus of protoplasts was collected and counted. An average of $6-8\times10^5$ protoplasts/ml could be isolated from 1 g of callus.

2.3 Factors Affecting Yield

Among the many factors which influenced protoplast yield, the culture medium and subculture routine used to establish uniform callus cultures were especially important. Leaf explants of *S. sinuata* would form calli and the calli would exhibit growth on MS medium supplemented with various auxins and cytokinins at various concentrations (0.1 – 10 mg/l) (Boyes and Sink 1981 a). However, calli appearance and protoplast yield was much improved when calli were maintained on UM basal medium which included the growth regulators 2,4-D and kinetin.

Callus age and time from last subculture strongly affected protoplast yield. As the total age of the calli increased, the calli became less amenable to conversion to protoplasts by the standard enzyme solution. Although calli subcultures older than 1 year could still be used for protoplast isolation, the yield was greatly decreased (Boyes and Sink unpubl.). However, regular subculture, every 2 – 3 weeks, of calli pieces slightly less than 1 cm^2, selected from the upper periphery of the calli, gave the best results in regard to subsequent callus growth. When sampled 10 days after subculture, these calli gave the best protoplast yield (greatest number of protoplasts released and surviving).

Essential to obtaining large numbers of viable protoplasts was the dispersment of the cells by pressing calli through a sieve, and treatment of the resulting cells and small cell clumps in a pre-plasmolysis solution. While sieve pore size probably was not critical, chopping the calli with a scalpel and placing them directly into enzyme solution did not result in a significant yield increase of protoplasts.

The enzyme solution also was a major factor affecting protoplast yield. Carrying out the above steps would not give acceptable yields if the calli were not incubated in CPW salts amended with 2% Cellulase R-10, 0.75% Pectinase, and 1% Driselase. Other enzymes tested at various concentrations and combinations resulted in fewer protoplasts and/or in non-viable protoplasts.

Table 1. Protoplast culture medium for Salpiglossis (mg/l)

a) Salts:
KNO_3 (1700), $MgSO_4 \cdot 7H_2O$ (370), $MnSO_4 \cdot 4H_2O$ (17), $CuSO_4 \cdot 5H_2O$ (0.025), $ZnSO_4 \cdot 7H_2O$ (8.6), $CaCl_2 \cdot 2H_2O$ (440), KI (0.83), $CoCl_2$ (0.025), KH_2PO_4 (170), H_3BO_3 (6.2), $Na_2MoO_4 \cdot 2H_2O$ (0.25)
b) Iron solution of 3.724 g of the disodium salt of ethylene diaminetetraacetic acid, and 2.784 g of $FeSO_4 \cdot 7H_2O$ (10 ml)
c) Vitamins:
Thiamine HCl (2.0)
Myoinositol (1.0 g)
d) Organic nitrate sources:
L-glutamine (250), L-serine (0.1)
e) Hormones:
IAA (0.5), 2,4-D (1.0), BA (0.5)
f) Carbon sources:
Sucrose (30000)
g) Osmoticum:
Mannitol (90000)

3 Culture of Protoplasts

Protoplasts collected by floatation were counted and plated in liquid protoplast culture media at a density of 5×10^4/ml resulted in 28% plating efficiency, whereas 2.5×10^4/ml resulted in only 12% efficiency. The nitrogen source was found to be critical to the survival of protoplasts. Initiation of protoplast division required MS medium devoid of ammonium nitrate and supplemented with L-glutamine (250 mg/l), L-serine (1 mg/l), thiamine HCl (2 mg/l) and myo-inositol (1000 mg/l) (Table 1).

After the protoplasts and culture medium were mixed to the proper density, they were plated in a thin layer (2–4 ml) in 60×15 mm Petri dishes. Incubation was carried out in diffuse light (11–15 $\mu E/m^2/s$) at 22 °C. During the first 48 h of culture, as many as 25% of the protoplasts died. The remaining protoplasts agglutinated, enlarged slightly, and after 24 more hours entered first division (Fig. 1 A, B). Initiation of first division was sometimes delayed for as long as 12 days after initiation of the culture when older calli were used as a cell source. Second division (Fig. 1 C) generally was observed about 6 days after the initiation of first division. Beginning at this time 0.3–0.6 ml of fresh culture medium were added to the protoplast suspension on a weekly basis. The osmoticum was reduced gradually by adding culture medium with 0.32 M, 0.16 M and 0.0 mannitol at 3-week intervals, respectively. Colonies were evident by 26 days post culture initiation (Fig. 1 D). These colonies attained sizes of 2–3 mm in diameter after 2 months' growth. The calli were light green in colour, indicating that some differentiation had begun. At this time the calli were transferred to shoot regeneration medium.

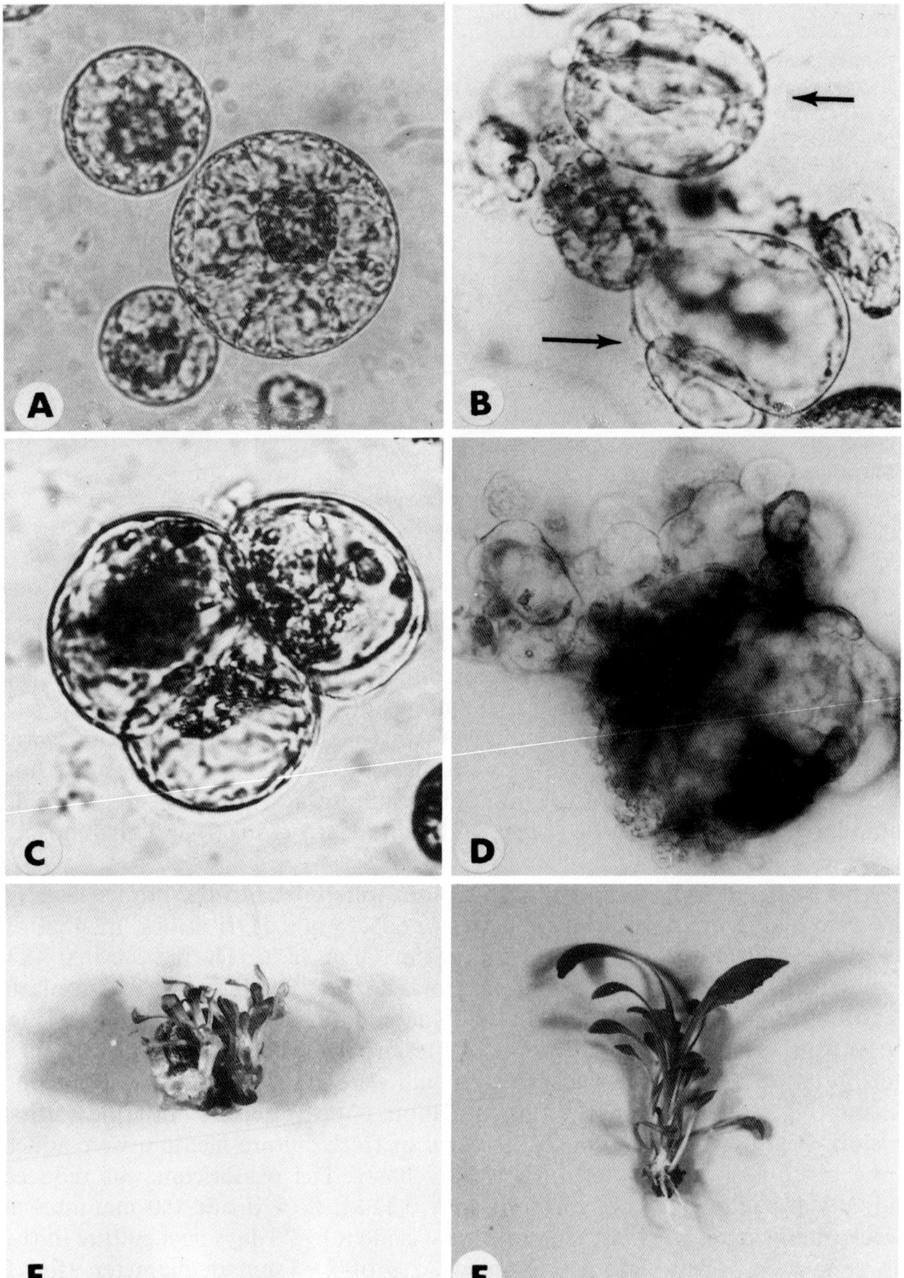

Fig. 1A–F. Regeneration of isolated callus protoplasts of *Salpiglossis sinuata*. **A** 48-h-old isolated protoplasts from 7-day-old callus subcultures (×145). **B** First division of isolated protoplasts, 6 days after plating in medium indicated in Table 1 (×92). **C** Second division, 14 days after plating in medium indicated in Table 1 (×92). **D** Two-month-old cell colony (×145). **E** Shoots produced from callus on MS + 1 mg/l 2ip (actual size). **F** Adventitious root initiation on 30-day-old shoot cultured on MS + 0.001 mg/l 2,4-D (actual size)

4 Regeneration of Plants

Calli 2–3 mm in diameter were cultured on shoot regeneration medium (MS+1 mg/l 2ip) at 22±2 °C in a 16-h photoperiod (74–84 $\mu E/m^2/s$). Smaller calli generally did not survive on this medium. The cultured calli rapidly increased in size and in approximately 21 days, green shoot buds could be observed under the dissecting microscope. These buds appeared as small shiny outgrowths to the unaided eye. Once shoots began to develop, the calli no longer increased in size (Fig. 1E). The shoots, however, continued to elongate and could be excised from the calli in about 14 days after their initial appearance. Any callus at the base of the shoots was gently removed at the time of excision.

Root initiation was carried out at the same temperature and light conditions used for shoot initiation. The excised shoot bases were dipped in filter-sterilized IBA (1000 ppm) and transferred to root induction medium (MS+0.001 mg/l 2,4-D). Approximately 75% of the shoots had developed root initials after 14 days in the root induction medium (Fig. 1F). Plantlets could be transferred to soil in 3–5 weeks at an efficiency of 85–90%. The plants were initially covered with plastic bags to prevent dehydration and gradually removed for longer periods of time each day until the plants were acclimated. Acclimatized plants were transferred to the greenhouse and fertilized weekly with 200 ppm N20/P8.6/K16.6.

5 Protoplast Fusion

Power and Chapman (1983) fused albino protoplasts of *Salpiglossis*, derived from cell suspension cultures, with *Petunia parodii* leaf mesophyll protoplasts. Both species were adjusted to 4×10^5 protoplasts/ml in CPW+13% mannitol and the pelleted protoplasts were fused with 25% PEG, MW 6000. The selection system was based on limited division of *P. parodii* and colourless callus formation of *S. sinuata*. Green putative hybrid calli were recovered, but most shoots regenerated were *Salpiglossis;* the others were green shoots like *Salpiglossis* or malformed. It may be that this intra-subfamilial somatic hybridization failed due to taxonomic relationships. Whether this is indeed the case, or an example of somatic cell level incompatibility that may be circumvented by asymmetric protoplast fusion (Gupta et al. 1984) remains to be determined.

6 Conclusions and Prospects

As mentioned in the introduction, *Salpiglossis* is a potential partner for protoplast fusion with other solanaceous species that are cultivated as ornamental plants. The authors know of several attempts, in addition to that reported by Power and Chapman (1983), to fuse *Salpiglossis* with *Petunia* in several laboratories, using selective systems that theoretically are applicable. To date, however, no report has appeared on protoplast fusions using *Salpiglossis* that resulted in somatic hybrid plants. Likewise, *Salpiglossis* has not yet been the focus of transformation, uptake, virus or other protoplast-based genetic manipulations. In summary, *Salpiglossis*, because of its minor economic importance, suffers from a lack of attention in biotechnology including protoplast methodologies. As gene transfer systems become more refined and readily adapted to more species, *Salpiglossis* should receive attention both as a donor and recipient species of genes.

7 Protocol

A. Isolation of Protoplasts

(1) Leaf-derived calli are cultured on UM medium and harvested 10 days after subculture.
(2) Dispersed calli are preplasmolyzed in CPW solution amended with 0.44 M mannitol for 30 min.
(3) Calli are incubated in CPW solution + 0.44 M mannitol + 2% Cellulase R-10 + 0.75% Pectinase + 1.0% Driselase for 4–5 h, at 27 °C, 50 rpm on a gyratory shaker, in diffuse light.
(4) Protoplasts are separated from the enzyme solution by centrifugation at 80 *g* for 2 min.
(5) Protoplasts are isolated by means of floatation in CPW salts + 0.44 M sucrose for 10 min at 100 *g*.

B Culture of Protoplasts

(1) Protoplasts are plated at a density of 5.0×10^4/ml.
(2) The protoplast culture medium is devoid of ammonium nitrate and includes L-glutamine (250 mg/l), L-serine (0.1 mg/l), thiamine HCl (2 mg/l), and myo-inositol (1000 mg/l).
(3) Droplets of fresh culture medium (0.3–0.6 ml) are added to each dish after the beginning of first division (3–5 days) and on a weekly basis thereafter.
(4) The osmoticum is reduced over a 9-week period by adding culture medium amended with 0.32 M, 0.16 M and 0 mannitol at 3-week intervals, respectively.,
(5) Macroscopic green calli, 2–3 mm in diameter (approx. 2 months in age) are transferred to regeneration medium.

C Regeneration of Plants

(1) Macroscopic calli are cultured on MS + 1 mg/l 2ip for shoot initiation.
(2) About 21 days after culture initiation, green shoot buds should be evident.
(3) About 14 days after the appearance of shoot buds, the shoots are elongated sufficiently to be dissected.
(4) Any callus at the base of the shoots is removed.
(5) Excised shoots are dipped in filter-sterilized aqueous IBA (1000 ppm) and transferred to MS + 0.001 mg/l 2,4-D.
(6) Roots develop in about 2 weeks and rooted plants can be transferred to soil in 3–5 more weeks.
(7) After 3 weeks in soil, the acclimatized plants are transferred to the greenhouse.

References

Boyes CJ, Sink KC (1981 a) Morphogenetic responses of *Salpiglossis sinuata* L. leaf explants and callus. Sci Hortic 15:53–60

Boyes CJ, Sink KC (1981 b) Regeneration of plants from callus-derived protoplasts of *Salpiglossis*. J Am Soc Hortic Sci 106:42–46

D'Arcy WG (1978) A preliminary synopsis of *Salpiglossis* and other Cestreae (Solanaceae). Ann Bot Gard 65:698–724

Freason EM, Power JB, Cocking EC (1973) The isolation, culture and regeneration of *Petunia* leaf protoplasts. Dev Biol 33:130–137

Gupta PP, Schieder O, Gupta M (1984) Intergeneric nuclear gene transfer between somatically and sexually incompatible plants through asymmetric protoplast fusion. Mol Gen Genet 197:30–35

Hughes HG, Lam SL, Janick J (1973) In vitro culture of *Salpiglossis sinuata* L. Hortic Sci 8:335–336

Lee CH, Skirvin RM, Soltero AI, Janick J (1977) Tissue culture of *Salpiglossis sinuata* L. from leaf discs. Hortic Sci 12:547–549

Murashige T, Skoog F (1962) A revised medium for rapid growth and bioassays with tobacco tissue cultures. Physiol Plant 15:473–497

Nitsch JP, Nitsch C (1969) Haploid plants from pollen grains. Science 163:85–87

Power JB, Chapman JV (1983) Intra- and intersubfamilial somatic hybridization within the Solanaceae, pp 276–277. In: Potrykus I et al. (eds) Protoplasts 1983 Poster Proc, Sixth Int Protoplast Symp, Basel. Birkhäuser, Basel

Uchimiya H, Murashige T (1974) Evaluation of parameters in the isolation of viable protoplasts from cultured tobacco cells. Plant Physiol 54:936–944

II.12 Isolation and Culture of Protoplasts of Forest Tree Species

EDWARD G. KIRBY[1], MICHAEL A. CAMPBELL and RICARDO M. PENCHEL

1 Introduction

The development of techniques in molecular biology and biotechnology and their application to plant genetic improvement promise to be of singular importance in the future of forest genetics (Karnosky 1981; Kirby 1982; Ahuja 1984a; Libby and Rauter 1984; Powledge 1984; Torrey 1985; Bajaj 1986; Kirby and David 1987). Mass clonal propagation, *in vitro* selection, somatic habridization, somaclonal and gametoclonal variation and genetic transformation using suitable vectors are only a few of the procedures having direct applications in crop improvement today. A key element in applying these techniques to forest tree improvement is establishment of successful procedures for isolation of protoplasts of important forest species and for their culture and the regeneration of whole plants from protoplasts. This chapter will (1) summarize reports of successful protoplast isolation and culture for forest species by focussing on recent advances, and (2) discuss long-range applications of protoplast techniques to tree improvement.

2 The Status of Protoplast Research with Forest Species

2.1 Starting Materials

Because of their potential impact on breading and production of novel phenotypes, protoplasts have been isolated from many different tissues of a number of species of forest trees, however, there are, only a limited number of reports of successful culture of protoplasts (Tables 1, 2). The starting material from which tree protoplasts are prepared is of major significance and affects both isolation efficiency and the ability of protoplasts to regenerate cell cultures, callus and plantlets.

In conifers, protoplasts have been isolated from seedling materials and from callus and cell suspension cultures derived both from seedlings and from needles of adult trees. The most notable success in regenerating dividing cells and callus from conifer protoplasts has been achieved using seedling source tissues (embryos or cotyledons) pre-conditioned *in vitro* by culture on a medium containing auxin

[1] Department of Biological Sciences, Rutgers University, Newark, New Jersey, USA

Biotechnology in Agriculture and Forestry, Vol. 8
Plant Protoplasts and Genetic Engineering I (ed. by Y. P. S. Bajaj)

Table 1. Isolation and successful culture of conifer protoplasts

Source material	Enzymes and osmoticum	Culture medium	Results	Reference
Biota orientalis				
Cotyledons	0.1% Onozuka 0.15% pectinase 0.05% Rhozyme (all desalted) 0.7 M glucose	0.5 X MS medium 0.7 M glucose 15 μM BAP 0.5 μM NAA	Mitosis	David et al. (1981)
Picea glauca				
Embryogenic cell lines	0.5% Cellulase R-10 0.25% Driselase 0.25% Rhozyme 0.5 M mannitol 5 mM $CaCl_2$[a]	VE medium 0.37 M glucose 5 mM gln 2 mg/l 2,4-D 1 mg/l BAP Agar bead culture	Mitosis; callus; embryos	Bekkaoui et al. (1987); Attree et al. (1987)
Pinus contorta				
Cell suspensions	0.5% Macerase 1% Cellulysin 0.4 M sorbitol overnight	NM medium 0.4 M glucose 0.5 μM 2,4-D 5.0 μM NAA 0.4 μM BAP	Colonies	Hakman and von Arnold (1983)
Pinus coulteri				
Cotyledons	2% Onozuka 0.1% Pectolyase 10% mannitol 500 mg/l $CaCl_2$ 6 h	B5 medium, 900 mg/l $CaCl_2$ 0.5 mg/l 2,4-D, NAA, GA, BAP and zeatin	Colonies	Patel et al. (1984)
Pinus lambertiana				
Callus and cell suspensions	2% Cellulysin 1% pectinase 1% Macerase 0.5% K-dextran-sulfate 12% mannitol 50 mM MES 3 – 5 h (callus) 16 – 18 h (needles)	DCR medium 1% glucose 2% sucrose 6% mannitol 50 mg/l gln 200 mg/l CH 1 mg/l 2,4-D 0.5 mg/l NAA 0.2 mg/l BAP 1% DMSO 5 mM MES	Colonies	Gupta and Durzan (1986b)
Pinus pinaster				
Cotyledons	0.15% Onozuka 0.05% Rhozyme 0.08% pectinase 0.7 M glucose overnight	0.5 X MS medium 10% FeEDTA 0.7 M glucose 15 μM BAP 0.5 μM NAA	Colonies	David and David (1979); David et al. (1982); David et al. (1986); Faye and David (1983)
Pinus taeda				
Cell suspensions	0.5% Cellulysin 0.5% Macerozyme 0.5% hemicellulase 0.5% Driselase 0.5% BSA 0.2 M sorbitol 0.4 M mannitol 9 h	LM medium 1.5 mM $CaCl_2$ 10 mM arg 5 mM asp 10 mM gln	Colonies	Teasdale and Rugini (1983)

Table 1 (continued)

Source material	Enzymes and osmoticum	Culture medium	Results	Reference
Pseudotsuga menziesii				
Cotyledons	2.0% Cellulysin 0.5% Macerase 0.55 M sorbitol 3.5 h	0.5 X modified MS 0.55 M sorbitol 15 µM NAA, 0.5 µM BAP 50 mM gln	Colonies Callus	Kirby (1980); Kirby and Cheng (1979)

Medium Abbreviations: B5, Gamborg (1977); DCR, Gupta and Durzan (1985); LM, Verma et al. (1982); MS, Murashige and Skoog (1962); NM, Nagy and Maliga (1976).
Medium Component Abbreviations: arg, arginine; asp, asparigine; BAP, N_6-benzylaminopurine; BSA, bovine serum albumin; CH, casein hydrolysate; 2,4-D, 2,4-dichlorophenoxyacetic acid; DMSO, dimethyl sulfoxide; GA, gibberellic acid; gln, glutamine; MES, 2(N-morpholino)ethane sulfonic acid; NAA, α-naphthaleneacetic acid.
[a] Supplied as $CaCl_2 \cdot 2H_2O$.

and cytokinin prior to protoplast isolation (Kirby and Cheng 1979; David and David 1979; Kirby 1980; David et al. 1982).

Although protoplasts have been successfully isolated from haploid cells of conifers, including pollen of *Cupressus arizonica* (Duhoux 1980) and female gametophytes of *Picea abies* (Hakman et al. 1986), regeneration of dividing cells from these protoplasts has not been reported. In addition, roots of seedlings of maritime pine (*Pinus pinaster*) have served as source material for protoplasts, but these root protoplasts have not been reported to divide in culture (Faye and David 1983).

The efficient isolation of protoplasts from suspension cultures of loblolly pine (*Pinus taeda*) (Teasdale and Rugini 1983) and lodge pole pine (*Pinus contorta*) (Hakman and Von Arnold 1983) required that cell cultures be in the exponential phase of growth at the time of protoplast isolation. Supporting these observations are preliminary studies from our laboratory indicating that during the early exponential phase of the culture cycle of cell suspensions of Douglas-fir maintained on a medium containing 30 mM glutamine, large numbers of metabolically active protoplasts can be easily isolated.

For angiosperm forest species, reports of protoplast isolation from leaves of juvenile and adult trees, cotyledons, cell suspensions and callus have appeared in the literature (Table 2). Recent work has shown that by carefully defining the starting materials from which protoplasts are prepared, the efficient regeneration of cell cultures from protoplasts of several angiosperm forest trees can be accomplished. Techniques for protoplast isolation from birch, aspen and poplar, as developed in McCown's laboratory, involve the initiation and maintenance of shoot cultures *in vitro* using a two-step procedure. First, nodal explants are multiplied on a medium containing high cytokinin, after which single shoots are transferred to medium containing low cytokinin in order to produce large, healthy leaves from which protoplasts are prepared (Russell and McCown 1986).

Perhaps one of the most exciting recent development in biotechnology of forest trees is the establishment of embryogenic tissue cultures (Hakman et al.

Table 2. Isolation and culture of protoplasts of angiosperm forest trees

Source material	Enzymes and osmoticum	Culture medium	Results	Reference
Alnus glutinosa				
Leaves	0.5% cellulase 0.2% Macerozyme 2% sucrose 0.5 M mannitol overnight	MS micronutrients N_6 macronutrients 2 mg/l 2,4-D 0.5 mg/l kinetin 2% sucrose 0.05 M mannitol 0.05 – 0.1 mM putrescine liquid microculture	Callus	Huhtinen et al. (1982)
Alnus glutinosa				
Juvenile leaves	2% Cellulysin 0.5% Macerase 11% mannitol overnight	CPW salts	Viability	Butt (1985)
Alnus incana				
Leaves	0.5% cellulase 0.2% Macerozyme 0.5 M mannitol 2% sucrose overnight	as for *Alnus glutinosa* (Huhtinen et al. 1982)	Callus	Huhtinen et al. (1982)
Alnus incana				
Cell suspensions	1% Onozuka 0.5% Rhozyme 0.03% Macerase 13% mannitol overnight	KM 8p medium 1 µM 2,4-D 2,4 µM zeatin 5 µM NAA	Callus	Tremblay et al. (1985)
Betula pendula				
Cotyledons; juvenile leaves	2% Cellulysin 0.5% Macerase 11% mannitol overnight	CPW salts	Viability	Butt (1985)
Betula platyphylla szechuanica				
Juvenile leaves; shoot cultures	0.1 – 0.5% Macerase 0.5 – 2% Cellulysin 0.7 M sucrose 0.1% BSA 0.1 X MS salts 16 – 18 h	HS medium over LS medium	Viability	Smith and McCown (1982)
Betula pubescens				
Cotyledons; leaves	2% Cellulysin 0.5% Macerase 11% mannitol overnight	CPW salts	Viability	Butt (1985)
Fagus sylvatica				
Leaves from adult trees	1% cellulase 0.5% Macerozyme 0.1% pectinase 0.7 M mannitol 20 – 22 h	WPM medium 100 mg/l casein 500 mg/l gln 1 mg/l ascorbic acid 1 mg/l Ca pentothenate 0.5 mg/l BAP 0.5 mg/l IAA 0.5 mg/l NAA	Mitosis	Ahuja (1984b)

Table 2 (continued)

Source material	Enzymes and osmoticum	Culture medium	Results	Reference
		0.02 mg/l 2,4-D 10% mannitol 1.8% glucose 1% sucrose liquid cultures		
Liriodendron tuliperifera				
Embryogenic cell lines	MeSo medium 2% Cellulysin 1% Macerase 100 mg/l MES 0.5 M mannitol 100 mg/l BSA 500 mg/l $CaCl_2$[a]	MeSo medium 500 mg/l $CaCl_2$[a] 250 mg/l xylose 250 mg/l fructose 40 g/l sucrose	Colonies; embryos; plantlets	Merkle and Sommer (1987)
Populus alba × *P. grandidentata*				
Shoot cultures	0.5% cellulase 0.1% Macerase 550 mM sucrose 0.1% BSA WPM salts 4.7 mM $Ca(NO_3)_2$ 1 µM NAA 0.1 µM BAP 4 h	WPM medium, 1.0 µM NAA 0.1 µM BAP 4.7 mM $Ca(NO_3)_2$ 366 mM sucrose 46 mM mannitol 46 mM sorbitol 46 mM xylitol 46 mM inositol liquid cultures	Colonies; callus; plantlets	Russell and McCown (1986)
Populus nigra var. "Butulifolia" × *P. trichocarpa*				
Shoot cultures	0.5% cellulase 0.1% Macerase 550 mM sucrose 0.1% BSA WPM salts 4.7 mM $Ca(NO_3)_2$ 1 µM NAA 0.1 µM BAP 4 h	WPM medium, 1.0 µM NAA 0.1 µM BAP 4.7 mM $Ca(NO_3)_2$ 366 mM sucrose 46 mM mannitol 46 mM sorbitol 46 mM xylitol 46 mM inositol liquid cultures	Colonies; callus; plantlets	Russell and McCown (1988)
Populus tacatriicho				
Leaves from adult trees	2% Cellulysin 0.5% Macerase 11% mannitol overnight	CPW salts	Viability	Butt (1985)
Populus tremula				
Leaves from adult trees	0.5% cellulase 0.1% Macerozyme 0.1% BSA 0.7 M mannitol CPW salts 20–22 h	WPM medium 100 mg/l casein 500 mg/l gln 1 mg/l ascorbic acid 1 mg/l Ca pantothenate 0.5 mg/l BAP 0.5 mg/l IAA 0.5 mg/l NAA 0.02 mg/l 2,4-D 10% mannitol	Mitosis	Ahuja (1983)

Table 2 (continued)

Source material	Enzymes and osmoticum	Culture medium	Results	Reference
		1.8% glucose 1% sucrose liquid cultures		
Populus tremula				
Shoot cultures	0.5% cellulase 0.1% Macerase 550 mM sucrose 0.1% BSA WPM salts 4.7 mM $Ca(NO_3)_2$ 1 µM NAA 0.1 µM BAP 4 h	WPM medium, 1.0 µM NAA 0.1 µM BAP 4.7 mM $Ca(NO_3)_2$ 366 mM sucrose 46 mM mannitol 46 mM sorbitol 46 mM xylitol 46 mM inositol liquid cultures	Colonies; callus; plantlets	Russell and McCown (1988)
Salix spp.				
Mature leaves	2% Cellulysin 0.5% Macerase 11% mannitol overnight	CPW salts	Viability	Butt (1985)
Santalum album				
Cell suspensions	1% cellulase 1% Macerozyme 0.5% Driselase 0.55 M mannitol 2–6 h	V47 medium 0.5 mg/l 2,4-D 0.5 mg/l BAP 3 mM MES 750 mOs/kg mannitol	Embryoids and plantlets	Rao and Ozias-Akins (1985)
Ulmus glabra var. "Camperdown"				
Leaves from adult trees	2% Cellulysin 0.5% Macerase 11% mannitol overnight	CPW salts	Viability	Butt (1985)
Ulmus spp.				
Callus	1.5% Cellulysin 0.5% Macerase 0.5% dextran 0.7 M mannitol overnight	KM 8p medium 1 µM 2,4-D 2.4 µM NAA 1 µM zeatin	Callus and plantlets	Sticklen et al. (1985a, 1985b)

Medium Abbreviations: CPW, Ahuja (1983); KM 8p, Kao and Michayluk (1975); MS, Murashige and Skoog (1962); MeSo, Merkle and Sommer (1986); N_6, Hu et al. (1978); HS (high salt), Shepard (1980); LS (low salt); Shepard (1980); WPM, Lloyd and McCown (1980); V47, Binding (1974).
Medium Component Abbreviations: arg, arginine; asp, asparigine; BAP, N_6-benzylaminopurine; BSA, bovine serum albumin; CH, casein hydrolysate; 2,4-D, 2,4-dichlorophenoxyacetic acid; DMSO, dimethyl sulfoxide; GA, gibberellic acid; gln, glutamine; IAA, indoleacetic acid; MES, 2(N-morpholino)ethane sulfonic acid; NAA, *a*-naphthaleneacetic acid.
[a] Supplied as $CaCl_2 \cdot 7H_2O_8$.

1985; Gupta and Durzan 1986a, 1987a; Merkle and Sommer 1986; Hakman and Fowke 1987). Utilization of embyrogenic cell cultures as starting material for protoplast isolation has been described for white spruce (Attree et al. 1987; Bekkaoui et al. 1987), loblolly pine (Gupta and Durzan 1987b) and yellow poplar (*Liriodendron tulipifera*) (Merkle and Sommer 1987). Regeneration of somatic embryos from protoplasts derived from these embryogenic cell lines has been reported. Further research using this approach is clearly indicated.

2.2 Protoplast Isolation and Purification

2.2.1 Cell Wall Degrading Enyzmes

Commercially available cell wall degrading enzymes, including cellulases, hemicellulases and pectinases, have been utilized for isolating tree protoplasts. Enzymes are widely available and many commercial enzyme preparations have been reported as effective in protoplast isolation from tree species. For some species, including maritime pine (David et al. 1984), an increase in protoplast yield and viability has been obtained by using desalted enzymes. In our experience with protoplasts isolated from cotyledons, needles and cell cultures of conifers and from cotyledons and young leaves of *Eucalyptus* species, it has not been necessary to desalt enzymes in order to obtain good protoplast release. It is important, however, to be aware that impurities in commercially available enzymes, including salts, phenolic substances nucleases and proteases, may have detrimental effects on viability and behavior of isolated protoplasts in culture. This is particularly true in protocols where long enzyme treatments are specified.

2.2.2 Isolation Mixtures

The use of plasmolyzing agents is important, not only to plasmolyze the cells and expose cell walls for enzymatic digestion, but also to maintain protoplast integrity during isolation and culture. Most frequently mannitol is used as the sole plasmolyzing agent. Sorbitol, either alone or in combination with mannitol, and glucose and sucrose have also been used successfully (Redenbaugh et al. 1981; Teasdale and Rugini 1983; David et al. 1982). Concentrations of plasmolyzing agents generally range between 0.4 and 0.7 M. It is important to note, however, that the exact level must be empirically determined for each starting material.

Buffer components, including 2-[N-morpholino]ethanesulfonic acid (MES), may be beneficial for the isolation of protoplasts of forest trees (David et al. 1984; Rao and Ozias-Akins 1985). Incorporating elevated levels of calcium chloride, usually in the range of 2–10 mM, has also been shown to help maintain protoplast integrity. In isolating protoplasts of *Pinus taeda*, Teasdale and Rugini (1983) reported that potassium chloride (0.7 mM) was beneficial in this regard. Teasdale and Rugine (1983) also reported an increase in viability of protoplasts of *Pinus taeda* when bovine serum albumin (BSA) (500 mg/ml) was incorporated into the isolated mixture, presumably to tie up phenolics and proteases present in the commercial enzyme preparations.

2.2.3 Isolation Times

The time required for release of protoplasts by cell wall degrading enzymes differs with the species utilized, the nature of the starting material and the enzymes employed. Protoplasts can be produced in as little as 1 h from leaves of cultured seedlings of *Eucalyptus grandis* using a combined level of 2.5% Cellulysin and Macerase (Behrndt and Kirby unpubl.), whereas 3.5 h are required to release protoplasts from Douglas-fir cotyledons using the same concentrations of enzymes (Kirby and Cheng 1979). Lower enzyme concentrations generally require longer incubation periods (up to 24 h), as has been reported for cotyledon protoplasts of maritime pine (David et al. 1982) and leaf and cell culture protoplasts of elm (Redenbaugh et al. 1981).

2.2.4 Purification of Protoplast Fractions

Purification of protoplasts is most frequently accomplished by repeated centrifugation and resuspension, usually in culture medium (Kirby and Cheng 1979; David and David 1979). Tree protoplasts, particularly those isolated from conifer cell suspensions, are extremely fragile. Centrifugation procedures frequently disrupt fragile protoplasts. Simple gravity sedimentation under proper osmotic conditions has been reported successful in purification of protoplasts from cell suspensions of loblolly pine (Teasdale and Rugine 1983).

2.2.5 Determination of Protoplast Viability

Isolation and culture of protoplasts from woody plants is not as easily accomplished as with many herbaceous plant species. It is appropriate to have a means of monitoring the relative success of each step during isolation and culture. One way in which this can be done is to assess viability. In our laboratory, cell and protoplast viability is routinely determined using the vital stain, Evans blue (0.1%, w/v, in culture medium for 1 min followed by rinsing with fresh medium). A similar procedure has been followed for cell cultures and protoplasts of loblolly pine (Gupta and Durzan 1987b).

2.3 Protoplast Culture

2.3.1 Medium Components

Protoplast culture media are based on cell culture media and contain basal medium components, auxin and cytokinin, sucrose or glucose, vitamins, organic supplements and basic amino acids, such as asparagine, glutamine and arginine (Tables 1, 2). Studies with protoplasts of Douglas-fir (Kirby 1980), maritime pine (David et al. 1984), white spruce (Bekkaoui et al. 1987) and other conifers have demonstrated that a limiting factor in obtaining dividing cells from conifer protoplasts is the presence of glutamine in the culture medium. In most cases glutamine can substitute for undefined organic supplements, such as malt and yeast extracts, or casein hydrolysate. From Table 1, it is apparent that addition of glutamine to the culture media for conifer protoplasts at levels between 5 and 50 mM is beneficial.

2.3.2 *Plating Density*

The plating densities for protoplasts of tree species vary considerably, from 5×10^3 protoplasts/ml in poplar and aspen (Russell and McCown 1988) to 2×10^6 protoplasts/ml in maritime pine (David and David 1979). In woody species, protoplast plating efficiencies are generally low and diffusion of phenolic compounds from non-viable cells may affect the growth of viable cells. Therefore, high plating densities are not recommended.

2.3.3 *Plating Technique*

The most frequent procedure employed with tree protoplasts is liquid culture. In this procedure, protoplasts are suspended in culture medium at a desirable density and then cultured as small droplets (David et al. 1982), or as cell suspensions, either with or without moderate agitation (Teasdale and Rugini 1983). We have investigated the use of polyester fleeces for support of protoplasts during culture (Kirby 1980). Use of these materials appears beneficial, perhaps because it allows free diffusion of inhibitory substances, including phenolics, away from living cells and provides a large surface area for diffusion of oxygen into the medium. For poplar and aspen protoplasts (Russell and McCown 1986) floating screen discs have been used, which offer many of the advantages of the polyester fleeces, while greatly facilitating protoplast observation. Recent experiments in our laboratory have shown that high purity low melting point agarose (e.g. Sea Plaque and Sea Prep from the Marine Colloids Division, FMC Corp., Rockland, Maine, USA) when used at low concentrations (in the range of 0.6%) can provide an adequate agar medium for plating of conifer cell cultures and protoplasts.

Regeneration of rapidly growing cell cultures from protoplasts is greatly faciliated by sequential lowering of the osmotic potential of the culture medium. This is achieved in liquid cultures by first allowing protoplasts and cells to settle to the bottom of the culture vessel. A known volume of the original culture medium is removed and replaced with new medium of lower osmotic potential. The frequency with which this is done must be determined empirically.

High plating efficiencies and rapid regeneration of somatic embryos from protoplasts isolated from embryogenic cell suspension cultures of white spruce (*Picea glauca*) have been reported (Attree et al. 1987) using a modification of the agarose bead procedure described by Shillito et al. (1983) in which protoplasts are suspended in agarose beads which are cultured as suspensions in liquid medium. This technique also facilitates the lowering of the osmoticum level of the culture medium without disturbing growing protoplast cultures.

3 Long-Range Application of Protoplast Techniques to Tree Improvement

3.1 Somaclonal Variation, Somatic Hybridization and Cybrids

Although techniques of somaclonal variation have yet to be applied to forest trees, for crop plants perhaps the most rewarding application of protoplasts has been their use in somaclonal variation. Somaclonal variation is currently being utilized as an alternative technique for the induction of genetic variability in a number of important crops including potato, tomato, carrot, tobacco, alfalfa, sugar cane, wheat, rapeseed, pepper, rice, oats, corn (Larkin and Scowcroft 1981; Evans et al. 1984). The application of this technique to forest trees will be forthcoming as new procedures become established allowing routine regeneration of plants from protoplasts and cell cultures.

Somatic hybridization is potentially an extremely powerful technique for the production of nuclear and cytoplasmic hybrids (cybrids). Applications of fusion technology is far from routine, however, and applications to forest trees await improved protoplast culture techniques and defined selection procedures.

3.2 Transformation Technology

Recent reports have demonstrated that the technology for genetic transformation of forest trees has arrived. Introduction of a herbicide resistance gene (*aro* A), which confers resistance to glyphosate, into poplar using a binary *Agrobacterium* vector coupled with a disk transformation scheme has been achieved (Fillatti et al. 1986). Parsons et al. (1986) have reported transformation of a superior hybrid poplar clone (*Popolus trichocarpa* × *deltoides* clone H11) using both octopine and agropine strains of *Agrobacterium tumefaciens*. Transformation was demonstrated by the presence of foreign T-DNA in derived tissues using Southern blotting and by agropine and octopine assays. In this study, wound site transformation, leaf disk transformation and transformation of stem segments all proved successful. Further work with clone H11 of hybrid poplar (Pythoud et al. 1987) has demonstrated that increased transformation efficiency can be achieved using a "supervirulent" strain of *Agrobacterium rhizogenes* containing pTVK291, a plasmid containing part of the *vir* region of pTiBo542.

Considerable effort has focussed on establishment of transformation procedures for conifer species using *A. tumefaciens*. Sederhoff et al. (1986) have demonstrated expression of bacterial genes for opine biosynthesis in crown gall tissue of loblolly pine (*Pinus taeda*). Two strains of *A. tumefaciens* (U3 and M2/73) proved effective. Although the presence of T-DNA in proliferating pine gall tissues was not determined, opine assays provide good evidence for transformation. Transformation of micropropagated shoots of Douglas-fir (*Pseudotsuga menziesii*) using derivatives of *A. tumefaciens* strain A6 has also been achieved (Dandekar et al. 1987). Transformed tissues demonstrated phytohormone-autotrophic growth *in vitro*, octopine production, the presence of foreign DNA sequences and expression of a chimeric kanamycin-resistance gene.

In addition to *Agrobacterium*-mediated transformation, additional transformation techniques that may play a role in transformation of forest trees include direct gene transfer (electroporation), microinjection and organellar transformation. Development of techniques for the isolation, culture and regeneration of forest trees from protoplasts, however, are prerequisites to the application of these efficient transformation protocols to tree improvement.

3.3 Summary and Conclusions

As has been shown in cereals, isolation of protoplasts from embryogenic cell cultures may lead to routine regeneration of forest trees from protoplasts. There certainly is good cause for optimism. Recent work has indicated that somatic embryos can be regenerated from protoplasts isolated from proembryonal cells of loblolly pine (Gupta and Durzan 1987b), from embryogenic cell suspensions of white spruce (Attree et al. 1987) and yellow poplar (Merkle and Sommer 1987). The direction of future work will clearly be on (1) establishment of embryogenic and morphogenically competent cultures of important forest species to serve as source materials for protoplast isolation, and (2) the efficient production of plants from protoplast-derived cell cultures. The integration of techniques such as somaclonal and gametoclonal variation, somatic hybridization and direct gene transfer into three breeding programs awaits routine regeneration of trees from protoplasts.

References

Ahuja MR (1983) Short note: Isolation and culture of mega and normal protoplasts in aspen. Silvae Genet 32:5–6

Ahuja MR (1984a) Protoplast research in woody plants. Silvae Genet 33:32–37

Ahuja MR (1984b) Isolation and culture of mesophyll protoplasts from mature beech trees. Silvae Genet 33:37–39

Attree SM, Bekkauoi F, Dunstan DI, Fowke LC (1987) Regeneration of somatic embryos from protoplasts from an embryonic suspension culture of white spruce (*Picea glauca*). Plant Cell Rep 6:480–483

Bajaj YPS (1986) Biotechnology in agriculture and forestry 1. Trees I. Springer, Berlin Heidelberg New York Tokyo

Bekkaoui F, Sexena PK, Attree SM, Fowke LC, Dunstan DI (1987) The isolation and culture of protoplasts from an embryogenic cell suspension culture of *Picea glauca*.. Plant Cell Rep 6:476–479

Binding H (1974) Regeneration von haploiden und diploiden Pflanzen aus Protoplasten von *Petunia hybrida*. Z Pflanzenphysiol 74:327–336

Butt AD (1985) A general method for the high-yield isolation of mesophyll protoplasts from deciduous tree species. Plant Sci 42:55–59

Dandekar AM, Gupta PK, Durzan DJ, Knauf V (1987) Transformation and foreign gene expression in micropropagated Douglas-fir. Biotechnology 5:587–590

David A, David H (1979) Isolation and callus formation from cotyledon protoplasts of pine (*Pinus pinaster*). Z Pflanzenphysiol 94:173–177

David H, David A, Mateille T (1981) Isolation and culture of protoplasts of two gymnosperms: *Pinus pinaster* and *Biota orientalis*. In: Boulay M (ed) Proc Int Colloq in vitro culture of tissues of forest trees, AFOCEL, Nangis, France, pp 339–347

David H, David H, Mateille T (1982) Evaluation of parameters affecting the yield, viability and cell division of *Pinus pinaster* protoplasts. Physiol Plant 56:108–113

David H, Jarlet E, David A (1984) Effects of nitrogen source, calcium concentration and osmotic stress on protoplasts and protoplast-derived cell cultures of *Pinus pinaster* cotyledons. Physiol Plant 61:477–481

David H, de Boucard MT, Gaultier JM, David A (1986) Sustained division of protoplast-derived cells from primary leaves of *Pinus pinaster*: factors affecting growth and nuclear DNA content. Tree Physiol 1:21–30

Duhoux E (1980) Protoplast isolation of gymnosperm pollen. Z Pflanzenphysiol 99:207–214

Evans DA, Sharp WR, Medina-Filho HP (1984) Somaclonal and gametoclonal variation. Am J Bot 71:759–774

Faye M, David A (1983) Isolation and culture of gymnosperm root protoplasts (*Pinus pinaster*). Physiol Plant 59:359–362

Fillatti JJ, Sellmer JC, McCown BH (1986) Regeneration and transformation of *Populus*. In: Sommers DA, Gengenbach BG, Biesboer DD, Hackett WP, Green CE (eds) Proc VI Congr Plant Tissue and Cell Culture, International Association for Plant Tissue Culture, Minneapolis, Minnesota, p 127

Gamborg O (1977) Culture media for plant protoplasts. In: Recheigl Jr. M (ed) CRC Handbook Series in Nutrition and Foods, Vol. IV, CRC Press, Cleveland, Ohio, pp 415–422

Gupta PK, Durzan DJ (1985) Shoot multiplication from mature trees of Douglas-fir (*Pseudotsuga menziesii*) and sugar pine (*Pinus lambertiana*). Plant Cell Rep 4:177–179

Gupta PK, Durzan DJ (1986a) Somatic embryogenesis from callus of mature sugar pine embryos. Biotechnology 4:643–645

Gupta PK, Durzan DJ (1986b) Isolation and cell regeneration of protoplasts from sugar pine (*Pinus lambertiana*). Plant Cell Rep 5:346–348

Gupta PK, Durzan DJ (1987a) Biotechnology of somatic polyembryogenesis and plantlet regeneration in loblolly pine. Biotechnology 5:147–151

Gupta PK, Durzan DJ (1987b) Somatic embyros from protoplasts of loblolly pine proembryonal cells. Biotechnology 5:710–712

Hakman IC, Fowke LC (1987) An embryogenic cell suspension culture of *Picea glauca* (white spruce). Plant Cell Rep 6:20–22

Hakman IC, von Arnold S (1983) Isolation and culture of protoplasts from cell suspensions of *Pinus contorta* Dougl. ex Loud. Plant Cell Rep 2:92–94

Hakman IC, Fowke LC, von Arnold S, Eriksson T (1985) The development of somatic embryos in tissue cultures initiated from immature embryos of *Picea abies* (Norway spruce). Plant Sci 38:53–59

Hakman IC, von Arnold S, Fellner-Feldegg H (1986) Isolation and DNA analysis of protoplasts from developing female gametophytes of *Picea abies* (Norway spruce). Can J Bot 64:108–112

Hu H, Hsi TY, Tseng CC, Ouyang TW, Ching CK (1978) Applications of anther culture to crop plants. In: Thorpe TA (ed) Frontiers of Plant Tissue Culture 1978. Univ Calgary Printing Service, Calgary, Canada, pp 123–130

Huhtinen O, Honkanen J, Simola LK (1982) Ornithine and putrescine-supported divisions in cell colony formation in leaf protoplasts of alders (*Alnus glutinosa* and *A. incana*). Plant Sci Lett 28:3–9

Kao KN, Michayluk MR (1975) Nutritional requirements for growth of *Vicia hajastana* cells and protoplasts at a very low population density in liquid medium. Planta 126:105–110

Karnosky DF (1981) Potential for forest tree improvement via tissue culture. Bioscience 31:114–120

Kirby EG (1980) Factors affecting proliferation of protoplasts and cell cultures of Douglas-fir. In: Sala F, Parisi B, Cella R, Ferri OC (eds) Plant cell cultures: results and perspectives, Elsevier/North Holland, Amsterdam, pp 289–293

Kirby EG (1982) The use of in vitro techniques for genetic modification of forest trees. In: Bonga JM, Durzan DJ (eds) Tissue culture in forestry. Martinus Nijhoff/Junk, Amsterdam, pp 369–386

Kirby EG, Cheng TY (1979) Colony formation from protoplasts derived from Douglas-fir cotyledons. Plant Sci Lett 14:145–154

Kirby EG, David A (1987) Use of protoplasts and cell cultures for physiological and genetic studies of conifers. In: Hanover J, Keathley D (eds) Genetic manipulation of woody plants. Plenum, New York, pp 185–197

Larkin PJ, Scowcroft W (1981) Somaclonal variation – a novel source of variability from cell cultures for plant improvement. Theor Appl Genet 60:197–214

Libby WJ, Rauter RM (1984) Advantages of clonal forestry. For Chron 60:145–149

Lloyd G, McCown BH (1980) Commerically feasible micropropagation of mountain laurel, *Kalmia latifolia*, by use of shoot tip culture. Prov Int Plant Prop Soc 30:421–427

Merkle SA, Sommer HE (1986) Somatic embryogenesis in tissue cultures of *Liriodendron tulipifera*. Can J For Res 16:420–422

Merkle SA, Sommer HE (1987) Regeneration of *Liriodendron tulipifera* (Family Magnoliaceae) from protoplast culture. Am J Bot 74:1317–1321

Murashige T, Skoog F (1962) A revised medium for rapid growth and bioassays with tobacco tissue cultures. Physiol Plant 15:473–497

Nagy JL, Maliga P (1976) Callus induction and plant regeneration from mesophyll protoplasts of *Nicotiana sylvestris*. Z Pflanzenphysiol 78:453–455

Parsons TJ, Sinkar VP, Stettler RF, Nester EW, Gordon MP (1986) Transformation of poplar by *Agrobacterium tumefaciens*. Biotechnology 4:533–536

Patel KR, Shekhawat NS, Berlyn GP, Thorpe TA (1984) Isolation and culture of protoplasts from cotyledons of *Pinus coulteri*. Plant Cell Tissue Organ Cult 3:85–90

Powledge TM (1984) Biotechnology touches the forest. Biotechnology 2:763–772

Pythoud F, Sinkar VP, Nester EW, Gordon MP (1987) Increased virulence of *Agrobacterium rhizogenes* conferred by the *vir* region of pTiBo542: Application to the genetic engineering of poplar. Biotechnology 5:1323–1327

Rao PS, Ozias-Akins P (1985) Plant regeneration through somatic embryogenesis in protoplast cultures of sandalwood (*Santalum album*). Protoplasma 124:80–86

Redenbaugh K, Karnosky DF, Westfall RD (1981) Protoplast isolation and fusion in three *Ulmus* species. Can J Bot 59:1436–1443

Russell JA, McCown BH (1986) Culture and regeneration of *Populus* leaf protoplasts isolated from non-seedling tissue. Plant Sci 46:133–142

Russell JA, McCown BH (1988) Recovery of plants from leaf protoplasts of hybrid-poplar and aspen clones. Plant Cell Rep 7:59–62

Sederhoff R, Stomp AM, Chilton WS, Moore LW (1986) Gene transfer into loblolly pine by *Agrobacterium tumefaciens*. Biotechnology 4:647–649

Shepard JF (1980) Mutant selection and plant regeneration from potato mesophyll protoplasts. In: Rubenstein I (ed) Genetic improvement of crops: emergent techniques, Univ Minnesota Press, Minneapolis, Minnesota, pp 186–192

Shillito RD, Paszkowski J, Portykus I (1983) Agarose plating and a bead type culture technique enable and stimulate development of protoplast-derived colonies in a number of plant species. Plant Cell Rep 2:244–247

Smith MA, McCown BH (1982) A comparison of source tissue for protoplast isolation from three woody plant species. Plant Sci Lett 28:149–156

Sticklen MB, Domir SC, Lineberger RD (1985a) Shoot regeneration from protoplasts of *Ulmus*× "pioneer". Plant Sci 41:29–34

Sticklen MB, Lineberger RD, Domir SC (1985b) Isolation and culture of protoplasts of *Ulmus*× "homestead". Plant Sci 41:117–120

Teasdale RD, Rugini R (1983) Preparation of viable protoplasts from suspension-cultured loblolly pine (*Pinus taeda*) cells and subsequent regeneration to callus. Plant Cell Tissue Organ Culture 2:253–261

Torrey JG (1985) The development of plant biotechnology. Amer Scientist 73:354–363

Tremblay FM, Power JB, Lalonde MA (1985) Callus regeneration from *Alnus incana* protoplasts isolated from cell suspensions. Plant Sci 41:211–216

Verma VC, Litvay JD, Johnson MA (1982) Media development for cell suspensions of conifers. Proc. 5th International Congress Plant Tissue and Cell Culture, Internatl Assoc Plant Tissue Culture, Tokyo, Japan, pp 59–60

Section III
Protoplast Fusion and Somatic Hybridization

III.1 Fusion of Plant Protoplasts – Techniques

K. N. KAO[1] and M. R. MICHAYLUK

1 Introduction

Hybridization of somatic cells has been suggested as a possible technique to overcome the cross-incompatibility of unrelated plant species. Somatic hybrids might be of great value for crop improvement (Schenk and Hildebrandt 1968). Hybridization of somatic plant cells requires the following steps:

a) Removal of the cell walls to form protoplasts.
b) Induction of close membrane contact at the molecular level and reversible membrane breakdown for fusion between two independent protoplasts (Poste and Allison 1973; Lucy 1978; Zimmermann and Vienken 1982).
c) Identification and selection of the somatic hybrids.
d) Culture of the hybrid protoplasts.
e) Regeneration of plants from the hybrids.

Attempts to hybridize somatic cells were made early in the century by Winkler, Küster and Michel (in Power et al. 1970; Melchers and Labib 1974). Michel (1937) demonstrated the fusion of homo- and heterospecific protoplasts after their treatment with a sodium nitrate solution. However, fusion was rare, and subsequent fusion products could not be cultured (Power et al. 1970).

Numerous plant somatic hybrids and cybrids have been produced in recent years. These advances have become possible because of the improvement in techniques for protoplast isolation, culture, fusion and selection of heterokaryocytes. There are numerous publications on these subjects. Among them are Gleba and Sytnik's book *Protoplast Fusion* (1984) and Harms's review article *Hybridization by Somatic Cell Fusion* (1985), see also Chap. I.1, this Vol. In this chapter, a number of protoplast fusion techniques are described.

[1] Plant Biotechnology Institute National Research Council of Canada, Saskatoon, Saskatochewan, Canada S7N OW9
NRCC No. 29582

Biotechnology in Agriculture and Forestry, Vol. 8
Plant Protoplasts and Genetic Engineering I (ed. by Y. P. S. Bajaj)

2 Methods of Fusion

2.1 Spontaneous Fusion

Fusion of plant protoplasts is known to occur spontaneously during enzymatic degradation of cell walls. This type of homoplasmic fusion has been attributed to the expansion of plasmodesmata between protoplasts derived from adjacent cells (Cocking 1972; Power and Cocking 1971).

Miller et al. (1971) observed that up to 25% of the protoplasts isolated from cultured soybean cells were multinucleate. The average number of nuclei per protoplast was 2.45 (ranging from 1–45) in *Vicia hajastana* and 2.51 (1–35) in brome grass. Mononucleate protoplasts isolated from cultured cells constituted only 38% of the *Vicia* population and 47% of the brome grass population (Kao et al. 1973).

Motoyoshi (1971) indicated that 50% of the total protoplasts enzymatically isolated from callus cells of maize endosperm were multinucleate, containing 2 to 28 nuclei. The frequency of spontaneous fusion could vary from 20 to 80%, but the higher frequencies were generally favoured by higher temperatures (35–40 °C) during protoplast isolation and by deplasmolyzing osmotic shocks (Keller et al. 1973).

2.2 Fusion of Cells by Mechanic Means

A. Fusion of the large-celled, filamentous marine red alga *Griffithsia tenuis* C. Agardh. (Waaland 1978). The method is as follows: Insert freshly isolated (male and/or female) intercalary filaments in a cylinder of cell wall from *Nitella* sp. filaments in such a way that the base of one filament of *G. tenuis* opposes the apex of the other filament. Allow enough time (ca. 4–8 h) for the upper filament to produce a rhizoid at its base and the lower filament to produce a specialized repair shoot cell at its apex. Fifty-four percent of the rhizoid and repair shoot can fuse by itself and form a single hybrid cell in 20 h; 84% of the hybrid cells regenerated new plants.

B. The giant protoplasts of *Acetabularia* have been fused by mechanically pushing together two protoplasts. This fusion does not depend upon the presence of fusion-inducing agents. The yield was more than 90% (Primke et al. 1978).

2.3 Fusion of Meiotic Protoplasts

Protoplasts released from meiocytes of *Lilium* and *Trillium* in an enzyme solution of 4% cellulase (Onozuka) and 2% Macerozyme in White's mineral solution with 0.4–0.45 M sucrose fused readily in a depression slide with gentle tapping (Ito and Maeda 1973; Ito 1973). Some of the di- and trinucleates completed the second division with the formation of typical tetrad configurations in a culture medium (Ito and Maeda 1974).

2.4 Fusion of Protoplasts with $NaNO_3$

Power et al. (1970) reported that protoplasts isolated from root tips of oats and other cereals in 0.56 M sucrose with 10% cellulase (Onozuka) and 5% Macerozyme were able to fuse in a 0.25 M sodium nitrate solution. Kameya and Takahashi (1972) indicated that isolation of protoplasts in enzyme solutions containing potassium salts and subsequently washed in solutions of sodium salts was effective for fusion of protoplasts. Carlson et al. (1972) reported that somatic hybrids of *Nicotiana glauca* + *N. langsdorffii* were obtained by the $NaNO_3$ method. However, Keller et al. (1973) indicated that a deplasmolyzing shock with sodium nitrate did not increase the frequency of interspecific fusion over that of a deplasmolyzing shock using sorbitol. Since most of the fusion treatments involved he removal of the enzymes which have been shown to contribute to osmolality, it is probable that a deplasmolyzing osmotic shock was associated with the treatment.

2.5 Fusion of Protoplasts by Fatty Acids and Esters

Ahkong et al. (1973) reported that 30 different fat-soluble substances caused the formation of multinucleated mouse and hen erythrocytes. Clones of viable hybrid cells were obtained upon treatment of mouse 3T3 Tk^- fibroblasts and Chinese hamster wg $31MP^-$ fibroblasts with glyceryl monooleate up to 32 times over untreated control cultures (Cramp and Lucy 1974). Very little work has been directed at the effects of fat-soluble substances on plant protoplast fusion. Keller et al. (1973) did not observe any substantial improvement in fusion frequency by adding lysolecithin. Addition of 0.01 mg glyceryl monooleate to 100 ml of PEG (polyethylene glycol) solution enhanced the frequency of heterokaryocyte formation (Kao unpubl.).

Nagata et al. (1979) reported that a positively charged phospholipid, 1,2-O-dipentadecylmethylidene-glycerol-3-phosphoryl-(N-ethylamino)-ethanolamine induced protoplast fusion. The fusion medium contained 1 mM phospholipid, 0.05 M $CaCl_2$, and 0.7 M mannitol. The fused cells were able to divide.

2.6 Fusion of Protoplasts with a High pH and High Calcium Ion Solution

Toister and Loyter (1973) reported that Ca^{2+} is able to induce fusion of chicken erythrocytes at pH 10.5. Soon afterwards, Keller and Melchers (1973) were able to obtain a high frequency (25%) of fusion between isolated leaf protoplasts when they were incubated at 37 °C in the presence of 0.05 M $CaCl_2$ and 0.05 M Na-glycine with 0.4 M mannitol at pH 10.5. According to Toister and Loyter (1973), a low percentage of the membrane phosphatidylethanolamine and lecithin were hydrolyzed and formed lyso compounds. The lyso compounds probably were responsible for the formation of homokaryons.

2.7 Fusion of Protoplasts by Osmotic Shock

Limited success in protoplast fusion from different species was achieved by a mild deplasmolyzing osmotic shock (Keller et al. 1973). The frequency of interspecific fusion was less than 1% and usually in the order of 0.1%. However, a high frequency of heteroprotoplast fusion by hypotonic osmotic shock could be achieved if the protoplasts were brought together by PEG (Kao and Saleem 1986; Takahashi et al. 1986).

2.8 Fusion of Protoplasts with Dextran and Dextran-Sulfate

Aggregation of protoplasts could be achieved with high molecular weight dextran sulfate or dextran (Kameya 1975). However, dextran sulfate was somewhat toxic to the protoplasts, while the dextrans were not. High MW dextrans (15%) in the presence of inorganic salts were able to induce protoplast aggregation and fusion (10%) (Kameya 1979). Kishinami and Widholm (1987) were able to regenerate cell colonies from dextran-treated protoplasts.

2.9 Fusion of Protoplasts with Polyvinyl Alcohol (PVA)

Protoplast adhesion and fusion could be induced by PVA (Nagata 1978). The most effective solution for cell fusion consisted of 15% (wt/wt) PVA (Polyscience, Code #2975), 0.05 M $CaCl_2$ and 0.3 M mannitol. The addition of this solution caused rapid adhesion of protoplasts. After this medium was washed out, a high percentage of cell fusion was observed. PVA has no harmful effect on the viability of protoplasts.

2.10 Fusion of Protoplasts with Polyethylene Glycol (PEG)

Kao and Michayluk (1974) reported that aggregation and fusion of free protoplasts could be induced by concentrated solutions of high-molecular weight polyethylene glycol (PEG, MW, 1500 to 6000) in the presence of Ca^{2+} ions. Soon this technique was confirmed by other researchers (Wallin et al. 1974; Constabel and Kao 1974; Kartha et al. 1974; Burgess and Fleming 1974; Fowke et al. 1975). The PEG-induced fusion is non-specific and has been effective for fusion of protoplasts from a large number of species, from different genera and families. Heterokaryocyte frequencies of up to 30% can occur. The procedure for the PEG (a thin layer) method (Kao 1975, 1982) is as follows:

1. Place ca. 150 µl of a 4–5% (v/v) protoplast suspension on a cover slip in a Petri dish and allow the protoplasts to settle in order to form a thin layer.
2. Slowly add ca. 300–450 ml of a PEG solution (Table 1) to the protoplast preparation.

Table 1. Solutions for protoplast fusion. (Kao 1982)

Compound	Solution	
	PEG	High pH-Ca^{2+} ion
Glucose (g)	–	5.4
PEG1540 (g)	50	–
$CaCl_2 \cdot 2H_2O$ (mg)	150	735
KH_2PO_4 (mg)	10	–
Glycine (mg)	–	375
H_2O (ml)	100	100
pH	5.5 – 6.0 (KOH)	10.5 (NaOH)

3. Incubate the protoplasts in the PEG solution at room temperature for 10 – 20 min.
4. Slowly elute the PEG with a protoplast culture medium.

Higher fusion frequencies were usually obtained if the PEG was eluted with a high pH and high calcium ion solution at pH 10.5 (Table 1), followed by washing the treated protoplasts with a protoplast culture medium (Kao et al. 1974). Up to 50% of the heterokaryocytes of soybean + pea were found in the protoplast population after the treatment. A high pH and high Ca^{2+} solution at pH 9 alone produced only 4% – 5% heterokaryocytes. The pea mesophyll protoplasts were unable to tolerate pH 10.5 (Kao and Wetter 1977).

Glimelius et al. (1978) indicated that the attachment of protoplasts to the Petri dishes induced by PEG, could be enhanced by using dishes coated with Concanavalin A. As a consequence of stronger attachment, more fusion products remained after the dilution and washing procedure. Haydu et al. (1977) reported that adding dimethylsulfoxide in PEG could increase the frequency of protoplast fusion.

A hypotonic solution with a proper osmotic strength at near neutral pH for eluting PEG and providing an osmotic shock could be just as effective as a high pH-Ca^{2+} ion solution to elute PEG to induce protoplast fusion (Kao and Saleem 1986; Takahashi et al. 1986). Fujita and Tabata (1986) reported that protoplasts from cultured cells of *Lithospermum, Coptis,* and *Duboisia* which were very difficult to fuse by PEG or electroshock, could, however, be effectively fused by a combination of PEG and a hypotonic treatment.

The only disadvantage to fusing protoplasts on a cover slip as described above is that the protoplasts tend to adhere tightly to the cover slip for a number of days. During this period, the differences in morphological character between parental cells and heterokaryocytes were observed to disappear, thus making it difficult to isolate heterokaryocytes by mechanical means.

Due to the high viscosity of high molecular weight PEG, fusion of protoplasts or mammalian cells with PEG in suspension is difficult. However, Hales (1977) overcame this problem by using a lower molecular weight PEG (PEG 1000) plus dimethylsulfoxide for the fusion of suspended mammalian cells. Power et al. (1976) simply reduced the PEG (PEG 6000) concentration to 15%, Schieder

Table 2. Solutions for fusion of protoplasts at interface. (Kao 1988)

Compound	Solution		
	Glucose	PEG	Hypotonic high pH
Glucose (g)	9	–	5.4
$CaCl_2 \cdot 2H_2O$ (mg)	100	150	294
KH_2PO_4 (mg)	10	10	–
PEG 4000 (g)	–	30	–
Sucrose (g)	–	17	–
CAPS[a] (mg)	–	–	66
H_2O (ml)	100	100	100
pH	6	6	10

[a] CAPS = cyclohexyl amino propane sulfonic acid (pKa = 10.4).

(1978) by employing only 4.5% PEG (PEG 6000) in a high pH-Ca^{2+} ion solution [0.2 M $Ca(NO_3)_2$ buffered with 0.05 M glycine and NaOH at pH 9] for fusion of protoplasts in suspension.

A better method for obtaining free floating fusion products by PEG treatment is by fusion of protoplasts at the interface of two solutions (Binding and Nehls 1978; Kao 1986). Fusion of protoplasts at the interface of a PEG-sucrose and a glucose solution with a high pH and hypotonic shock was able to produce up to 30% free floating heterokaryocytes with good viability in the treated population (Kao 1986). The method is a follows:

1. Pipette ca. 300 µl of a PEG-sucrose solution (Table 2) onto a 22×22 mm cover slip. (The cover slip was placed in a 60×15 mm plastic Petri dish).
2. Slowly add ca. 150 µl of a 10% (v/v) protoplast suspension in a glucose solution (Table 2) onto the centre of the PEG layer. Allow the protoplasts to settle at the interface (ca. 5–10 min). Add two 0.5-ml aliquots of a hypotonic high pH solution (Table 2) and then 3 ml of a protoplast culture medium at 5-min intervals.
3. Incubate protoplasts in this mixture for 30–60 min then collect them in a centrifuge tube using a Pasteur pipette and centrifuge at 50–100 *g* for 6 min to sediment protoplasts. Remove the supernatant with a Pasteur pipette and wash the protoplasts twice with 10-ml aliquots of a protoplast culture medium.

PEG ($HOCH_2$ $(CH_2-O-CH_2)_n$ CH_2OH) is very water-soluble. The ether linkage makes the molecule slightly negative in polarity and capable of forming hydrogen bonds with the positively polarized group of membrane substances. When the chain of the PEG molecule is long enough, it is possible that it acts as a molecular bridge between the surfaces of adjacent protoplasts, and adhesion occurs. PEG also can bind Ca^{2+} as well as any other cations. The Ca^{2+} perhaps forms a bridge between the negatively polarized groups of the membrane and PEG, thus enhancing adhesion. Although the protoplasts became tightly adpressed to each other when incubated in a PEG solution, only a very low frequency of heteroplasmic fusion occurred. However, soon after the PEG solution was diluted with a protoplast culture medium or a hypotonic solution, many heterokaryocytes were observed (Kao and Michayluk 1974). Thus most of the fusion occurred during washing of the protoplasts. It is likely that those PEG molecules which bind to the protoplast membranes either directly or indirectly through Ca^{2+} ions may be eluted during the washing process. This could result in disturbance and redistribution of electric charges. Since the membranes of two tightly adhering protoplasts are in intimate contact over large surface areas, such disturbances could be strong enough to cause membrane fusion. Osmotic shock also plays an important role in fusion (Keller et al. 1973; Kao and Michayluk 1974; Kao and Saleem 1986; Takahaski et al. 1986).

Many factors can affect PEG-induced adhesion, fusion and viability of heterokaryocytes (Kao and Michayluk 1974; Kao et al. 1974; Constabel and Kao 1974; Ferenczy et al. 1975; Anné and Peberdy 1975; Weber et al. 1976; Kao and Wetter 1977; Kao 1978; Kao and Saleem 1986).

1. Tight adhesion of protoplasts and a high frequency of protoplast fusion occurs only in a concentrated solution of PEG (MW over 1000).
2. PEG-induced adhesion and fusion are enhanced by enriching the PEG solution with Ca^{2+} ions, while they are inhibited by K^+ or Na^+ ions at high pH. However, if the Ca^{2+} ion concentration is high enough, then adhesion and fusion can occur in a PEG solution containing high Na^+ at a high pH.
3. The osmolality of various solutions has a profound effect on protoplast fusion and viability of the protoplasts.
4. Protoplasts from young leaves or fast-growing cultures are the best material for fusion. Protoplasts from old mesophyll cells or old cultured cells are not suitable for fusion. Old cells presumably have secondary cell walls which are not digested by the enzymes. Rapid wall regeneration on the protoplasts before they are exposed to the fusogenic agent can also reduce fusion.
5. Deleterious effects of PEG to protoplasts have been reported; however, this is most likely caused by using impure PEG rather than PEG per se. Kao and Saleem (1986) reported that desalted PEG is far less harmful to protoplasts than commercial pharmaceutical grade PEG. The polyethylene glycol 4000 used for gas chromatography (Merck) was equally good or better than desalted PEG (Kao unpubl.; Fazekas de St. Groth and Scheidegger 1980).
6. The concentration and type of enzymes have profound effects on protoplast fusion and viability of the heterokaryocytes (Kao et al. 1974; Fitzsimons and Weyers 1985). Treatment of soybean and pea cells with enzyme solutions containing Driselase usually resulted in a higher frequency of fusion than enzyme solutions with other cellulases. However, the heterokaryocytes from Driselase-treated protoplasts were usually much weaker, perhaps because of the high activity of phospholipase in Driselase. The phospholipase may convert lecithin to lysolecithin, thus affecting the configuration of the membrane, and increase the chances of membrane fusion (Lucy 1970).
7. Protoplast population density also plays an important role in protoplast fusion. The protoplast population density should be regulated in such a way that most of the fusion occurs between two protoplasts and multiple fusions are reduced to a minimum. A 5–10% (v/v) protoplast suspension was used for fu-

sion of protoplasts at the interface of a PEG-sucrose and a glucose solution (Kao 1986), while a 4–5% (v/v) protoplast suspension was used for fusion of protoplasts settled on a solid surface by PEG methods (Kao 1982). A protoplast population density of $2-8\times10^4$/ml was found to be optimum for electrofusion (Watts and King 1984).

8. There are large differences in the ability of the protoplasts from different species to tolerate enzyme and fusogenic treatment. Protoplasts from cultured cells, especially soybean, could tolerate the enzymes, PEG, high pH and Ca^{2+} treatment very well, while the protoplasts from young leaves of *Nicotiana langsdorffii, N. glauca* and pea are much more sensitive to these conditions. Preincubation of the mesophyll cells of leaves for 2–3 days in a nutritional solution could greatly increase the stability and cell division of potato protoplasts (Shepard and Totten 1977). The *N. glauca, N. langsdorffii* and pea protoplasts reacted in a similar manner. Their resistance to PEG and high pH and Ca^{2+} solution treatments could also be greatly increased. An increase in tolerance was usually accompanied by a decrease in fusion rate. Barley mesophyll protoplasts are usually very weak. In order to increase the tolerance of these protoplasts to PEG treatment, it is essential to reduce the enzyme concentration to 1/5 or 1/10 of that used for pea leaves. For fusion of barley with soybean, a short treatment with the PEG solution usually produced healthy heterokaryocytes. The barley protoplasts were very sensitive to high pH and Ca^{2+} solution treatment. Very little gain in heterokaryocyte formation between soybean and barley was observed by eluting the PEG with the high pH and Ca^{2+} solution.

2.11 Electrofusion

Senda et al. (1979) with the aid of a micromanipulator brought two electrodes into contact with adjoining protoplasts and, after applying a brief electrical pulse, induced fusion. Subsequently, Zimmermann and Scheurich (1981) reported that mesophyll protoplasts of *Vicia faba* in a salt-free mannitol solution formed aggregates or "pearl chain" in a highly inhomogeneous alternating electric field (sine wave, 5 to 10 V peak-to-peak value, 500 KH_z, electrode distance 200 μm). This "pearl chain" aggregate was stable only for the duration of the applied field. By the additional application of a high single field pulse (Square wave, 15 V, 50 μs), it was possible to induce cell fusion with the aggregates. This electrically stimulated fusion of cells proceeded at room temperature and under physiological pH conditions without the use of chemical reagents, and gave a high yield. During the dielectrophoretically induced adhesion of the protoplasts to one another, the field strength must be chosen so that the dielectric breakdown of the membrane is avoided, but at the same time, the strength of the subsequently applied single pulse must be high enough to induce dielectric breakdown at the sites of contact between the protoplasts membranes. The products of electrofusion were viable (Verhock-Köhler et al. 1983). Zachrisson and Bornman (1984) reported that protoplasts subjected to fusogenic conditions, that is, application of voltage that induces reversible membrane breakdown, were capable not only of survival but also of cell wall resynthesis, cell division and subsequent growth and development. Bates and Hasenkampf (1985) confirmed that calli and clonally derived plantlets recovered from electrofusion were hybrids.

Watts and King (1984) developed a simple method for large-scale electrofusion of protoplasts. The protoplasts were aggregated in a ratio-frequency field (10 V RMS, 0.5 MH_z) for 15–30 s with an inter-electrode distance of 5 mm, and were then fused with a 300 V DC pulse. They indicated that a prolonged exposure to salt-free mannitol solutions resulted in protoplasts bursting rather than fusing when the DC pulse was applied. The presence of 0.1–0.5 mM $CaCl_2$ in the electrofusion mixture virtually eliminated breakage during fusion and improved the efficiency of the process. The protoplasts were able to divide after this treatment. A transferable electrode permitted electrofusion of 1-ml volumes of culture in standard tissue-culture dishes in about 20 s. Electrofusion of individually selected protoplasts with a computer-directed system is possible (Schweiger et al. 1987).

The maximum electrofusion value obtained depended on a number of factors including protoplast origin, size and chain length. Leaf mesophyll protoplast fused much more readily than suspension-culture protoplasts. Large protoplasts tended to fuse more readily than small protoplasts. In short chains ($\leq$ five protoplasts), the fusion frequency was lower, but the proportion of one-to-one products was greater than in long chain ($\geq$ 10 protoplasts) (Tempelaar and Jones 1985; Bates 1985).

Nea et al. (1987) reported that lysophosphatidylcholine, dimethylsulfoxide, calcium chloride and pronase, which are known to enhance PEG-induced fusion of animal or plant cells, were all found to stimulate the electrofusion of protoplasts, while low temperature reduced fusion. However, lysophosphatidylcholine and dimethylsulfoxide treatments also increased cell lysis.

3 Summary

Spontaneous fusion may occur between protoplasts from cells which are connected by plasmodesmata. Heteroplasmic fusion must be induced. This can be achieved in various ways, by pushing two protoplasts together using mechanical means, by adding fat-soluble substances such as glyceryl monooleate or water-soluble polymers such as polyethylene glycol or polyvinyl alcohol to the protoplast mixture, or by induction of lyso compounds in the cell membrane of the protoplasts with a high pH and Ca^{2+} ion solution, or by induction of reversible membrane breakdown using an electric current or with a hypotonic osmotic solution.

Fusion of portoplasts at the interface of a PEG-sucrose and a glucose solution with a high pH and hypotonic shock or electrofusion in weak ionic solution (0.1–0.5 mM $CaCl_2$) may be two of the most desirable methods.

References

Ahkong QF, Fisher D, Tampion W, Lucy JA (1973) The fusion of erythrocytes by fatty acids, esters, retinol and α-tocopherol. Biochem J 136:147–155

Anné J, Peberdy JF (1975) Conditions for induced fusion of fungal protoplast in polyethylene glycol solution. Arch Microbiol 105:201–205

Bates GW (1985) Electrical fusion for optimal formation of protoplast heterokaryons in *Nicotiana.* Planta 165:217–224

Bates GW, Hasenkampf CA (1985) Culture of plant somatic hybrids following electrical fusion. Theor Appl Genet 70:227–233

Binding H, Nehls R (1978) Somatic cell hybridization of *Vicia faba* + *Petunia hybrida.* Mol Gen Genet 164:137–143

Burgess J, Fleming EN (1974) Ultrastructural studies of the aggregation and fusion of plant protoplasts. Planta 118:183–193

Carlson PS, Smith HH, Dearing RD (1972) Parasexual interspecific plant hybridization. Proc Natl Acad Sci USA 69:2292–2294

Cocking EC (1972) Plant cell protoplast – isolation and development. Annu Rev Plant Physiol 23:29–50

Constabel F, Kao KN (1974) Agglutination and fusion of plant protoplasts by polyethylene glycol. Can J Bot 52:1603–1606

Cramp FC, Lucy JA (1974) Glyceryl monooleate as a fusogen for the formation of heterokaryons and interspecific hybrid cells. Exp Cell Res 87:107–110

Fazekas de St. Groth S, Scheidegger D (1980) Production of monoclonal antibodies: Strategy and tactics. J Immunol Methods 35:1–21

Ferenczy L, Kevei F, Szegedi M (1975) High-frequency fusion of fungal protoplasts. Experientia 31:1028–1030

Fitzsimons PJ, Weyers JDB (1985) Properties of some enzymes used for protoplast isolation. In: Pilet P-E (ed) The physiological properties of plant protoplasts. Springer, Berlin Heidelberg New York Tokyo, pp 12–23

Fowke LC, Rennie PJ, Kirkpatrick JW, Constabel F (1975) Ultrastructural characteristics of intergeneric protoplast fusion. Can J Bot 53:272–278

Fujita Y, Tabata M (1986) Secondary metabolites from plant cells – pharmaceutical applications and progress in commercial production. In: Green CE, Somers DA, Hackett WP, Biesboer DD, Alan R (eds) Plant tissue and cell culture. Liss, Inc, New York, pp 169–185

Gleba YY, Sytnik KM (1984) Protoplast fusion, genetic engineering in higher plants. In: Shoeman R (ed). Springer, Berlin Heidelberg New York Tokyo

Glimelius K, Wallin A, Eriksson T (1978) Concanavalin A improves the polyethylene glycol method for fusing plant protoplasts. Physiol Plant 44:92–96

Hales A (1977) A procedure for the fusion of cells in suspension by means of polyethylene glycol. Somatic Cell Genet 3:227–230

Harms CT (1985) Hybridization by somatic cell fusion. In: Fowke LC, Constabel F (eds) Plant Protoplasts. CRC Press, Boca Raton, Florida, pp 169–203

Haydu Z, Lazär G, Dudits D (1977) Increased frequency of polyethylene glycol-induced protoplast fusion by dimethylsulfoxide. Plant Sci Lett 10:357–360

Ito M (1973) Studies on the behavior of meiotic protoplasts II. Induction of a high fusion frequency in protoplasts from liliaceous plants. Plant Cell Physiol 14:865–872

Ito M, Maeda M (1973) Fusion of meiotic protoplasts in liliaceous plants. Exp Cell Res 80:453–456

Ito M, Maeda M (1974) Meiotic division and fusion of nuclei in multinucleate cells from the induced fusion of meiotic protoplasts of liliaceous plants. Bot Mag Tokyo 87:219–228

Kameya T (1975) Induction of hybrids through somatic cell fusion with dextran sulfate and gelatin. Jpn J Genet 50:235–246

Kameya T (1979) Studies on plant cell fusion: Effects of dextran and pronase E on fusion. Cytologia 44:449–456

Kameya T, Takahashi N (1972) The effects of inorganic salts on fusion of protoplasts from roots and leaves of *Brassica* species. Jpn J Genet 47:215–217

Kao KN (1975) A method for fusion of plant protoplasts with polyethylene glycol. In: Gamborg OL, Wetter LR (eds) Plant tissue culture methods. National Research Council of Canada, pp 22–27

Kao KN (1978) Plant protoplast fusion and somatic hybrids. In: Proceedings of Symposium on Plant Tissue Culture. Science Press, Peking. pp 331–339

Kao KN (1982) Protoplast fusion and isolation of heterokaryocytes. In: Wetter LR, Constabel F (eds) Plant Tissue Culture Methods. Second edition. National Research Council of Canada, pp 49–56

Kao KN (1986) Fusion of plant protoplasts at the interface of a glucose and a sucrose-polyethylene glycol solution. J Plant Physiol 126:55–58

Kao KN, Michayluk MR (1974) A method for high-frequency intergeneric fusion of plant protoplasts. Planta (Berl.) 115:355–367

Kao KN, Saleem M (1986) Improved fusion of mesophyll and cotyledon protoplasts with PEG and high pH-Ca^{2+} solutions. J Plant Physiol 122:217–225

Kao KN, Wetter LR (1977) Advances in techniques of plant protoplast fusion and culture of heterokaryocytes. In: Brinkley BR, Porter KR (eds) International Cell Biology 1976–1977. Rockfeller University Press, New York, pp 216–224

Kao KN, Gamborg OL, Michayluk MR, Keller WA, Miller RA (1972) The effect of sugars and inorganic salts on cell regeneration and sustained division in plant protoplasts. In: Tempe J (ed) Protoplastes et fusion de Cellules Somatiques Végétales. CNRS, Paris, pp 207–213

Kao KN, Constabel F, Michayluk MR, Gamborg OL (1974) Plant protoplast fusion and growth of intergeneric hybrid cells. Planta (Berl.) 120:215–227

Kartha KK, Gamborg OL, Constabel F, Kao KN (1974) Fusion of rapeseed and soybean protoplasts and subsequent division of heterokaryocytes. Can J Bot 52:2435–2436

Keller WA, Melchers G (1973) The effect of high pH and calcium on tobacco leaf protoplast fusion. Z Naturforsch 28c:737–741

Keller WA, Harvey BL, Kao KN, Miller RA, Gamborg OL (1973) Determination of the frequency of interspecific protoplast fusion by differential staining. In: Tempe J (ed) Protoplastes et fusion de cellules somatiques végétales. CNRS, Paris, pp 455–463

Kishinami I, Widholm JM (1987) Auxotrophic complementation in intergeneric hybrid cells obtained by electrical and dextran-induced protoplast fusion. Plant Cell Physiol 28:211–218

Lucy JA (1970) The fusion of biological membranes. Nature 227:814–817

Lucy JA (1978) Mechanisms of chemically induced cell fusion. In: Poste G, Nicolson GL (eds) Membrane fusion. Elsevier/North-Holland Biomedical Press, Amsterdam, pp 267–304

Melchers G, Labib G (1974) Somatic hybridization of plants by fusion of protoplasts. I. Selection of light resistant hybrids of "haploid" light sensitive varieties of tobacco. Mol Gen Genet 135:277–294

Michel W (1937) Über die experimentelle Fusion pflanzlicher Protoplasten. Arch Exp Zellforsch 20:230–252

Miller RA, Gamborg OL, Keller WA, Kao KN (1971) Fusion and division of nuclei in multinucleated soybean protoplasts. Can J Genet Cytol 13:347–353

Motoyoshi F (1971) Protoplasts isolated from callus cells of maize endosperm. Formation of multinucleate protoplasts and nuclear division. Exp Cell Res 68:452–456

Nagata T (1978) A novel cell-fusion method of protoplasts by polyvinyl alcohol. Naturwissenschaften 65:263–264

Nagata T, Eibel H, Melchers G (1979) Fusion of plant protoplasts induced by a positively charged synthetic phospholipid. Z Naturforsch 34c:460–462

Nea NJ, Bates GW, Gilmer PJ (1987) Facilitation of electrofusion of plant protoplasts by membrane-active agents. Biochim Biophys Acta 897:293–301

Poste G, Allison AC (1973) Membrane fusion. Biochim Biophys Acta 300:421–465

Power JB, Cocking EC (1971) Fusion of plant protoplasts. Sci Prog Oxf 59:181–198

Power JB, Cummins SE, Cocking EC (1970) Fusion of isolated plant protoplasts. Nature 225:1016–1018

Power JB, Frearson EM, Hayward C, George D, Evans PK, Berry SF, Cocking EC (1976) Somatic hybridization of *Petunia hybrida* and *P. parodii.* Nature 263:500–502

Primke M, Berger S, Schweiger H-G (1978) Protoplasts from Acetabularia: Isolation and fusion. Cytobiologie 16:375–380

Schenk RU, Hildebrandt AC (1968) Somatic hybridization: a new approach to genetic change. Am J Bot 55:731

Schieder O (1978) Somatic hybrids of *Datura innoxia* Mill + *Datura discolor* Bernh. and of *Datura innoxia* Mill + *Datura stramonium* L. var. tatula L. I. Selection and characterisation. Mol Gen Genet 162:113–119

Schweiger H-G, Dirk J, Koop H-U, Kranz E, Neuhaus G, Spangenberg G, Walff D (1987) Individual selection, culture and manipulation of higher plant cells. Theor Appl Genet 73:769–783

Senda M, Takeda J, Abe S, Nakamura T (1979) Induction of cell fusion of plant protoplasts by electrical stimulation. Plant Cell Physiol 20:1441–1443

Shepard JF, Totten RE (1977) Mesophyll cell protoplasts of potato; Isolation, proliferation and plant regeneration. Plant Physiol 60:313–316

Takahashi S, Fujita Y, Yamada Y (1986) Effective method for protoplast fusion. In: VI International Congress of Plant Tissue and Cell Culture, Abstracts, Aug. 3–8, 1986. University of Minnesota, Minneapolis, Minn., USA, p 317

Tempelaar MJ, Jones MGK (1985) Fusion characteristics of plant protoplasts in electric fields. Planta 165:205–216

Toister Z, Loyter A (1973) The mechanism of cell fusion. II. Formation of chicken erythrocyte polykaryons. J Biol Chem 248:422–432

Verhock-Köhler B, Hampp R, Ziegler H, Zimmermann U (1983) Electro-fusion of mesophyll protoplasts of *Avena sativa.* Planta 158:199–204

Waaland SD (1978) Parasexually produced hybrids between female and male plants of *Griffithsia tenuis* C. Agardh. a red alga. Planta 138:65–68

Wallin A, Glimelius K, Eriksson T (1974) The induction of aggregation and fusion of *Daucus carota* protoplasts by polyethylene glycol. Z. Pflanzenphysiol 74:64–80

Watts JW, King JM (1984) A simple method for large-scale electrofusion and culture of plant protoplasts. Biosci Rep 4:335–342

Weber G, Constabel F, Williamson F, Fowke L, Gamborg OL (1976) Effect of pre-incubation of protoplasts on PEG-induced fusion of plant cells. Z Pflanzenphysiol 79:459–464

Zachrisson A, Bornman CH (1984) Application of electric field fusion in plant tissue culture. Physiol Plant 61:314–320

Zimmermann U, Scheurich P (1981) High frequency fusion of plant protoplasts by electric fields. Planta 151:26–32

Zimmermann U, Vienken J (1982) Electric field-induced cell-to-cell fusion. J Membr Biol 67:165–182

III.2 Ultrastructural Studies of Plant Protoplast Fusion

L. C. FOWKE[1]

1 Introduction

A wide variety of plant protoplasts have been fused and in many cases hybrid plants have been regenerated (see reviews by Fowke and Gamborg 1980; Schieder and Vasil 1980; Gleba and Sytnik 1984; Constabel and Cutler 1985; Bajaj et al. 1988). A number of different fusion techniques have been reported; however, the most effective involves the use of polyethylene glycol (PEG) (Kao and Michayluk 1974; Wallin et al. 1974). PEG-induced fusion is non-specific and has been used to fuse protoplasts of widely separated plant species and genera, animal cells and combinations of plant protoplasts and animal cells. Recently, electrofusion has also been used successfully to fuse protoplasts, and regeneration of hybrids has been achieved (e.g. Zimmerman et al. 1984; Schieder and Kohn 1986; De Vries and Tempelaar 1987). The major emphasis of the fusion work to date has been the development of techniques to produce hybrid plants which can be used in plant breeding programmes.

This chapter examines the morphology of protoplast fusion with emphasis on ultrastructural studies. Table 1 lists plant protoplast fusions which have been studied by electron microscopy. The major focus of this chapter is fusion of higher plant protoplasts, but a brief discussion of the interaction of algal protoplasts with higher plant protoplasts is also included. Fusions between higher plant protoplasts and animal cells or fungal protoplasts are listed in Table 1 but not discussed further. All fusions represented pictorially have been induced with PEG.

2 Higher Plants

2.1 Protoplasts

Structural studies of the fusion of a range of higher plant protoplasts have been reported. In most cases cell suspension culture protoplasts have been used as one partner and leaf protoplasts as the other. This choice permits the visual identification of fusion products at both the light and electron microscope level. Proto-

[1] Department of Biology, University of Saskatchewan, Saskatoon, Canada S7N OWO

Biotechnology in Agriculture and Forestry, Vol. 8
Plant Protoplasts and Genetic Engineering I (ed. by Y. P. S. Bajaj)

Table 1. Ultrastructural studies of higher plant protoplast fusion

Fusion partners	Features illustrated by electron microscopy	Reference
Nicotiana × *Nicotiana*	Agglutination and fusion Role of plasmodesmata in fusion	Withers and Cocking (1972)
Avena × *Avena*	Agglutination and fusion	Withers (1972)
Nicotiana × *Nicotiana*	Role of plasmodesmata in fusion	
Pisum × *Centaurea*	Agglutination and fusion	Davey and Short (1972)
Daucus × *Daucus*	Agglutination and fusion	Wallin et al. (1974)
Nicotiana × *Nicotiana*	Agglutination and fusion	Burgess and Flemming (1974)
Parthenocissus × *Saccharomyces*	Agglutination and uptake	Davey and Power (1975)
Pisum × *Vicia*	Agglutination and fusion	Fowke et al. (1975)
Melilotus × *Glycine*	Structure and development of fusion products Chloroplast degeneration	Fowke et al. (1976)
Nicotiana × *Nicotiana*	Agglutination of labelled protoplasts	Burgess et al. (1977a)
Pisum × *Glycine*	Structure and development of fusion products Nuclear fusion Chloroplast degeneration	Fowke et al. (1977)
Nicotiana × *Nicotiana*	Identification of nuclei	Nagy et al. (1977)
Nicotiana × *Gallus*	Agglutination and fusion Structure of fusion products	Willis et al. (1977)
Skimmia × *Skimmia*	Agglutination and fusion Plasma membrane structure	Robenek and Peveling (1978)
Daucus × *Xenopus*	Fusion Structure and development of fusion products	Davey et al. (1978)
Nicotiana × *Arabidopsis*	Agglutination and fusion Structure of fusion products	Sidorov et al. (1978)
Pisum × *Nicotiana*	Agglutination and fusion	Syono et al. (1979)
Daucus × *Chlamydomonas*	Fusion Structure and development of fusion products	Fowke et al. (1979)
Vicia × *Vicia*	Dedifferentiation of chloroplasts	Rennie et al. (1980)
Arabidopsis × *Brassica*	Structure of hybrid nuclei	Nagl and Hoffmann (1980)
Arabidopsis × *Brassica*	Trichome structure in hybrids	Gleba and Hoffmann (1980)
Pisum × *Nicotiana*	Structure of fusion products	Kajita et al. (1980)
Glycine and Drosophila	Agglutination	Hadlaczky et al. (1980)
Daucus × *Stigeoclonium*	Fusion and structure of fusion products	Fowke et al. (1981)
Nicotiana × *Glycine*	Structure and development of fusion products Fate of chloroplasts	Chien et al. (1983)
Spinacea × *Daucus*	Structure of protoplasts and fusion products Degeneration of chloroplasts	Hodgson and Rose (1984)
Daucus and Xenopus	Structure of fusion products	Ward (1984)
Apium × *Aspergillus*	Agglutination and uptake	Lynch et al. (1986)

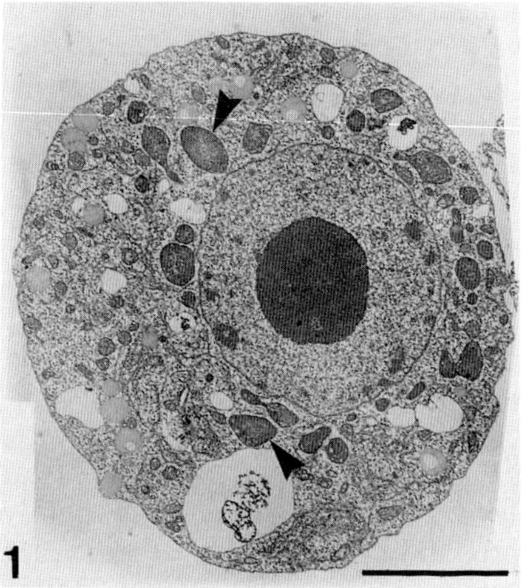

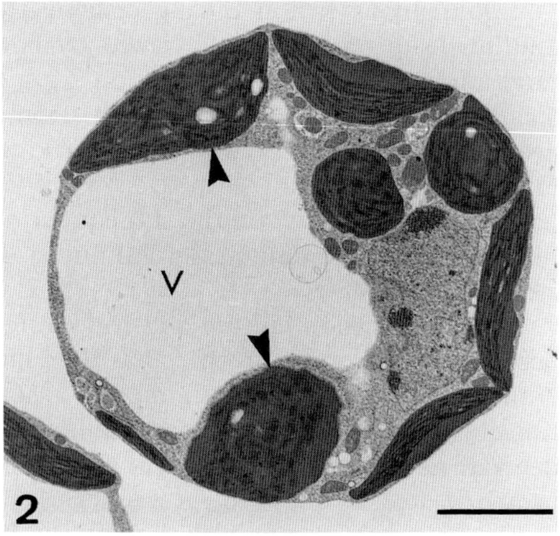

Fig. 1. Electron micrograph showing pinto bean cell culture protoplast. Note the undifferentiated leucoplasts (*arrows*). Bar = 5 μm

Fig. 2. Electron micrograph showing turnip leaf protoplast. Note the prominent chloroplasts (*arrows*). *V* vacuole. Bar = 5 μm

plasts derived from cell suspension cultures contain undifferentiated leucoplasts (Fig. 1), while leaf protoplasts are characterized by well-developed chloroplasts (Fig. 2). The hybrids thus contain two clearly recognizable plastid types. Leaf protoplasts have been combined with other types of protoplasts containing leucoplasts (e.g. carrot root – Hodgson and Rose 1984) or in some cases easily identifiable bacteroids (e.g. pea root nodule – Kajita et al. 1980).

2.2 Ultrastructure of Isolated Protoplasts

Cell suspension protoplasts are characterized by the presence of numerous small vacuoles rather than the large central vacuole of leaf protoplasts (Figs. 1, 2). The leucoplasts of cell suspension protoplasts are small, undifferentiated, contain few internal lamellae and are usually randomly distributed in the cytoplasm. In contrast, the chloroplasts of leaf protoplasts are much larger and contain well-developed grana and stroma lamellae. The chloroplasts are typically flattened against the plasma membrane (Fig. 2).

Freshly isolated protoplasts are characterized by the absence of a cell wall as demonstrated by conventional transmission (Figs. 1, 2) and scanning electron microscopy (Burgess and Linstead 1976; Burgess et al. 1977a, b), freeze etching (Grout 1975; Willison and Klein 1982) and high resolution replica techniques (Williamson et al. 1977). The quantity of cell wall material present on the surface of protoplasts has an important effect on the frequency of protoplast fusion (Weber et al. 1976). In some cases the surface of protoplasts is characterized by spherical membrane projections (e.g. Williamson et al. 1977).

Spontaneous fusion during enzyme treatment results in the formation of multinucleate protoplasts (reviewed by Fowke and Gamborg 1980). Ultrastructural studies of protoplast isolation from tobacco leaf and oat root tissues indicate that spontaneous fusion results from an expansion of plasmodesmata connecting adjacent cells (Withers and Cocking 1972). The number of nuclei present in protoplasts does not seem to interfere with the actual fusion process, but appears to have an effect on the subsequent development of fusion products. Large fusion products containing several nuclei usually deteriorate and die (Constabel et al. 1975).

Both suspension culture and leaf protoplasts are characterized by a cortical network of microtubules which has been illustrated by immunofluorescence techniques (Lloyd et al. 1980; Simmonds 1986; Galway and Hardham 1986) and by transmission electron microscopy of large patches of the plasma membrane (van der Valk et al. 1980). The cortical microtubules appear to be primarily randomly arranged. Recent results suggest a negative correlation between the presence of a cortical network of microtubules and the frequency of protoplast fusion (Hahne and Hoffmann 1986).

Protoplasts of two algae (*Chlamydomonas* – Fowke et al. 1979; *Stigeoclonium* – Fowke et al. 1981) have been successfully fused with higher plant protoplasts (see Sect. 3 below). The *Chlamydomonas* used is a mutant which lacks both flagellae and cell walls. Protoplasts of this alga are easily recognized by the presence of a large U-shaped chloroplast. Protoplasts of the filamentous alga *Stigeoclonium* are also characterized by large easily recognizable chloroplasts (Marchant and Fowke 1977).

2.3 Protoplast Agglutination

The initial step in the fusion process is the adhesion or agglutination of protoplasts. Agglutination is achieved by incubating mixtures of protoplasts in high concentrations (25–33%, w/v) of high molecular weight PEG (e.g. PEG 1570, PEG 4000, PEG 6000) either in test tubes or on glass or plastic surfaces. Under ideal conditions, many pairs of protoplasts adhere tightly in the PEG solution (Fig. 3). Ultrastructural studies indicate that in some cases membrane contact between the protoplasts is uniformly tight (e.g. Burgess and Flemming 1974; Wallin

Fig. 3. Scanning electron micrograph showing agglutinated pea (*P*) leaf and soybean (*S*) cell culture protoplasts. The outlines of the chloroplasts are visible at the surface of the pea protoplast. Bar = 10 μm. (Fowke et al. 1977)

Fig. 4. Electron micrograph showing agglutination between pea (*P*) leaf and *Vicia* (*Vi*) cell culture protoplasts. Note the areas of tight adhesion (*arrows*) between adjacent plasma membranes. *Ch* chloroplast. Bar = 1 μm. (Fowke et al. 1975)

Fig. 5. a Light micrograph of partially fused pea (*P*) leaf protoplast and *Vicia* (*Vi*) cell culture protoplast. **b** An electron micrograph of the membranes indicated by the *arrows* in **a**. An area of cytoplasmic continuity exists between the pea (*P*) and *Vicia* (*Vi*) cytoplasm. Note the agglutinated plasma membranes (*arrows*) and the *Vicia* leucoplasts (*L*) and nucleus (*N*). *V* vacuole. Bar = 2 μm. (Fowke et al. 1975)

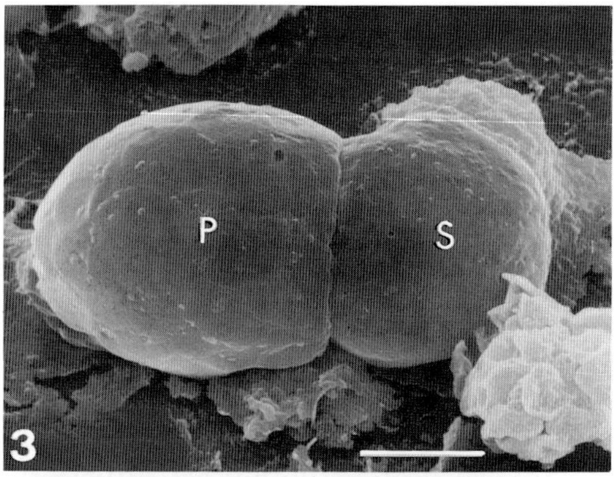
3
P
S

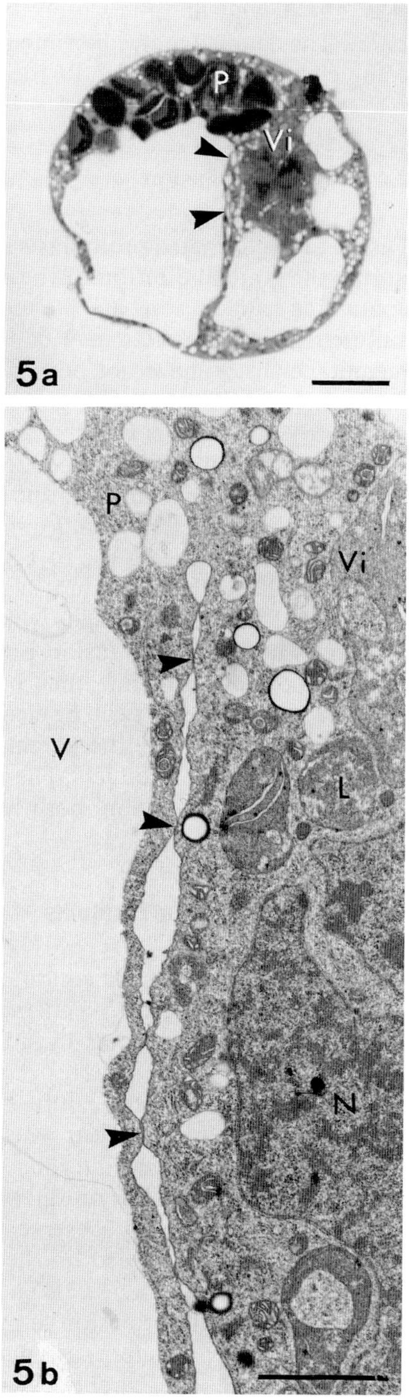
5a
P
Vi
5b
P
Vi
V
L
N

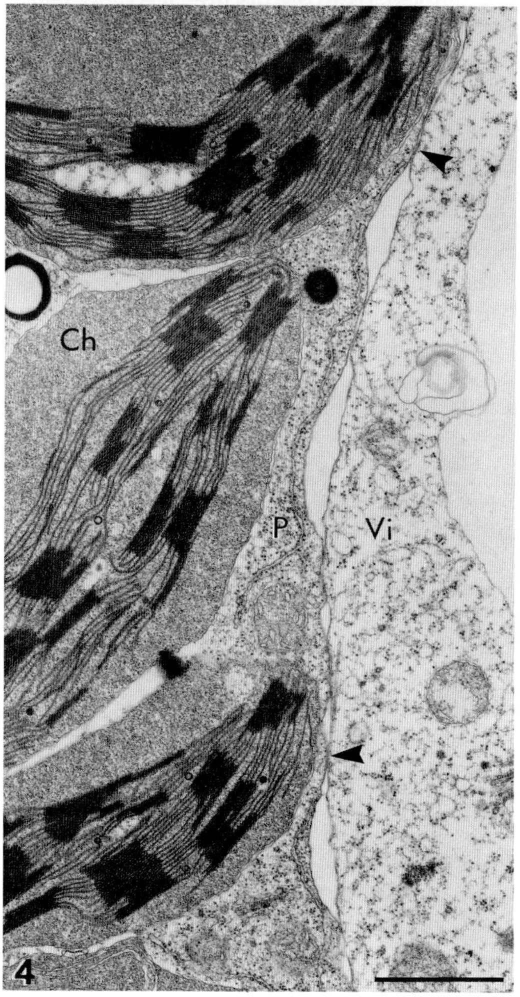
4
Ch
P
Vi

et al. 1974), while in others it occurs in localized regions (Fig. 4 and Fowke et al. 1975). The detailed molecular interactions occurring during agglutination are not completely understood (for further discussion see Constabel and Cutler 1985).

2.4 Protoplast Fusion

Fusion of agglutinated protoplasts usually does not occur in the PEG solution but during the gradual dilution of the PEG with elution medium. Alkaline solutions containing calcium ions are the most effective for elution, although viability of cultured fusion products may be reduced (Binding 1974). The actual events of membrane fusion have not been described morphologically. Fusion most likely occurs very rapidly and conventional chemical fixatives do not act sufficiently quickly to capture this event.

Despite the problems associated with conventional fixatives, partial fusion products have been examined and described (Figs. 5, 6 and Wallin et al. 1974; Fowke et al. 1975). The fusing protoplasts contain regions of cytoplasmic continuity as well as large areas of agglutinated plasma membrane. The fate of the agglutinated membranes is not clear; however, the ultrastructural observations suggest that they form vesicles which disperse in the cytoplasm. Within the first few hours after elution, fusion products can be easily identified with the light microscope, but with time they become more difficult to recognize. However, putative fusion products can be clearly distinguished with the electron microscope (Fig. 6). In cases where the parental nuclei differ markedly in size, they can be recognized in fusion products. The clearest markers, however, are the plastids with fusion products containing both leucoplasts and chloroplasts (Fig. 6).

2.5 Cultured Fusion Products

Under appropriate culture conditions, fusion products will form new cell walls, divide, and in some cases regenerate to plants. It has been possible to follow the early stages of this process by microscopical methods. The formation of true hybrid cells by nuclear fusion occurs early in the culture period. Ultrastructural studies indicate that nuclei can fuse during interphase by the formation of nuclear bridges (Fig. 7). It is not clear whether this type of fusion results in healthy cells; Kao (1978) suggests that interphase nuclear fusions of this type do not develop normally. Healthy fusion products may arise by nuclear fusion during the first mitosis, as is commonly observed with animal fusions. Newly formed hybrid nuclei can be distinguished on the basis of chromatin content at both the light and electron microscope level (Fig. 8 and Constabel et al. 1975). With prolonged culture and continued division of hybrid cells, it is no longer possible to recognize the two genomes in interphase nuclei. In fact with many hybrids, most of the chromosomes of one parent are eventually lost (e.g. soybean + tobacco – Kao 1977). In other hybrid cell lines, little or no chromosome loss occurs (e.g. *Arabadopsis + Brassica* – Gleba and Hoffmann 1978).

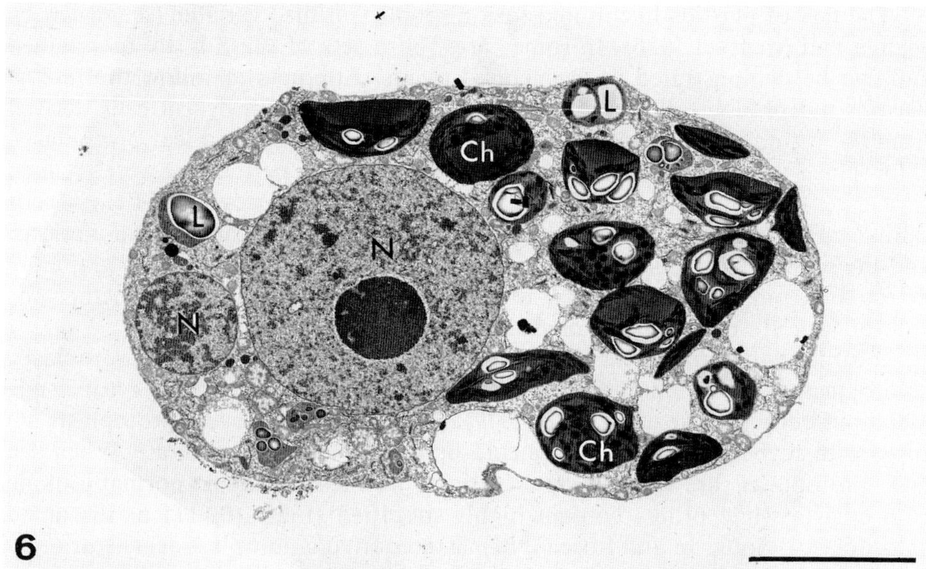

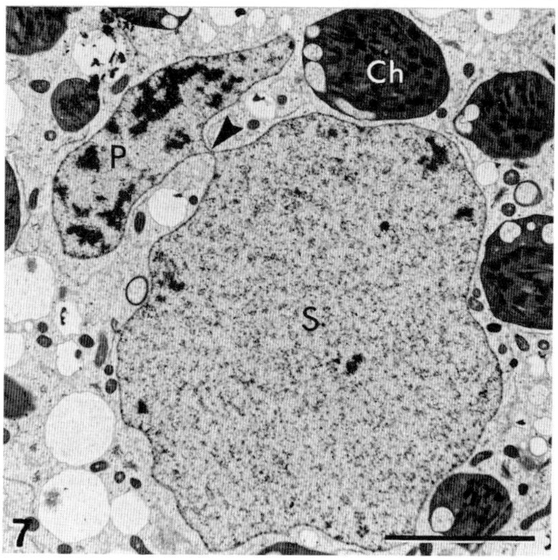

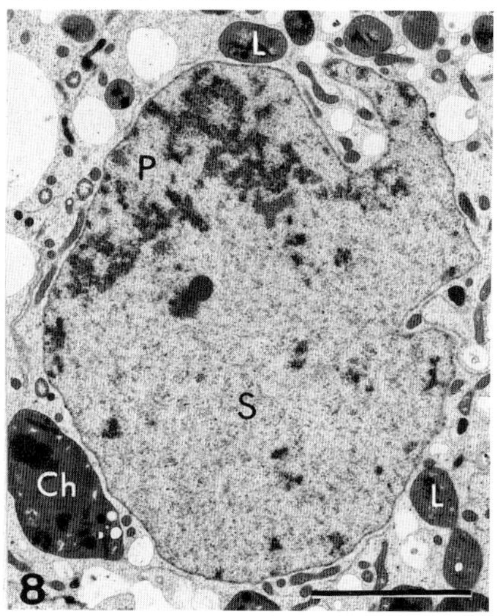

Fig. 6. Electron micrograph showing fusion product of *Vicia hajastana* cell culture protoplast and *Vicia narbonensis* leaf protoplast. The two nuclei (*N*) have not fused. Both leucoplasts (*L*) and chloroplasts (*Ch*) are present. Bar = 10 μm. (Rennie et al. 1980)

Fig. 7. Electron micrograph showing nuclear bridge (*arrow*) between interphase nuclei from pea (*P*) leaf and soybean (*S*) cell culture in a fusion product. *Ch* chloroplast. Bar = 5 μm. (Fowke et al. 1977)

Fig. 8. Electron micrograph showing nucleus containing both pea (*P*) leaf chromatin and soybean (*S*) cell culture chromatin in a fusion product. *Ch* chloroplasts; *L* leucoplasts. Bar = 5 μm. (Fowke et al. 1977)

The fate of plastids in cultured fusion products is also variable (see review by Maliga and Menczel 1986). In some cases, both sets of plastids are maintained and can be demonstrated in regenerated plants. Attempts to follow the fate of plastids morphologically have met with limited success. In hybrid cells derived from the fusion of soybean cell suspension culture and tobacco leaf protoplasts, chloroplasts are recognized by electron microscopy for a period of 2 weeks (Chien et al. 1983). During the first few days of culture, the chloroplasts of these cells exhibit rather unusual shapes, possibly as a result of large starch accumulations, and the number of granal stacks in reduced (Fig. 9a, 9b). By 2 weeks grana are detected in only a few plastids. It is not clear from this study whether the chloroplasts are degenerating or dedifferentiating.

In other wide crosses involving soybean cell suspension culture protoplasts, chloroplasts exhibit a number of alterations suggesting some type of cytoplasmic incompatibility (Fowke et al. 1976, 1977). When sweet clover leaf protoplasts are fused with soybean protoplasts, the chloroplasts appear to degenerate. When pea leaf protoplasts are used, some fusion products contain normal-looking chloroplasts while others contain highly modified (Figs. 10, 11) or disrupted chloroplasts. Hodgson and Rose (1984) also reported chloroplast degeneration in cultured hybrid cells derived from fused protoplasts of spinach leaves and carrot roots.

In all fusion studies to date, it appears that the leucoplasts from the cell suspension culture parent remain unaltered and apparently reproduce normally in the developing hybrid callus.

Very limited information is available regarding the fate of plastids in fusions involving closely related protoplasts. In a fusion study using leaf and cell culture protoplasts of the same or closely related species of *Vicia*, the leaf chloroplasts appear to dedifferentiate rather than degenerate (Rennie et al. 1980). After 7 days of culture, the chloroplasts in fusion products are characterized by a reduction in size and number of grana membranes (Fig. 12). In some cases grana are almost completely absent. In contrast, chloroplasts in leaf protoplasts cultured for the same period of time do not dedifferentiate and resemble chloroplasts in leaf cells. The rather drastic plastid changes characteristic of wide crosses are not observed in any of the fusion products. These results suggest that there is little, if any, incompatibility between closely related fusion partners.

Fig. 9a. Light micrograph showing 4.5-day-old cell cluster resulting from the fusion of tobacco leaf and soybean cell culture protoplasts. An electron micrograph of this cluster is shown in **9b**. The hybrid cells contain tobacco chloroplasts (*single arrows*) and soybean leucoplasts (*double arrows*). *N* nucleus; *V* vacuole. Bar = 10 μm. (Chien et al. 1983)

Fig. 10. Electron micrograph showing chloroplast in cultured fusion product from pea leaf and soybean cell culture. The internal membranes of the chloroplast are swollen. Bar = 1 μm. (Fowke et al. 1977)

Fig. 11. Electron micrograph showing chloroplast in cultured fusion product from pea leaf and soybean cell culture. Note the enlarged disoriented grana (*arrows*). Bar = 1 μm. (Fowke et al. 1977)

Fig. 12. Electron micrograph showing chloroplast in cultured fusion product from *Vicia hajastana* leaf and cell culture protoplasts. The reduction in size and number of grana (*arrows*) may indicate dedifferentiation of the plastid. Bar = 1 μm. (Rennie et al. 1980)

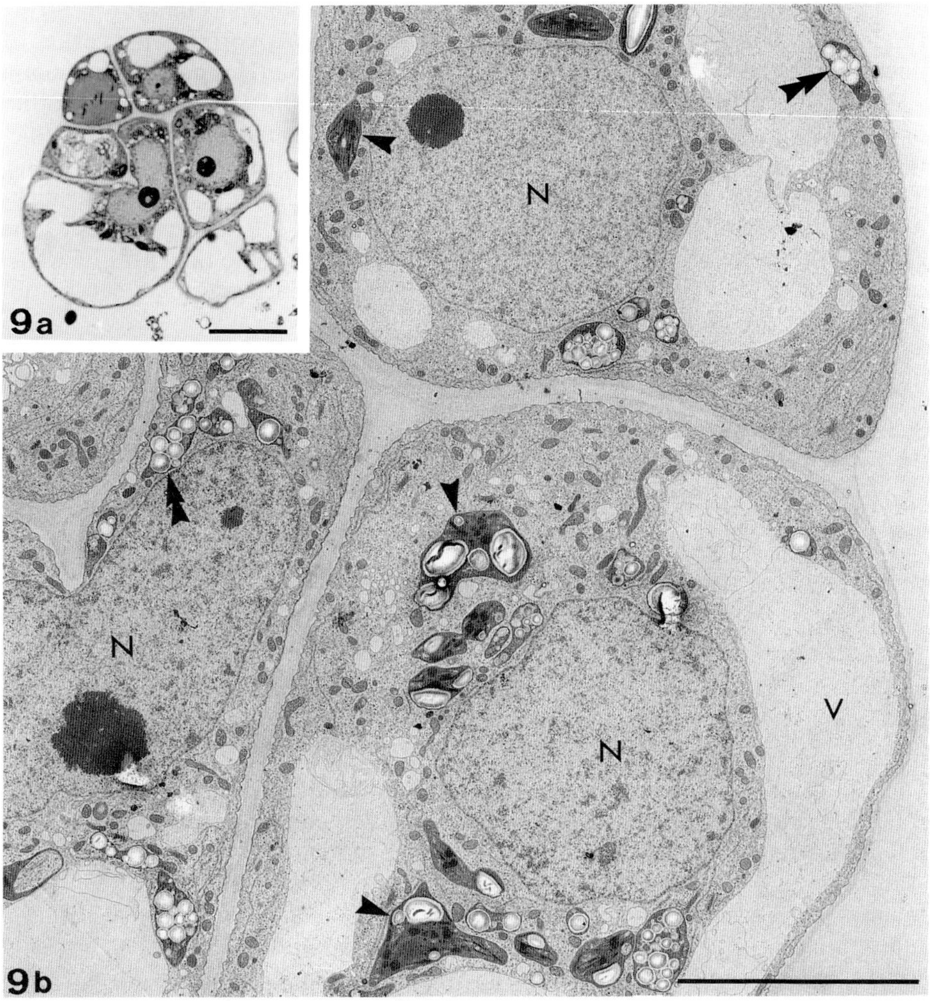
9a
N
N
N
V
9b

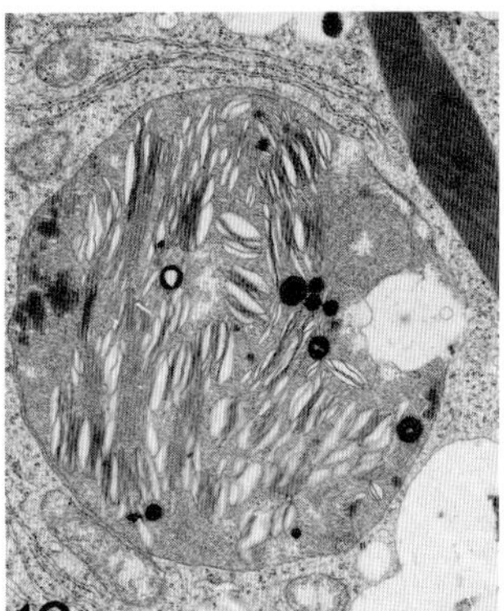
10

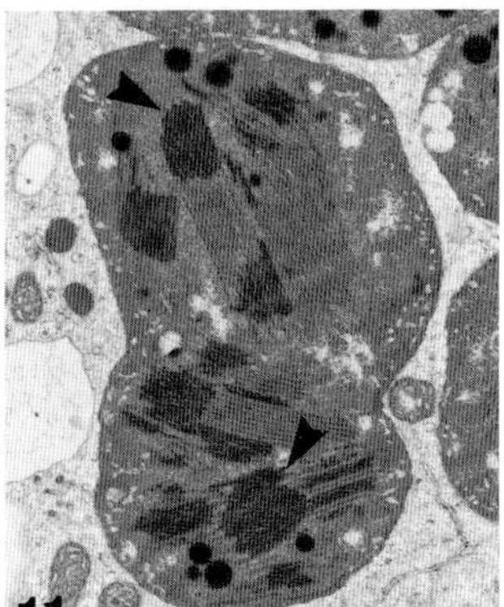
11

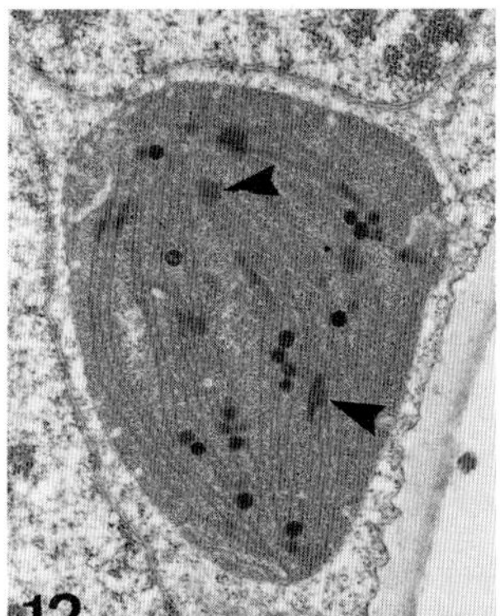
12

The fate of most other cell organelles in fusion products is unknown. Structural information is available which indicates that cortical networks of microtubules become integrated during protoplast fusion (Hahne and Hoffmann 1986) and vacuoles can fuse (Constabel et al. 1980). For studies of other organelles such as mitochondria, ribosomes, endoplasmic reticulum, and dictyosomes, the light and electron microscopes will most likely be of limited value. The fate of these organelles will be determined by biochemical and genetic techniques. Electron microscopy should, however, be useful for further studies of the cell wall in regenerating hybrids. Information concerning the thickness of walls, orientation of cellulose microfibrils and numbers of plasmodesmata can be obtained using this instrument.

3 Algae and Higher Plants

Protoplasts have been isolated from a variety of algae (e.g. Marchant and Fowke 1977; Berliner 1983) and some have been fused with protoplasts of higher plants (Fowke et al. 1979, 1981). Experiments of this nature provide the opportunity to study the interaction of plasma membranes of different cells as well as the interaction of cell organelles in mixed cytoplasm from widely separated plant groups.

3.1 *Chlamydomonas* + Carrot

One of the most interesting fusions studied to date is the combination of wild carrot protoplasts and protoplasts of the unicellular green alga *Chlamydomonas reinhardii* (mutant CW 15-9). This mutant is particulary well suited for such studies because it lacks cell walls and flagella but retains basal bodies and is easily maintained in culture. Agglutination and rapid fusion are induced by suspending protoplasts in a PEG solution. The *Chlamydomonas* chloroplast is easily recognized within the carrot cytoplasm with both light and electron microscopes (Figs. 13, 14). A true membrane fusion event occurs, as evidenced by the incorporation of the algal plasma membrane (marked by the basal body – Fig. 14)

Fig. 13. Light micrograph showing fusion between carrot and *Chlamydomonas* (*arrow*) protoplasts. Nomarski optics. Bar = 10 μm. (Fowke et al. 1979)

Fig. 14. Electron micrograph showing the *Chlamydomonas* chloroplast (*Ch*) within the cytoplasm of the carrot protoplast. Note the algal basal body (*arrow*) at the surface of the carrot protoplast. Bar = 3 μm. (Fowke et al. 1979)

Fig. 15. Light micrograph showing the agglutination of a *Stigeoclonium* protoplast (*arrow*) to the surface of the carrot protoplast. Bar = 10 μm. (Fowke et al. 1981)

Fig. 16. Light micrograph showing fusion between carrot and *Stigeoclonium* (*arrow*) protoplasts. Bar = 10 μm. (Fowke et al. 1981)

Fig. 17. Electron micrograph showing a *Stigeoclonium* chloroplast (*Ch*) and nucleus (*N*) in the cytoplasm of the carrot protoplast. *V* vacuole. Bar = 5 μm. (Fowke et al. 1981)

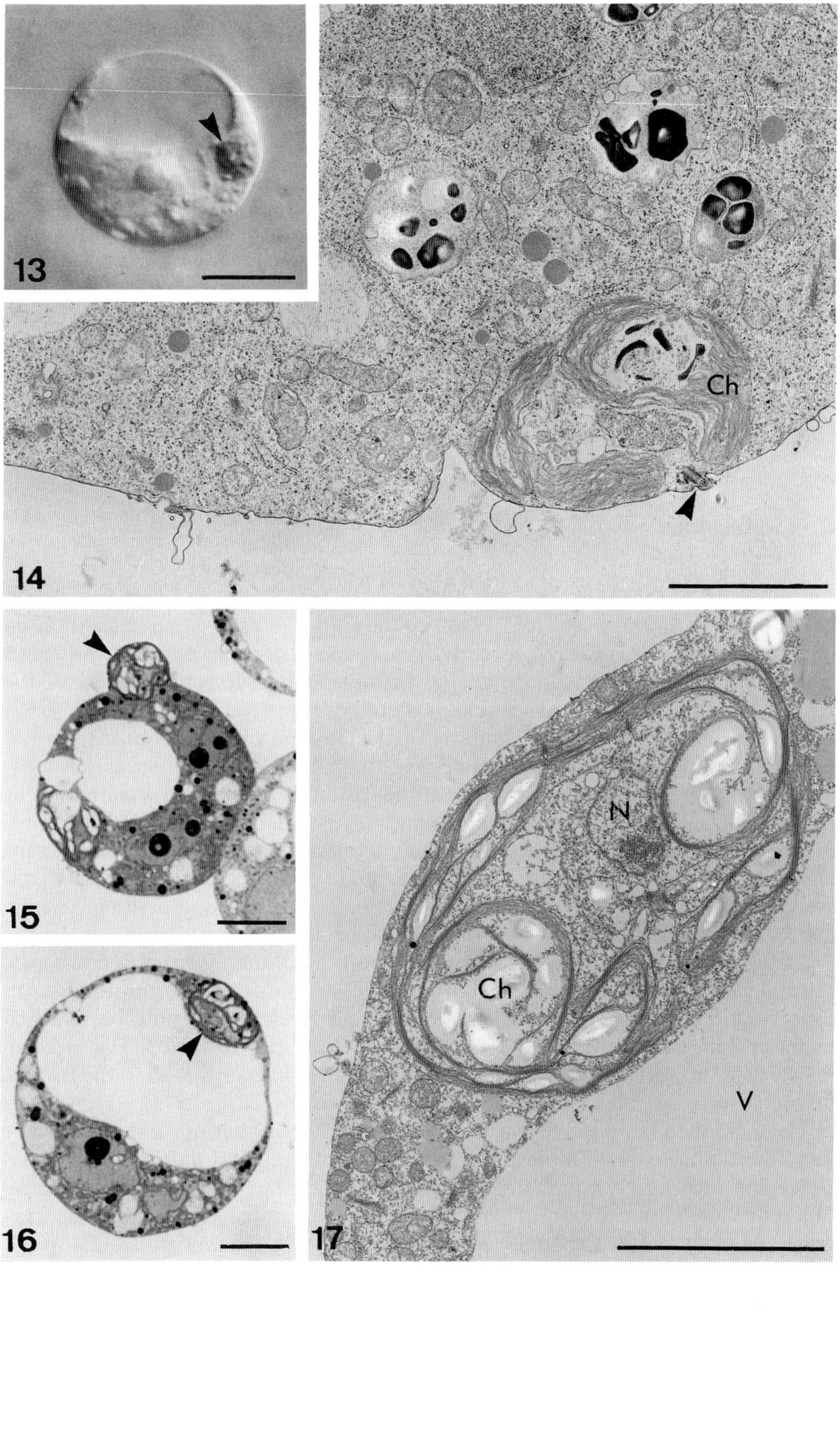
13
Ch
14
15
16
N
Ch
V
17

into the carrot plasma membrane. The algal cytoplasm is released directly into the carrot protoplast. Such fusion products will regenerate a cell wall and divide in culture, and remnants of the algal chloroplast are recognizable for a period of up to 10 days.

3.2 *Stigeoclonium* + Carrot

Protoplasts from filamentous green algae can also be fused with higher plant protoplasts. Using techniques similar to those used with *Chlamydomonas*, protoplasts of the green alga *Stigeoclonium* have been fused with carrot protoplasts (Figs. 15–17). The agglutination and fusion process can be followed with both light and electron microscopes. Algal organelles devoid of a plasma membrane are observed within the carrot cytoplasm, indicating that a true fusion event has occurred. The fate of the algal organelles has not been followed in cultured fusion products.

4 Conclusions

The majority of the fine structural studies of fusing plant protoplasts were published in the period from the mid 1970's to the early 1980's and there has not been much interest in this research recently. Protoplast fusion has by no means been completely explored, and the following examples illustrate some aspects of the process which deserve further cytological study.

The nature of the actual membrane fusion between plant protoplasts has not been determined and would be interesting to explore further. Freeze etch techniques have provided information about the fusion of membranes in animal cells (e.g. Orci et al. 1977), and similar techniques could be applied to plant protoplasts. The very rapid freezing techniques developed recently for animal systems could also be tried (e.g. Heuser 1980). The fate of the agglutinated plasma membranes following fusion could also be examined by labelling individual plasma membranes and following the distribution of the labels in fusion products. Labelling experiments could be carried out with both light and electron microscopes. For light microscopy, lectins which are available commercially or antibodies directed against the plasma membrane (Villanueva et al. 1986) could be attached to fluorescent probes and used to label protoplasts prior to fusion. For electron microscopy, uniform populations of latex spheres could be attached to the surface of protoplasts by either lectins or antibodies (Molday et al. 1975). The various labels could then be followed during and after protoplast fusion. It might also be very interesting to examine more fully the fate of the algal plasma membranes which are incorporated into the carrot plant protoplast surface during fusion. Studies of this nature might indicate whether the integrated algal membrane fragment can manufacture new cell wall material and, if so, the type of wall. It would be intriguing to establish whether fusion products would be motile if flagellated *Chlamydomonas* were used in such fusion studies. Finally, it might be interesting to compare the processes of electrofusion and PEG-induced fusion of protoplasts. Such a study might include a look at the development of fusion products as well.

5 Protocols

Protoplasts are isolated by treating cells or tissues with a buffered solution containing enzymes (e.g. cellulase, pectinase, hemicellulase) and an osmoticum (e.g. sorbitol, mannitol). Detailed procedures vary with the source plant material and usually must be empirically determined (see examples in Constabel 1982; Vasil 1984; Eriksson 1985).

The best preservation of protoplasts is achieved with fixation in glutaraldehyde and postfixation in osmium tetroxide. The critical step in the fixation process appears to be the initial treatment with glutaraldehyde. Best results are obtained with a low concentration of glutaraldehyde (1%) prepared in the medium used to culture protoplasts or in a mannitol or sorbitol solution of the same osmolarity (Fowke 1982).

Protoplasts are dehydrated stepwise with either alcohol or acetone. Very gradual dehydration is required for particularly fragile protoplasts.

For transmission electron microscopy, dehydrated protoplasts are slowly infiltrated with one of a variety of resins (e.g. araldite, epon, Spurr) and pelleted into embedding capsules. Alternatively, they can be flat embedded to permit selection of individual protoplasts for sectioning (Fowke 1982). Protoplasts are sectioned with glass or diamond knives, mounted on grids, stained with uranyl acetate and lead citrate and examined in the electron microscope.

For scanning electron microscopy, dehydrated specimens are dried in a critical point drier, coated with a thin layer of gold by shadowing or sputter coating and examined in the scanning electron microscope (Fowke 1982).

Acknowledgements. I wish to express my appreciation to all those who were involved in the original work described in this chapter. A special thanks to Pat Rennie for her continued excellent technical assistance. The financial support of the Natural Sciences and Engineering Research Council of Canada is gratefully acknowledged.

References

Bajaj YPS, Gosal SS, Gill MS (1987) Somatic hybridization – a biotechnological approach to plant breeding. In: Ganguli PN, Mandal AK (ed) Plant breeding. Oxford + IBH Publ., New Delhi

Berliner MD (1983) Protoplasts of nonvascular plants. In: Giles KL (ed) Plant protoplasts. Academic Press, New York, pp 21–31

Binding H (1974) Fusion experiments with isolated protoplasts of *Petunia hybrida* L. Z Pflanzenphysiol 72:422–426

Burgess J, Flemming EN (1974) Ultrastructural studies of the aggregation and fusion of plant protoplasts. Planta 118:183–193

Burgess J, Linstead PJ (1976) Scanning electron microscopy of cell wall formation around isolated plant protoplasts. Planta 131:173–178

Burgess J, Linstead PJ, Fisher VEL (1977a) Studies on higher plant protoplasts by scanning electron microscopy. Micron 8:113–122

Burgess J, Linstead PJ, Harnden JM (1977b) The interpretation of scanning electron micrographs. Micron 8:181–191

Chien YC, Fowke LC, Rennie PJ, Wetter LR (1983) Ultrastructure and isozyme analysis of cultured soybean-*Nicotiana* fusion products. In: Potrykus I, Harms CT, Hinnen A, Hutter R, King PJ, Shillito RD (eds) Protoplasts 1983 – Poster Proc. Birkhäuser, Basel, pp 84–86

Constabel F (1982) Isolation and culture of plant protoplasts. In: Wetter LR, Constabel F (eds) Plant tissue culture methods, N.R.C., Saskatoon, pp 38–48
Constabel F, Cutler AJ (1985) Protoplast fusion. In: Fowke LC, Constabel F (eds) Plant protoplasts. C.R.C. Press, Boca Raton, pp 54–65
Constabel F, Dudits D, Gamborg OL, Kao KN (1975) Nuclear fusion in intergeneric heterokaryons. Can J Bot 53:2092–2095
Constabel F, Koblitz H, Kirkpatrick JW, Rambold S (1980) Fusion of cell sap vacuoles subsequent to protoplast fusion. Can J Bot 58:1032–1034
Davey MR, Power JB (1975) Polyethylene glycol-induced uptake of micro-organisms into higher plant protoplasts: an ultrastructural study. Plant Sci Lett 5:269–274
Davey MR, Short KC (1972) The isolation, culture and fusion of legume and cornflower leaf protoplasts. Colloq Int Cent Natl Rech Sci 212:437–453
Davey MR, Clothier RH, Balls M, Cocking EC (1978) An ultrastructural study of the fusion of cultured amphibian cells with higher plant protoplasts. Protoplasma 96:157–172
De Vries SE, Tempelaar MJ (1987) Electrofusion and analysis of potato somatic hybrids. In: Bajaj YPS (ed) Biotechnology in agriculture and forestry 3. Potato. Springer, Berlin Heidelberg New York Tokyo
Eriksson TR (1985) Protoplast isolation and culture. In: Fowke LC, Constabel F (eds) Plant protoplasts. CRC Press, Boca Raton, pp 1–20
Fowke LC (1982) Electron microscopy of protoplasts. In: Wetter LR, Constabel F (eds) Plant tissue culture methods. N.R.C. Saskatoon, pp 72–81
Fowke LC, Gamborg OL (1980) Applications of protoplasts to the study of plant cells. Int Rev Cytol 68:9–51
Fowke LC, Rennie PJ, Kirkpatrick JW, Constabel F (1975) Ultrastructural characteristics of intergeneric protoplast fusion. Can J Bot 53:272–278
Fowke LC, Rennie PJ, Kirkpatrick JW, Constabel F (1976) Ultrastructure of fusion products from soybean cell culture and sweet clover leaf protoplasts. Planta 130:39–45
Fowke LC, Constabel F, Gamborg OL (1977) Fine structure of fusion products from soybean cell culture and pea leaf protoplasts. Planta 135:257–266
Fowke LC, Gresshoff PM, Marchant HJ (1979) Transfer of organelles of the alga *Chlamydomonas rheinhardii* into carrot cells by protoplast fusion. Planta 144:341–347
Fowke LC, Marchant HJ, Gresshoff PM (1981) Fusion of protoplasts from carrot cell cultures and the green alga *Stigeoclonium*. Can J Bot 59:1021–1025
Galway ME, Hardham AR (1986) Microtubule reorganization, cell wall synthesis and establishment of the axis of elongation in regenerating protoplasts of the alga *Mougeotia*. Protoplasma 135:130–143
Gleba YY, Hoffmann F (1978) Hybrid cell lines of *Arabidopsis thaliana + Brassica campestris*: no evidence for specific chromosome elimination. Mol Gen Genet 165:257–264
Gleba YY, Hoffmann F (1980) „Arabidobrassica“: a novel plant obtained by protoplast fusion. Planta 149:112–117
Gleba YY, Sytnik KM (1984) Protoplast fusion – Genetic engineering in higher plants. Monographs on theoretical and applied genetics. Vol 8. Springer, Berlin Heidelberg New York Tokyo
Grout BWW (1975) Cellulose microfibril deposition at the plasmalemma surface of regenerating tobacco mesophyll protoplasts: a deep etch study. Planta 123:275–282
Hadlaczky GY, Burg K, Maroy P, Dudits D (1980) DNA synthesis and division in interkingdom heterokaryons. In Vitro 16:647–650
Hahne B, Hoffmann F (1986) Cortical microtubules and protoplast fusion: effect and fate of microtubular lattices. Plant Sci 47:199–206
Heuser J (1980) Three-dimensional visualization of coated vesicle formation in fibroblasts. J Cell Biol 84:560–583
Hodgson RAJ, Rose RJ (1984) Fusion of spinach mesophyll protoplasts with carrot root parenchyma protoplasts and the effect on spinach chloroplasts. J Plant Physiol 115:69–78
Kajita S, Matsui K, Syono M, Suzuki M, Nagata T (1980) Fine structure of fusion bodies formed between pea root nodule and tobacco mesophyll protoplasts. Z Pflanzenphysiol 97:233–240
Kao KN (1977) Chromosomal behaviour in somatic hybrids of soybean – *Nicotiana glauca*. Mol Gen Genet 150:225–230
Kao KN (1978) Plant protoplast fusion and somatic hybrids. In: Proc Symp Plant tissue culture. Science Press, Peking, pp 331–339

Kao KN, Michayluk MR (1974) A method for high-frequency intergeneric fusion of plant protoplasts. Planta 115:355–367

Lloyd CW, Slabas AR, Powell AJ, Lowe SB (1980) Microtubules, protoplasts and plant cell shape. Planta 147:500–506

Lynch PT, Isaac S, Collin HA (1986) Uptake of fungal protoplasts by plant protoplasts. Proc VI Int Congr Plant Tissue Cell Cult, Minneapolis, Abstr, pp 390

Maliga P, Menczel L (1986) Chloroplast transfer and recombination through protoplast fusion. In: Vasil IK (ed) Cell culture and somatic cell genetics of plants, Vol 3. Academic Press, New York, pp 601–612

Marchant HJ, Fowke LC (1977) Preparation, culture and regeneration of protoplasts from filamentous green algae. Can J Bot 55:3080–3086

Molday RS, Dryer WJ, Rembaum A, Yen SPS (1975) New immunolatex spheres: visual markers of antigens on lymphocytes for scanning electron microscopy. J Cell Biol 64:75–88

Nagl W, Hoffmann F (1980) "Arabidobrassica": evidence for intergeneric somatic hybrid nature from electron microscopic morphometry of chromatin. Eur J Cell Biol 21:227–228

Nagy JI, Paszkowski J, Joo F (1977) Identification of parental nuclei in *Nicotiana* heterokaryons. Z Pflanzenphysiol 84:295–301

Orci L, Perrelet A, Friend DS (1977) Freeze-fracture of membrane fusions during exocytosis in pancreatic B-cells. J Cell Biol 75:23–30

Rennie PJ, Weber G, Constabel F, Fowke LC (1980) Dedifferentiation of chloroplasts in interspecific and homospecific protoplast fusion products. Protoplasma 103:253–262

Robenek H, Peveling E (1978) Beobachtungen am Plasmalemma während der Fusion isolierter Protoplasten von *Skimmia japonica* Thunb. Ber Dtsch Bot Ges 91:351–359

Schieder O, Kohn H (1986) Protoplast fusion and generation of somatic bybrids. In: Vasil IK (ed) Cell Culture and Somatic Cell Genetics of Plants, Vol 3. Academic Press, New York, pp 569–588

Schieder O, Vasil IK (1980) Protoplast fusion and somatic hybridization. Int Rev Cytol Suppl 11B:21–46

Sidorov VA, Gleba Y, Sytnik KM (1978) Ultrastructural study of *Arabidopsis thaliana* (L.) Heynh. cultured protoplasts and arabidopsis+tobacco fusion products. Arabidopsis Inf Serv 15:70–77

Simmonds DH (1986) Prophase bands of microtubules occur in protoplast cultures of *Vicia hajastana* Grossh. Planta 167:469–472

Syono K, Nagata T, Suzuki M, Kajita S, Matsui C (1979) Fusion of pea root nodule protoplasts with tobacco mesophyll protoplasts. Z Pflanzenphysiol 95:449–457

van der Valk P, Rennie PJ, Connolly JA, Fowke LC (1980) Distribution of cortical microtubules in tobacco protoplasts. An immunofluorescence microscopic and ultrastructural study. Protoplasma 105:27–43

Vasil IK (ed) (1984) Cell culture and somatic cell genetics of plants, Vol. 1. Academic Press, New York

Villanueva MA, Metcalf TN, Wang JL (1986) Monoclonal antibodies directed against protoplasts of soybean cells. Planta 168:503–511

Wallin A, Glimelius K, Eriksson T (1974) The induction of aggregation and fusion of *Daucus carota* protoplasts by polyethylene glycol. Z Pflanzenphysiol 74:64–80

Ward M (1984) Fusion of plant protoplasts with animal cells. In: Beers RF, Bassett EG (eds) Cell fusion: gene transfer and transformation. Raven, New York, pp 189–205

Weber G, Constabel F, Williamson F, Fowke LC, Gamborg OL (1976) Effect of preincubation of protoplasts on PEG-induced fusion of plant cells. Z Pflanzenphysiol 79:459–464

Williamson FA, Fowke LC, Weber GW, Constabel F, Gamborg O (1977) Microfibril deposition on cultured protoplasts of *Vicia hajastana*. Protoplasma 91:213–219

Willis GE, Hartmann JX, Lamater ED (1977) Electron microscopic study of plant-animal cell fusion. Protoplasma 91:1–14

Willison JHM, Klein AS (1982) Cell wall regeneration by protoplasts isolated from higher plants. In: Brown RM (ed) Cellulose and other natural polymer systems. Plenum, New York, pp 61–85

Withers LA (1972) Plant protoplast fusion: methods and mechanisms. Colloq Int Cent Natl Rech SCI 212:517–545

Withers LA, Cocking EC (1972) Fine-structural studies on spontaneous and induced fusion of higher plant protoplasts. J Cell Sci 11:59–75

Zimmermann U, Vienken J, Pilwat G (1984) Electrofusion of cells. In: Chayen J, Bitensky L (eds) Investigative microtechniques in medicine and biology. Dekker, New York, pp 89–166

III.3 Symmetric Versus Asymmetric Fusion Combinations in Higher Plants

I. NEGRUTIU[1], A. MOURAS[2], Y. Y. GLEBA[3], V. SIDOROV[3], S. HINNISDAELS[1], Y. FAMELAER[1], and M. JACOBS[1]

1 Introduction

The terms symmetric and asymmetric hybrids refer to the nuclear constitution of fusion products. Symmetric hybrids are defined as consisting of complete sets of chromosomes from both the parents, while asymmetric ones result from a preferential or unidirectional loss of chromosomes belonging to one of the fusion partners. Since protoplasts can be isolated from almost any plant species and readily induced to fuse in all possible combinations, it was expected and even accepted that fusion technology was going to make important contributions in both basic and applied research. This matter was critically evaluted by Harms (1983a), Burgess (1984) and reviewed in detail by Gleba and Sytnik (1984) and Horn et al. (1986).

Our intention is to present several recent developments in this area and discuss them with respect to previous considerations, associate them to novel technical advances in plant research, and place them in the current prospectives on crop improvement.

2 Wide Crosses by Means of Protoplast Fusion

First successful experiments on somatic hybridization were reconstruction experiments: fused protoplasts from species that interbreed in nature produced plants that resembled the control, sexual hybrids (Melchers 1977). This substantiated the initial enthusiasm which during more than 10 years produced sustained efforts in obtaining novel hybrid clones by fusion of protoplasts belonging to distinct genera, tribes, families etc. Among them *Arabidobrassica,* soybean + *Nicotiana*, soybean + barley, *Atropa* + *Nicotiana, Datura* + *Atropa,* tomato + potato, carrot + tobacco and many others (reviewed by Gleba and Sytnik 1984; Schieder et al. 1985; Glimelius et al. 1986). [Even plant protoplasts and animal cell fusions have been studied, such as cell lines of carrot + *Xenopus* (Ward 1984), and tobacco + bull or human sperms (sic) (Lima-de-Faria et al. 1983).]

[1] Laboratorium voor Plantengenetica, Vrije Universiteit Brussel, Instituut voor Moleculaire Biologie, Sint-Genesius-Rode (Belgium)
[2] Laboratoire de Biologie Cellulaire, Université Bordeaux II, Talence (France)
[3] Academy of Sciences, Institute of Botany, Repina 2, Kiev (USSR)

Biotechnology in Agriculture and Forestry, Vol. 8
Plant Protoplasts and Genetic Engineering I (ed. by Y. P. S. Bajaj)

When disparate partners are fused, the complexity of nuclear and cytoplasmic interactions is such that genomic (somatic) incompatibilities [at their various levels, but basically of the zygotic and postzygotic type (Harms 1983b)], are expected to be well beyond those acting in sexual crosses. As a matter of fact, a variable degree of success has been encountered with wide somatic hybridizations: synkaryons are formed in many fusion combinations, they undergo a few initial cell divisions and then fail to proliferate (Harms 1983b). Thus in fusion experiments with several Solanaceous species (Binding 1976) no viable fusion products were obtained among pairwise combinations of tobacco, tomato and *Petunia* even when subprotoplasts of one partner species were used. Unexpectedly, the combination *Vicia faba* + *Petunia* allowed the recovery and prolonged maintenance in culture of fusion products (Binding and Nehls 1978). In several other reports, wide crosses turned out to be successful in that proliferating colonies and callus cultures were produced (see above and Constabel et al. 1976 and references therein; Poulsen et al. 1980; Gleba et al. 1982; Pental et al. 1986; O'Connell and Hanson 1986).

Is there any rationale behind such results? First, in all cases except those reported by Binding (1976), Nehls (1978) and Binding and Nehls (1978), where fusions were performed between protoplasts directly isolated from differentiated tissue (leaf, stem, fruit), use was made as one fusion partner of protoplasts from established in vitro culture cells. Second, contrary to any phylogenetic criteria or considerations, certain fusion combinations appeared more successful compared to others, for example whenever soybean was present (Constabel et al. 1976).

Going back to the first observation, the fact that fusion products from wide crosses, where cultured cells were used as one fusion partner, proliferate actively suggests that somatic incompatibility may be bypassed mainly because of intrinsic characteristics of in-vitro cultured cells. Such cells evolve in a permissive and destabilizing environment (Bayliss 1975) where the tight regulatory controls acting upon cells within the whole plant, organ, etc. cease operating, the cells undergoing the in-vitro dedifferentiation process. This erases, suppresses (more or less reversibly?) at least certain "domains" involved in somatic incompatibility processes. Such loss of genomic "identity" is most likely a gradual phenomenon (this could be associated with the loss of morphogenetic potential in long-term cultured cells); therefore, in short-lived abortive intergeneric synkaryons, produced by the fusion of protoplasts derived from differentiated tissues, the somatic incompatibility reactions still act within cells that have not been exposed to the cell culture environment before. As a matter of fact, proliferating wide cross-hybrid cells are usually highly polyploid, genetically unstable or unbalanced. Random chromosome elimination from hybrid cells has been observed and frequently associated with structural rearrangements (Gleba and Syntnik 1984 and references therein; Poulsen et al. 1980; Hoffmann and Adachi 1981; Gleba et al. 1983; Endo et al. 1987). Such cultures are usually non-morphogenic, which can result from either loss of regeneration capacity during the in-vitro passaging, or because the induction of differentiation and differentiation per se in such hybrid cells reveals or engages otherwise repressed somatic incompatibility processes (also see Dudits et al. 1980). If, however, plants were regenerated, they are abnormal and/or sterile, and could not be incorporated into breeding schemes (Burgess 1984) (see further Sect. 3).

The typical example is the tomato + potato combination (Poulsen et al. 1980). Nuclear genomes of both species were retained in the hybrid nucleus, while cpDNA's segregated mitotically. The large variation in chromosome number (attributed to the hyperploidy and karyotype instability within both the callus culture used for potato protoplast production and in the regenerants) made it impossible to evaluate whether the somatic segregation of the chloroplasts was due to random sorting out or to more specific nuclear-cytoplasm interactions in various hybrid clones (also see O'Connell and Hanson 1986). Therefore, the use of somatic hybridization to restore the morphogenetic potential of established cultured cells (Maliga et al. 1977) is probably of limited use, not only because it is unlikely to be the rule (e.g. O'Connell and Hanson 1985), but also because the genetic constitution of long-term cultured cells will not allow the recovery of fertile plants.

3 Spontaneous Asymmetrization of Fusion Products

The "final" product of fusions between taxonomically unrelated species is usually an asymmetric combination of parental genomes, with parts of one or both genomes being lost during the in-vitro stage (Gleba and Sytnik 1984; Imamura et al. 1987 and references therein). The extent to which the two genomes undergo ploidization, mitotic segregation, chromosome elimination or rearrangements is unpredictable.

The difficulty in controlling the asymmetrization process stems from at least four considerations: (1) as already mentioned, most of the available wide somatic hybrids were produced by the fusion of stable euploid leaf protoplasts with genetically unstable protoplasts from cultured cells. In such fusions the "leading" partner is the most rapidly cycling nucleus, usually the cultured cell parent (Harms 1983b); (2) the cell cycle phase can affect the interaction of plant nuclei at the earliest stages after fusion. Heterophasic synkaryons result in premature chromosome condensation (PCC) induced by the mitotic nucleus to the fusion partners: if in G_1 phase, the chromosomal threads remain single-stranded, if in S-phase, the chromatin suffers "pulverization" to small fragments (Gould and Daines 1986) (note that the signal from mitotic nuclei inducing PCC, is recognized by nuclei from intergeneric and interkingdom cells); (3) the asymmetrization being basically random or in favour of the rapidly cycling partner, it is difficult to estimate the role of somatic incompatibility mechanisms in the speed and level of asymmetrization; (4) more specifically, nuclear-cytoplasmic interactions in wide hybridizations being poorly understood, the parallel asymmetrization of cytoplasmic genetic determinants ("cybridization") remains as unpredictable as that of the nuclear parts (Pelletier et al. 1985). Organelle segregation seems to occur very rapidly only under defined selection pressures (Fluhr et al. 1983; Flick et al. 1985; Pental et al. 1986; Zuckerman et al. 1986; Morgan and Maliga 1987 and references therein). Two cases of intergeneric cybridization are well documented so far, one containing the nucleus of *Petunia* and chloroplasts of tobacco (Pental et al. 1986), the other containing the nucleus of tobacco and the chloroplasts of *Atropa* (Kushnir et al. 1987). On the other hand, rearrangements following recombination between parental mitochondrial DNA in regions not normally undergoing intragenomic recombination, as well as sorting out, have been observed (Rothenberg and Hanson 1987; Morgan and Maliga 1987).

A certain asymmetrization has to take place before differentiation processes can occur. This could be of particular importance in both understanding the bases of wide hybridization conditions together with the role of somatic incompatibility reactions (by dissecting them into sub-domains), and also in terms of applied breeding objectives. The fact that the asymmetrization is a gradual process, where morphogenesis resumes after prolonged callus proliferation, reduces – in our view – the potential interest of the spontaneous asymmetrization process (also see Sects. 6 and 7).

Thus, variation in morphology and ploidy was reported in subclonal regenerants of somatic hybrids between *Lycopersicon esculentum* and *Solanum rickii* (O'Connell and Hanson 1986). Even in the best-documented asymmetric hybrids of *Datura innoxia* + *Atropa belladonna* obtained after spontaneous chromosome elimination following screening for morphogenic events and where the chromosome complement of *Atropa* was strongly reduced from 72 to 6 or 2 chromosomes, the regenerated shoots were inviable in soil or upon grafting (Schieder et al. 1985). In *Arabidobrassica* (Hoffmann and Adachi 1981), asymmetric hybrids with obvious chromosomal rearrangements produced plants with predominantly regular vegetative morphology. However, abnormalities were recorded at flower level, indicating that reproductive organs were particularly exposed to the joint impact of the in-vitro culture passage, chromosome rearrangement, and somatic incompatibility processes in the asymmetric hybrids. Last but not least, regenerants with a *Hyoscyamus*-type phenotype (and cp-DNA pattern) from a tobacco+henbane combination produced viable and abortive seeds (Jia et al. 1983); no detectable tobacco signal for species-specific repetitive DNA could be identified in such plants, suggesting to us that almost complete asymmetrization may have taken place and resulted in parental-type individuals (also see Pental et al. 1986). It is interesting to note that completion of elimination of the tobacco complement apparently took place after regeneration of shoot structures and their proliferation through grafting. This indicates that somatic incompatibility may still operate in somatic hybrids with a given degree of asymmetrization when such tissues are challenged through developmental pathways.

4 Symmetric Hybrids Between Unrelated Plant Species – the Disappointment

Excellent, empirically devised fusion techniques were developed during the last two decades and used in the production, under in-vitro culture conditions, of a variety of species combinations with no counterpart in nature. Despite the academic interest of the matter, it became obvious that no novel hybrid plants among distantly related species would be engineered by means of fusion technology. Furthermore, there is only circumstantial evidence to suggest that protoplast fusion can result in introgression, via illegitimate recombination or homeologous pairing, of limited genetic information from one fusion partner into the other (Jia et al. 1983; Dudits et al. 1979, 1980; Schieder et al. 1985). Gene introgression, if any, is not the direct result of the fusion process, but mainly the unpredictable consequence of the in-vitro passage, as discussed in Sections 2 and 3. At most, the fusion technique holds promise as an alternative (Harms 1983b and references therein, Hamill et al. 1985 and references therein) or a complement (Glimelius 1985) to sexual cross-programmes for interspecies combinations (cultivars and wild species), with emphasis on a more efficient transfer and/or recombination of cytoplasmic genetic determinants (cybrids). As a matter of fact, the availability of selectable chloroplast mutants enabled Medgyesy et al. (1986) to demonstrate transmission of paternal chloroplasts in Nicotiana's, e.g. a sexual cybridization method.

5 Technical Advances with Relevance to Fusion Studies

In an attempt to further understand the various cellular processes associated with the fusion event per se, but also and mainly to explore further the fusion technology as a tool of chromosome transfer and recombination, as a tool to study somatic incompatibility with respect to hybrid genomic complexity and developmental constraints, we have identified a series of recent technical developments in plant biology which should confer to this area the appropriate level of sophistication imposed by the above objectives. These technical achievements are discussed below.

5.1 The Fusion Technique

First of all, fusion techniques per se have constantly evolved (see Chap. III.1, this Vol.). They have been simplified and made more flexible and reproducible for both chemical (PEG) and electric field-mediated fusion. Such techniques can be adapted either for mass-fusion (Watts and King 1984; Tempelaar and Jones 1985; Puite et al. 1985; Negrutiu et al. 1986) or for the micromanipulation of individual plant cells (Schweiger et al. 1987). In the latter case, the one-to-one electrofusion of preselected protoplasts and subprotoplasts makes it possible to control the fusion partners a priori and also to combine different cell compartments (cell reconstitution), and this in association with microinjection (Spangenberg and Schweiger 1986; Spangenberg et al. 1986).

5.2 Dominant and Recessive Mutant Plants. The Universal Hybridizer

Among the various ways to identify and isolate fusion products, the availability of biochemical, selectable marker traits is of particular interest. A series of biochemical selectable mutants have been selected via haploid or diploid protoplast isolation and culture (reviewed by Negrutiu et al. 1984). Many of them have been

Table 1. Linkage test for the Universal Hybridizer T116-A1 for the recessive (NR^-) and dominant (kanamycin-resistance) traits. Ratios of growing: non-growing seedlings are given in simple segregation (on 10 mM KNO_3 as sole nitrogen source) and double segregation (on 10mM KNO_3 + 100 mg per 1 kanamycin) tests

Progeny	Segregation	
	Observed	Expected
F1 : KNO_3	No segregation	–
F1 : KNO_3 + km	0.99 : 1 (544)	1 : 1
F2 : KNO_3	3.3 : 1 (537)	3 : 1
F2 : KNO_3 + km	1.37 : 1 (879)	1.28 : 1[a]

Note: Figures in brackets represent total of seedlings counted.
[a] – the result shows that the NR^- and km^R are unlinked; the expected ratio in case of linkage is 1 : 1.

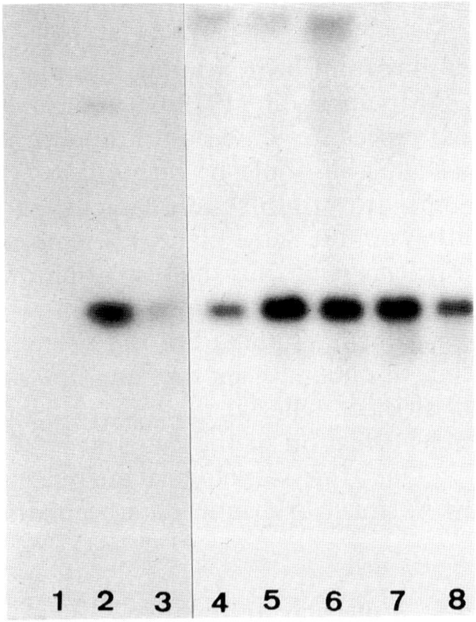

Fig. 1. Southern blot analysis showing the physical presence of the kanamycin gene in several Universal Hybridizer plants. DNA was digested with EcoR5, 10 µg loaded on the gel, and hybridized to the 1.2 kb EcoR5 fragment of the kanamycin gene from Tn-5. *Lane 1* wild type; *lane 2* 10 copies reconstruction; *lane 3* one copy reconstruction; *lanes 4 to 8* transformed plants T116-B4, T116-B2, T116-B12, T116-A1, and T116-A9

regenerated into plants and analyzed biochemically and genetically. Several of them (nitrate reductase-less, amino acid auxotrophs) were used as euploid, leaf protoplast partners in fusion experiments (Negrutiu et al. 1986).

Since mutants have not yet been identified in numerous plant species, one possible solution is to construct universal hybridizers (Lo Schiavo et al. 1983; Brunold et al. 1987; Ye et al. 1987). In our case, a nitrate reductase-less (NR^-) mutant defective in the cofactor function (cnx20) and for which no reversion was obtained (Dirks et al. 1986) was further exploited to produce a multimarker universal hybridizer (UH.). Briefly, the homozygous recessive cnx20 mutant was transformed by direct gene transfer for kanamycin resistance (Negrutiu et al. 1987 and references therein). Several NR^-, kanamycin-resistant clones were regenerated into plants and analyzed for the marker traits. In T_{116}–A_1 and T_{116}–A_9 clones, for example, the foreign gene was inherited as a single dominant trait, unlinked to the NR deficiency (Table 1 and Fig. 1). Protoplasts from the UH were used in fusion experiments with partners lacking any selectable trait (see Sect. 7).

To further sophisticate the UH, we transformed leaf disks with *Agrobacterium rhizogenes* (strain A_4) and from the isolated transformed roots plants were regenerated again. Thus, several screenable markers from the T-DNA of the bacteria were associated to the already available selectable markers NR^- and kanamycin resistance. In addition, a periclinal albino sector was identified and multiplied in one of the UH plants, thus providing a visual marker as well.

5.3 Fine Chromosomal Analysis of Plant Material – the Protoplast Approach

Fine caryological analysis with plant material was developed in association with classical breeding methods in cereals (Brar and Khush 1986), but such techniques are far from being routine in many other species.

Mouras et al. (1978, 1986; also see Ambros et al. 1986) described karyotyping procedures based on protoplast isolation from root meristems. Three major advantages should be mentioned: (1) protoplast preparations produce chromosomes two- to three-fold larger than by root squashes; (2) the chromosomes are always in the same plane; (3) removal of the cell wall dramatically increases the efficiency and thus the resolution of in situ hybridization protocols (Mouras et al. 1987).

To better standardize the preparation of root protoplasts and speed up the time of obtaining appropriate root material, we coupled the karyological analysis of all dicot material, asymmetric hybrid plants etc. (see Sect. 7) to the preliminary transformation of the material by *Agrobacterium rhizogenes.*

Thus, fine karyological analysis and localization of genes onto chromosomes by in situ hybridization are becoming simple tools of analysis with plant cells.

5.4 Molecular Techniques

Among the many techniques on recombinant DNA developed or adapted to plant material, we shall refer to only a few in order to familiarize the inexperienced reader with the new dimension and solutions the experimental plant biology can bring to old objectives, such as the protoplast fusion matter.

Thus, simple miniaturized DNA extraction procedures are by now in use (Grimsley et al. 1987) and automated extraction will be made available in a near future. Species-specific DNA probes consisting of repetitive DNA sequences and monoclonal antibodies to plasmamembrane antigens are being identified and exploited in fusion experiments (Saul and Potrykus 1984; Imamura et al. 1987; Fitter et al. 1987).

Furthermore, bacterial genes (antibiotic resistance, T-DNA genes from *Agrobacterium* etc.) and a rapidly increasing collection of cloned plant genes (Horn et al. 1986) represent additional selectable and screenable markers in fusion experiments and powerful tools to analyze the fate (physical presence, localization, level of expression) of defined genes or gene products.

Progress is being made in refining techniques on organelle DNA analysis, an essential requirement in elucidating organelle behaviour and heredity (Medgyesy et al. 1985; Morgan and Maliga 1987; Rothenberg and Hanson 1987). In human cells and cell hybrids, mt-DNA analysis by RFLP's can now detect single nucleotide substitutions, and contribute to reconsidering errors in constructing restriction maps and estimating sequence diversity (Singh et al. 1987), as well as in establishing codominant or dominant/recessive relationship among mitochondrial genomes (Zuckerman et al. 1986; Hayashi et al. 1987), which governs the segregation pattern of the organelle.

6 Asymmetric Hybrids in Wide Crosses by Gamma-Fusion: Towards Chromosome Transfer and Gene Introgression

Gamma-fusion stands for asymmetric hybrids obtained following inactivation of donor protoplasts. It follows that the abbreviation can be employed to designate gamma-hybrids.

From results described in Sections 2 and 3 it became obvious that there was a need (1) to direct the elimination of chromosomes in various fusion combination and (2) to maximize the chance of regenerating gamma-hybrids by a more careful device of fusion partners, protoplast source, culture conditions and timing etc.

Unidirectional Elimination of Donor Chromosomes. This has been attempted almost exclusively by the irradiation with sub- or lethal doses of the donor fusion partner. Little is known about the effects of irradiation on the donor protoplast itself and even less on the behaviour of gamma-fusion products in very early stages after fusion (Negrutiu et al. 1984). Menczel et al. (1982) showed that there was a decreased frequency of somatic hybrids and an increased frequency of cybrids with increased radiation doses (in the range of 5 to 30 krad).

Nicoll et al. (1987), in a study on endosperm responses to irradiated pollen, described several nuclear abnormalities, including enhanced numbers of polyploid restitution nuclei, bridges between nuclei, excluded metaphase chromosome fragments, disrupted mitotic synchrony, spindle malfunction. They underlined the importance of the state of the nucleus (before or after DNA synthesis) at irradiation, mainly in association with doses above 100 krad, and it was suggested that the exclusion of donor DNA occurs very early after "gamma-fertilization". With respect to the state of the donor nucleus at irradiation one should remember the reference made to cell cycle effects during and after protoplast fusion (Sect. 3). If PCC has more destructive effects in G_1 and mainly S-phase nuclei than in other stages of the cell cycle, one can imagine that in function of the choice of the protoplast source one can reinforce or even simulate irradiation effects in donor-recipient interactions after fusion.

At this stage, it is worthwhile mentioning that one major phenomenon associated with irradiated pollen is mutational damage, as shown by us and others in "egg-transformation"-type experiments (Famelaer et al. 1986 and references therein; Nicoll et al. 1987 and references therein), and this was also suggested to occur at high rates in protoplast systems (Negrutiu et al. 1984, 1987).

Highly Asymmetric Hybrids by Means of Gamma-Fusion. Combination of irradiation and fusion of protoplasts (exhibiting selectable phenotypes and almost systematically isolated directly from plant material) produced an interesting series of asymmetric hybrid clones, such as carrot + parsley (Dudits et al. 1980), *Datura* + *Physalis* (Gupta et al. 1984), *Lycopersicon* species (O'Connell and Hanson 1985), *Solanum* species (Sidorov et al. 1987), several *Solanaceae* (Glimelius et al. 1986), henbane + tobacco (Imamura et al. 1987), tobacco + barley (Somers et al. 1986). Different combinations of isozyme, species-specific DNA, organelle DNA, and immunological probes, as well as chromosome counts and morphology were employed to measure the level of asymmetry, and show the preferential loss of the irradiated DNA. With one exception (Imamura et al. 1987), where in fact both fusion partners were of cell culture origin, regeneration was reported in all combinations tested.

Selection/screening pressures via selectable or visible markers appeared as an efficient parameter in the system. The regeneration "activity" per se can be considered as a further selection pressure to reveal or accentuate the asymmetry; for example,Sidorov et al. (1987) noticed that the asymmetrization of the clones in the *Solanum* combination was not complete at the plant regeneration step. Control experiments with the non-irradiated combination (Gupta et al. 1984) demonstrated the role of the "gamma-asymmetrization" in stabilizing the clones and favouring plant regeneration.

In all listed reports the irradiation doses used were below 50 krad (usually 10–20 krad). At sublethal doses (1.5 krad) the irradiated nucleus could be completely rescued by the recipient to produce symmetric hybrid lines (O'Connell and Hanson 1985). At lethal doses the amounts of nuclear donor DNA were relatively large (Imamura et al. 1987) and only rarely were highly asymmetric clones clearly described (Gupta et al. 1984; Dudits et al. 1980). This suggests that the number of chromosome breaks per cell is relatively small, and that there is a strong selection for centromere within the irradiated partner, and no real proof so far of chromosome fragment introgression into the recipient.

The somatic incompatibility barriers did not seem to be active in the reported combinations following gamma-fusion, or it was precisely in symmetric fusions between protoplasts directly isolated from plant tissues that the incompatibility reactions were the strongest (see Sect. 2). This may indicate that irradiation alone can suppress or strongly reduce such incompatibility processes. The work of Glimelius et al. (1986), where gamma-cybrids were produced within Solanaceae, suggests that the nuclear incompatibility "border" between the five tribes of Solanaceae was set between the subfamilies Solanoideae and Cestroideae.

Concerning asymmetrization at organelle level, the irradiation condition does not seem to establish a clear-cut situation: there is a feeling that chloroplasts from the irradiated partner can be preferentially transferred; selective pressures or selective disadvantages (albino, cms) at organelle level are obviously interacting as well (Morgan and Maliga 1987 and references therein; Pental et al. 1986; Glimelius et al. 1986; Sidorov et al. 1987).

7 Gamma-Fusion in the *N. plumbaginifolia* System

In this species we developed a gamma-fusion strategy based on the following considerations:

1. *N. plumbaginifolia* is a true diploid species ($2n = 2x = 20$) with a distinct karyotype (Mouras et al. 1986), all chromosomes being telocentrics.
2. The protoplast-whole plant cycle is short and the regeneration potential high (Installé et al. 1985). The in-vitro passage was reduced in time to the period required to visualize selected fusion products, and proliferate them in view of plant regeneration.
3. Among the numerous recessive and dominant selectable markers available in this species, the mutant cnx 20 was chosen for two main reasons: it is a well-characterized NR^- mutation for which no spontaneous or induced reversions are known; the cofactor function (cnx A type, Molybdenum restorable) for which it is coding is most likely highly conserved within various plant species and even other organisms. It therefore makes an ideal recipient in gammma-fusion experiments. The UH (Sect. 4), a derivative of this mutant, was used as a valuable control in fusions with the non-irradiated partners.
4. Gamma-fusions were performed in the 10 to 200 krad range in an attempt to provoke a maximum of radiation damage; the somatic incompatibility reactions were assessed by providing a wide range of fusion combinations, from phylogenetically close to unrelated species.

The results are summarized below.

The Incompatibility Barriers. The recipient *N. plumbaginifolia* protoplasts (cnx 20 and UH) were fused respectively with gamma-irradiated and non-irradiated protoplasts of *N. tabacum, N. sylvestris, Datura innoxia, Atropa belladonna, Petunia hybrida, Lycopersicon esculentum, Solanum tuberosum, Hyoscyamus muticus, Brassica napus, Arabidopsis thaliana, Medicago sativa* and *Beta vulgaris.* Leaf protoplasts were isolated from both recipient and donor species, with the exception of *Beta* protoplasts (cell suspension). In the case of *Petunia* and tobacco, both leaf and cell suspension protoplasts were employed. Only the combinations with tobacco, *N. sylvestris, Atropa* and sugarbeet produced stable growing clones.

Thus, the success of a given fusion combination did not seem to depend on phylogenetic criteria only. Moreover, in our experimental system, gamma-fusion did not allow us to extend the range of successful fusion combinations; the irradiation damage alone does not seem to be sufficient in breaking down incompatibility barriers.

Irradiation, Fragmentation of Donor Chromosomes, and Regeneration of Asymmetric Hybrid Plants. Irradiation favoured a rapid and unidirectional elimination of most or part of the donor chromosomes. The donor genome underwent fragmentation, the level of which increased with the dose (Fig. 2). Deleted or minichromosomes were maintained mitotically and meiotically. By analyzing sufficient clones one could identify individuals with few donor chromosomes. In general, the extent of chromosome elimination was higher in wider hybrid combinations (Fig. 3).

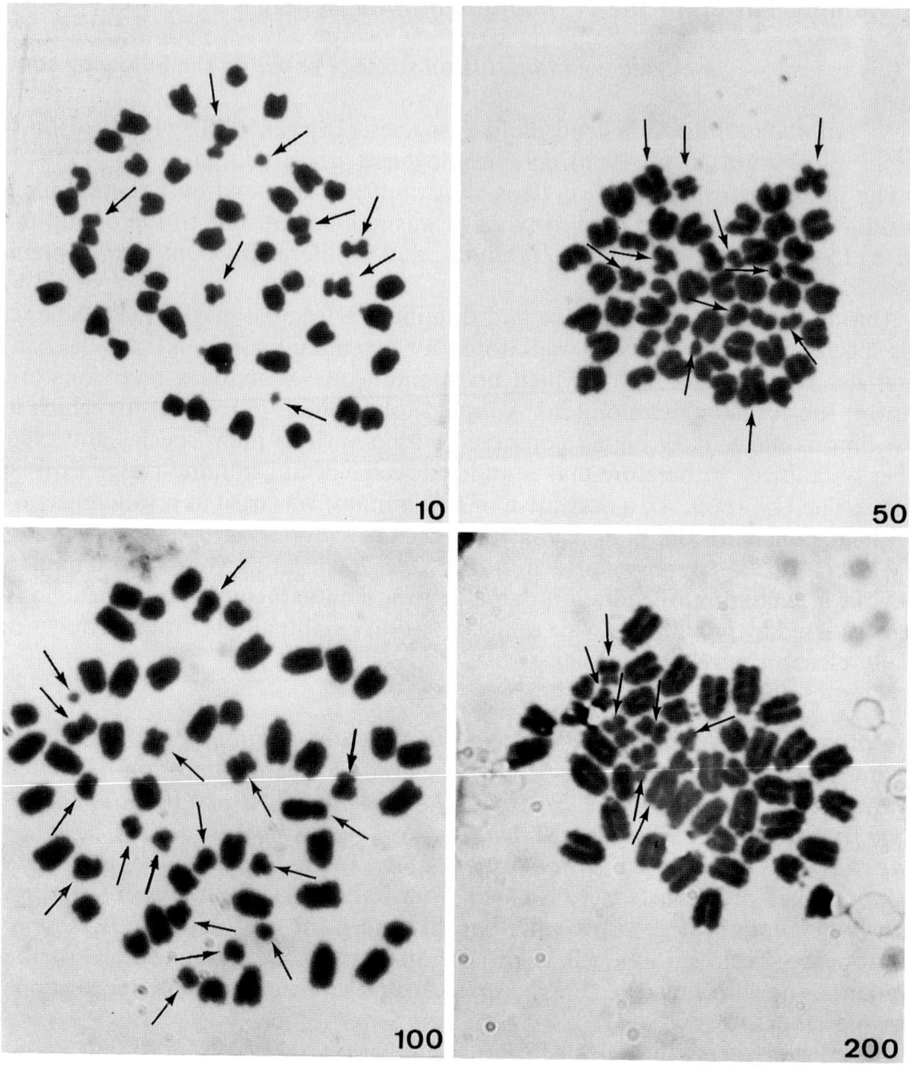

Fig. 2. Effect of the irradiation dose on the level of fragmentation and asymmetrization in the fusion combination *N. plumbaginifolia (nia 26)* + *N. sylvestris* V42. The recipient *N. plumbaginifolia* has telocentric and the donor *N. sylvestris* has metacentric and minichromosomes (*arrows*). Figures at the *right* of the pictures represent krad.

Fig. 3. Metaphase plates from root meristems in three highly asymmetric hybrids of **A** *N. plumbaginifolia* + *N. sylvestris;* **B** *N. plumbaginifolia* + *Beta vulgaris;* **C** *N. plumbaginifolia* + *Atropa belladonna.* Donor chromosomes are marked with *arrows*

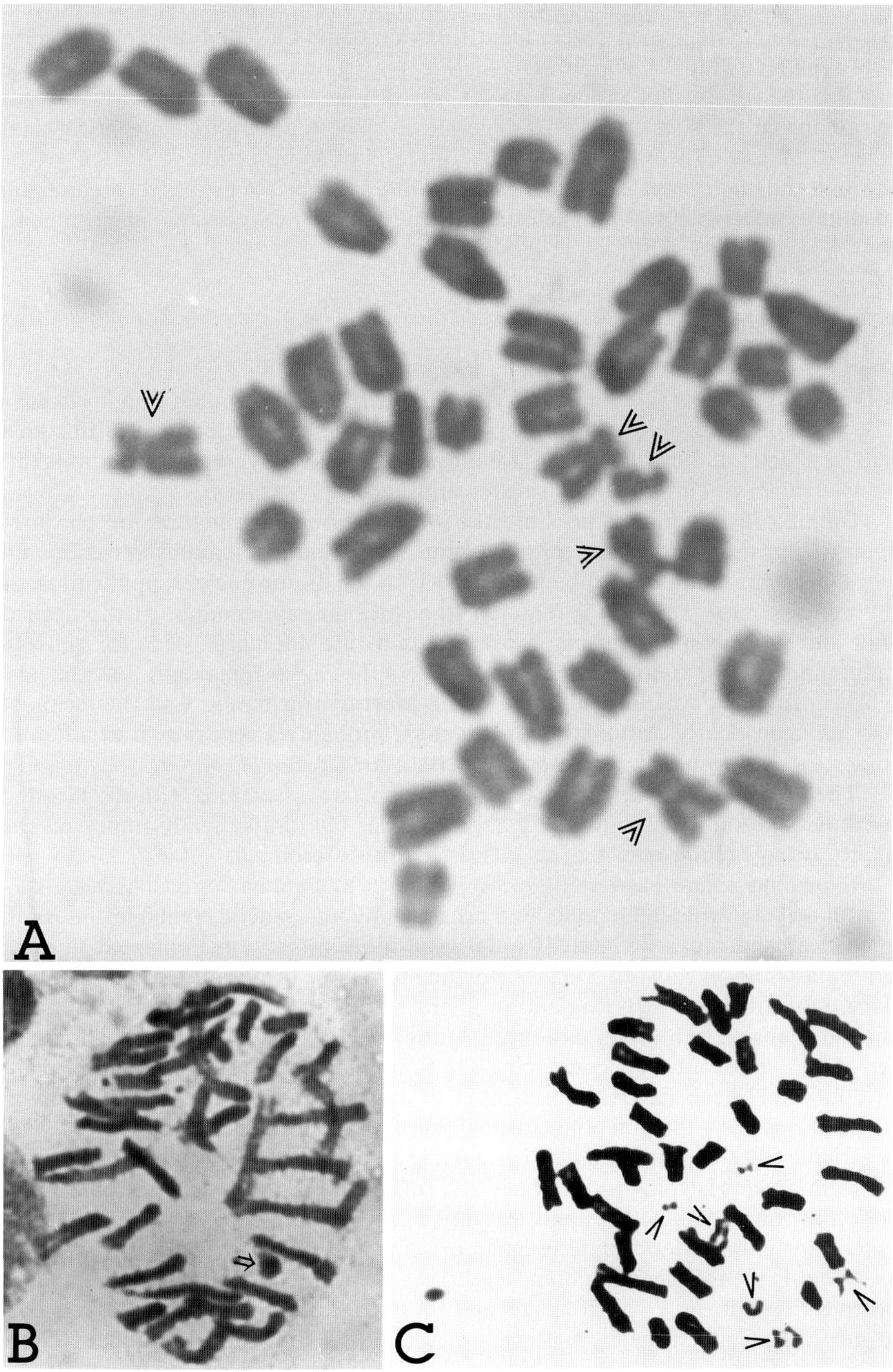
A
B
C

The regenerated plants frequently exhibited a recipient type phenotype. Among them, fertile plants were identified that could be selfed or backcrossed to wild-type recipient. Whenever seed set was uncertain or too low, embryo rescue enabled the recovery of progeny plants. Fertility of the asymmetric hybrids kept improving in the successive cross-generations. These generations are being analyzed to establish the evolution in time and level of stabilization of the karyotype through the meiotic cycle, and to evaluate with appropriate probes the expression of donor, selectable and non-selectable, genes in the new genomic environment.

8 Conclusion

Steady progress has been made in developing fusion techniques and strategies. They shed light on a series of new problems we are confronted with in this area and in which the direct interests have moved from novel hybrid combinations (cf. Sects. 1 to 3) to chromosome transfer and gene introgression.

Gamma-fusion makes possible the unidirectional elimination of chromosomes from the donor partner; the extent of elimination remains variable, and since elimination appears to be random within the donor nucleus, application of defined selection pressures is required to ensure the maintenance of the desired trait and its corresponding chromosome. As a rule, the parts of donor genome maintained in the asymmetric hybrids possess their own functional centromeres, while introgression of limited amounts of genetic information into the recipient has not been clearly demonstrated. It is most probably a very rare, if at all, even under the experimental conditions tested so far.

The use of euploid cells as source of protoplasts, associated with the directed asymmetrization process of the gamma-fusion, has dramatically improved the ability to regenerate plants from identified fusion products.

Bypassing sexual barriers of incompatibility by gamma-fusion has been successful up till now in interspecies and occasionally intergeneric combinations, both at nuclear and organelle level. If novel genomic combinations, obtained through protoplasts fusion between species belonging to remote taxonomic groups, can produce only structures surviving and proliferating under in vitro culture conditions, the use of fusion technology as a breeding tool is basically compromised. As a matter of fact, most interests concerning the introgression of valuable traits in plant breeding can be satisfied within the range of interspecies and intergeneric combinations. If this is so, the potential classical breeding methods, reinforced with several new tissue culture techniques (such as embryo rescue, hormone application, in-vitro fertilization etc.) is going to be largely self-sufficient. The area of immediate application for fusion should be in those groups of species where the reproductive biology is poorly understood or technically extremely difficult.

What is the real need in going beyond the above-mentioned taxonomic range? Engineering nitrogen fixation, storage proteins etc. across various taxonomic groups? For such strategic matters it is very likely that the recombinant DNA technology, associated to more and more powerful breeding methods, will be the chosen way.

The major contribution of protoplast fusion is first expected to cover several areas of fundamental research, without excluding some limited achievements in the applied field.

References

Ambros PF, Matzke MA, Matzke AJM (1986) Detection of 17 kb unique sequence (T-DNA) in plant chromosomes by in situ hybridization. Chromosoma 94:11–18

Bayliss MW (1975) The effect of growth in vitro on the chromosome complement of *Daucus carota* suspension cultures. Chromosoma 51:401–411

Binding H (1976) Somatic hybridization experiments in Solanaceous species. Mol Gen Genet 144:171–175

Binding H, Nehls R (1978) Somatic cell hybridization of *Vicia faba* + *Petunia hybrida*. Mol Gen Genet 164:137–143

Brar DS, Khush GS (1986) Wide hybridization and chromosome manipulation in cereals. In: Evans DA, Sharp WR, Ammirato PV (eds) Handbook of plant cell culture, Vol. 4. Macmillan, New York, pp 221–263

Brunold C, Krüger-Lebus C, Saul MW, Wegmüller S, Potrykus I (1987) Combination of kanamycin resistance and nitrate reductase deficiency as selectable markers in one nuclear genome provides a universal somatic hybridizer in plants. Mol Gen Genet 208:469–473

Burgess J (1984) The revolution that failed. New Sci 104/1428:26–29

Constabel F, Weber G, Kirkpatrick JW, Pahl K (1976) Cell division of intergeneric protoplast fusion products. Z Pflanzenphysiol 79:1–7

Dirks R, Negrutiu I, Heinderickx M, Jacobs M (1986) Genetic analysis of revertants for the nitrate reductase function in *Nicotiana plumbaginifolia*. Mol Gen Genet 202:309–311

Dudits D, Hadlaczky G, Bajszar GY, Kouczc C, Lazar G, Horvath G (1979) Plant regeneration from intergeneric cell hybrids. Plant Sci Lett 15:101–112

Dudits D, Fejer O, Hadlaczky F, Koucz C, Lapar G, Horvath G (1980) Intergeneric gene transfer mediated by plant protoplast fusion. Mol Gen Genet 179:283–288

Endo T, Komiya T, Masumitsu Y, Morikawa H, Yamada Y (1987) An intergeneric hybrid cell line of *Duboisia hopwoodii* and *N. tabacum* by protoplast fusion. J Plant Physiol 129:453–459

Famelaer Y, Cammaerts D, Karp A, Sidorov V, De Brouwer D, Negrutiu I, Jacobs M (1986) Cellular engineering by "gamma fusion" and "egg transformation". In: Nuclear techniques and in vitro culture for plant improvement. International Atomic Energy Agency Vienna, IAEA-SM-282/44, pp 453–461

Fitter MS, Normann PM, Hahn MG, Wingate VPM, Lamb C (1987). Identification of somatic hybrids in plant protoplast fusion with monoclonal antibodies to plasma-membrane antigens. Planta 170:49–54

Flick CE, Kut SA, Bravo JE, Gleba Yu, Evans DA (1985) Segregation of organelle traits following protoplast fusion in *Nicotiana*. Biotechnology 3:555–560

Fluhr R, Aviv D, Edelman M, Grahon E (1983) Cybrids containing mixed and sorted-out chloroplasts following interspecific somatic fusions in *Nicotiana*. Theor Appl Genet 65:289–294

Gleba YuYu, Sytnik KM (1984) Protoplast fusion – Genetic engineering in higher plants. Monographs on theoretical and applied genetics, Vol 8. Springer, Berlin Heidelberg New York Tokyo

Gleba YuYu, Momot VP, Cherep NN, Skarzhynskaya MV (1982). Intertribal hybrid cell lines of *Atropa belladonna* × *Nicotiana chinensis* obtained by cloning individual protoplast fusion products. Theor Appl Genet 62:75–79

Gleba YuYu, Momot VP, Okolot AN, Cherep NN, Skarzhynskaya MV, Kotov V (1983) Genetic processes in intergeneric cell hybrids *Atropa* + *Nicotiana*. Theor Appl Genet 65:269–276

Glimelius K (1985) Sexual and somatic hybridization. Hereditas Suppl 3:41–47

Glimelius K, Fahlesson J, Sjödin C, Sundberg E, Djupsjöbacka M, Fellner-Feldegg H, Bonnet HT (1986) Somatic hybridization and cybridization as potential methods for widening of the gene-pools of crops within Brassicaceae and Solanaceae. In: Horn W, Jensen CJ, Odenbach W, Schieder O (eds) Genetic Manipulation in plant breeding. Proceedings International Symposium Eucarpia, Sept. 8–13, 1985, Berlin. de Gruyter, Berlin, pp 663–682

Gould AR, Daines RJ (1986) Plant protoplasts and the cell cycle. In: Fowke LC, Constabel F (eds) Plant Protoplasts. CRC Press Inc, Boca Raton Florida, pp 67–77

Grimsley N, Hohn T, Davies W, Hohn B (1987) *Agrobacterium*-mediated delivery of infectious maize streak virus into maize. Nature 325:177–179

Gupta PP, Schieder O, Gupta M (1984) Intergeneric nuclear gene transfer between somatically and sexually incompatible plants through asymmetric protoplast fusion. Mol Gen Genet 197:30–35

Hamill JD, Pental D, Cocking EC (1985) Analysis of fertility in somatic hybrids of *Nicotiana rustica* and *N. tabacum* and progeny over two sexual generations. Theor Appl Genet 71:486–490

Harms CT (1983a) Somatic hybridization by plant protoplast fusion. In: Potrykus I, Harms CT, Hinnen A, Hütter R, King PJ, Shillito RD (eds) Protoplasts 1983. Birkhäuser, Basel, pp 69–84

Harms CT (1983b). Somatic incompatibility in the development of higher plant somatic hybrids. Q Rev Biol 58:325–351

Hayashi J, Yonekawa H, Murakami J, Tagashira Y, Perreirra-Smith OM, Shay JW (1987) Mitochondrial genomes in intraspecies mammalian cell hybrids display codominant or dominant/recessive behaviour. Exp Cell Res 172:218–227

Hoffmann F, Adachi T (1981) *Arabidobrassica*: chromosomal recombination and morphogenesis in asymmetric intergeneric hybrid cells. Planta 153:586–593

Horn W, Jensen CJ, Odenbach W, Schieder O (eds) (1986) Genetic manipulation in plant breeding. Proceedings International Symposium Eucarpia, Sept. 8–13, 1985, Berlin. de Gruyter, Berlin

Imamura J, Saul MW, Potrykus J (1987) X-ray irradiation promoted asymmetric somatic hybridization and molecular analysis of the products. Theor Appl Genet 74:445–450

Installé P, Negrutiu I, Jacobs M (1985) Protoplast-derived plants in *Nicotiana plumbaginifolia*: improving the regeneration response of wild type and mutant cultures. J Plant Physiol 119:443–454

Jia J, Potrykus I, Lazar GB, Saul M (1983) *Hyoscyamus-Nicotiana* fusion hybrids selected via auxotroph complementation and verified by species-specific DNA hybridization. In: Potrykus I, Harms CT, Hinnen A, Hütter R, King PJ, Shillito RD (eds) Protoplasts 1983. Birkhäuser, Basel, pp 110–111

Kushnir SG, Schlumukov LR, Pogrebnyak NJ, Berger S, Gelba Yu (1987) Functional cybrid plants possessing a *Nicotiana* genome and an *Atropa* plastome. Mol Gen Genet 209:159–163

Lima-de-Faria A, Hiraoka T, Jarl C, Bornman CH, Joworska H (1983) Fusion of bull and human with *Nicotiana* protoplasts. Hereditas 99:191–201

Lo Schiavo F, Giovinazzo G, Terzi M (1983) 8-Azaguanine resistant carrot cell mutants and their use as universal hybridizers. Mol Gen Genet 192:326–329

Maliga P, Lazar G, Joo' I, Nagy AH, Menczel L (1977) Restoration of morphogenic potential in *Nicotiana* by somatic hybridization. Mol Gen Genet 157:291–296

Medgyesy P, Feyes E, Maliga P (1985) Interspecific chloroplast recombination in a *Nicotiana* somatic hybrid. PNAS 82:6960–6964

Medgyesy P, Pay A, Marton L (1986) Transmission of parental chloroplasts in *Nicotiana*. Mol Gen Genet 204:195–198

Melchers G (1977) Plant hybrids by fusion of protoplasts. In: Beers RF, Bassett EG (eds) Recombinant molecules: impact on science and society. Raven, New York, pp 209–227

Menczel L, Galiba G, Nagy F, Maliga P (1982) Effect of radiation dosage on efficiency of chloroplast transfer by protoplast fusion in *Nicotiana*. Genetics 100:487–495

Morgan A, Maliga P (1987) Rapid chloroplast segregation and recombination of mitochondrial DNA in *Brassica* cybrids. Mol Gen Genet 209:240–246

Mouras A, Salessess G, Lutz A (1978) Sur l'utilisation de protoplastes en cytologie: amélioration d'une méthode récente en vue de l'identification des chromosomes mitotiques des genres *Nicotiana* et *Prunus*. Caryologia 31:117–127

Mouras A, Wildenstein C, Salesses G (1986) Analysis of karyotype and C-banding pattern of *N. plumbaginifolia* using two techniques. Genetica 68:197–203

Mouras A, Saul MW, Essad S, Potrykus I (1987) Localization by in situ hybridization of a low copy chimaeric resistance gene introduced into plants by direct gene transfer. Mol Gen Genet 207:204–209

Negrutiu I, Jacobs M, Cattoir-Reynaerts A (1984) Progress in cellular engineering of plants: biochemical and genetic assessment of selectable markers from cultured cells. Plant Mol Biol 3:289–302

Negrutiu I, De Brouwer D, Watts JW, Sidorov VI, Dirks R, Jacobs M (1986) Fusion of plant protoplasts: a study using auxotrophic mutants of *Nicotiana plumbaginifolia.* Theor Appl Genet 72:279–286

Negrutiu I, Mouras A, Horth M, Jacobs M (1987) Direct gene transfer to plants: present developments and some future prospectives. Plant Physiol Biochem 25:493–503

Nehls R (1978) The use of metabolic inhibitors for the selection of fusion products of higher plant protoplasts. Mol Gen Genet 166:117–118

Nicoll MF, Chapman GP, James DJ (1987) Endosperm responses to irradiated pollen in apples. Theor Appl Genet 74:508–515

O'Connell MA, Hanson MR (1985) Regeneration of somatic hybrid plants formed between *Lycopersicon aesculentum* and *Solanum rickii.* Theor Appl Genet 72:59–65

Pelletier G, Vedel F, Belliard G (1985) Cybrids in genetics and breeding. Hereditas Suppl 3:49–56

Pental D, Hamill DJ, Pirrie A, Cocking EC (1986) Somatic hybridization of *Nicotiana tabacum* and *Petunia hybrida.* Recovery of plants with *P. hybrida* nuclear genome and *Nicotiana tabacum* chloroplast genome. Mol Gen Genet 202:342–347

Poulsen C, Porath D, Sacristan MD, Melchers G (1980) Peptide mapping of the ribulose biphosphate carboxylase small subunit from the somatic hybrid of tomato and potato. Carlsberg Res Commun 45:249–267

Puite KJ, Van Wikselaar P, Verhoeven H (1985) Electrofusion, a simple and reproducible technique in somatic hybridization of *Nicotiana plumbaginifolia* mutants. Plant Cell Rep 4:274–276

Rothenberg M, Hanson MR (1987) Recombination between parental mitochondrial DNA following protoplast fusion can occur in a region which normally does not undergo intragenomic recombination in parental plants. Curr Genet 12:235–240

Saul MW, Potrykus I (1984) Species-specific repetitive DNA used to identify interspecific somatic hybrids. Plant Cell Rep 3:65–67

Schieder O, Gupta PP, Krumbiegel-Schroener G, Hein T, Steffen A (1985) Novel techniques in handling and manipulating cells. Hereditas Suppl 3:65–75

Schweiger HG, Dirk J, Koop HU, Kranz E, Neuhaus G, Spangenberg G, Wolff D (1987) Individual selection, culture, and manipulation of higher plant cells. Theor Appl Genet 73:769–783

Sidorov VA, Zubco MK, Kuchko AA, Komanitsky IK, Gleba YuYu (1987) Somatic hybridization in potato: use of gamma-irradiated protoplasts of *Solanum pinnatisectum* in genetic reconstruction. Theor Appl Genet 74:346–368

Singh G, Neckelmann N, Wallace DC (1987) Conformational mutations in human mitochondrial DNA. Nature 329:270–272

Somers DA, Narayanan KR, Kleinhofs A, Cooper-Blaud S, Cocking EC (1986) Immunological evidence for transfer of the barley nitrate reductase structural gene to *Nicotiana tabacum* by protoplast fusion. Mol Gen Genet 204:296–301

Spangenberg G, Schweiger HG (1986) Controlled electrofusion of different types of protoplasts including cell reconstitution in *Brassica napus.* Eur J Cell Biol 41:51–56

Spangenberg G, Neuhaus G, Schweiger HG (1986) Expression of foreign genes in a higher plant cell after electrofusion-mediated cell reconstitution of a microinjected karyoplast and a cytoplast. Eur J Cell Biol 42:236–238

Tempelaar MJ, Jones MGK (1985) Directed electrofusion between protoplasts with different responses in a mass function system. Plant Cell Rep 4:92–95

Ward M (1984) Fusion of plant protoplasts with animal cells. In: Beers RF, Bassett EG (eds) Cell fusion: gene transfer and transformation. Raven, New York, pp 189–197

Watts JW, King JM (1984) A simple method for large-scale electrofusion and culture of plant protoplasts. Biosci Rep 4:335–342

Ye J, Hauptmann RM, Smith AG, Widholm JM (1987) Selection of a *Nicotiana plumbaginifolia* universal hybridizer and its use in intergeneric somatic hybrid formation. Mol Gen Genet 208:474–480

Zuckerman SH, Gillespie FP, Solus JF, Rybczynski R, Eisenstadt JM (1986) Mitochondrial protein synthesis in interspecific somatic cell hybrids. Somatic Cell Mol Genet 12:449–458

III.4 Somatic Hybridization in Tomato[4]

MAUREEN R. HANSON[1], MARY A. O'CONNELL[2], and CHARLES VIDAIR[3]

1 Introduction

The reasons for subjecting the tomato cultivar, *Lycopersicon esculentum*, to numerous cell and tissue culture studies have been twofold: agronomic improvement of this important crop species and exploitation of tomato's abundant classical genetics for basic research purposes. Several years ago we began experiments utilizing protoplasts of *Lycopersicon* species, with the goals of developing methods to culture and regenerate plants, influencing transmission of genomes following cell fusion, and genetically analyzing plants resulting from protoplast or fusion product regeneration. At the time, little information was available about the requirements for successful culture and regeneration of members of the tomato genus. We looked to the facile protoplast culture and fusion of other Solanaceous species, *Nicotiana* and *Petunia*, as guides, as well as the arduous but effective methods for potato protoplasts developed by Shepard et al. (1980). Here we will summarize our knowledge of *Lycopersicon* protoplast culture and fusion and the analysis of regenerated plants.

In our experience, protoplast culture of the tomato cultivar is greatly affected by local environmental conditions for plant growth, as well as by genotype. Comparisons of protocols from different laboratories are thus difficult. Several other laboratories have also described regeneration of tomato protoplasts (Koblitz and Koblitz 1982; Morgan and Cocking 1982; Shahin 1984, 1985; Niedz et al. 1985) or production of somatic hybrid plants (Melchers et al. 1978; Shepard et al. 1983; Zapata and Sink 1981; Tabaeizadeh et al. 1985; Handley et al. 1986; Adams and Quiros 1985; Sink and Reynolds 1986) (also see Chap. III.5, this Vol.).

2 Protoplast Isolation

Growth conditions of the source plants are critical for protoplast yield and viability. For example, leaves watered with $1/10 \times$MS salts (Murashige and Skoog 1962)

[1] Section of Genetics and Development, Cornell University, Biotechnology Building, Ithaca, N.Y. 14853, USA
[2] Dept. Agronomy, Horticulture and PGEL, Box 3GL, New Mexico State Univ., Las Cruces, New Mexico 88003, USA
[3] Dept. of Radiation Oncology, CED-200, Univ. California San Francisco, San Francisco, CA 94143
[4] A portion of this article has been derived (with permission) from one previously published in the *Iowa State Journal for Research* (Hanson et al. 1987). The literature review for this article was completed in 1986

Biotechnology in Agriculture and Forestry, Vol. 8
Plant Protoplasts and Genetic Engineering I (ed. by Y. P. S. Bajaj)

Table 1. Media composition. (O'Connell and Hanson 1985)

	JSCL 2.5	JSR[a]	JSC-12	TR 1	N 13
Macroelements (g/l)					
KNO_3	7.6	1.9	1.9	1.9	0.41
$CaCl_2-2H_2O$	1.76	0.44	0.44	0.44	–
$MgSO_4-7H_2O$	1.48	0.37	0.37	0.37	0.55
KH_2PO_4	0.68	0.17	0.17	0.17	0.14
NH_4	–	0.028	0.11	–	–
$Ca(NO_3)-4H_2O$	–	–	–	–	0.96
$(NH_4)_2SO_4$	–	–	–	–	0.14
KCl	–	–	–	–	0.0027
NH_4NO_3	–	–	–	1.65	–
Microelements (mg/l)					
Na_2ethylenediaminetetra-acetic acid	18.5	18.5	18.5	15	15
$FeSO_4-7H_2O$	13.9	13.9	13.9	11	11
H_3BO_3	3.1	3.1	3.1	6	6
$MnCl_2-4H_2O$	9.9	9.9	9.9	–	–
$ZnSo_4-zH_2O$	4.6	4.6	4.6	4	4
Kl	0.42	0.42	0.42	0.8	0.8
$Na_2MoO_4-2H_2O$	0.13	0.13	0.13	0.25	0.25
$CuSO_4-5H_2O$	0.013	0.013	0.13	0.025	0.025
$CoSO_4-7H_2O$	0.015	0.15	0.015	–	–
$MnSO_4-H_2O$	–	–	–	18	18
$CoCl_2-6H_2O$	–	–	–	0.025	0.025
Organics (mg/l)					
Thiamine	0.05	0.5	0.5	5	2
Glycine	2	2	2	–	–
Nicotinic acid	5	5	5	3	0.5
Pyridoxine	–	–	–	0.5	–
Folic acid	0.5	0.5	0.5	–	–
Biotin	0.5	0.5	0.5	–	–
Casein hydrolysate	50	100	100	–	–
Myo-inositol	4400	–	100	100	1000
Adenine-SO_4	–	–	40	–	–
Sugars (g/l)					
Sucrose	68	34	4.7	–	5
Mannitol	4.4	8.8	109.2	–	–
Sorbitol	4.4	–	–	–	–
Xylitol	3.8	–	–	–	–
Glucose	–	–	–	30	–
Other (mg/l)					
6-Benzyladenine	0.4	0.2	–	–	–
Zeatin	–	–	2	1	–
2,4-Dichlorophenoxy-acetic acid	0.6	–	–	–	–
Indole-3-acetic acid-phenylalanine	–	–	0.1	0.1[7]	–
2-(N-morphilino)ethane sulfonic acid	–	–	970	–	–
Phytagar	–	–	7000	7000	7000
pH	5.6	5.6	5.8	6.0	6.0

[a] 1×R salts is a stock solution of the macro and micro elements in JSR

Table 2. Leaf mesophyll protoplast yields from 12 consecutive experiments. Red Cherry leaves were peeled to cover 3 – 4 60-mm dishes of enzyme solution

Experiment date	Protoplast yield
12/16/83	0.81×10^6
12/20/83	1.79×10^6
12/22/83	1.10×10^6
12/28/83	1.66×10^6
1/4/84	2.10×10^6
1/5/84	1.82×10^6
1/9/84	1.79×10^6
1/11/84	1.78×10^6
1/16/84	2.56×10^6
1/18/84	2.19×10^6
1/26/84	1.28×10^6
1/30/84	2.44×10^6

released few protoplasts, while leaves from plants watered with Nitsch (1965) salts were well digested. A mixture of 1 : 10 incandescent to fluorescent light (12 h day, 1500 – 2000 ft candles) in growth chambers was found superior to 100% fluorescent light. Our plants were usually maintained at 22 °C day, 18 °C night at 60 – 80% relative humidity and watered with Nitsch (1965) salts, modified so that $(NH_4)S_2O_4$ was reduced to 0.1 mM (see Table 1). Greenhouse-grown plants were unsuitable as protoplast donors due to high variability in culture response and frequent systemic infections. Variability in protoplast yield depending on environmental conditions of the source plant have also been noted by Cassells and Barlass (1976, 1978). By defining growth conditions of plants in vermiculite (O'Connell and Hanson 1986) or shoot cultures in artificial media (Hosticka and Hanson 1984; Shahin 1984), a reproducible source of leaf protoplasts can be obtained. Protoplast yields from 12 consecutive experiments are listed in Table 2.

Suspension cultures are generally more recalcitrant to digestion than leaf tissue grown under appropriate conditions. Rapidly growing, recently transferred suspension cultures are essential for good digestion. By using large amounts of starting material, satisfactory amounts of suspension culture protoplasts can be obtained even if only a small percentage of cells are digested.

We found *Lycopersicon* tissues to be, in general, more sensitive to toxic enzyme effects than other Solanaceous species such as tobacco or petunia. Desalting the commercial enzyme preparations could improve protoplast yield and appearance from certain plant material. However, we were able to identify commercial enzyme preparations and incubation conditions which resulted in good protoplast yield and viability (Sect. 7).

3 Protoplast Culture

For culture of mesophyll protoplasts of tomato cultivars, media were based on those of Shepard (1980), with modification as described (Table 1; O'Connell and

Table 3. Irradiated *L. pennellii* suspension culture protoplasts can act as feeders for Red Cherry leaf protoplasts. Presence of feeder protoplasts has caused colony formation by Red Cherry protoplasts to be independent of their density. Irradiated protoplasts cultured separately as controls do not divide

Density of Red Cherry leaf protoplast/ml	Density of irradiated *L. pennellii* protoplasts/ml	Average number of colonies after 12 days (in 5 random fields)
500	0	0
1000	0	0
5000	0	5.8
500	80000	6.0
1000	80000	14.2
5000	80000	10.2

Hanson 1985; Vidair and Hanson 1985). Two key factors in successful cultivar protoplast culture were low NH_4^+ ion concentrations in initial media and rapid dilution of protoplasts.

In addition to tomato, we have regenerated protoplasts from *L. pennellii*, the F1 hybrid of *L. esculentum* × *L. pennellii*, and L2, a *L. peruvianum* × *L. esculentum* hybrid produced by embryo culture (Thomas and Pratt 1981). All three of these genotypes respond much more readily than the cultivar in explant and protoplast culture to a variety of media. Ease of culturing these genotypes is comparable to many *Nicotiana* and *Petunia* lines, while tomato cultivars are substantially more difficult to handle.

Developing from L2 a line sexually compatible with *L. esculentum* cultivars but retaining good culturability, as proposed by Thomas and Pratt (1981), would indeed be a boon to research requiring the use of protoplasts. However, perhaps such a line could be more simply developed from the *L. esculentum* × *L. pennellii* F1 hybrid, since sexual crosses are more readily performed with this genotype.

Our *Lycopersicon* protoplast culture experiments were undertaken as a prelude to fusion studies. While performing control co-cultivation experiments with *L. esculentum* mesophyll protoplasts and a potential fusion partner, protoplasts from a hormone-autotrophic *Agrobacterium*-transformed *L. pennellii* line, it was noticed that the mesophyll protoplasts could divide in hormone-free medium containing transformed protoplasts (Table 3). Utilizing this observation, Vidair and Hanson (1985) described a quantitative assay to characterize transformed lines' phenotypes based on their ability to replace the auxin and cytokinin requirement of normal tomato mesophyll protoplasts. Different lines differed in their ability to feed normal tomato protoplasts. This work also suggested that nurse feeder protoplasts from transformed lines might find a use as a protoplast culture method for recalcitrant tomato cultivars. Further experiments would be required to examine whether tomato cultivar protoplast-derived colonies induced by transformed feeder protoplasts have greater or lesser regenerative capacity in comparison to those induced by commercial preparations of hormones provided in media.

4 Regeneration of Plants

4.1 Explant Culture

As a simple way to gain some information concerning appropriate regeneration media for *Lycopersicon* cells, we examined the response of hypocotyl, leaf, and stem explants of various genotypes on media, a large number of which had been tested on tomato previously (Padmanabhan et al. 1974; Behki and Lesley 1976; Kartha et al. 1976; Tal et al. 1977; Herman and Haas 1978; Ohki et al. 1978; Hangarter et al. 1980; Frankenberger et al. 1981; Kut and Evans 1982; Locy 1983). We observed that the cultivated tomato, *L. esculentum*, is a particularly poor species in culture compared to many of its wild relatives. Of those tested, most wild *L. pennellii*, *L. hirsutum*, and *L. peruvianum* accessions produced abundant callus and prolific shoots on a range of typical plant culture media. In contrast, most of these media resulted in slow-growing callus and few shoots when tomato cultivars were examined. Among the tomato cultivars, Tiny Tim proved to produce the most abundant, friable, and rapidly growing callus (Hosticka and Hanson 1982), while Red Cherry was superior to others tested in number of shoots formed from leaf explants (Hanson 1981).

Effective media for shoot formation from tomato cultivar and wild species tissue explants are either MS505 (MS medium containing 9.2 μM zeatin; B. Thomas pers. comm.), MST5 (MS medium containing 0.11 μM IAA and 4.6 μM zeatin) or a MS medium containing 10 μM IAA-L-phenylalanine and 4.6 μM zeatin (Hangarter et al. 1980; Hanson 1982), the latter two containing 2% glucose instead of sucrose. Our preferred callus media are the 2D/1P medium of Thomas and Pratt (1981) supplemented with 10% coconut milk or a MS medium containing 10 μM IAA-L-alanine and 4.6 μM zeatin (Hangarter et al. 1980; Hanson 1982; Hosticka and Hanson 1982). These callus and shooting media were not the ones ultimately chosen for protoplast-derived calli (O'Connell and Hanson 1985, 1986; Vidair and Hanson 1985; Sect. 7). However, these media may find use in transformation experiments involving *Lycopersicon* species, now that infection and transformation of leaf explant cells of Solanaceous species by *Agrobacterium* strains carrying antibiotic resistance markers on Ti plasmids have been described (Horsch et al. 1985).

4.2 Phenotype and Genotypes of Protoplast-Derived Tomato Plants

Thirteen different cultivars were induced to form callus from protoplasts. Most of the regenerated cultivar plants obtained were derived from UC82B and Petoseed No. 46 leaf protoplasts (O'Connell and Hanson 1985). Some variations in leaf shape or growth habit were noted, but studies have not been carried out with these plants to determine whether they carry heritable genetic alterations, as were reported for explant regenerates (Evans and Sharp 1983). However, a number of protoplast-derived plants of the *L. esculentum* × *L. pennellii* F1 hybrid were examined more thoroughly. One such plant exhibited variegation typical of tissue sorting-out for a plastid mutation (Hosticka and Hanson 1984). Three sets of pro-

Fig. 1. Increasing roundness of leaves in protoplast regenerates of *L. esculentum* Red Cherry × *L. pennellii* F1 hybrid associated with increasing ploidy. *2N* leaf from diploid protoplast donor plant; *4N* tetraploid protoplast regenerate; *8N* octoploid protoplast regenerate. (O'Connell et al. 1986)

toplast-derived plants, obtained from three different experiments, appeared to have much rounder leaves than the parents. It was observed that the round shape was related to ploidy, with roundness increasing from diploid to tetraploid to octoploid (Fig. 1; O'Connell et al. 1986). Of 100 plants, 89 regenerated from protoplasts were tetraploid or octoploid, while only of 23 F1 hybrid plants regenerated from leaf slices on MS505 were tetraploid (O'Connell et al. 1986). Either spontaneous fusion or doubling during protoplast culture evidently resulted in plants of higher ploidy. The relatively low number of non-diploid plants from leaf slices indicates that MS505 may be a good choice for regeneration of *Agrobacterium*-transformed explants.

The genetic stability of the regenerated F1 hybrids was examined, to the limited degree possible, by testing for isozyme loci characteristic of the parental plant. No variability at the tested loci was detected (O'Connell et al. 1986).

5 Protoplast Fusion

5.1 Optimum Fusion Protocol

Our laboratory compared electrical and chemical fusion methods for *Lycopersicon* protoplasts. Using a home-made electrical fusion device, W. H. Loh (unpubl.) typically obtained fusion rates of approximately 5% (Fig. 2). A number of published chemical fusion methods were tried in the course of these studies, such at the methods of Gamborg et al. (1981) Kao and Michayluk (1974), and

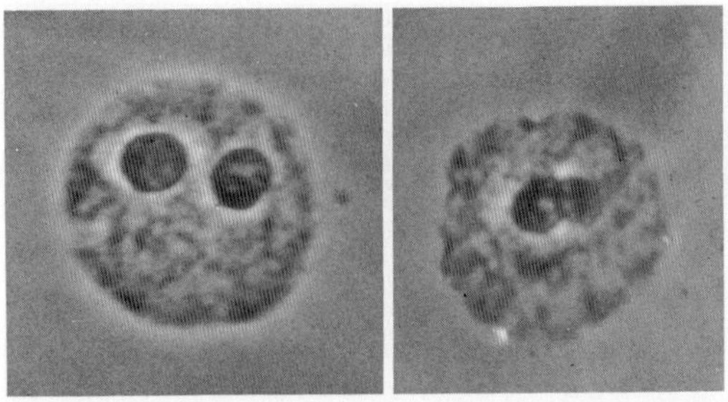

A B

Fig. 2. Fixed and stained *Lycopersicon* protoplasts following electrical fusion (W.H. Loh and M. Hanson unpubl.). **A** Two nuclei within protoplast after electrical treatment. Two nuclei in the same protoplast were seen in approximately 5% of protoplasts after fusion treatment, rarely in control untreated protoplasts. **B** Following electrical treatment, two nuclei appear to be fusing in a protoplast

Medgyesy et al. (1980), along with some modifications to these methods. We were successful in observing fusion products using fluorescent microscopy only when we had modified the methods of Medgyesy et al. (1980) to reduce the rate of dilution of the fusogen solutions (O'Connell and Hanson 1985). The critical difference among the published fusion procedures is the incorporation of a washing step with a solution containing NaCl, W5, prior to the addition of the fusion solutions (Medgyesy et al. 1980). Because comparable or better rates could be obtained by chemical fusion than those achieved with electrical fusion, all of our somatic hybrid calli and plants produced thus far have been derived following a polyethylene glycol fusion method (Sect. 7).

Fusion frequency can be monitored by fluorescence microscopy by determining the number of cells exhibiting red autofluorescence of chlorophyll from parental mesophyll protoplasts as well as green fluorescence from fluorescein isothiocyanate provided to suspension protoplasts. We determined the number of doubly labeled protoplasts 24 h after fusion following transfer of the viable released cells to fresh media (Sect. 7.2) and observed up to 25% doubly fluorescent protoplasts in randomly selected microscope fields. Only those fusion cultures in which fusion frequencies of 10% or more were observed were cultured further. It should be realized that a great many protoplasts remain attached to the fusion culture dish and are not cultured further or included in the calculation of fusion frequency. The protoplasts which remain attached to the dish appear damaged and unresponsive to the culture medium.

The protoplasts from the fusion experiments are cultured essentially as described in Sect. 7.2. Regeneration frequencies among somatic hybrid clones are dependent on the genotype of the somatic hybrid as well as the culture media (Sect. 5.2.2). In those constructions where regeneration was possible, virtually all of the somatic hybrid callus cultures regenerated shoots on the media described in Sect. 7.2. For example, in three separate fusion combinations between three different tomato cultivars and a regenerable cell line of *L. pennellii*, an average shoot regeneration frequency of 79% was obtained for the somatic hybrid calli cultured (Early Girl, 12/18; UC82b, 9/13; Manapal, 4/4) (O'Connell and Hanson unpubl.).

5.2 Cell Lines and Plants Arising from Protoplast Fusion

Nuclear loci were studied by examining isozyme markers. Electrophoretic mobilities of isozymes coded for by parental lines were known to differ at a number of

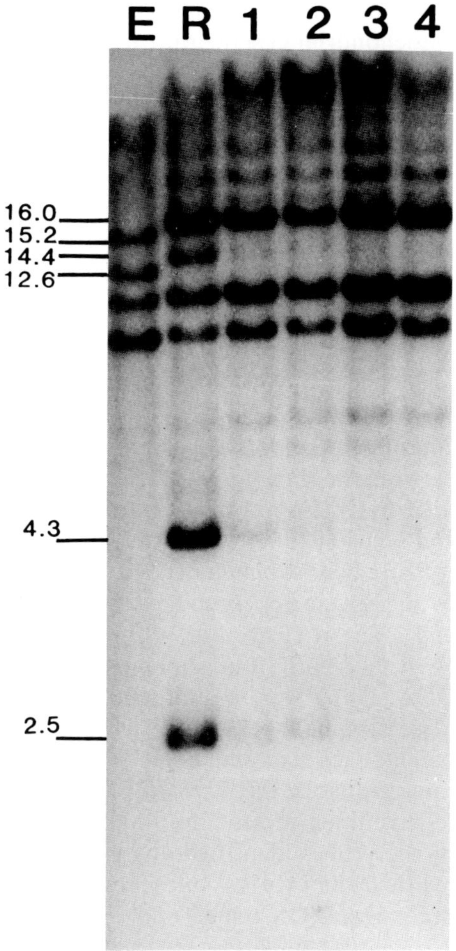

Fig. 3. Identification of mitochondrial ribosomal RNA genes in somatic hybrid plants. Lane *E* contains DNA from *L. esculentum*, *R* contain DNA from *S. rickii*. Lanes *1–4* contains DNA from somatic hybrid plants no. 4, 5, 6, and 7, respectively. The Southern blot of the agarose gel containing Nrul-digested DNA was probed with a labeled soybean mitochondrial DNA clone containing 18S and 5S mitochondrial rDNA. (O'Connell and Hanson 1986)

loci (Tanksley and Rick 1980). Somatic hybrids carrying chromosomes from both parents would be expected to exhibit isozymic forms characteristic of the fusion partners (Fig. 3). A dimeric isozyme should also be present as a third, heterodimeric form if both alleles are present.

Organelle genomes were studied by hybridizing labeled chloroplast DNA (Hanson et al. 1986) or labeled mitochondrial DNA (mtDNA) clones (McClean and Hanson 1986) to total DNA preparations. Because chloroplast genomes in somatic hybrids have been found to recombine only rarely (Clark et al. 1985; Medgyesy et al. 1985), the parental source of the chloroplast genome in hybrid lines could be effectively determined by observing at least one parental-specific fragment. However, because mitochondrial genomes in fusion products are known to recombine (Belliard et al. 1979; Boeshore et al. 1983; Rothenberg et al. 1985), mtDNA probes can only give information about a limited portion of the somatic hybrid mitochondrial genome. Finding restriction fragments characteristic of one parent in a somatic hybrid line does not preclude the presence of mitochondrial genetic information from the other parent. Only thorough examination of a somatic hybrid's mtDNA can reveal whether or not recombination has occurred (Hanson 1983).

5.2.1 L. esculentum + Solanum rickii

S. rickii was chosen as a fusion partner because it is one of the *Solanum* species most closely related to *Lycopersicon* (Rick 1979). Most of the calli arising from a fusion of San Marzano tomato mesophyll protoplasts and suspension culture protoplasts of *S. rickii* LA1974 rapidly regenerated shoots typical of *S. rickii*, an easily regenerable wild species. One callus did not regenerate immediately and was carried through transfers on shoot-inducing media for 11 months, at which time a number of shoots appeared. These shoots exhibited morphologies intermediate between those of the two fusion partners, and their somatic hybrid origin was verified by finding both parental phosphoglumutase isozymes present (O'Connell and Hanson 1986). At least some of the variation in leaf shape of the individual somatic hybrid plants derived from the same callus is likely to be due to ploidy, because some appear to be tetraploid and others approach hexaploidy (O'Connell and Hanson 1986).

Analyses of four shoots revealed that they all carried *S. rickii* chloroplast DNA. Using soybean mitochondrial ribosomal RNA and maize cytochrome oxidase subunit II gene probes, mtDNA fragments characteristic of only the *S. rickii* parental genome were detected. However, not all of the fragments hybridizing to these clones in the *S. rickii* parent were present in the somatic hybrids, and the relative intensities of the fragments were also different (Fig. 3). Although none of the probes identified any *L. esculentum* mtDNA fragments in these somatic hybrids, the hybrids' mitochondrial genomes are clearly unique with respect to the parents (O'Connell and Hanson 1986).

5.2.2 L. esculentum + L. pennellii

Most of our fusion experiments have utilized *L. pennellii* LA716, a wild species which produces rapidly growing suspension cultures, regenerates plants readily, can sexually hybridize with cultivars when used as the pollen parent, but carries a mitochondrial genome that has significantly diverged from *L. esculentum* (McClean and Hanson 1986). Using these two fusion parents, experiments were conducted with the ultimate goal of creating (1) cybrids carrying the *L. esculentum* nuclear genome and organelle genetic information from *L. pennellii* and (2) asymmetric somatic hybrids carrying some but not all nuclear information from *L. pennellii* in a background of an intact *L. esculentum* nucleus. We based our experimental strategies on ones which had been successful in creating asymmetric hybrids and cybrids in *Nicotiana* (Gupta et al. 1982; Menczel et al. 1982), and the results obtained, however, were quite different from previously published results with *Nicotiana*.

In an early set of fusion experiments, either UC82B or Petoseed No. 46 leaf protoplasts were fused with protoplasts from an old non-regenerating suspension culture of *L. pennellii.* After either unirradiated or 1.5 krad-treated *L. pennellii* protoplasts were used as a fusion partner, screening of subsequent calli for both parental isozymic forms of phosphoglucoisomerase isozyme indicated that 70% were of fusion product origin. This high percentage of somatic hybrid calli was probably due to enhanced growth and division frequency of fused cells relative to that of parental lines.

None of over 100 such calli regenerated shoots. In this experiment, lack of regeneration of one parental line was apparently dominant in the somatic hybrids, unlike findings with *Nicotiana* fusions (Maliga et al. 1977; Glimelius and Bonnett 1981; Gleba and Evans 1983). We have since shown that *L. esculentum + L. pennellii* somatic hybrids can be regenerated into plants provided that a regenerating suspension culture parent is used (O'Connell and Hanson 1987).

Of a sample of nine of the non-regenerating somatic hybrid calli examined for parental chloroplast genome, four carried one or the other parental genome, while five exhibited restriction fragments characteristic of both. When limited portions of the mitochondrial genomes were examined with a set of probes known to distinguish the two parents (McClean and Hanson 1986), only *L. pennellii*-specific fragments were detected in eight of the nine lines sampled. These data on organelle genomes are consistent with other observations of organelle genome sorting-out following protoplast fusion (Belliard et al. 1979; Menczel et al. 1982; Boeshore et al. 1983; Clark et al. 1985).

The somatic hybrid calli described by O'Connell and Hanson (1985) were screened and assigned as somatic hybrid only on the basis of their *Pgi*-1 isozymes. Because both parental cells were irradiated in some of our experiments, we extended our analysis to three different isozyme loci to find whether irradiation might have affected transmission of nuclear genes. When a sample of calli from *L. esculentum* + *L. pennellii* fusion experiments was analyzed, it was discovered that not all calli were hybrid at all loci. The *L. pennellii* parent was more likely to be represented in the hybrid callus. Twelve of 37 analyzed somatic hybrid calli arising following fusion of parental protoplasts 2.5 krad-irradiated exhibited *L. pennellii*-type isozymes of *Got*-2, *Pgi-1, and Pgm*-2. Seventeen of the 37 calli were hybrid at one locus, but *L. pennellii*-type at the other two loci. Four were hybrid at two loci and *L. pennellii* at one, and only four were hybrid at all three loci (O'Connell et al. 1986).

Preferential presence of the *L. pennellii* allele was detected even in calli sampled from an experiment in which neither parent was irradiated (O'Connell et al. 1986). Eleven of 38 calli analyzed were hybrid at one locus and *L. pennellii*-type at two others. Nineteen of the 38 were hybrid at two loci and *L. pennellii* at one. Eight calli were hybrid at all three loci. How these genotypes arose remains to be investigated. Now that Bernatsky and Tanksley (1986) have developed a set of cDNA markers for tomato chromosomes, the chromosomal composition of future tomato cell culture and fusion regenerates can be more extensively characterized.

5.2.3 *L. esculentum* + *L. pennellii* Tumor Line

In an attempt to produce an asymmetric somatic hybrid, we fused 10 krad-irradiated protoplasts from an *Agrobacterium* A6-transformed *L. pennellii* tumor line with Red Cherry mesophyll protoplasts (C. Vidair and M. Hanson unpubl.). The fused protoplasts were grown in liquid Shepard's (1980) CL medium without hormones for 1 day. The next day liquid Shepard's (1980) R medium without hormones was added. Seven days later, the medium was replaced with Shepard's (1981) liquid C-medium without hormones. After 5 more days, colonies scraped from the dish as well as free-floating colonies were put in a centrifuge tube and spun at 500 *g* for 5 min. The cells were resuspended in C-medium without hormones containing 0.5% low-melting agarose and spread onto 100 mm dishes containing hardened C-medium without hormones. After 6–8 weeks, about 200 calli appeared. Three calli were selected for further study.

Because these three calli could grow on hormone-free medium, they evidently carried T-DNA specifying hormone autotrophy. When phosphoglucoisomerase isozymes were examined, the calli were found to carry the two homodimeric forms specified by parental alleles, as well as the heterodimer. Thus, these three calli clearly contained cells of somatic hybrid origin. Because the fusion frequency was low in this experiment (a few percent), finding somatic hybrid cells in a small sam-

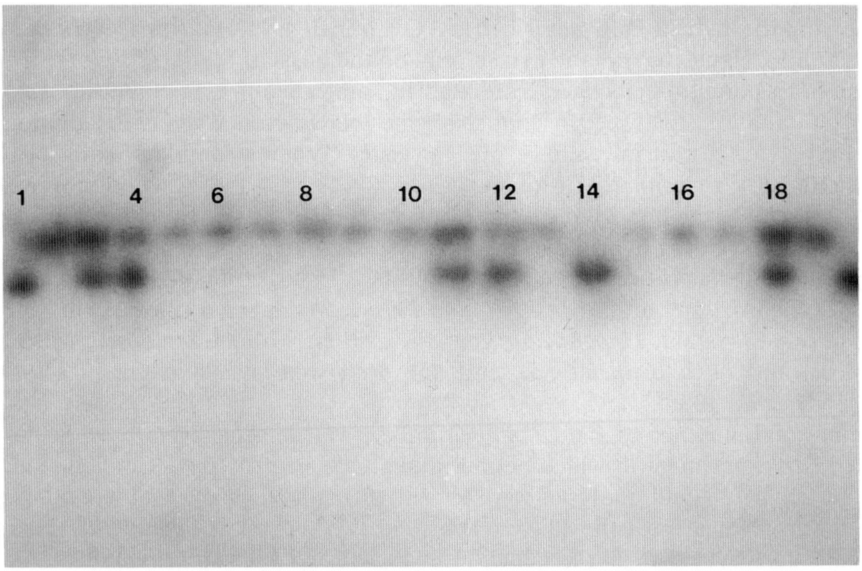

Fig. 4. Isozyme gel stained for phosphoglucomutase activity: lane 1, *L. esculentum;* 2, *L. pennellii* tumor line; 3, F1 hybrid *L. esculentum* × *L. pennellii*; 4, mixture of *L. pennellii* and *L. esculentum* tissue; 5, *L. esculentum* + *L. pennellii* tumor somatic hybrid line 1(SH1); 6, *L. esculentum* + *L. pennellii* tumor somatic hybrid line 2 (SH2); 7, *L. esculentum* + *L. pennellii* tumor somatic hybrid line 3 (SH3); 8, SH3; 9, SH2; 10, SH1; 11, F1 hybrid; 12, mixture; 13, tumor; 14, *L. esculentum*; 15, SH1; 16, SH2; 17, SH3; 18, F1 hybrid; 19, tumor; 20, *L. esculentum* (C. Vidair, M. Hanson, M. O'Connell, L. Hosticka unpubl.)

ple was probably due to preferential growth relative to parental lines. The irradiation may have inhibited division of tumor protoplasts, and *L. esculentum* protoplasts tend to be damaged by fusion treatments.

Since the *L. pennellii* parent was irradiated at a high level, we expected that some of the *L. pennellii* alleles may not have been transferred to the fusion nucleus. However, when *Pgm*-2 isozymes were examined, we observed the form characteristic of the *L. pennellii* parent only – the *L. esculentum* form was missing (Fig. 4). Like the experiments with non-transformed *L. pennellii* fusion partners, the irradiated parent's allelic form was present instead of the non-irradiated parent's isozyme.

6 Conclusions and Prospects

At the time our fusion experiments were initiated, *Agrobacterium* lines carrying engineered T-DNA which could confer antibiotic resistances to plant cells were not available. Now that transformed antibiotic-resistant regenerable lines can be obtained by using engineered Ti plasmids, these lines can provide more suitable material for attempts to produce asymmetric hybrids. Our experiences to date indicate, however, that even 10 krad of irradiation of *L. pennellii* protoplasts, which prevents any cell division, is still insufficient to prevent transmission of markers present on several different chromosomes. Other methods for fragmentation of nuclear chromosomes may need to be utilized for transferring small regions of the nuclear genomes from wild *Lycopersicon* relatives to the tomato cultivar.

Among plants of economic importance, tomato is one of the more easily culturable species. Within the *Lycopersicon* genus are wild species which are as readily regenerated from protoplasts as model species such as *Nicotiana plumbaginofolia* or *Petunia hybrida*. Tomato is likely to find increasing use as a target for introduction of engineered genes. However, tissue explants (McCormick et al. 1986) rather than protoplasts are the method of choice for introduction of DNA via Ti plasmid-based vectors. Protoplasts will find use for transient gene expression studies, non-*Agrobacterium*-based DNA transformation, and for production of somatic hybrids and cybrids.

7 Protocols

7.1 Protoplast Isolation

Packed cells from rapidly growing suspension cultures were incubated for 8 h in 4–7 vol 1/10 × Nitsch salts, 10 nM MES, 0.5 M mannitol, 10 g/l cellulase RS, 5 g/l pectinol, 2.5 g/l Rhozyme HP150, pH 5.6 at 25C on a gyratory shaker. If fluorescent labeling was desired, 12.5 mg/l fluoresceinisothiocynate (FITC) was added to the incubation mixture. Protoplasts can be collected by filtering the digestion mixture through a 37-μm nylon mesh and pelleting at 500 *g* for 5 min. Alternatively, the protoplasts can be purified as described below.

The leaf protoplast isolation procedure used by O'Connell and Hanson (1985, 1986) was based on that of Shepard (1980). Young fully expanded leaves were surface-sterilized and placed upside-down on 1/4 Nitsch watering solution (Nitsch 1965) for 5 days at room temperature in the dark. The lower epidermis was peeled off and the peeled side exposed to 0.3 M sucrose, 5 mM MES, 10 mg/l casein hydrolysate, 1 mM polyvinylpyrrolidone, 1/4 × R salts, 2 g/l Pectinol Ac (Rohm and Haas), 15 g/l Cellulysin (Calbiochem), pH 5.6, for 18 h in the dark. The protoplasts were layered over 2.5 ml 15% Percoll (Pharmacia) in JS Rinse (0.3 M sucrose, 1 × R salts, 10 mg/l casein hydrolysate (ICN), pH 5.6) and centrifuged for 10 min at 500 *g*. The protoplasts that floated and those at the Percoll-enzyme interface were removed into a new tube, 7–10 vol of JS Rinse were added, and the mixture centrifuged at 500 *g* for 10 min. The floating protoplasts were collected and counted in a hemocytometer. In general, optimum plating density for unfused protoplasts ranges from 10^4 to 10^5. Higher initial densities of protoplasts must be used in protoplast fusion experiments, since substantial cell breakage and death occurs.

In the nurse culture experiments with tumor suspension protoplasts described by Vidair and Hanson (1985), leaf protoplasts from the Large Red Cherry cultivar were obtained by pretreating leaves from young plants upside down on sterile water overnight. Leaves were incubated peeled side down on an enzyme solution containing Nitsch salts, 9% mannitol, 15 mM MES, 0.7% Cellulysin, 0.1% Pectinol AC, pH 5.8 for 6–12 h, gently shaking in the dark. The incubation mixture was centrifuged for 5 min at 500 *g*. The pellet was resuspended in 1/10 × Nitsch salts, 10% sucrose, 20% Percoll and 15 mM MES, pH 5.8 before centrifuging at 500 *g* for 10 min. Intact protoplasts were collected from the top, mixed, with 9% mannitol, 1/10 × Nitsch salts, 15 mM MES, pH 5.8, and pelleted before suspending in culture medium. Protoplast yields are shown in Table 2. This isolation protocol was also found suitable for preparation of regenerable leaf protoplasts from *L. pennellii* and L2, a *L. peruvianum-L. esculentum* hybrid produced by embryo culture (Thomas and Pratt 1981).

7.2 Protoplast Culture and Regeneration

The media we used for protoplast culture and regeneration of *L. esculentum, S. rickii, L. pennellii*, and somatic hybrids are shown in Table 1 (taken from O'Connell and Hanson 1985). Alternative media appropriate for culture of *Agrobacterium*-transformed protoplasts or for culture of tomato leaf protoplasts with nurse tumor protoplasts can be found in Vidair and Hanson (1985). Observation of the specific protoplast type cultured is needed to determine the best time for transfer of cells to subsequent media, as such features as genotype and tissue source can affect response to culture.

In our experiments with *L. pennellii* LA716 or *S. rickii* suspension cultures and *L. esculentum* UC82B or Petoseed no. 46 leaves, protoplasts were initially plated in JSCL 2.5 medium. After 4 days, the cells were diluted 2:1 with a 1:1 mixture of JSCL 2.5 and JSR media. Three days later the culture was diluted 4:1 with the same 1:1 mixture. Five days later, 5 ml of the cultures were placed on solid JSC-12 medium and grown until the microcalli were 1–2 mm in diameter. Microcalli could then be individually transferred to a new JSC-12 plate. After illumination 12 h/day, 1500 lx, for about 1 month, calli that turned green and reached at least 0.5 cm in size were transferred to the TR-1 medium. After a month, the green calli were transferred to new TR-1 medium. As shoots became sufficiently large, they were transferred to N13 media in larger vessels such as Magenta (Chicago, IL) boxes for root induction. Rooted plantlets were acclimatized after planting in vermiculite and enclosure in plastic bags for a week.

7.3 Protoplast Fusion

Of the chemical fusion methods attempted, Medgyesy et al.'s (1980) procedure, which incorporates DMSO, proved to be the optimum after some modification. Parental protoplasts of 0.3 ml (4×10^5 were placed in 60-mm plastic Petri dishes for 5 min of settling; 0.5 ml of 30% polyethylene glycol (MW 6000, Koch-Light), 0.3 M glucose, and 50 mM CaCl was added slowly; 0.2 ml of W10 (Medgyesy et al. 1980) was added after 15 min). After 20 min, 1.5 ml JSCL 2.5 medium was added. Then 5 min later, the fluid above the surface-attached protoplasts was gently removed. Three ml of JSCL 2.5 medium was added and cultures incubated at 20 °C in the dark. Twenty-four hours later, the dishes were gently shaken to dislodge the protoplasts, and the cells were transferred to 60-mm dishes with the addition of 2 ml of JSC1 2.5 media. At this time it is possible to determine the fusion frequency among those cells which have been released from the bottom of the dish. We routinely observed between 10 and 25% doubly fluorescent protoplasts in randomly selected microscope fields.

Acknowledgements. Most of the work described here was carried out while the authors were at the Dept. of Biology, University of Virginia, Charlottesville. Excellent technical assistance was provided by Mary Ellen Wessinger, Christine Snyder, and Linda Hosticka. This research was supported by Agrigenetics Research Associates, the U.S.D.A. Competitive Grants Program (Genetic Mechanisms), and the McKnight Foundation.

References

Adams TL, Quiros CF (1985) Somatic hybridization between *Lycopersicon peruvianum* and *L. pennellii*: regenerating ability and antibiotic resistance as selection systems. Plant Sci 40:209–219

Behki RM, Lesley SM (1976) In vitro plant regeneration from leaf explants of *Lycopersicon esculentum* (tomato). Can J Bot 54:2409–2414

Belliard G, Vedel F, Pelletier G (1979) Mitochondrial recombination in cytoplasmic hybrids of *Nicotiana tabacum* by protoplast fusion. Nature 218:401–403

BernatskyR, Tanksley SD (1986) Towards a saturated linkage map in tomato based on isozyme and random cDNA sequences. Genetics 112:887–893

Boeshore ML Lifshitz I, Hanson MR, Izhar S (1983) Novel composition of mitochondrial genomes in *Petunia* somatic hybrids derived from cytoplasmic male sterile and fertile plants. Mol Gen Genet 190:459–467

Cassells AC, Barlass M (1976) Environmentally induced changes in the cell walls of tomato leaves in relation to cell and protoplast release. Physiol Plant 37:239–246

Cassells AC, Barlass M (1978) A method for the isolation of stable mesophyll protoplasts from tomato leaves throughout the year under standard conditions. Physiol Plant 42:236–242

Clark EM, Izhar S, Hanson MR (1985) Independent segregation of the plastic genome and cytoplasmic male sterility in *Petunia* somatic hybrids. Mol Gen Genet 199:440–445

Evans DA, Sharp WR (1983) Single gene mutations in tomato plants regenerated from tissue culture. Science 221:949–951

Frankenberger EA, Hasegawa PM, Tigcelaar EC (1981) Influence of environment and developmental state on the shoot-forming capacity of tomato genotypes. Z Pflanzenphysiol 102:221–232

Fromm M, Taylor LP, Walbot V (1985) Expression of genes transferred into monocot and dicot plant cells by electroporation. Proc Natl Acad Sci USA 82:5824–5828

Gamborg OL, Shyluk JP, Shahin EA (1981) Isolation, fusion and culture of plant protoplasts. In: Thorpe TA (ed) Plant tissue culture. Academic Press, New York, pp 115–153

Gleba YY, Evans DA (1983) Genetic analysis of somatic hybrid plants. In: Evans DA, Sharp WR, Ammirato PV, Yamada Y (eds) Handbook of plant cell culture Vol 1. MacMillan, New York, pp 322–357

Glimelius K, Bonnett HT (1981) Somatic hybridization in *Nicotiana*: restoration of photoautotrophy to an albino mutant with defective plastids. Planta 153:497–503

Gupta PP, Gupta M, Schieder O (1982) Correction of nitrate reductase defect in auxotrophic plant cells through protoplast-mediated intergeneric gene transfers. Mol Gen Genet 188:378–383

Handley LW, Nickels RL, Cameron MW, Moore PP, Sink KC (1986) Somatic hybrids plants between *Lycopersicon esculentum* and *Solanum lycopersicoides*. Theor Appl Genet 71:691–697

Hangarter RP, Peterson MD, Good NE (1980) Biological activities of indoleactylamino acids and their use as auxin in tissue culture. Plant Physiol 65:761–767

Hanson MR (1981) Regeneration of tomato plants in vitro from leaf and hypocotyl explants. Tomato Gen Coop Rep 32:15–16

Hanson MR (1982) Cell and tissue culture of *Lycopersicon*. In: Fujiwara A (ed) Plant Tissue Culture 1982. Maruzen, Tokyo, pp 193–194

Hanson MR (1983) Stability, variation and recombination in plant mitochondrial genomes via cell culture and somatic hybridization. Oxford Surv Plant Mol Cell Biol 1:33–52

Hanson MR, Boeshore ML, McClean PE, O'Connell MA, Nivison HT (1986) The isolation of mitochondria and mitochondrial DNA. Methods Enzymol 118:437–453

Hanson MR, O'Connell MA, Vidair C (1987) Regeneration of calli and plants following protoplast fusion in *Lycopersicon*. Iowa State J Res 61:445–458

Herman EB, Hass GJ (1978) Shoot formation in tissue cultures of *Lycopersicon esculentum* Mill. Z Pflanzenphysiol 89:467–470

Horsch RB, Fry JE, Hoffmann NL, Eichholtz D, Rogers SG, Fraley RT (1985) A simple and general method for transferring genes into plants. Science 227:1229–1231

Hosticka L, Hanson MR (1982) A tomato cultivar exhibiting rapid callus growth. Tomato Gen Coop Rep 34:9

Hosticka L, Hanson MR (1984) Induction of plastid mutations by nitrosomethyl urea. J Hered 75:242–246

Kao KN, Michayluk MR (1974) A method for high-frequency intergeneric fusion of plant protoplasts. Planta 115:355–367

Kartha K, Gamborg OL, Shyluk JP, Constabel F (1976) Morphogenetic investigations on in vitro leaf culture of tomato (*Lycopersicon esculentum* Mill. cv. Starfire) and high frequency plant regeneration. Z Pflanzenphysiol 77:292–301

Koblitz H, Koblitz D (1982) Experiments on tissue culture in the genus *Lycopersicon* Miller: mesophyll protoplast regeneration to plants in *Lycopersicon esculentum* cv. 'Nadja'. Plant Cell Rep 1:143–146

Kut SA, Evans DA (1982) Plant regeneration from cultured leaf explants of eight wild tomato species and two related *Solanum* species. In Vitro 18:593–598

Locy RD (1983) Callus formation and organogenesis by explants of six *Lycopersicon* species. Can J Bot 61:1072–1079

Maliga PG, Lazar G, Joo F, Nagy AH, Menczel L (1977) Restoration of morphogenic potential in *Nicotiana* by somatic hybridization. Mol Gen Genet 157:291–296

McClean PE, Hanson MR (1986) Mitochondrial DNA sequence divergence among *Lycopersicon* and related *Solanum* species. Genetics 112:649–667

McCormick S, Niedermeyer J, Fry J, Barnason A, Horsch R, Fraley R (1986) Leaf disc transformation of cultivated tomato (*L. esculentum*) using *Agrobacterium tumefaciens*. Plant Cell Rep 5:81–84

Medgyesy P, Menczel L, Maliga P (1980) The use of cytoplasmic streptomycin resistance; chloroplast transfer from *Nicotiana tabacum* into *Nicotiana sylvestris*, and isolation of their somatic hybrids. Mol Gen Genet 179:693–698

Medgyesy P, Fejes E, Maliga P (1985) Interspecific chloroplast recombination in a *Nicotiana* somatic hybrid. Proc Natl Acad Sci 82:6960–6964

Melchers G, Sacristan MD, Holder AA (1978) Somatic hybrid plants of potato and tomato regenerated from fused protoplasts. Carlsberg Res Commun 43:203–218

Menczel L, Galiba G, Nagy F, Maliga P (1982) Effect of radiation dosage on efficiency of chloroplast transfer by protoplast fusion in *Nicotiana*. Genetics 100:487–495

Morgan A, Cocking EC (1982) Plant regeneration from protoplasts of *Lycopersicon esculentum* Mill. Z Pflanzenphysiol 106:97–104

Murashige T, Skoog F (1962) A revised medium for rapid growth and bioassays with tobacco tissue cultures. Physiol Plant 15:473–497

Niedz RP, Ruter SM, Handley LW, Sink KC (1985) Plant regeneration from leaf protoplasts of six tomato cultivars. Plant Sci 39:199–204

Nitsch JP (1965) Deux espèces photopériodiques de jours courts: *Plumbago indica* L. et *Plumbago zeylanica*. Bull Soc Bot Fr 112:517–522

O'Connell MA, Hanson MR (1985) Somatic hybridization between *Lycopersicon esculentum* and *Lycopersicon pennellii*. Theor Appl Genet 70:1–12

O'Connell MA, Hanson MR (1986) Regeneration of somatic hybrid plants formed between *Lycopersicon esculentum* and *Solanum rickii*. Theor Appl Genet 72:59–65

O'Connell MA, Hosticka LR, Hanson MR (1986) Use of isozymes to examine genome stability in cultured *Lycopersicon*. Plant Cell Rept 5:276–279

O'Connell MA, Hanson MR (1987) Regeneration of somatic hybrids formed between *Lycopersicon esculautum* and *L. pennellii* Theor Appl Genet 75:83–89

Ohki S, Bigot C, Mousseau J (1978) Analysis of shoot-forming capacity in vitro in two lines of tomato (*Lycopersicon esculentum* Mill.) and their hybrids. Plant Cell Physiol 19:27–42

Padmanabhan V, Paddock EF, Sharp WR (1974) Plantlet formation from *Lycopersicon esculentum* leaf callus. Can J Bot 52:1429–1432

Rick CM (1979) Biosystematic studies in *Lycopersicon* and closely related species of *Solanum*. In: Hawkes JG, Lester RN, Skelding AD (eds) The biology and taxonomy of the Solanaceae. Academic Press, London, pp 667–677

Rothenberg M, Boeshore ML, Hanson MR, Izhar S (1985) Intergenomic recombination of mitochondrial genomes in a somatic hybrid plant. Curr Genet 9:615–618

Shahin EA (1984) Isolation and culture of protoplasts: tomato. In: Vasil IK (ed) In: Cell Culture and Somatic Cell Genetics of Plants Vol 1. Academic Press, New York, pp 370–380

Shahin EA (1985) Totipotency of tomato protoplasts. Theor Appl Genet 69:235–240

Shephard JF (1980) Abscisic acid-enhanced shoot initiation in protoplast-derived calli of potato. Plant Sci Lett 18:327–333

Shepard JF, Bidney D, Shahin E (1980) Potato protoplasts in crop improvement. Science 208:17–24

Shepard JF, Bidney D, Barsby T, Kemble R (1983) Genetic transfer in plants through interspecific protoplast fusion. Science 219:683–688

Sink KC, Reynolds JF (1986) Tomato (*Lycopersicon esculentum*). In: Bajaj YPS (ed) Biotechnology in agriculture and forestry 2. Crops I. Springer, Berlin Heidelberg New York, pp 319–344

Tabaeizadeh Z, Perennes C, Bergounioux C (1985) Increasing the variability of *Lycopersicon peruvianum* Mill. by protoplast fusion with *Petunia hybrida* L. Plant Cell Rep 4:7–11

Tal M, Dehan K, Heikin H (1977) Morphogenetic potential of cultured leaf sections of cultivated and wild species of tomato. Ann Bot 41:937–941

Tanksley SD, Rick CM (1980) Isozymic gene linkage map of the tomato: applications in genetics and breeding. Theor Appl Genet 57:161–170

Thomas BR, Pratt B (1981) Efficient hybridization between *Lycopersicon esculentum* and *L. peruvianum* via embryo callus. Theor. Appl Genet 59:215–219

Vidair C, Hanson MR (1985) *Agrobacterium*-transformed tomato cells replace the hormone requirements for growth of tomato leaf protoplasts. Plant Sci 41:185–192

Zapata FJ, Sink KC (1981) Somatic embryogenesis from *Lycopersicon peruvianum* leaf mesophyll protoplasts. Theor Appl Genet 59:265–268

III.5 *Lycopersicon + Petunia* Somatic Hybrids

Z. TABAEIZADEH[1], C. BERGOUNIOUX[2], and Y. DEMARLY[3]

1 Introduction

Interspecific and intergeneric hybridization has been successfully employed for plant improvement. However, due to incompatibility barriers, many sexual crosses are not possible. Moreover, cytoplasmic genetic elements are known to be transmitted maternally in most higher plants (Gillham 1978). This limits, to some extent, the induction of variability in the cytoplasmic genome after crossing. With recent advancement in genetic engineering and biotechnology research (see Bajaj 1986), plant breeders are now able to combine classical and new technologies for plant improvement. Somatic hybridization overcomes the sexual incompatibility and uniparental inheritance of cytoplasmic traits, and provides a wide range of nuclear cytoplasmic genetic combinations.

We used protoplast fusion technology in order to transfer cytoplasmic male sterility (CMS) from *Petunia* to the genus *Lycopersicon*. This type of male sterility, which is found in 140 species (Lazer and Lersten 1972), does not exist in tomato. A cytoplasmic male sterile line could have a great economical impact on the production of hybrid seeds in tomato. The frequent production of plants with cytoplasmic male sterility, after interspecific cross-hybridization, indicates that CMS is due to an interaction between the nuclear genome of one species with the cytoplasmic genome of another species (Gerstel et al. 1980). Several lines of circumstantial evidence generally suggest that the mitochondrial genome is a possible site of determinants conditioning this trait (Levings and Pring 1976; Pring et al. 1982; Powling and Gliss 1983; for a review see Hanson and Conde 1985). Transfer of CMS by protoplast fusion has been demonstrated in *Petunia* (Izhar and Power 1979; Bunisset-Bergounioux and Perennes 1980), in *Nicotiana* (Belliard et al. 1977; Zelcer et al. 1978) and in *Brassica* (Pelletier et al. 1983). Research on the induction of cytoplasmic male sterility in the genus *Lycopersicon* was initiated by Rick (1959) when he grafted tomato on the cytoplasmic male sterile line of *Petunia*. However, his efforts and those of several others who took the same approach were unsuccessful. Somatic hybridization could be a rational strategy for the induction of this trait in tomato.

[1] Département des Sciences biologiques, Université du Québec à Montréal, Montréal, H3C 3P8, Canada.
[2] Laboratoire de Physiologie Végétale Moléculaire, Université de Paris-Sud, Orsay, 91405, France.
[3] Laboratoire d'Amélioration des Plantes, Université de Paris-Sud, Orsay, 91405, France.

Biotechnology in Agriculture and Forestry, Vol. 8
Plant Protoplasts and Genetic Engineering I (ed. by Y.P.S. Bajaj)

Table 1. Constituents and concentrations of growth regulators used in culture media for protoplast division, callus growth and plant regeneration in the genus *Petunia*

Species	Hormones for induction of division		Hormones for callus growth (mg/l)		Hormones for plant regeneration (mg/l)		Reference
P. hybrida	2,4-D	0.5	Kinetin	2.56	Kinetin	2.56	Frearson et al. (1973)
	NAA	1.5	IAA	4.00	IAA	4.00	
	BAP	1.0					
P. hybrida	2.4-D	1.4	2,4-D	1.4	NAA	2.0	Durand et al. (1976)
	BAP	0.4	BAP	0.4	BAP	0.5	
P. parodii	2,4-D	0.5	2,4-D	0.5	Kinetin	2.56	Power et al. (1976)
	NAA	1.5	NAA	1.5	IAA	4.00	
	BAP	1.0	BAP	1.0			
P. inflata	NAA	2.0	NAA	2.0	Kinetin	0.05	Power et al. (1976)
	BAP	0.5	BAP	0.5	or		
					Zeatin	1.00	
P. voilaceae	NAA	5.0	NAA	5.0	Zeatin	1.0	Power et al. (1976)
	BAP	1.0	BAP	1.0	or		
					BAP	1.0	
P. axillaris	2,4-D	1.0	2,4-D	1.0	Zeatin	1.0	Power et al. (1976)
	BAP	0.5	BAP	0.5			
P. hybrida × *P. parodii*	NAA	5.0	NAA	5.0	IAA	2.0	Power et al. (1976)
	BAP	0.5	BAP	0.5	BAP	1.0	

The techniques for protoplast culture and plant regeneration in *Petunia* have been developed for several years (Frearson et al. 1973; Durand et al. 1976; Power et al. 1976). The constituents and concentrations of growth regulators which have been successfully used for induction of protoplast division, callus growth and plant regeneration in this genus are summarized in Table 1. With respect to interspecific somatic hybridization in *Petunia*, interesting results have been produced by Power et al. (1979, 1980) and Bunisset-Bergounioux and Perennes (1980).

In contrast to *Petunia*, development of protoplast technology has been rather slow in *Lycopersicon*, although it was the first plant material used for the enzymatic isolation of protoplasts (Cocking 1960). Reports on plant regeneration from cultivated tomato protoplasts are quite recent. This is due to several problematic factors affecting protoplast manipulation, plant regeneration and somatic hybridization. Our data on somatic hybridization of *Lycopersicon peruvianum* (2n = 24) and *Petunia hybrida* (2n = 14) will be more meaningful with an overview of these factors.

2 Important Factors in Protoplast Culture and Somatic Hybridization of the Genus *Lycopersicon*

2.1 Source of Protoplasts: Growth Condition and Pretreatment

A survey of the published reports on tomato protoplasts reveals that substantial efforts have been made in testing different culture media and different genotypes. However, only few authors have investigated the effect of growth conditions and treatment of source plants on yield, viability and division of leaf mesophyll protoplasts. Cassells and Barlass (1976, 1978) recommended that plants be grown under low light intensity, to increase protoplast yield and viability. Increased levels of calcium pectate in plants grown under high light intensity were suggested to be a barrier to cell wall degradation. Tal and Watts (1979) obtained a greater yield of leaf mesophyll protoplasts when the source plants were grown under relatively high humidity. The effects of different photoperiod applied to source plants on subsequent protoplast division and callus formation were investigated in *Lycopersicon esculentum* and *L. peruvianum* (Tabaeizadeh et al. 1984). In *L. esculentum*, protoplasts isolated from plants grown under 16 h, daylength followed by a 9-h daylength period for 5 days produced more cell colonies than protoplasts obtained from plants kept only under a 16-h daylength period. This difference was not significant in *L. peruvianum*. In both species, incubation of whole plant at 4 °C in the dark for 15 h before protoplast isolation had a significant positive effect on cell colony formation. The type of watering solution has also been reported to influence the yield of protoplasts isolated from leaf mesophyll in different cultivars of *L. esculentum* (Hanson 1982). From several reports it is evident that the use of the first leaves of young (4–6 weeks old) plants is one of the initial determinants for protoplast viability from different species of *Lycopersicon* (Tabaeizadeh et al. 1984, 1985; Imanishi and Hiura 1983; Shahin 1985; Niedz et al. 1985; O'Connell and Hanson 1985). Precise differences in day and night temperatures do not seem to be important as they vary in different reports. In addition to leaf mesophyll, cell suspensions (O'Connell and Hanson 1985; Hassanpour-Estahbanati and Demarly 1986) and Calli (Morgan and Cocking 1982) have also been successfully used as source of protoplasts in different species of *Lycopersicon*.

2.2 Protoplast Culture, Induction of Division and Plant Regeneration

From all of the reports it is obvious that after isolation, tomato leaf mesophyll protoplasts need a dark period of 4–10 days followed by low light intensity (800–1500 lx). Even in those cases where tomato protoplasts were directly incubated under low light intensity, the whole plant (Shahin 1985) or detached leaves (O'Connell and Hanson 1985) were placed in the dark for a few days before protoplast isolation. The reports indicate that a temperature of 25–30 °C is optimum for initiation and maintenance of tomato protoplast division, as first suggested by Zapata et al. (1977).

Various basal culture media have been used, and due to the genetic diversity of different species or cultivars from one species, it is impossible to select an optimum basal medium. MS (Murashige and Skoog 1962) and B5 (Gamborg et al. 1968) media have been successfully used either in the exact composition or with slight modifications. For the achievement of high plating efficiency 2,4-D (2,4-dichlorophenoxy acetic acid) has been shown to be essential in the initial culture medium (Table 2). However, Shahin (1985) used 2,4-D-supplemented medium only after the induction of cell division and formation of microcolonies. Table 2 indicates that no exact rule has been followed with respect to auxin/cytokinin levels for protoplast division in *Lycopersicon*.

After induction of division, osmotic pressure has been gradually decreased by dilution of the cultures with a medium containing reduced amounts of sugar (Muhlbach 1980; Tabaeizadeh et al. 1985). In some cases the frequent replacement of the original medium by fresh medium has been suggested to be essential for callus formation (Niedz et al. 1985).

Zeatin alone or with gibberellic acid has been successfully used for plant regeneration of *L. esculentum* protoplasts (Table 2). *L. peruvianum* protoplasts regenerate in a medium containing BAP (benzylaminopurine) alone or in combination with IAA (indoleacetic acid).

No direct correlation between plating efficiency and plant regeneration could be found, the work of Shahin (1985) being an example. He obtained a plating efficiency of 1.4–50% from 14 different cultivars of *L. esculentum*; in the cv. Floradade protoplasts with 22% plating efficiency, the frequency of plant regeneration was less that 25%, while UC-82 cv. with only 1.4% plating efficiency had a shoot regeneration capacity of more than 50%. At this point it might be necessary to distinguish two different pathways of plant regeneration in tomato protoplasts: (1) plant regeneration through somatic embryogenesis, and (2) plant regeneration through de novo organization of shoot meristems in callus. Zapata and Sink (1981) observed embryoid-like structures from *L. peruvianum* proto-

Table 2. Constituents and concentration of growth regulators used in culture media for protoplast division, callus growth and plant regeneration in the genus *Lycopersion*

Species	Hormones for induction of divison[a]		Hormones for callus growth[a]		Hormones for plant regeneration[a]		Reference
L. peruvianum	2,4-D	1	[b]		BAP	10 μM	Muhlbach (1980)
	NAA	0.5	–		IAA	0.1 – 10 μM	
	BAP	0.5					
L. peruvianum	NAA	1	–		Zeatin	1	Zapata and Sink (1981)
	Kinetin	1					
	or 2,4-D	0.5					
L. esculentum	2,4-D	1	2,4-D	1	Zeatin	1	Morgan and Cocking (1982)
	NAA	0.5	NAA	0.5			
	BAP	0.5	BAP	0.5			
L. esculentum	2,4-D	4,53 μM	2,4-D	4.50 μM	BAP	10 – 40 μM	Koblitz and Koblitz (1982, 1983)
	Kinetin	0.93 μM	Kinetin	1 μM	IAA	1 μM	
	BAP	2.00 μM			or IAA	1 μM	
					Zeatin	10 – 40 μM	
L. esculentum	2,4-D	1	NAA	0.5	BAP	10 μM	Tabaeizadeh (1983 a)
and	NAA	0.5	BAP	0.5	IAA	0.1 μM	Tabaeizadeh et al. (1984, 1985)
L. peruvianum	BAP	0.5					
L. peruvianum	NAA	2.5	–		BAP	1 – 5	Imanishi and Hiura (1983)
	BAP	0.25 – 0.75					
L. esculentum	NAA	1	2,4-D	0.2	Zeatin	1	Shahin (1985)
	Zeatin	0.5	BAP	0.5	GA3	0.2	
L. esculentum	2,4-D	1	2,4-D	1	Zeatin	2	Niedz et al. (1985)
	NAA	1	IAA	1			
	BAP	0.5	BAP	0.5			
L. Chilense	2,4-D	0.4	Zeatin	0.4	Zeatin	1	Hassanpour-Estahbanati and
	BAP	0.6	BAP	0.4	GA3	0.6	Demarly (1986)

[a] Hormones concentrations are in mg/l except where indicated.
[b] – : no information is available

plasts in a medium where the plating efficiency was 1–5%. In a medium where the plating efficiency was 20–30%, no embryoids were formed. However, the calli formed in the second medium were able to regenerate shoots. Therefore tomato protoplasts might have the capacity for both types of regeneration and, depending on the culture medium, one or the other type is expressed. It is necessary to mention that no other report has indicated the formation of somatic embryos or embryoid-like structures from tomato protoplasts.

2.3 Influence of Genotype on Plant Regeneration

Plant regeneration from *Lycopersicon peruvianum* leaf mesophyll protoplasts was first reported by Muhlbach (1980) followed by Zapata and Sink (1981), Tabaeizadeh (1983), Tabaeizadeh et al. (1985) and Imanishi and Hiura (1983). However, attempts to induce shoot formation from protoplasts of *L. esculentum* remained unsuccessful for several years. Although somatic hybrid plants of *L. esculentum + Solanum tuberosum* were produced the control unfused tomato protoplasts were never regenerated in these experiments (Melchers et al. 1978). It was only in 1982 that regeneration of two different cultivars (Lukullus and Nadga) of *L. esculentum* was reported (Koblitz and Koblitz 1982; Morgan and Cocking 1982). In the last few years there has been much progress in plant regeneration from tomato protoplasts (Table 3) and so far plants have been regenerated from 23 different cultivars (Koblitz and Koblitz 1983; Shahin 1985; Niedz et al. 1985). In all of the above-cited reports significant differences between different species or different cultivars have been observed with respect to plant regeneration capacity. The same phenomenon has already been shown in *Petunia* (Bergounioux 1974).

Transfer of regeneration capacity to cultivated tomato by interspecific cross-hybridization of *L. esculentum* with *L. peruvianum* was suggested by Thomas and Pratt (1981) and was demonstrated by Koornneff et al. (1986). A tomato genotype superior in plant regeneration was obtained by classical plant breeding, emphasizing the importance of the simultaneous use of traditional and new technologies for improvement of plant genome.

2.4 Protoplast Fusion: Selection of the Somatic Hybrids

Lack of suitable selectable markers for identification of the somatic hybrids from the homologous fusion products and unfused protoplasts is a major problem in somatic hybridization. Selection based on the use of auxotrophic or dominant resistant mutants has been successful. However, these types of mutants are developed in only few species. Comparison of morphological characters of regenerated plants can be used as a first selection (followed by cytological, biochemical and molecular analysis). However, this comparison is efficient mostly in those somatic hybridizations where a sexual hybrid of the parental species already exists. This is not the case in intergeneric somatic hybridization. Therefore it is very difficult to predict morphological characters of such a hybrid. Never-

Table 3. Plating efficiency and morphogenetic response of tomato protoplasts

Genotype	Plating efficiency[a] (%)	Morphogenetic response[b]	Reference
L. peruvianum	22	+	Muhlbach (1980)
L. peruvianum	1–30	+	Zapata and Sink (1981)
L. esculentum (Lukullus)	14	+	Morgan and Cocking (1982)
L. peruvianum (T45)	42	+	Tabaeizadeh (1983 a)
L. esculentum (2NI)	14	–	Tabaeizadeh et al. (1984, 1985)
L. peruvianum	35	+	Imanishi and Hiura (1983)
L. peruvianum	33	+	Shahin (1985)
L. esculentum:			
Red Cherry	43	+	
Cocktail Cherry	50	+	
VFNT-Cherry	9	+	
VF-36	21	+	
Manapal	2	+	
Floradade	22	+	
UC-82	1.4–2	+	
Red Ace-VF Type	3	+	
Roma	2	+	
Beefsteak	3	+	
San Marzano	9	+	
Improved Pearson	2	+	
Heinz 733	7	+	
Heinz 2152	28	+	
L. esculentum:			Niedz et al. (1985)
LA 1268	87	+	
Lukullus	28	+	
Red Cherry	65	+	
P.I. 126914	83	–	
Ventura	58	+	
Walter	35	+	
Sub-Arctic Maxi	75	+	
LA 1622	0.1%	–	
L. Chilense (LA-130)	17	+	Hassanpour-Estambanati and Demarly (1986)

[a] Exact definition of plating efficiency is variable in different reports. It has generally been defined as % of toal number of cultured protoplasts or % of viable protoplasts 24 h after isolation, which divide.

[b] +, shoots have been produced from protoplasts; –, no shoot formation.

theless, a somatic hybrid plant of *L. esculentum* and *Solanum rickii* was identified on the basis of morphology and isozyme analysis (O'Connell and Hanson 1986).

Screening based on differential growth requirements and culture media led to the identification of somatic hybrids in *L. esculentum* + *Solanum tuberosum* (Shepard et al. 1983). Inactivation of protoplasts of one partner by metabolic inhibitors (Medgyesy et al. 1980) or irradiation (O'Connell and Hanson 1985) can lead to the growth of only unfused protoplasts and homologous fusion products of one partner plus somatic hybrids. Identified heterokaryons following fusion of mesophyll protoplasts with cell suspension protoplasts or fusion of fluorescence-stained protoplasts can be selected by a micromanipulator (Galbraith and Galbraith 1979; Galbraith and Mauch 1980). Somatic hybrids can also be sorted out by a cell sorter (Redenbaugh et al. 1982). In this case fluorescence-stained protoplasts can be used, however marked protoplasts with fluorescence-labelled polyclonal (Bergounioux et al. 1983) or monoclonal (Fitter et al. 1987) antibodies are more specific. The efficiency of this technique could be up to 10^4 sorted protoplasts/h (Afonso et al. 1985).

3 An Overview of Somatic Hybridization with the Genus *Lycopersicon*

The first successful experiments on somatic hybridization with the genus *Lycopersicon* were reported by Melchers et al. (1978). Somatic hybrid plants were produced after protoplasts fusion of tomato (*Lycopersicon esculentum*) and dihaploid strain of potato (*Solanum tuberosum*). Somatic hybrid plants with a slightly thickened rhizome had a leaf shape and flower morphology different from tomato and potato. Analysis of Fraction-I protein revealed that the small subunit polypeptides in these plants were from both parents; however, the large subunit polypeptides were exclusively from either tomato or potato. None of these hybrids had fertile flowers. This work was successfully repeated by Shepard et al. (1983), the chlorophyll-deficient protoplasts of a variegating protoclone (protoplast-derived clone) of potato being fused with protoplasts of two different cultivars of tomato. Restriction pattern of mitochondria and chloroplast DNA of these hybrids was identical to potato. Initially, these somatic hybrids showed the expected 72 chromosomes; however, the cuttings from these hybrids displayed mitotic instability.

In another study, fusion of *L. esculentum* mesophyll protoplasts with protoplasts isolated from suspension cultures of *Solanum pennellii* led to the production of somatic hybrid calli (O'Connell and Hanson 1985). These calli predominantly contained either only one of the parental chloroplast DNA type or a mixture of both types. Out of the nine calli which were analyzed for mitochondrial DNA, only one exhibited the presence of both parental mitochondrial DNA and the eight others were identical with those of *S. pennellii*. However, this result was not conclusive, since only a small part of the mitochondrial genome was examined. The same authors (O'Connell and Hanson 1986) identified a single somatic hybrid callus clone following protoplast fusion of *L. esculentum + Solanum rickii*. Several plants regenerated from this callus clone, varying from one another with respect to leaf morphology and ploidy level. These plants exhibited the expression of both parental genomes for phosphoglucomutase. However, their chloroplast and mitochondrial genome resembled that of *S. rickii*.

Somatic hybrid plants were also produced following protoplast fusion of *L. esculentum + Solanum lycopersicoides* (Handley et al. 1986). These plants, which were phenotypically similar to *S. lycopersicoides*, showed the addition of parental bands for several isozymes as well as for the small subunit of Fraction-I protein. The cytoplasmic genome of these plants has not been studied.

The preceding paragraphs summarize the salient developments on somatic hybridizations with the genus *Lycopersicon*. However, the present chapter focusses attention on somatic hybridization of *Lycopersicon*+*Petunia*.

4 Somatic Hybridization in *Lycopersicon*+*Petunia*

At the beginning of this study several genotypes of wild and cultivated species of *Lycopersicon* were tested with respect to division of protoplast, using different basal media, growth regulators and growth conditions. Sustained cell division was obtained in one genotype of *L. peruvianum* (T45) and one genotype of *L. esculentum* (2NI). Intergeneric somatic hybridization was conducted with these two genotypes: *L. esculentum*+*Petunia hybrida* and *L. peruvianum*+ *P. hybrida*. However, plants were regenerated only from the second combination, for which the results are presented below.

4.1 Plant Material

A pure line of *Petunia hybrida* (sfla) was used as one of the fusion partners (Bergounioux et al. 1986). *Lycopersicon peruvianum* var. T45 was chosen as the second partner. This genotype has been selfed since 1964. Seeds were generously supplied by Dr. Chr. Lehman (Genetic Institute, Academy of Science, GDR). Plants were grown in a greenhouse with a light intensity of 6000–8000 lx. Four-week-old plants were used for protoplast isolation. Based on our studies, cited in the previous section (2.1.), *L. peruvianum* plants were grown in 16 h light for 3 weeks, transferred to 9 h light for 5 days and finally were subjected to a 15-h cold treatment (4 °C in the dark) prior to protoplast isolation (Tabaeizadeh et al. 1984). Growth conditions for *Petunia hybrida* have been previously described (Bunisset-Bergounioux and Perennes 1980).

4.2 Protoplast Isolation

The youngest leaves were surface sterilized in 6% calcium hypochlorite for 15 min and were subsequently washed three times with sterile distilled water. The lower epidermis was peeled off and leaf fragments were floated (with exposed lower surface down) on the enzyme solution. The enzyme solution consisted of 0.3% Cellulase Onozuka R10 (Yakult Honsha Co.), 0.15% Macerozyme R10 (Yakult Honsha Co.) and 0.5 M mannitol. The enzyme-protoplast mixtures were filtered through a 100-μm stainless steel sieve. The protoplasts were removed from the enzyme solution by centrifugation at 40 *g* for 2 min and were washed once with a solution containing 0.1 mM $CaCl_2$ and 0.5 M mannitol, and once with the culture medium. The culture medium consisted of B5 (Gamborg et al. 1968) salts and vitamins, 2,4-D (1 mg/l), NAA (0.5 mg/l), BAP (0.5 mg/l), glucose (8.4%) and sucrose (1%).

4.3 Protoplast Fusion, Culture and Induction of Division

Protoplast fusion was induced by using a modified fusion technique of Bunisset-Bergounioux and Perennes (1980). The fusion solution consisted of 40% polyethylene-glycol (MW = 1500), 0.16% $CaCl_2$, 0.0095% KH_2PO_4 and 3.6% glucose. The pH was adjusted to 5.8. The choice of this concentration of PEG was based on microscopic observations of the relatively high rate of fusion (Fig. 1 a) and low number of dead protoplasts. The protoplasts of the parental lines were mixed (1 : 1) with a density of 5×10^5 protoplasts/ml. Five drops of fusion solution were placed (using a 0.5 ml pipette) in a 6-cm Petri dish, then one drop of protoplast suspension was added (using a 0.2-ml pipette) to each drop of fusion solution. After 15 min three successive dilutions were performed at 5-min inter-

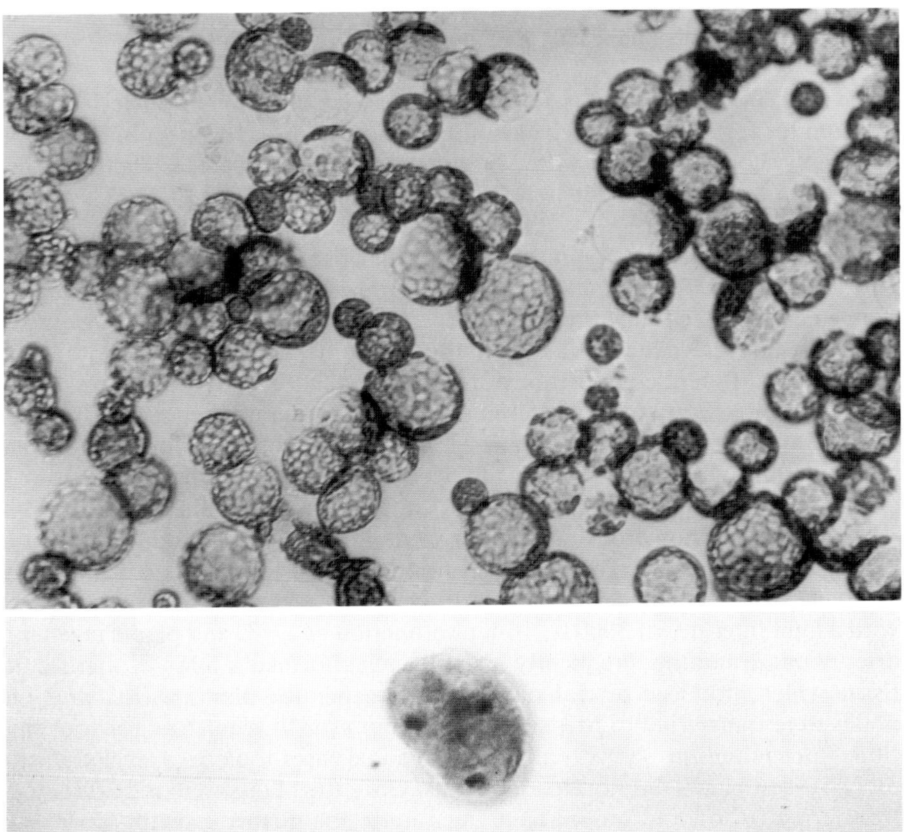

Fig. 1. Protoplast fusion of *Lycopersicon peruvianum + Petunia hybrida* induced with polyethylene glycol. **a** The adhesion phase of the protoplasts. Protoplasts of *Petunia hybrida* are larger than those of *L. peruvianum*. **b** A fixed protoplast after fusion, containing three nuclei. (Tabaeizadeh et al. 1983)

vals by adding one drop of the culture medium to each drop of protoplast-PEG solution. Subsequently, 4 ml of culture medium was added to each Petri dish. After 15 min the Petri dishes were sealed and incubated at 29 °C in the dark. In each fusion experiment unfused protoplasts as well as homologous fusions of each species were cultured as controls. After each fusion experiment, the frequency of fusion was investigated based on the number of fixed protoplasts containing two or more nuclei (Fig. 1 b). Fixation of the protoplasts and staining of the nuclei was according to a procedure described by Dutuit and Orcival (1979). The first division was observed after 4–5 days. When colonies of five to ten cells were formed, the cultures were diluted by adding 1–2 ml of the culture medium with a reduced concentration (0.3 M) of glucose. This dilution was performed once a week until the cell colonies were transferred to solid medium.

In *L. peruvianum*, protoplasts resulting from homologous fusion formed a higher number of cell colonies than unfused protoplasts. We did not find out whether this difference was due to a direct positive effect of PEG or to the increased ploidy level after fusion treatment.

4.4 Callus Growth and Plant Regeneration

Once the cell colonies were visible to the naked eye, they were transferred to solid medium for further growth. The medium was the same as for the initial protoplast culture, without 2,4-D and with 3% glucose and 1.5% sucrose. The Petri dishes were kept in a growth chamber at 27 °C with a light intensity of 1000 lx and 16-h photoperiod.

There was heterogeneity in the growth rate of calli resulting from fusion of *L. peruvianum* and *P. hybrida*. Once the calli reached the size of about 5 mm in diameter, they were transferred to the regeneration medium. At this stage most of the calli were green and friable. In the preliminary experiments several basal media with different compositions of growth regulators (BAP or zeatin alone or in combination with IAA) were used to test the regeneration capacity of *L. peruvianum* protoplasts. Plants were regenerated on MS basic medium with different combinations of BAP and IAA. Calli derived from *Petunia* protoplasts were not able to regenerate in this medium. This could therefore eliminate the possibility of plant regeneration from unfused or homologous fusion products of *Petunia* protoplasts among the protoplasts resulting from fusion of *L. peruvianum* and *Petunia hybrida.* Calli on the regeneration medium were kept in a growth chamber with a light intensity of 3000 lx, 16-h photoperiod at 27 °C and plants regenerated after about 2 months. Frequently a single callus formed a mass of meristematic tissue which produced several shoots. The regenerated plants with 2 or 3 internodes were transferred to MS medium without growth regulators for root initiation. Root initiation was very slow and in some cases took 5–8 weeks. Furthermore, several plants did not develop roots. Once the plants were rooted they were transferred to soil and moved to a greenhouse for further growth.

4.5 Characterization

Ninety eight plants were regenerated from 30 calli orginating from fusion experiments of *L. peruvianum* and *Petunia hybrida*. All the plants regenerated from the same calli were designated by the same number. The plants were characterized by cytological observation of the chromosomes at mitosis and meiosis, by morphological characters and finally by analysis of Fraction-I Protein.

4.5.1 Cytological Study of Mitosis and Meiosis

The procedure for observation of chromosomes at mitosis and meiosis has already been described (Tabaeizadeh et al. 1985). *Petunia hybrida* chromosomes (2n = 14) can be easily identified from those of *L. peruvianum* (2n = 24), as they are significantly larger in size (Fig. 2a, b). Most of the plants regenerated from fusion of *L. peruvianum* and *P. hybrida* were mixoploid. The chromosome numbers of some of the regenerated plants are shown in Table 4. Three plants, R3, R5 und R6 had 38 chromosomes, which is the sum of parental chromosome

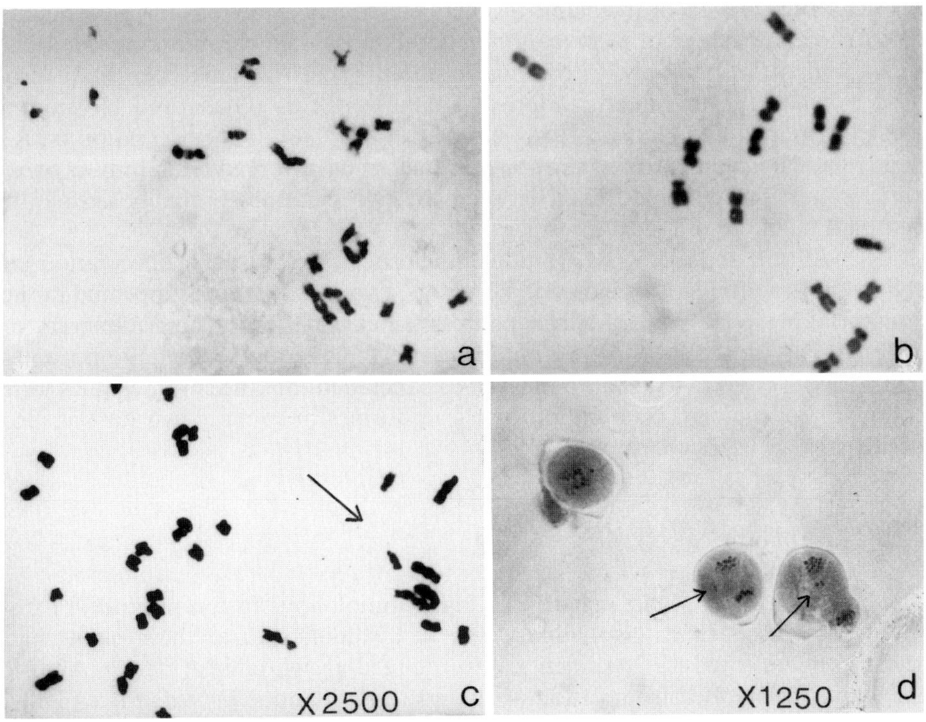

Fig. 2. a and **b** Mitotic metaphase of *L. peruvianum* and *P. hybrida*, respectively. **c** Mitotic metaphase of plant R3. The *arrow* marks the larger chromosomes of the size of those of *P. hybrida*. **d** Anaphase of meiosis of plant R3. The *arrow* marks lagging chromosomes at the equatorial plane. (Tabaeizadeh et al. 1983, 1985)

Table 4. Chromosome counts in root tip preparations from regenerated plants. (Tabaeizadeh et al. 1985)

Plant category	Plant number	Chromosome number
L. peruvianum +	R_2	24, 28, 32
P. hybrida	R_3	24, 30, 32, 38, 42, 46
	R_5	24, 38, 40
	R_6	38, 50
	R_7	32
	R_9	48
	R_{10}	24
	R_{15}	48
L. peruvianum +	R_1	48
L. peruvianum	R_{16}	48
	R_{27}	48
L. peruvianum	R_{31}	24
	R_{32}	24

numbers (24+14). In the metaphase plate of R3 plants, 24 chromosomes of the size of *L. peruvianum* chromosomes were observed along with 6–8 others which were unambiguously the size of *Petunia hybrida* chromosomes, off to one side of the field (Fig. 2c). All of the plants regenerated from homologous fusion products or unfused protoplasts of *L. peruvianum* had the expected chromosome numbers, i.e. 48 and 24, respectively. Chromosome number and size have been frequently used as a marker for identification of somatic hybrids (Gleba and Hoffmann 1980; O'Connell and Hanson 1986; Handley et al. 1986). Our observation on R3 plants indicates that chromosomes were transferred from *P. hybrida* into *L. peruvianum* but the incompatibility between the two genomes probably led to the elimination of certain chromosomes.

In addition to mitosis, we studied meiosis by microscope observation of stained pollen grains. At meiosis of R3 plants, two chromosomes appeared in the equatorial plane (Fig. 2d). Meiosis permits the elimination of supernumerary or different chromosomes. Since chromosomes of the size of *Petunia* chromosomes had been observed in the same plant, it was concluded that the chromosomes were eliminated at meiosis. Elimination of chromosomes in somatic hybrids has also been reported by Shepard et al. (1983).

4.5.2 Morphological Characters

The morphology of plants regenerated from homologous fusion products or unfused protoplasts of *L. peruvianum* was first studied. All of these plants had seven to nine dented leaflets, identical to the seed-grown plants (Fig. 3). Among the plants regenerated from fusion of *L. peruvianum* and *P. hybrida* three plants, R3, R6 and R7, had a leaf morphology distinctly different from *L. peruvianum* and mostly intermediate between this species and *Petunia*. The leaflets were round and fewer than *L. peruvianum* (Fig. 3). The flowers of R3 and R6 plants resembled those of *L. peruvianum*. Both plants were partially (R3) or completely (R6)

Fig. 3. Comparison of leaf morphology: **a** *L. peruvianum* seed-grown plant. **b** *L. peruvianum* regenerated from protoplasts. **c** *P. hybrida*. **d, e** Plant R3. **f, g** Plant R6. **h, i** Plant R7 in vitro and in vivo. (Tabaeizadeh et al. 1983, 1985)

male sterile. The backcross with *L. peruvianum* as the male parent was not successful and we could not find out whether this male sterility was nuclear or cytoplasmic. Plant R7, once transferred to soil, grew poorly and never produced flowers.

Leaf hair is another morphological character which has been studied in somatic hybrids (Gleba and Hoffmann 1980). Initially we observed simple hair in *L. peruvianum* and glandular hair in *Petunia hybrida* leaves of both seed-grown and protoplast-regenerated plants. However, a closer examination revealed that in *L. peruvianum* leaf hair morphology changes during plant growth; cotyledons have simple hair, the first leaves have a mixture of glandular and simple hair and only the later leaves of fully grown plants have exclusively simple hair. Several plants regenerated from fusion of *L. peruvianum* and *P. hybrida* exhibited the two types of leaf hair in different periods of their growth. However, due to the instability of this character in *L. peruvianum*, it was not possible to reach a clear conclusion.

4.5.3 Analysis of Fraction-I Protein (Ribulose-1,5-bisphosphate carboxylase/oxygenase)

This enzyme is particularly interesting because it consists of small and large subunit polypeptides encoded by the nuclear and chloroplast genomes, respectively (Kung et al. 1975). Purification of the enzyme by immunoprecipitation was carried out according to Hirai (1982). The antibodies were kindly provided by Dr. J. P. Jaquot (Laboratoire de Physiologie végétale Moléculaire, Université de Paris-Sud, Orsay, 91405 France). Purified enzyme was subjected to isoelectrofocussing in 5% polyacrylamide gel containing 2% of ampholine (LKB) pH 5–8 and 8 M urea.

Petunia hybrida and *L. peruvianum* both displayed a distinct pattern of the enzyme with respect to the isoelectric points of large and small subunit polypeptides (Fig. 4a). Although the stability of this enzyme during plant growth has already been reported (Chen et al. 1976), we verified this with our experimental plants. There was no difference in the isoelectric focussing pattern of seed-grown plants in two different stages of development (Fig. 4a). Plants regenerated from protoplasts or from homologous fusion had the same isoelectric focussing pattern as the seed-grown plants (Fig. 4b). Among the plants regenerated from fusion of *L. peruvianum* and *P. hybrida*, the large subunit of plant R7 had a more basic isoelectric point than either of the parents (Fig. 4a). Since the isoelectric focussing pattern of this enzyme was identical in seed-grown plant and plants regenerated from protoplast, the possibility that the new form of the large subunit in plant R7 was an artifact due to tissue culture could be excluded. The result could not be due to experimental conditions either since the same pattern was obtained in two independent isoelectrofocussing experiments. Therefore, the new form of the large subunit in plant R7 could be explained in two ways: (1) expression of *Lycopersicon* genetic information encoded by the chloroplast genome might have been modified by the presence of information originating from *Petunia* or (2) recombination of chloroplast DNA could have occurred. This latter

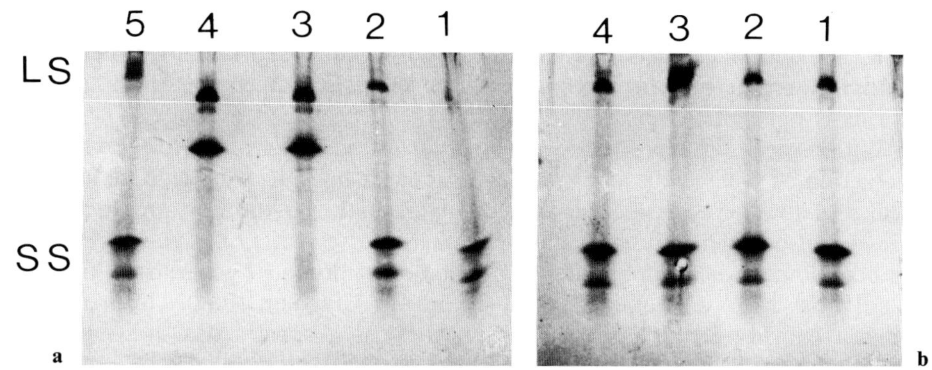

Fig. 4. Isoelectric focussing of large and small subunits of Fraction-I protein. *SS* small subunit. *LS* large subunit. **a***1* and **a***2* *L peruvianum* flowering and seedling plants, respectively. **a***3* and **a***4* *P. hybrid* flowering and seedling plant, respectively. **a***5* plant R7 with a large subunit which has a more basic isoelectric point than *L. peruvianum* and *P. hybrida*. **b***1* *L. peruvianum* seed-grown plant. **b***2* *L. peruvianum* plant regenerated from protoplasts. **b***3* Plant R3. **b***4* Plant R6. (Tabaeizadeh et al. 1983, 1985)

hypothesis was first suggested by Conde (1981), and it was demonstrated in interspecific *Nicotiana* somatic hybrid plants (Medgyesy et al. 1985). In most previous analysis of Fraction-I protein in somatic hybrid plants, one parental type of large subunit was found (Chen et al. 1977; Aviv et al. 1980). Reports on the presence of both parental types of large subunit are rare (Glimelius and Bonnett 1981). This shows that when fusion involves two different populations of chloroplasts, there is a rapid segregation to either one of the organelle types. It has also been suggested (Chen et al. 1977) that after protoplast fusion there is an active selection against cells containing a mixture of parental chloroplasts so that those cells with exclusively one or the other type of chloroplast DNA are competitively more vigorous. This view might explain the poor growth of plant R7. In spite of our efforts, we could not obtain enough leaf tissue to isolate chloroplast DNA from this plant.

Isoelectricfocussing pattern of Fraction-I protein of several other presumptive hybrid plants were the same as of *L. peruvianum*. We did not find out whether these plants lack the genetic information of *Petunia hybrida* related to this enzyme or whether the gene(s) was transferred but not expressed.

5 Conclusion and Prospects

Somatic hybridization of *Lycopersicon peruvianum + Petunia hybrida* led to the production of plants with nuclear and cytoplasmic variability. One plant contained *L. peruvianum* chromosomes plus six to eight chromosomes identical to those of *Petunia hybrida*. The same plant, along with one other, had intermediate leaf morphology and was male sterile. We could not find out the origin of this male sterility, since the backcross with *L. peruvianum* was not successful. Analysis of mitochondrial DNA of these two plants and its comparison with seed-grown and protoplast-regenerated parental plants could lead to a clear conclusion. Another plant (R7) contained a new form of the large subunit of Fraction-I protein. This could mean that the direct expression of the parental characters in somatic hybrids is not systematic.

With respect to the production of cytoplasmic male sterility in tomato through interaction of the nucleus and cytoplasm, somatic hybridization with wild related species of tomato such as *Solanum pennellii* could be another approach. This species can be used only as a male parent in crosses with *L. esculentum*. There has been some indication of cytoplasmic male sterility in F2 generations from the sexual cross of *L. esculentum* and *S. pennellii* (Anderson 1964). Unfortunately, somatic hybrid calli produced between these two species did not regenerate (O'Connell and Hanson 1985).

Direct or *Agrobacterium*-mediated transformation could be another approach for the induction of variability in *Lycopersicon* (Shahin et al. 1986). Rapid progress has been made in this area, and herbicide-tolerant plants (Shah et al. 1986) and plants resistant to virus infection (Abel et al. 1986) have been produced. Attempts to induce cytoplasmic male sterility through direct transfer of mitochondrial DNA fragments is an active area of research; however, no success has yet been reported. Therefore protoplast fusion still provides a unique tool, to introduce foreign DNA into a plant's mitochondrial genome (Rothenberg et al. 1985; Tabaeizadeh et al. 1986, 1987). Since plant regeneration from *L. esculentum* protoplasts is no longer a limiting factor (Shahin 1985; Niedz et al. 1985), production of a wide range of somatic hybrid plants in this species is expected in the near future.

6 Protocol for Isolation, Fusion and Culture of *Lycopersicon peruvianum* + *Petunia hybrida* Protoplasts

Protoplasts are isolated by incubation of leaf fragments in an enzyme solution containing Cellulase Onozuka R10 (0.3%), Macerozyme R10 (0.15%) and mannitol (0.5 M) for 3–4 h. The digested leaf tissues are filtered through a 100-μm stainless steel sieve. The enzyme protoplast mixture is then centrifuged at 40 *g* for 2 min. The protoplasts are washed once in a solution containing 0.1 mM $CaCl_2$ and 0.5 mM mannitol and once by the culture medium.

For the fusion, the protoplasts of both partners are mixed (1:1) at a density of 5×10^5 protoplasts/ml. Five drops of fusion solution (40% polyethyleneglycol, 0.17% $CaCl_2$, 0.0095% KH_2PO_4 and 3.6% glucose) are placed in a 6-cm Petri dish. One drop of protoplast mixture is added to each drop of fusion solution. After 15 min three successive dilutions are performed at 5-min intervals by adding one drop of the culture medium to each drop of PEG-protoplast solution. Subsequently, 4 ml of the culture medium is added to each Petri dish. Petri dishes are then sealed and incubated at 24 °C in the dark. The best medium for culture of fusion products of *Lycopersicon* and *Petunia* was recognized to be B5 salts and vitamins (Gamborg et al. 1968) supplemented with glucose (8.4%), sucrose (1%), 2,4-D (1 mg/l) and NAA (0.5 mg/l).

References

Abel PP, Nelson RS, De B, Hoffman N, Rogers SG, Fraley RT, Beachy RN (1986) Delay of disease development in transgenic plants that express the tobacco mosaic virus coat protein gene. Science 232:738–743

Afonso CL, Harkins KR, Thomas-Conpton MA, Krejci AE, Galbraith DW (1985) Selection of somatic hybrid plants in *Nicotiana* through fluorescence activated sorting of protoplasts. Biotechnology 3:811–818

Anderson WR (1964) Evidence of plasmon differentiation in *Lycopersicon*. Tomato Genetic Coop Rep 14:4–6

Aviv D, Fluhr R, Edelman M, Galun E (1980) Progeny analysis of interspecific somatic hybrids of *Nicotiana tabacum* (CMS) + *Nicotiana sylvestris* with respect to nuclear and chloroplast markers. Theor Appl Genet 56:145–150

Bajaj YPS (ed) (1986) Biotechnology in agriculture and forestry, Vols 1, 2. Springer, Berlin Heidelberg New York Tokyo

Belliard G, Pelletier G, Ferault M (1977) Fusion de protoplastes de *Nicotiana tabacum* à cytoplasmes différentes, Etudes des hybrides cytoplasmique néoformes. CR Acad Sci Ser D 284:749–752

Bergounioux C (1974) Déclanchement des phénomènes de régénération sur les pétales de *Petunia hybrida*. Ann Amélior Plant 24:55–62

Bergounioux C, Perennes C, Tabaeizadeh Z, Prevot P, Gadal P (1983) Characterization and separation of *Petunia hybrida* and *Lycopersicon esculentum* protoplasts based on natural fluorescence or FITC added to label antibodies in immunoreaction with the protoplast membrane. In: Potrykus I et al. (ed) Poster Proc. 6th Int. Protoplast Symposium. Birkhäuser, Basel, pp 262–263

Bergounioux C, Perennes C, Miege C, Gadal P (1986) The effect of male sterility on protoplast division in *Petunia hybrida*. Cell cycle comparison by flow cytometry. Protoplasma 130:138–144

Bunisset-Bergounioux C, Perennes C (1980) Transfert de facteur cytoplasmique de la fertilité mâle entre deux lignées de *Petunia hybrida* par fusion de protoplastes. Plant Sci Lett 19:143–149

Cassells AC, Barlass M (1976) Environmentally induced changes in the cell walls of tomato leaves in relation to cell and protoplast release. Physiol Plant 37:239–246

Cassells AC, Barlass M (1978) A method for the isolation of stable mesophyll protoplasts from tomato leaves throughout the year under standard conditions. Physiol Plant 42:236–242

Chen K, Kung SD, Gray JC, Wildman SG (1976) Subunit polypeptide composition of Fraction-I protein from various plant species. Plant Sci Lett 7:429–443

Chen K, Wildman SG, Smith H (1977) Chloroplast DNA distribution in parasexual hybrids as shown by polypeptide composition of Fraction-I protein. Proc Natl Acad Sci USA 74:5109–5112

Cocking EC (1960) A method for the isolation of plant protoplasts and vacuoles. Nature 187:962–963

Conde MR (1981) Chloroplast DNA recombination in *Nicotiana* somatic hybrids. Genetics 97:S26

Durand J, Potrykus I, Donn G (1976) Plantes issues de protoplastes de *Petunia*. Z Pflanzenphysiol 69:26–34

Dutuit P, Orcival J (1979) Mise en évidence d'une recoloration atypique du réactif de Schiff au niveau de structures cytoplasmique dans des protoplastes nouvellement formés à partir de feuilles de *Petunia hybrida*. C R Acad Sci 288:1163–1166

Fitter MS, Norman PM, Hahn MG, Wingata VPM, Lamb CJ (1987) Identification of somatic hybrids in plant protoplast fusions with monoclonal antibodies to plasma-membrane antigens. Planta 170:49–54

Frearson EM, Power JB, Cocking EC (1973) The isolation, culture and regeneration of *Petunia* leaf protoplasts. Dev Biol 33:130–137

Galbraith DW, Galbraith JEC (1979) A method for identification of fusion of protoplasts derived from tissue culture. J Plant Physiol 93:149–158

Galbraith DW, Mauch TJ (1980) Identification of fusion of plant protoplasts. II. Conditions for the reproducible fluorescence labelling of protoplasts derived from mesophyll tissue. J Plant Physiol 98:120–140

Gamborg OL, Miller RA, Ojima K (1968) Nutrient requirements of suspension cultures of soybean root cells. Exp Cell Res 50:151–158

Gerstel DV, Burns JA, Sand SA (1980) Cytogenetics of flower modification of a cytoplasmic male sterile tobacco. Genetics 96:223–235

Gillham NW (1978) Organelle heredity. Raven, New York, p 602

Gleba YY, Hoffmann F (1980) "Arabidobrassica", A novel plant obtained by protoplast fusion. Planta 149:112–117

Glimelius K, Bonnett HT (1981) Somatic hybridization in *Nicotiana*. Restoration of photoautotrophy to an albino mutant with defective plastids. Planta 153:497–503

Handley LW, Nickels RS, Cameron MW, Moore PP, Sink KC (1986) Somatic hybrid plants between *Lycopersicon esculentum* and *Solanum lycopersicoides*. Theor Appl Genet 71:691–697

Hanson MR (1982) Cell and tissue culture of *Lycopersicon*. In: Fujiwara A (ed) Plant Tissue Cult. 1982 Maruzen, Tokyo, pp 262–263

Hanson MR, Conde M (1985) Functioning and variation of cytoplasmic genomes: Lessons from cytoplasmic nuclear interactions effecting male fertility in plants. Int Rev Cytol 94:214–267

Hassanpour-Estahbanati A, Demarly Y (1986) Plant regeneration from protoplasts of *Lycopersicon chilense*. Physiol Vég 24:391–396

Hirai A (1982) Isoelectrofocusing of non-carboxymethylated Fraction-I protein from green callus. Plant Sci Lett 25:37–41

Imahishi S, Hiura I (1983) Culture and regeneration of *Lycopersicon peruvianum* leaf protoplasts. Jpn J Breed 33:359–368

Izhar S, Power JB (1979) Somatic hybridization in *Petunia*; a male sterile cytoplasmic hybrid. Plant Sci Lett 14:49–55

Koblitz H, Koblitz D (1982) Experiment on tissue culture in the genus *Lycopersicon* Miller. Mesophyll protoplast regeneration to plants *Lycopersicon esculentum* cv. Nadja. Plant Cell Rep 1:143–146

Koblitz H, Koblitz D (1983) Tissue and protoplast culture studies in *Lycopersicon esculentum* Miller var. Flammatum Lehm. cv. Bonner Beste and its mutant chloronerna. Plant Cell Rep 2:194–197

Koornneef M, Hanhart C, Jongsma M, Toma I, Weide R, Zabel P, Hille J (1986) Breeding of tomato genotype readily accessible to genetic manipulation. Plant Sci 45:201–208

Kung SD, Gray JC, Carlson PS (1975) Polypeptide composition of Fraction-I protein from parasexual hybrid plants in the genus *Nicotiana*. Science 187:353–355

Lazer KD, Lersten NR (1972) Anatomy and cytology of microsporogenesis in cytoplasmic male sterile angiosperms. Bot Rev 38:425–454

Levings CS III, Pring DR (1976) Restriction endonuclease analysis of mitochondrial DNA from normal and Texas cytoplasmic male sterile maize. Science 193:158–160

Medgyesy P, Menczel L, Maliga P (1980) The use of cytoplasmic streptomycin resistance: chloroplast transfer from *Nicotiana tabacum* into *Nicotiana sylvestris*, and isolation of their somatic hybrids. Mol Gen Genet 179:693–698

Medgyesy P, Fejes E, Maliga P (1985) Interspecific chloroplast recombination in a *Nicotiana* somatic hybrid. Proc Natl Acad Sci USA 82:6960–6964

Melchers G, Sacristan MD, Holder AA (1978) Somatic hybrid plants of potato and tomato regenerated from fused protoplasts. Carlsberg Res Commun 43:203–218

Morgan A, Cocking EC (1982) Plant regeneration from protoplasts of *Lycopersicon esculentum* Mill. J Plant Physiol 106:97–104

Muhlbach HP (1980) Plant regeneration potentials of mesophyll protoplasts from cultivated and a wild species of tomato. Planta 148:89–96

Murashige T, Skoog F (1962) A revised medium for rapid growth and bioassays with tobacco tissue cultures. Physiol Plant 15:473–497

Niedz ZRP, Rutter SM, Handley LW, Sink KC (1985) Plant regeneration from leaf protoplasts of six tomato cultivars. Plant Sci 39:199–204

O'Connell MA, Hanson MR (1985) Somatic hybridization between *Lycopersicon esculentum* and *Lycopersicon pennellii*. Theor Appl Genet 70:1–12

O'Connell MA, Hanson MR (1986) Regeneration of somatic bybrid plants formed between *Lycopersicon esculentum* and *Solanum rickii*. Theor Appl Genet 72:59–65

Pelletier G, Primard C, Vedel F, Chetrit P, Remy R, Rousselle P, Renard M (1983) Intergeneric cytoplasmic hybridization in Crucifereae by protoplast fusion. Mol Gen Genet 191:244–250

Power JB, Frearson EM, George D, Evans PK, Berry SF, Hayward C, Cocking EC (1976) The isolation, culture and regeneration of leaf protoplasts in the genus *Petunia*. Sci Lett 7:51–55

Power JB, Berry SF, Chapman JV, Cocking EC (1979) Somatic hybrids between unilateral cross incompatible *Petunia* species. Theor Appl Genet 55:97–99

Power JB, Berry SF, Chapman JV, Cocking EC (1980) Somatic hybridization of sexually incompatible petunias: *Petunia parodii, Petunia parviflora*. Theor Appl Genet 57:1–4

Powling A, Gliss THN (1983) Studies on the organelle genomes of sugarbeet with male fertile and male sterile cytoplasms. Theor Appl Genet 65:323–328

Pring DR, Conde MF, Schertz KF, Levings CS III (1982) Plasmid like DNAs associated with mitochondria of cytoplasmic male sterile sorghum. Mol Gen Genet 186:180–184

Redenbaugh K, Ruzin S, Bartholomew J, Bassham JA (1982) Characterization and separation of plant protoplasts via flow cytometry and cell sorting. J Plant Physiol 107:65–80

Rick CM (1959) Response of tomatoes grafted on cytoplasmically male sterile petunia. TGC Rep 9:43–44

Rothenberg M, Boeshore ML, Hanson MR, Izhar S (1985) Intergenomic recombination of mitochondrial genomes in a somatic hybrid plant. Curr Genet 9:615–618

Shah DM, Horsh RB, Klee HJ, Kishore GM, Winter JA, Tumer NF, Hironaka CM, Sanders PR, Gasser CS, Aykent S, Siegle NR, Rogers SG, Fraley RT (1986) Engineering herbicide tolerance in transgenic plants. Science 233:478–481

Shahin EA (1985) Totipotency of tomato protoplasts. Theor Appl Genet 69:235–240

Shahin EA, Sukhapinda K, Simpson RB, Spivey R (1986) Transformation of cultivated tomato by a binary vector in *Agrobacterium rhizogenes*: transgenic plants with normal phenotypes harbor binary vector T-DNA, but no Ri-plasmid T-DNA. Theor Appl Genet 72:770–777

Shepard JF, Bidney D, Barsby T, Kemble R (1983) Genetic transfer in plants through interspecific protoplast fusion. Science 219:683–688

Tabaeizadeh Z (1983) Etude d'accroissement de la variabilité nucléaire et cytoplasmique chez *Lycopersicon* Mill. par hybridation somatique intergénéraique et interspécifique. Dr. Ing. Thesis, University of Paris-Sud, Orsay, France

Tabaeizadeh Z, Bergounioux C, Perennes G (1983) Increasing the variability of *Lycopersicon* Mill. by protoplast fusion with *Petunia* L. In: Potrykus I et al. (ed) Poster Proc. 6th Int Protoplast Symposium. Birkhäuser, Basel, pp 90–91

Tabaeizadeh Z, Bunisset-Bergounioux C, Perennes C (1984) Environmental growth conditions of protoplasts source plants: Effect of subsequent protoplast division in two tomato species. Physiol Vég 22:223–229

Tabaeizadeh Z, Perennes C, Bergounioux C (1985) Increasing the variability of *Lycopersicon peruvianum* Mill. by protoplast fusion with *Petunia hybrida* L. Plant Cell Rep 4:7–11

Tabaeizadeh Z, Ozias-Akins P, Pring DR, Vasil IK (1986) Amplification of mitochondrial DNA in intergeneric somatic hybrids of the Gramineae. In: Proc. 1st Int. Workshop on Higher Plant Mitochondrial DNA. Airlie, p 58

Tabaeizadeh Z, Pring DR, Vasil IK (1987) Analysis of mitochondrial DNA from somatic hybrid cell lines of *Saccharum officinarum* (sugarcane) and *Pennisetum americanum* (peral millet). Plant Mol Biol 8:509–513

Tal M, Watts JW (1979) Plant growth conditions and yield of viable protoplasts isolated from leaves of *Lycopersicon esculentum* and *Lycopersicon peruvianum*. J Plant Physiol 92:207–214

Thomas BR, Pratt D (1981) Breeding tomato strains for use in cell culture research. Plant Mol Biol Newslett 2:102–1065

Zapata FJ, Sink KC (1981) Somatic embryogenesis from *Lycopersicon peruvianum* leaf mesophyll protoplasts. Theor Appl Genet 59:265–268

Zapata FJ, Evans PK, Power JB, Cocking EC (1977) The effect of temperature on the division of leaf protoplasts of *Lycopersicon esculentum* and *Lycopersicon peruvianum*. Plant Sci Lett 8:119–124

Zelcer A, Aviv D, Galun E (1978) Interspecific transfer of cytoplasmic male sterility by fusion between protoplasts of normal *Nicotiana sylvestris* and X-ray irradiated protoplasts of male sterile *Nicotiana tabacum*. J Plant Physiol 90:397–407

III.6 Somatic Hybridization in *Hyoscyamus* × *Nicotiana*

G. LAZAR[1]

1 Introduction

1.1 Importance and Distribution of *Nicotiana* and *Hyoscyamus* Species

The majority of the approximately 1500 species of the Solanaceae family comprise herbaceous or woody plants and they are distributed mostly in tropical or temperate regions of the Americas. There are a number of useful genera in this family, such as: *Solanum, Lycopersicum, Nicotiana,* and the important drug plants of Hyoscyamus, Atropa and *Datura.* The 76 species of the genus *Nicotiana* are distributed as follows: North America (12), Central America (14), South America (43) and Australia (14). Thus the centre of the area of this genus must be regarded as South America (Wulff 1943). *Nicotiana tabacum* is economically the most important species of the genus and the most widely grown commercial non-food plant in the world. It was introduced into Europe in 1560. The herbaceous *Hyoscyamus* species have their origin in Africa and southwest Asia. The best known is the poisonous weed henbane, *Hyoscyamus niger*, a natural source of hyoscyamine, hyoscine and atropine.

The allotetraploid *N. tabacum* is a relatively young species, having arisen less than 6 million years ago (Okamura and Goldberg 1985) from the diploid progenitors *N. sylvestris* and *N. tomentosiformis*, the female and male parents, respectively (Gray et al. 1974).

1.2 Significance and Use of in-Vitro Methods in *Hyoscyamus* and *Nicotiana*

No model plant species are easier to handle in in-vitro culture than tobacco and *Hyoscyamus*. Much of our understanding of plant protoplasts and tissue culture derives from *Nicotiana*. The classical studies of Skoog and Miller (1956) on plant morphogenesis were performed on tobacco stem pith explants. The totipotency of single plant cells (Vasil and Hildebrandt 1965) or isolated plant protoplasts (Takebe et al. 1971) was first demonstrated with tobacco. *Nicotiana* species have figured largely in advances in anther culture technology (Bajaj 1983). Tobacco

[1] Institute of Genetics, Biological Research Center, Hungarian Academy of Sciences, H-6701 Szeged, P.O.B. 521, Hungary
Present address: Department of Molecular Biology, Massachusetts General Hospital and Department of Genetics, Harvard Medical School, Massachusetts General Hospital, Boston, MA 02114, USA

Biotechnology in Agriculture and Forestry, Vol. 8
Plant Protoplasts and Genetic Engineering I (ed. by Y. P. S. Bajaj)

was used in the first reports describing successful single cell cloning (Bergmann 1960), isolation of auxotrophic mutants from tissue culture (Muller and Grafe 1978), somatic hybridization (Carlson et al 1972), *Agrobacterium* coculture transformation of plant protoplasts (Marton et al. 1979) and high efficiency DNA-mediated transformation (Paszkowski et al. 1984).

Hyoscyamus can be credited as having one of the best biochemical mutant collections available for higher plants (Strauss et al. 1981; Gebhardt et al. 1981, 1983), a reliable tissue culture (Lorz et al. 1979), a protoplast-plant regeneration system (Lorz et al. 1977; Wernicke et al. 1979; Wernicke and Thomas 1980), established methods for haploid production (Raghavan 1975, 1978; Sunderland and Wildon 1979) and cell genetic manipulation via somatic hybridization (Lazar et al. 1983; Gebhardt et al. 1983; Potrykus et al. 1984).

The lists above, although incomplete, highlight the importance of these species in the development of plant protoplast-tissue culture technology and their significant contribution to our understanding of the basic problems of plant cell biology. Protoplast-tissue culture and cell genetic modification methods (mutant isolation, somatic hybridization and transformation) for *Nicotiana* and *Hyoscyamus* have been refined to such an extent that these species have gained the reputation of being the easiest to handle and most reliable for experimental and practical purposes.

2 Isolation of Protoplasts

2.1 Plant Material

Protoplasts can be isolated from any organ or tissue of the plant (e.g. shoots, root tips, petals), but leaf or suspension culture cells are most frequently used. Rigorous maintenance of standardized growth conditions is crucial for successful, reproducible protoplast culture. Field-grown plants provide inconsistent material. For greenhouse-grown *Nicotiana* plants the ideal conditions are 22 °C, 15 h daylight, approximately 10000–20000 lx, and weekly feeding with high nitrogen fertilizer (Watts et al. 1974). These conditions can be used to grow *H. muticus* plants as well. Plants to be used for protoplast isolation are placed in the dark for 24 h.

Sterility is a major concern, and a controlled environment, which can be provided under in vitro conditions, can alleviate some of the variations associated with greenhouse-grown plants. For these reasons shoot cultures are the most reliable source for the isolation of protoplasts. *N. tabacum* shoot cultures are routinely maintained on MS hormone-free medium (Murashige and Skoog 1962) at 26 °C with continuous light of 1000–3000 lx. The optimum shoot culture conditions for protoplast isolation in other *Nicotiana* species, however, may differ from those required for vegetative propagation. *N. sylvestris* shoot culture, for instance, should be maintained on MS medium with 0.01 μM NAA and 0.03 M sucrose (Negrutiu and Mousseau 1980).

Since cultured cells are already conditioned to grow in vitro, they offer certain advantages over leaf mesophyll tissue. On the other hand, cultured cells are prone to accumulate chromosomal abnormalities (Fig. 2) and exhibit a gradual loss of morphogenetic ability. Suspension cultures, which are maintained in log phase by frequent subculturing (e.g. 3–4 days), have a higher growth rate, are more homogenous, and are generally superior as a source of protoplasts to callus cultures. When protoplasts are prepared from suspension cultures for fusion experiments, it is important to establish and maintain the cell cultures of both fusion partners under the same culture conditions.

2.2 Factors Influencing Protoplasts Yield

Protoplasts are isolated by treating tissues with mixture of cell wall degrading enzymes under osmotically adjusted conditions. Inconsistency between batches of enzymes is often due to such impurities as salts, phenolics, proteases, lipases and various contaminating enzymes. Desalting (Patnaik et al. 1981) of repurification (Schenk and Hildebrandt 1969) of commercial enzymes can improve protoplast yield and viability.

The age and physiological condition of the plant material is crucial, affecting not only the digestability of cell wall material, but also the osmotic properties of the cells and released plasts. The first three fully expanded leaves of 3–4 week old shoot cultures provide protoplasts with the highest plating efficiency. Greenhouse-grown plants should be transferred to an incubator (500 lx) for 2 days at 26 °C, before protoplast isolation. Suspension cultures should be in the logarithmic phase, 1–2 days after the last subculture.

It is necessary to determine osmotic conditions under which both partners may be maintained. For instance, for protoplasts derived from suspension cultures of *Nicotiana* and *Hyoscyamus*, the optimal isoosmotic conditions is 400 mOs/kg H_2O (Lazar et al. 1983).

For mesophyll protoplasts, preplasmolysis of leaf tissue prior to enzyme treatment can enhance yield and recovery (Cocking 1972). Leaves are always sliced into fine strips (1 mm), and may be evacuated for 2–3 min in the enzyme solution. *Nicotiana* leaves are incubated in K3 medium (Nagy and Maliga 1976) containing 0.4 M sucrose as osmotic stabilizer and the digestive enzymes cellulase (Onozuka R10, 1%) and Macerozyme R10 (0.5%). *Hyoscyamus* leaves are incubated in an enzyme solution containing cellulase (Onozuka R10, 0.2%) and Macerozyme R10 (0.5%) in mannitol (0.25 M) and $CaCl_2$ (0.1 M) as osmotic stabilizer. Protoplast isolation from suspension cultures of *Nicotiana* and *Hyoscyamus* are carried out as described in Section 7.1.

The incubation conditions of protoplast isolation (pH, temperature, light intensity, type of osmoticum, duration of treatment) are also critical, and should be established empirically. In general, the following conditions provide consistently high yield: suspension culture cells are agitated continuously (50 rpm) and leaf tissue swirled occasionally for 5–8 h with digestive enzymes, pH 5.4, in the dark at room temperature (max 30 °C). For overnight incubation the enzyme concentration, the temperature (<20 °C), or both should be reduced to avoid overtreatment (Strauss et al. 1981).

3 Protoplast Culture

Several properties of the donor tissue, including the genotype, physiological state, and the organ or tissue type, can influence the developmental response and nutritional requirements of the cultured protoplasts.

Protoplast culture media (Table 1) often support the growth requirement of protoplasts from a wide variety of species. The NT medium (Nagata and Takebe 1971) was originally used for mesophyll protoplasts, and the AA medium (Glimelius et al. 1978) used for protoplasts derived from the nitrate reductase-de-

Table 1. Various media used in this study

	AA cell culture medium/l	AA protoplast culture medium/l	Nitrate selection medium/l
$MgSO_4 \times 7H_2O$	370 mg	370 mg	370 mg
KH_2PO_4	340 mg	340 mg	340 mg
$CaCl_2 \times 2H_2O$	440 mg	440 mg	440 mg
KCl	2950 mg	2950 mg	–
KNO_3	–	–	3800 mg
$FeSO_4-Na_2EDTA$	5 ml	5 ml	5 ml
MS micro (1000×conc.)	1 ml	1 ml	1 ml
Nicotinic acid	0.5 mg	0.5 mg	0.5 mg
Thiamine-HCl	0.4 mg	0.4 mg	0.4 mg
Pyridoxine	0.1 mg	0.1 mg	0.1 mg
Myo-inositol	80 mg	80 mg	20 mg
Glutamate	876 mg	876 mg	–
Aspartate	226 mg	226 mg	–
Arginine	174 mg	174 mg	–
Glycine	7.5 mg	7.5 mg	–
Sucrose	20 g	68.46 g	20 g
Xylose	–	150 mg	–
Arabinose	–	150 mg	–
Glucose	–	150 mg	–
NAA	–	1 mg	–
BAP	–	0.5 mg	–
2,4-D	1 mg	–	1 mg
Kin	0.2 mg	–	0.2 mg
Gibberellic acid	0.1 mg	–	0.1 mg
	pH 5.7	pH 5.7	pH 5.7
	Autoclaved	Filter-sterilized	Autoclaved

$FeSO_4$-EDTA: for 100 ml, 557 mg $FeSO_4 \times 7H_2O$ and 745 mg Na_2EDTA

MS micro: for 100 ml (1000×conc.)

$MnSO_4$	2.23 g
H_3BO_3	0.62 g
$ZnSO_4 \times 7H_2O$	0.86 g
$Na_2MoO_4 \times 2H_2O$	25 mg
$CuSO_4$	2.5 mg
$CoCl_2 \times 6H_2O$	2.5 mg
KI	83 mg

ficient cell lines of *N. tabacum*. Both media work as well for the corresponding protoplasts cultures of *H. muticus* (Strauss et al. 1981; Lazar et al. 1983), supporting division frequencies up to 70%.

Protoplasts can be grown in suspension culture, in hanging droplets (Potrykus et al. 1976) or microdroplets (Kao 1977), or directly plated in soft agar (Nagata and Takebe 1971), and may be maintained with or without the presence of feeder cells (Cella and Galun 1980; Firoozabady 1986). However, bead culture, the growth in agitated liquid medium of protoplasts embedded in agarose beads, seems to be superior both for *Nicotiana* and *Hyoscyamus* in terms of division frequencies and plating efficiencies (Shillito et al. 1983).

The optimum protoplast density is between 10^4-10^5 per ml, depending on the source of protoplasts, medium composition and culture method. Protoplasts may burst upon direct plating into agarose. It is best to culture the protoplasts in liquid medium for 4–5 days at a density of 10^5 protoplasts/ml, then plate them into agarose solidified (Seakem agarose, 0.5%) medium at a density of 2×10^4 protoplasts/ml, and initiate agarose bead cultures. High plating efficiencies, 50–70%, can be obtained from both *H. muticus* and *N. tabacum* mesophyll protoplasts cultures using the NTC medium (Strauss et al. 1981) under the conditions described above.

Since freshly isolated protoplasts can burst upon exposure to light, cultures are incubated in the dark for the first 2–3 days, then transferred to low light (500 lx). After approximately 10 days in culture, the concentration of osmotic stabilizer should be reduced in a stepwise manner, over a period of 10–14 days. Protoplasts, however, can be grown to callus in the presence of high osmotic pressure, although the growth rate is significantly slower.

4 Plant Regeneration

The application of protoplast technology for practical purposes depends on the morphogenic ability of protoplast-derived cells. In both *Nicotiana* and *Hyoscyamus* morphogenesis can be induced fairly easily. High efficiency plant regeneration can be achieved through shoot formation from protoplast-derived calli (Wernicke and Thomas 1980), or through somatic embryogenesis directly from the protoplast stage (Lorz et al. 1977, 1979). With the application of double filter feeder plates the time required for plant regeneration from mesophyll protoplasts can be reduced (Firoozabady 1986).

5 Somatic Hybridization

Fusion treatment of two genetically different protoplast populations always results in a mixture of homokaryotic, heterokaryotic and unfused protoplasts. Therefore the overall efficiency of somatic hybridization depends primarily on the efficiency of selection for the low number of fusion products. The selection strate-

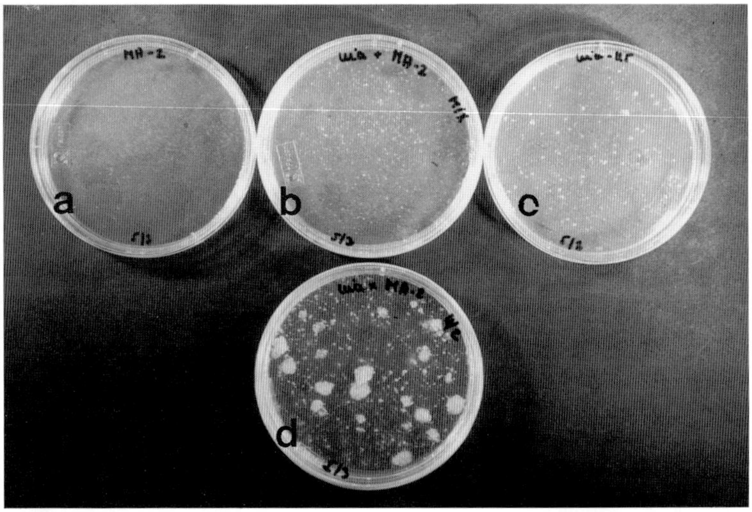

Fig. 1. Complementation selection for intergeneric somatic hybrids after fusion of nitrate reductase-deficient lines of *N. tabacum* (nia-115) and *H. muticus* (MA-2). Three weeks after the fusion treatment cell aggregates were plated to selective nitrate medium and incubated for 4 weeks **a** *H. muticus* MA-2 homokaryotic fusion control. **b** Parental coculture "cross feeding" control. **c** *N. tabacum* nia-115 homokaryotic fusion control. **d** Prototrophic intergeneric somatic hybrids selected after protoplast fusion of *N. tabacum* (nia-115)× *H. muticus* (MA-2) (G. Lazar unpubl.)

gy depends upon the aim one wishes to achieve (e.g. somatic hybrid or cybrid) and the availability of selectable markers. Selection system based on genetic complementation are the most reliable. Fortunately, for *Nicotiana* and *Hyoscyamus* species a large number of auxotrophic lines are available (see review by Bourgin 1986), with an abundant representation of nitrate reductase-deficient variants (89 cell lines). If the fusion partners have no selectable markers, one may select by physical means (see review Lazar 1983) or introduce selectable prokaryotic dominant resistance genes by *Agrobacterium*-mediated transformation (Horsch et al. 1985).

As a rule parental homokaryotic fusion controls and mixed parental coculture "cross-feeding" controls should always be established under the same culture conditions and density as for the heterofusion treatment (Fig. 1).

When recovery of somatic hybrid plants is required, the morphogenic ability of fusion partners as well as the timing of morphogenic induction after protoplast fusion are critical. Achieving morphogenesis at the early phase in the development of fusion products may ensure plant regeneration before extensive aneuploidization takes place. Protoplast fusion between different isogenic auxotrophic lines of *N. tabacum* and *H. muticus* has resulted in either non-morphogenic (Lazar et al. 1983) or morphogenic (Potrykus et al. 1984) intergeneric somatic hybrids. The *N. tabacum* cell lines used for somatic hybridization were the nitrate reductase mutants nia-115 and cnx-68, deficient in the apoenzyme or the molybdenum cofactor of the nitrate reductase respectively (Muller and Grafe 1978). The fusion partners were the *H. muticus* MA-2 mutant, deficient in the

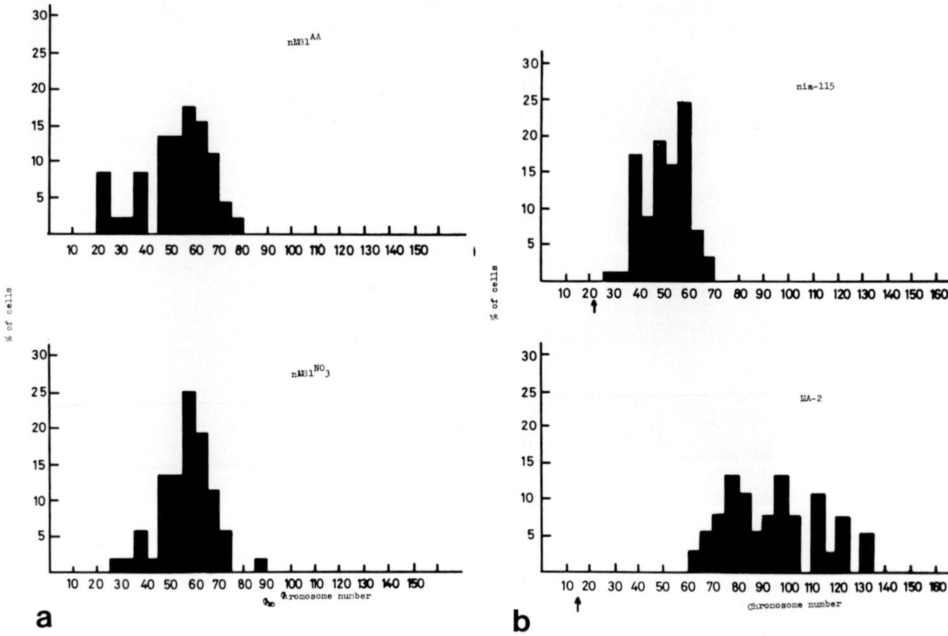

Fig. 2. **a** Chromosome number variation in suspension cultures of nM81 somatic hybrid line. The $nM81^{NO_3}$ subline is maintained under continuous selection, whereas the $nM81^{AA}$ subline is maintained under non-selective conditions. **b** Chromosome number variations in the nitrate reductase-deficient parental *N. tabacum* (nia-115) and *H. muticus* (MA-2) lines is cell suspension cultures. *Arrows* indicate the initial chromosome numbers. (G. Lazar unpubl.)

molybdenum cofactor of the nitrate reductase (Strauss et al. 1981), or the nic-IV2 nicotinamide auxotroph mutant (Gebhardt et al. 1981) respectively. All of the mutant lines were isolated from haploid cells, with chromosome numbers in *Nicotiana* n = x = 24 and in *Hyoscyamus* n = x = 28.

In the first set of fusion experiments the extensive chromosome number fluctuation in the intergeneric somatic hybrids resulted in a complete loss of the morphogenic response. This chromosome number variation, however, was not due to intergeneric genomic inbalance. Rather it was the manifestation of the intrinsic polyploidization and aneuploidization of *N. tabacum* (nia-115) and *H. muticus* (MA-2) parental lines prior to somatic hybridization (Fig. 2). During 2–4 years in culture, both parental cell lines developed extensive cytogenetic heterogenity. The nia-115 line exhibited a chromosome number variation between 25–70 (Fig. 2b). The extent of cytogenetic instability, however, was more dramatic in the MA-2 line, which exhibited aneuploidization at the higher ploidy levels, displaying chromosome number variations between 60 and 130 (Fig. 2b) This intrinsic cytogenetic instability, however, had no significant effect on the viability of somatic hybrids. The comparable complementation frequencies in the intraspecific *Nicotiana* nia-115 × cnx-68 and in the intergeneric *N. tabacum* nia-115 ×*H. muticus* MA-2-fusion combinations ($1-3\times10^{-3}$, calculated from total protoplast number) indicated compatibility between the genomes of the two species

Table 2. Complementation frequencies following different fusion combinations of NR deficient cell lines. (Lazar et al. 1983)

Fusion combinations	No. of independent fusion treatments	Total no. of selected hybrid clones	Complementation frequencies
nia-115×MA-2	18	200	$0-3.1\times10^{-3}$
nia-115×cnx-68	9	206	1.5×10^{-4} to 1.9×10^{-3}
nia-115 – nia-115	8	0	0
Ma-2×MA-2	8	0	0
cnx-60×cnx-68	8	0	0

in the selected calli (Table 2). Isoenzyme studies with seven enzymes (esterase, peroxidase, glucose-6-phosphate dehydrogenase, alcohol dehydrogenase, glutamate oxaloacetate transaminase, acid phosphatase, malate dehydrogenase) indicated that both fusion partners contributed equally to the biochemical makeup of the somatic hybrids (Lazar unpublished). This suggested the existence of a "symmetric genetic constitution" and the absence of somatic incompatibility at the dedifferentiated cell level. Metabolic cooperation between the two parental genomes, however, exists only at the level of housekeeping functions.

When undifferentiated tissue was forced to differentiate by morphogenic induction, the somatic incompatibility between *Nicotiana* and *Hyoscyamus* became apparent (Potrykus et al. 1984). Induction of morphogenesis and recovery of somatic hybrid plants have been successfully achieved only when the fusion partners were derived from young cell lines and when the morphogenesis was induced shortly (4–6 weeks) after the fusion treatments (Potrykus et al. 1984). The different clones expressed various degrees of morphogenic potential ranging from non-morphogenic, over various types of distorted shoot formation, to perfectly normal shoots and leaves (Fig. 3A, B). Further development to flowering plants was only possible from those clones which produced normal-looking shoots. Direct rooting of shoots was possible only in three clones after repeated subculture on agarose solidified medium (Fig. 3C). The routine procedure for plant regeneration was grafting onto *Nicotiana tabacum* (Fig. 3D).

These shoots could be easily rooted with commercial rooting powder (Fig. 3E, F). From seven clones ca. 100 plants were grown to maturity.

Shoot regeneration from hybrid cells selected for those genotypes which were able to express the regular morphogenetic program without interference from the other genome. As a result, all regenerants recovered from *N. tabacum* × *H. muticus* somatic hybrid cells showed extensive unidirectional loss of the *N. tabacum* genome (Saul and Potrykus 1984). The regenerated hybrid plants resembled *Hyoscyamus* in their morphology (Fig. 3), but retained and expressed the complementing gene from *Nicotiana,* along with a limited amount of DNA.

Spontaneous random bidirectional or unidirectional chromosome elimination in intergeneric somatic hybrids is a general phenomenon (see review by Harms 1983). Unidirectional chromosome elimination is more advantageous than random bidirectional elimination, because it establishes a donor-recipient relation-

Fig. 3. Regeneration of intergeneric somatic hybrid plants after the fusion of a nicotinamide-requiring *H. muticus* cell line (nic^{-}IVH2) with a nitrate reductase-deficient *N. tabacum* cell line (cnx-68). (Potrykus et al. 1984). **A** Variation in the morphogenic response among the selected clones. **B** Irregular morphogenesis. In this type of clone development beyond the teratoma-like stage has never been observed. **C** Rooting of primary shoot on agarose solidified hormone-free selective medium. **D** Multiple shoot formation from an intergeneric hybrid after grafting to *N. tabacum.* **E** Rooting of a previously grafted shoot results in normal plant development. **F** Flowering somatic hybrid plant

ship between the fusion partners. This is of great importance for practical genetic manipulations, since cotransfer of undesired genetic information is limited or eliminated. Furthermore, it can be exploited for gene mapping or linkage analysis, as was done with somatic hybrids of mammalian cells (McKusick and Ruddle 1977).

Fig. 3E, F

6 Conclusions and Prospects

Somatic hybridization was originally proposed as a technology with the potential to overcome some of the natural incompatibility barriers existing between non-crossable species. In *Hyoscyamus* × *Nicotiana* intergeneric combination, the results partially fulfilled these expectations. *Hyoscyamus*-like somatic hybrids carrying limited nuclear genetic information from the fusion partner *N. tabacum* could be regenerated and grown to maturity. Meiotic abnormalites, which give rise to a predominance of abortive seeds in self-crossed intergeneric hybrids, still hamper large-scale sexual propagation and the introduction of the new genotype into breeding programs.

The biochemical potential of somatic hybrids, including the ability to control the production of desirable secondary metabolities, has yet to be exploited. Biosynthetic pathways characteristic of parental species have been recapitulated in somatic hybrids (Nagao 1982), but the more interesting possibility of obtaining the synthesis and accumulation of new alkaloid species has yet to be explored.

7 Protocol

The following protocol is the best of a number of methods we have tried in our laboratory, based on a number of independent experiments.

Cell lines:

N. tabacum	nia-115	nitrate reductase apoenzyme-deficient.
	cnx-68	nitrate reductase Mo cofactor-deficient (Muller and Grafe 1978)
H. muticus	MA-2	nitrate reductase Mo cofactor-deficient. (Strauss et al. 1981)
	nic-IVH2	nicotinamide auxotroph. (Gebhardt et al. 1981)

7.1 Optimal Conditions/Medium for the Isolation and Purification of Protoplasts

Suspension cultures of the parental lines are maintained in AA cell culture medium (Muller and Grafe 1978) with weekly subcultivation (1:2 dilution with fresh medium) at 26 °C in the dark. For the nic-IVH2 auxotrophic line, the medium should be supplemented with 5 mg/l nicotinic acid. Protoplasts are isolated 1–2 days after the last subculture. Cell suspensions are sedimented, then 1 ml settled volume of the cells is resuspended in 10 ml NH enzyme and incubated for 5–6 h with gentle agitation (<50 rpm) at room temperature in the dark.

Protoplasts are separated from the undigested cell clumps by filtration through a 50-µm pore size sieve, and diluted with equal volume of washing solution to reduce the density of the enzyme solution for the subsequent pelleting (<100 g/3 min, in swinging bucket centrifuge). The protoplasts are resuspended in 1-ml washing solution and further purified by overlaying a 0.5 M sucrose cushion. After sedimentation for 10 min/100 *g* the protoplasts are collected from the interphase, washed once with SCT buffer and resuspended in SCT buffer at the final density of 10^5 protoplasts per ml.

NH enzyme solution for protoplast isolation from suspension cultures of H. muticus and N. tabacum:

cellulase (Onozuka R10)	2% w/v
Macerozyme R10	1% w/v
dissolved in:	
mannitol	0.3 M
$CaCl_2 \times 2H_2O$	0.04 M
MES	0.5%
pH 5.4 (KOH), filter sterilized	

Protoplast washing solution:

$CaCl_2 \times 2H_2O$	0.16 M
MES	0.5%
pH 5.6, autoclaved	

SCT solution for resuspending protoplasts before fusion:

sorbitol	0.3 M
$CaCl_2 \times 2H_2O$	0.05 M
Tris-HCl pH 7.0	0.002 M
autoclaved	

7.2 Optimal Conditions/Medium for Protoplast Fusion

The two parental protoplast suspensions are mixed (1:1) in a sterile centrifuge tube, and 1 ml aliquots are pipetted in 150 µl droplets into plastic Petri dishes (60×15 mm). Protoplasts are allowed to sediment for 10–15 min to create a uniform single layer on the bottom of the Petri dish. Roughly the same volume (150 μl) of 50% PEG 4000 is added to each droplet and incubated for 10 min. Thereafter the Petri dish is gradually filled with 3 ml of post-fusion buffer in a period of 10 min, and washed once with the same buffer for 5 min. Finally protoplasts are washed twice for 5 min with AA protoplast culture medium (Glimelius et al. 1978).

PEG solution for fusion induction:

Polyethylene glycol (PEG 4000)	50% w/v
dissolved in	
sorbitol	0.3 M
$CaCl_2 \times 2H_2O$	0.05 M
filter sterilized	

Post-fusion buffer:

sorbitol	0.15 M
$CCl_2 \times 2H_2O$	0.1 M
Tis-HCl, pH 9.0	0.0025 M
autoclaved	

7.3 Optimal Conditions/Medium for Protoplast Culture and Plant Regeneration

Three ml AA protoplasts culture medium is added to the protoplasts, then the Petri dish is sealed with PARAFILM and incubated in the dark at 26 °C. Two or three days later the cultures can be transferred to dim light (500 lx). After approximately 10 days in culture, the osmotic pressure of the medium is gradually decreased over a period of 2 weeks by sequential dilution with the appropriate selection medium (e.g. nitrate cell culture medium if the parental lines have nitrate reductase deficiency). Developing cell clumps are collected by sedimentation (200 *g*/5 min), washed twice with selection medium, then plated onto 0.7% agar solidified selection medium and incubated at 26 °C with a 12 h light period of 3000 lx. Proliferating clones are transferred to shoot regeneration medium (selection medium containing 0.01 mg/l NAA [1-naphthalene-acetic acid) and 0.25 mg/l BAP (6-benzylaminopurine)]. Rooting of primary shoots can be achieved by serial subculturing in agarose solidified hormone-free selection medium. An alternative possibility is grafting to *N. tabacum*, which can lead to vigorous shoot development. These shoots can then be easily rooted with commercial rooting powder.

Acknowledgements. This report summarizes a part of the somatic hybridization, cell genetic analysis programme which was carried out over the past few years in the group of Profesor Ingo Potrykus at the Friedrich Miescher Institut, Basel. The critical reading of the manuscript by Alan Dow is acknowledged.

References

Bajaj YPS (1983) In vitro production of haploids. In: Evans DA, Sharp WR, Ammirato PV, Yamada Y (eds) Handbook of plant cell culture, Vol 1. Macmillan, New York, pp 228–287

Bergmann L (1960) Growth and division of single cells of higher plants in vitro. J Gen Physiol 43:841–851

Bourgin JP (1986) Isolation and characterisation of mutant cell lines and plants: auxotrophs and conditional lethal mutants. In: Vasil IK (ed) Cell culture and somatic cell genetics of plants, Vol. 3. Plant regeneration and genetic variability, Academic Press, New York, pp 475–512

Carlson PS, Smith HH, Dearing RD (1972): Parasexual interspecific plant hybridisation. Proc Natl Acad Sci USA 69:2292–2294

Cella R, Galun E (1980) Utilization of irradiated carrot cell suspensions as feeder layer for cultured *Nicotiana* cells and protoplasts. Plant Sci Lett 19:243–252

Cocking EC (1972) Plant cell protoplasts–isolation and development. Annu Rev Plant Physiol 23:29–50

Firoozabady E (1986) Rapid plant regeneration from *Nicotiana* mesophyll protoplasts. Plant Sci 46:127–131

Gebhardt C, Schnebli V, King PJ (1981) Isolation of biochemical mutants using haploid mesophyll protoplasts of *Hyoscyamus muticus* II. Planta 153:81–89

Gebhardt C, Shimamoto K, Lazar G, Schnebli V, King PJ (1983) Isolation of biochemical mutants using haploid mesophyll protoplasts of *Hyoscyamus muticus* III. Planta 159:18–24

Glimelius K, Eriksson T, Grafe R, Muller A (1978) Somatic hybridisation of nitrate reductase deficient mutants of *Nicotiana tabacum* by protoplast fusion. Physiol Plant 44:273–277

Gray JC, Kung SD, Wildman SG, Sheen SJ (1974) Origin of *Nicotiana tabacum* detected by polypeptide composition of fraction 1 protein. Nature 252:226–227

Harms CT (1983) Somatic hybridization by plant protoplast fusion. In: Potrykus I, Harms CT, Hinnen A, Hutter R, King PJ, Shillito RD (eds) Protoplasts 1983, Experientia Suppl. Vol. 46. Birkhäuser, Basel, pp 69–84

Horsch RB, Fry JE, Hoffmann NL, Eizhholtz D, Rogers SG, Fraley RT (1985) A simple and general method for transferring genes into plants. Science 227:1229–1231

Lazar G (1983) Recent development in plant protoplast fusion and selection technology. In: Potrykus I, Harms CT, Hinnen A, Hutter R, King PJ, Shillito RD (eds) Protoplasts 1983, Experientia Suppl Vol. 46. Birkhäuser, Basel, pp 61–67

Lazar G, Fankhauser H, Potrykus I (1983) Complementation analysis of a nitrate reductase-deficient *Hyoscyamus muticus* cell line by somatic hybridisation. Mol Gen Genet 189:359–364

Lorz H, Potrykus I, Thomas E (1977) Somatic embryogenesis from tobacco protoplasts. Naturwissenschaften 64:439–440

Lorz H, Wernicke W, Potrykus I (1979) Culture and plant regeneration of *Hyoscyamus* protoplasts. Planta Medicico 36:21–29

Marton L, Wullems GJ, Molendijk L, Schilperoort RA (1979) In vitro transformation of cultured cells from *Nicotiana tabacum* by *Agrobacterium tumefaciens.* Nature 277:129–131

McKusick VA, Ruddle FM (1977) The status of the gene map of the human chromosome. Science 196:390–405

Muller AJ, Grafe R (1978) Isolation and characterisation of cell lines of *Nicotiana tabacum* lacking nitrate reductase. Mol Gen Genet 161:67–76

Murashige F, Skoog F (1962) A revised medium for rapid growth and bioassays with tobacco tissue cultures. Physiol Plant 15:473–497

Nagao T (1982) Somatic hybridisation by fusion of protoplasts III. Somatic hybrids of sexually incompatible combinations of *Nicotiana tabacum* + *N. repanda* and *N. tabacum* + *Salpiglossis sinuata.* Jap J Crop Sci 51:35—42

Nagata T, Takebe I (1971) Plating of isolated tobacco mesophyll protoplasts on agar medium. Planta 99:12–20

Nagy JI, Maliga P (1976) Callus induction and plant regeneration from mesophyll protoplasts of *N. sylvestris.* Z Pflanzenphysiol 78:453–455

Negrutiu I, Mousseau J (1980) Protoplast culture from in vitro grown plants of *Nicotiana sylvestris* Spegg. and Comes. Z Pflanzenphysiol 100:373–376

Okamuro JK, Goldberg RB (1985) Tobacco single copy DNA is highly homologous to sequences present in the genomes of its diploid progenitors. Mol Gen Genet 198:290–298

Paszkowski J, Shillito RD, Saul M, Mandak V, Hohn T, Hohn B, Potrykus I (1984) Direct gene transfer to plants. EMBO J 3:2717–2722

Patnaik G, Wilson D, Cocking EC (1981) Importance of enzyme purification for increased plating efficiency and plant regeneration from single protoplasts of *Petunia parodii.* Z Pflanzenphysiol 102:199–205

Potrykus I, Harms CT, Lorz H (1976) Problems in culturing cereal protoplasts. In: Dudits D, Farkas G, Maliga P (eds) Cell genetics in higher plants. Akademiai Kiado Budapest, pp 129–140

Potrykus I, Jia J, Lazar G, Saul M (1984) *Hyoscyamus muticus* + *Nicotiana tabacum* fusion hybrids selected via auxotroph complementation. Plant Cell Rep 3:68–71

Raghavan V (1975) Induction of haploid plants from anther culture of henbane. Z Pflanzenphysiol 76:89–92

Raghavan V (1978) Origin and development of pollen embryoids and pollen calluses in cultured anther segments of *Hyoscyamus niger* (Henbane). Am J Bot 65:982–1002

Saul MW, Potrykus I (1984) Species specific repetitive DNA used to identify interspecific somatic hybrids. Plant Cell Rep 3:65–67

Schenk RU, Hildebrandt AC (1969) Production of protoplasts from plant cells in liquid culture using purified commercial cellulases. Crop Sci 9:629–631

Shillito RD, Paszkowski J, Potrykus I (1983) Agarose plating and a bead-type culture technique enable and stimulate development of protoplast derived colonies in a number of plant species. Plant Cell Rep 2:244–247

Skoog F, Miller CO (1956) Chemical regulation of growth and organ formation in plant tissue cultures in vitro. Symp Soc Exp Biol 9:118–131

Strauss A, Bucher F, King PJ (1981) Isolation of biochemical mutants using haploid mesophyll protoplasts of *Hyoscyamus muticus*. Planta 153:75–80

Sunderland N, Wildon DC (1979) A note on the pretreatment of excised flower buds in float cultures of *Hyoscyamus* anthers. Plant Sci Lett 15:169–175

Takebe I, Labib G, Melchers G (1971) Regeneration of whole plants from isolated mesophyll protoplasts of tobacco. Naturwissenschaften 58:318–320

Vasil V, Hildebrandt AC (1965) Differentiation of tobacco plants from single isolated cells in microcultures. Science 150:889–892

Watts JM, Motoyoshi F, King JM (1974) Problems associated with the production of stable protoplasts of cells of tobacco mesophyll. Ann Bot 38:667–671

Wernicke W, Thomas E (1980) Studies on morphogenesis from isolated plant protoplasts: shoot formation from mesophyll protoplasts of *Hyoscyamus muticus* and *Nicotiana tabacum*. Plant Sci Lett 17:401–407

Wernicke W, Lorz H, Thomas E (1979) Plant regeneration from leaf protoplasts of haploid *Hyoscyamus muticus* L. produced via anther culture. Plant Sci Lett 15:239–249

Wulff EV (ed) (1943) An introduction to hystorycal plant geography. Chronica Botanyca Co, Waltham, Mass. USA, p 32

III.7 *Medicago* Protoplasts: Fusion, Culture and Plant Regeneration

D.M. Gilmour, T.J. Golds and M.R. Davey[1]

1 Introduction

Forage legumes are recognized for their nutritional value as animal feed and for their contribution to the nitrogen economy of the soil. Alfalfa or lucerne (*Medicago sativa* L.) was the first forage to be domesticated, and has been recognized as an agronomically useful crop plant for centuries. The evolution of a symbiotic relationship with *Rhizobium* and the development of a deep taproot are major factors which have contributed to the success of this plant. Currently, alfalfa is one of the most important forage species in the world, and is grown on about 35 million ha worldwide, with Argentina and the United States being the main producers.

The agronomic value of alfalfa could be further improved by genetic manipulation, including hybridization with other species (Barnes et al. 1977). Thus, desirable characteristics such as disease, frost and drought resistance, increased protein content and leaf production, and a creeping habit with grazing resistance, could be introduced from wild perennial *Medicago* species into the cultivated crop. In addition, attention has been given to several annual species of *Medicago,* including *M. disciformis, M. minima, M. polymorpha, M. rugosa, M. scutellata* and *M. truncatula.* These species offer the breeder germplasm resistant to alfalfa weevil and potato leafhopper, both major pests of alfalfa. Introduction of useful characteristics from wild species into the cultivated crop has not been possible by conventional breeding, and these efforts have also been hindered by lack of information on crossing relationships within the genus as a whole (Barnes and Ratcliffe 1969; Barnes et al. 1977; Arcioni et al. 1982; McCoy and Smith 1984). The ability to isolate protoplasts from both cultivated alfalfa and wild species, to fuse protoplasts both chemically and electrically, and to regenerate plants from protoplast-derived tissues, may circumvent some of these restrictions.

To date, the cultivated crop has been regenerated from protoplasts derived from leaves (Kao and Michayluk 1980; Santos et al. 1980; Johnson et al. 1981; Pezzotti et al. 1984; Atanassov and Brown 1984), seedling cotyledons (Lu et al. 1982), seedling roots (Xu et al. 1982; Pezzotti et al. 1984), and cell suspension cultures (Atanassov and Brown 1984, Pezzotti et al. 1984). In the case of wild *Medicago* species, plants have been regenerated from cell suspension protoplasts

[1] Plant Genetic Manipulation Group, Department of Botany, University of Nottingham, University Park Nottingham, NG7 2RD, UK

Biotechnology in Agriculture and Forestry, Vol. 8
Plant Protoplasts and Genetic Engineering I (ed. by Y.P.S. Bajaj)

of *M. coerulea, M. glutinosa* (Arcioni et al. 1982), and *M. falcata* (Gilmour and Davey unpubl.), from root and leaf protoplasts of *M. arborea* (Mariotti et al. 1984), and from seedling cotyledon protoplasts of *M. difalcata, M. falcata, M. glutinosa, M. hemicycla* and *M. varia* (Gilmour et al. 1987a).

2 Isolation of Protoplasts

Leaves of pot-grown plants, cotyledons and roots of seedlings, and cell suspension cultures are sources of protoplasts.

2.1 Leaf Protoplasts

Plants for the isolation of leaf protoplasts are to be grown under controlled conditions, preferably in soil-less compost (EFF Products Ltd., Guilford, Surrey UK), and maintained in a growth cabinet (Fisons Scientific Apparatus, Loughborough, UK) with day and night temperatures of 18 °C and 16 °C respectively and an 18 h daylength of 5000 lx, provided by daylight fluorescent tubes. The first fully expanded leaves are excised from 30–35-day-old plants and surface sterilized in 5% v/v Domestos bleach (Lever Bros. UK) for 10 min, followed by washing with sterile tap water (six changes). The lower epidermis is removed, and the peeled leaflets floated on the surface of a plasmolyzing solution (Frearson et al. 1973) containing CPW salts (KH_2PO_4 27.2, KNO_3 101, $CaCl_2 \cdot 2H_2O$ 1480, $MgSO_4 \cdot 7H_2O$ 246, KI 0.16, $CuSO_4 \cdot 5H_2O$ 0.025 mg/l) with the addition of 13% w/v mannitol (CPW13M) for 1 h, before incubating 1 g fresh weight of leaf tissue in 12 ml of cell wall degrading enzyme mixture for 6–16 h at 25 °C in the dark. The enzyme solution consists of 0.4 g Cellulase R10, 0.1 g Macerozyme R10 (Kinki Yakult MFG. Co. Ltd., Nishinomya, Japan) and 0.4 g Rhozyme HP150 (Röhm and Hass Co. Ltd., Philadelphia, USA) dissolved in 20 ml of distilled water. The solution is purified by passage through a Biogel P6 column (Biorad; 80×150 μm size) and the first enzyme peak collected (Santos et al. 1980). To this collected volume (approx. 35 ml) is added 4 ml of a×10 stock solution of the same salts as used in the plasmolysing solution. The pH is adjusted to 5.6 before filter sterilization (Satorius membrane filter, 0.2 μm pore size). Protoplasts are released by gentle agitation of the digested leaf pieces, and the enzyme-protoplast mixture filtered through a nylon sieve (64 μm mesh). Protoplasts are pelleted by centrifugation (100 *g*, 5 min) and purified by resuspension in 21% w/v sucrose solution with CPW salts (CPW 21S), and centrifuged (100 *g*, 5 min). Intact, floating protoplasts are transferred to the protoplast culture medium of Kao (1977) or Kao and Michayluk (1975), washed twice, and counted.

2.2 Cotyledon Protoplasts

Seeds of *Medicago* species are sterilized with a solution of 0.1% w/v mercuric chloride and 0.1% w/v solution lauryl sulphate for 15 min followed by 20% v/v

Domestos bleach (30 min). Seeds are washed with five changes of sterile tap water, before sowing on the surface of 50 ml aliquots of agar solidified (0.6% w/v; Sigma) hormone-free Murashige and Skoog (1962) based medium with 2% w/v sucrose (MSO). Seedlings are grown at 25 °C under continuous fluorescent light (2000 lx).

Cotyledons are excised from 4–7-day-old seedlings, cut transversely into 1 mm strips, and plasmolyzed for 1 h in CPW13M. Thirty cotyledon pairs are incubated in 10 ml of enzyme consisting of 2% w/v Rhozyme HP150, 2% w/v Meicelase (Meiji Seika Kaiska Ltd., Tokyo, Japan) and 0.03% w/v Macerozyme R10 in CPW13M, for 16 h at 25 °C on a rotary shaker (30 rpm). Protoplasts are released by gently squeezing the cotyledon strips, the enzyme protoplast mixture filtered (64 μm nylon mesh), and the protoplasts pelleted (80 *g*, 5 min). Cotyledon protoplasts cannot be purified from cellular debris by flotation in sucrose solution (CPW 21S). Consequently, protoplasts are washed by centrifuging through CPW13M (80 *g*, 5 min) followed by K8P protoplast culture medium (two changes), to remove particulate debris. Clean, intact protoplasts are resuspended in fresh KP8 medium and counted.

2.3 Root Protoplasts

Seeds are surface sterilized and sown on agar medium as in Section 2.2, but germinated in the dark. Roots (approx. 1 cm long) are excised from 2–3-day-old seedlings, and cut transversely into 0.5–1 mm segments. The latter are plasmolyzed in CPW13M for 1 h. The enzyme mixture, incubation conditions, and washing procedures are the same as those for cotyledon protoplasts.

2.4 Cell Suspension Protoplasts

Protoplasts can be obtained from cell suspensions derived from callus of leaf (Arcioni et al. 1982) and seedling explant origin (Atanassov and Brown 1984; Gilmour et al. 1987c) maintained in UM liquid medium (Uchimiya and Murashige 1974). Cells are harvested 3–4 days after subculture, and protoplasts released (Fig. 1) using the enzyme mixture described for the isolation of cotyledon

Fig. 1. Freshly isolated *Medicago sativa* cell suspension protoplasts. ×250

Fig. 2. *M. sativa* cell suspension protoplasts undergoing division after 72 h in culture. ×250

Fig. 3. Somatic embryos developing from protoplast-derived callus. The callus was maintained on UM medium with 2 mg/l 2,4-D and 0.25 mg/l kinetin for 4 weeks, before transfer to MS medium with 0.05 mg/l NAA and 0.5 mg/l BAP. ×10.5

Fig. 4. Leaf emergence from a germinating somatic embryo, 7 days after transfer to hormone-free MS medium. ×7.5

Fig. 5. A young *Medicago* plantlet with a well-developed root system after 2 weeks culture on hormone-free MS medium. ×2.5

Fig. 6. A flowering plant of *M. sativa* cv. Vertus regenerated from protoplast-derived callus, shown 4 months after transfer to the glasshouse. ×0.15

1
2
3
4
5
6

protoplasts (2–3 g fresh weight of cells in 20 ml enzyme mixture). Washing of cell suspension protoplasts is as described in Section 2.1.

3 Culture of Protoplasts

The highly enriched protoplast culture media of Kao (1977) and Kao and Michayluk (1975) have been used extensively for *Medicago* protoplasts isolated from different tissues. At Nottingham, these media are prepared according to the formulations given in the original references, but with some modifications. The protoplast culture medium 8P of Kao and Michayluk (1975), designated KM8P at Nottingham, has glucose increased from 68.4 g/l to 100 g/l. Similarly, in the case of the protoplast culture medium of Kao (1977), designated K8P at Nottingham, glucose is also increased from 68.4 g/l to 100 g/l, sucrose is increased from 125 mg/l to 250 mg/l, while thiamine HCl is reduced from 10 mg/l to 1 mg/l. Plating densities of 1×10^5/ml for mesophyll and cell suspension protoplasts cultured in KM8P, and 5×10^4/ml for root and cotyledon protoplasts in K8P result in optimum division frequencies for these protoplast systems, with cell division commencing within 72 h of culture (Fig. 2).

Protoplasts are cultured in 56×14 mm Nunclon dishes (A/S Nunc, Kamstrupnej 99, Kamstrup, DK-4000 Roskilde, Denmark), each containing 3 ml of culture medium. The osmoticum in the cultures is reduced gradually by diluting the protoplast culture medium with 1 ml volumes of protoplast medium mixed with cell culture medium at intervals of 7, 14 and 21 days. The proportion of protoplast to cell medium in the mixtures is 2:1, 1:1 and 0:1 at the three dilution stages (Lu et al. 1983). The protoplast medium KM8P is mixed with a cell medium (designated KM8 at Nottingham), prepared according to the formulation given for medium 8 by Kao and Michayluk (1975), but with thiamine HCl reduced from 10 mg/l to 1 mg/l. Similarly, the protoplast medium K8P is mixed with its respective cell medium, designated K8 in this laboratory. In the latter, the original formulation of Kao (1977) is modified as follows. Thiamine HCl is reduced from 10 mg/l to 1 mg/l, casamino acids from 250 to 125 mg/l, coconut milk from 2% to 1% v/v, and sucrose from 30 to 20 g/l. The concentration of the growth regulators, 2,4-D and zeatin are also changed from 0.2 mg/l to 0.1 mg/l and 0.5 mg/l to 0.2 mg/l respectively. Initial culture of protoplasts in the dark at 25 °C followed by transfer to the light (700 lx, 25 °C) after 7 days (cotyledon and cell suspension protoplasts), 14 days (root protoplasts) or 21–28 days (leaf protoplasts), stimulates cell colony formation. The division frequency (number of dividing protoplasts expressed as a percentage of the total population) is assessed after 7–10 days of culture.

Leaf protoplasts can also be cultured in 56×14 mm Nunclon dishes containing 4 ml of agar-solidified (0.5% w/v; Sigma) medium overlaid with 1 ml of liquid medium. The isolated protoplasts are plated in the liquid layer (Santos et al. 1980). Inclusion of a filter paper layer at the agar-liquid interface stimulates protoplast division and facilitates transfer of protoplast-derived colonies to fresh medium. Low-melting temperature agarose is a useful alternative to agar, and also increases the plating efficiency of *M. sativa* cell suspension protoplasts (Holbrook et al. 1985) and cotyledon protoplasts from wild Medicago species (Gilmour et al. 1987a) compared to agar and liquid media. In one application of the agarose culture technique, protoplasts are suspended at a density of 5×10^4/ml in a mixture of equal volumes of 1.2% w/v Sea Plaque agarose (FMC Corporation, Marine Colloids Division, Rockland, ME 04841, USA) and double strength K8P medium. Twenty to thirty droplets, each approximately 100 µl in volume, are allowed to gel on the bottom of 9-cm diameter plastic dishes (Sterilin Ltd. Middlesex, UK). The protoplast-containing droplets are then bathed in 6 ml of K8P medium. Plating of protoplasts in agarose droplets also facilitates estimation of the number of protoplasts undergoing division. Subsequent counting of colony throughput and its expression as a percentage of the number of dividing protoplasts enables an estimate to be obtained of the plating efficiency, i.e. the number of protoplasts which undergo sustained division.

High plating efficiencies of cotyledon protoplasts from wild *Medicago* species can be obtained when the protoplasts are mixed with nurse protoplasts from an albino cell line of *M. sativa* (Gilmour et al. 1987 a). K8P liquid medium containing the wild *Medicago* protoplasts at 5×10^4/ml, is mixed with 2 ml of the same medium containing the albino protoplasts, also at 5×10^4/ml. The albino protoplasts are cultured at 24 °C for 1 to 2 days prior to mixing with wild-type protoplasts. Cell colonies of the wild species can be recognized by their green colour compared to the white albino colonies after 4–6 weeks of culture, and are transferred manually, using fine forceps, to fresh medium.

Small-scale culture chambers are useful when culturing limited numbers of protoplasts, e.g. heterokaryons resulting from fusion experiments (see Sect. 5). In preparing such chambers, as described by Gilmour et al. (1987 c), pieces of Sartorius membrane filters (Sartorius Membrane Filter GmbH, P.O. Box 19, 3400 Göttingen, FRG), 3×0.75 cm in size, 12-µm pore size, are formed into cylinders and the ends sealed by carefully applying acetone. Cylinders are autoclaved and washed three times with K8P medium before being embedded on the bottom of 56×14 mm Nunclon dishes in 2 ml of a mixture of equal volumes of 1.2% w/v Sea Plaque agarose and double strength K8P medium. *Medicago* protoplasts from a readily available source, e.g. those from cell suspensions or seedling cotyledons, are plated as a nurse tissue at 5×10^4/ml outside the culture chamber, in 2 ml of K8P liquid medium overlaying the agarose layer. After 1–2 days, a low number of protoplasts of the species of interest is introduced into the centre of each chamber. Reduction of the osmoticum during culture is achieved by addition of fresh medium of lower osmotic potential to the nurse protoplasts. Using this technique, as few as 37 *M. sativa* cv. Regen-S cell suspension protoplasts have been cultured to callus (Gilmour et al. 1987c) the optical clarity of the agarose gelling agent permitting microscopic examination of protoplasts plated in the centre of the cellulose nitrate cylinder.

Table 1. Protoplast culture, somaclonal variation, somatic hybridization and transformation of *Medicago* species

Species	Protoplast source	Enzyme mixture	Protoplast culture medium	Culture response	Reference
Protoplast culture					
Medicago sativa cv. Canadian No 1, No 2	Leaf mesophyll	1% Cellulase R10 1% Rhozyme HP150 0.5% Sigma Pectinase 5.4% sorbitol 5.4% mannitol	Kao (1977)	Somatic embryos, plants	Kao and Michayluk (1980)
cv. Europe	Leaf mesophyll	5% Cellulase R10 1.25% Macerozyme R10 11% mannitol (Biogel P6 purified)	Kao and Michayluk (1975)	Somatic embryos, plants	Santos et al. (1980, 1983)
cv. Regen S	Leaf mesophyll	2% Cellulase R10 2% Rhozyme HP150 1% Macerozyme R10 5.5% sorbitol 5.5% mannitol	Modified Kao and Wetter (1977)	Somatic embryos, plants	Johnson et al. (1981)
cv. Europe	Root tips	4% Meicelase 2% Rhozyme HP150 0.3% Macerozyme R10 13% mannitol	Kao (1977); Kao and Michayluk (1975)	Somatic embryos, plants	Xu et al. (1982)
cv. Europe	Seedling cotyledons	As Xu et al. (1982)	Kao (1977)	Shoots, plants	Lu et al. (1982)
cv. Europe	Leaf mesophyll	As Santos et al. (1980)	Modified Kao (1977)	Somatic embryos, plants	Lu et al. (1983)
	Seedling cotyledons; Root tips	As Xu et al. (1982)			
cv. Adriana	Leaf mesophyll	As Kao and Michayluk (1980)	Kao and Michayluk (1975)	Somatic embryos, plants	Pezzotti et al. (1984)
	Root tips; Cell suspension	As Xu et al. (1982)			
cvs. Citation, Answer, Regen S	Leaf mesophyll Cell suspension	0.1% Cellulase R10 1% Driselase 1% Rhozyme HP150 7.3% mannitol	Modified Kao (1977)	Somatic embryos, plants	Atanassov and Brown (1984)

cv. Rambler	Cell suspension	As Atanassov and Brown (1984)	Modified Kao (1977) and plated in Sea Plaque agarose	Callus (at higher plating efficiency than in liquid medium)	Holbrook et al. (1985)
cvs. Rangelander, Regen S	Leaf mesophyll	0.5% Cellulase 0.5% Rhozyme HP 150 0.25% Macerozyme 11% mannitol	Modified Kao (1977) with electrical stimulation	Somatic embryos	Dijak et al. (1986)
cvs. Rambler, Regen S, Rangelander	Leaf mesophyll	As Dijak et al. (1986)	Modified Kao (1977)	Somatic embryos	Dijak and Brown (1987)
Medicago coerulea *M. glutinosa*	Leaf mesophyll Cell suspension	As Santos et al. (1980) As Xu et al. (1982)	Kao and Michayluk (1975)	Somatic embryos, plants	Arcioni et al. (1982)
M. arborea	Root tips Leaf mesophyll	As Xu et al. (1982) 2% Cellulase R 10 2% Cellulysin 1% Driselase 5.5% mannitol 5.5% sorbitol	Kao and Michayluk (1975)	Shoots, plants	Mariotti et al. (1984)
M. difalcata *M. falcata* *M. glutinosa* *M. hemicycla* *M. varia*	Seedling cotyledons	2% Meicelase 2% Rhozyme HP 150 0.03% Macerozyme R 10 13% mannitol	Kao (1977)	Somatic embryos, plants	Gilmour et al. (1987 c)
M. falcata *M. sativa*	Cell suspension Seedling cotyledons	As Gilmour et al. (1987 a)	Kao (1977) with membrane culture chambers embedded in agarose	Microcolonies from low numbers (<50) of plated protoplasts	Gilmour et al. (1988)

Table 1 (continued)

Species	Protoplast source	Enzyme mixture	Protoplast culture medium	Characteristics of regenerated plants	Reference
Somaclonal variation					
M. sativa cv. Europe	Leaf mesophyll	As Santos et al. (1980)	Kao and Michayluk (1975)	Phenotypic and cytological variation; *Verticillium* resistance	Latunde-Dada and Lucas (1983)
cv. Regen S	Leaf mesophyll	As Johnson et al. (1981)	Modified Kao and Wetter (1977)	Phenotypic and cytological variation. Resistance to *Fusarium solani* and/or *F. roseum* and *Rhizoctonia solani*	Johnson et al. (1984)

Fusion partners	Protoplast source	Fusion method	Heterokaryon culture	Culture response	Reference
Somatic hybridisation					
M. falcata + *M. sativa*	Leaf mesophyll	High pH/Ca^{2+} (Keller and Melchers 1973)	Kao (1977)	Somatic hybrid plants	Téoulé (1983)
M. arborea + *M. sativa*	Leaf mesophyll Root; cell suspension	Electrofusion	Kao and Michayluk (1975) in agarose layers	Putative hybrid colonies	Arcioni et al. (1986)
M. falcata + *M. sativa*	Cell suspension Seedling cotyledons	PEG with high pH/Ca^{2+}	Modified Kao (1977); Heterokaryons mixed with albino *M. sativa* cell suspension protoplasts as nurse	Hybrid callus	Gilmour et al. (1987b)
M. quasifalcata + *M. sativa*	Seedling cotyledons	PEG with high pH/Ca^{2+}			

Species	Protoplast source	Enzyme mixture	Protoplast culture medium	Culture response	Reference
Transformation					
M. sativa cv. Rambler	Cell suspension	0.1% Cellulase RS 1.0% Driselase 1.0% Rhozyme HP150 7.3% Mannitol	Modified Kao (1977)	Transformed callus following microinjection of Ti plasmid	Reich et al. (1986a, b)

4 Regeneration of Plants

Regeneration from protoplast-derived tissues of *Medicago* occurs mainly by embryogenesis (Fig. 3), with one report of organogenesis and somatic embryogenesis occurring simultaneously in *M. sativa* (Santos et al. 1980), and regeneration via organogenesis in *M. arborea* (Mariotti et al. 1984a). Regeneration has been shown to be highly genotype-dependent. Indeed, this is not surprising in an open-pollinated plant such as alfalfa, since each cultivar represents a highly heterogeneous population (Kao and Michayluk 1980). Such variation in regeneration capability has been reported in tissues derived from mesophyll protoplasts (Kao and Michayluk 1981) and callus from seedling explants (Mitten et al. 1984; Brown and Atanassov 1985; Nagarajan et al. 1986). Initially, studies of plant regeneration in *Medicago* were restricted to a selected line of cv. Saranac called Regen-S, superior in its in vitro embryogenic response (Bingham et al. 1975). However, the creeping-rooted *Medicago* cultivars, which contain a high input of germplasm from *M. falcata* and Ladak, have now been shown to have the highest regeneration capability (Brown and Atanassov 1985).

Embryogenesis was first reported from alfalfa mesophyll protoplasts by Kao and Michayluk (1980), globular embryos being formed in liquid medium after 3–4 weeks of culture. Protoplasts from one genotype formed embryos in liquid medium containing 0.1 mg/l benzyladenine and 1 mg/l naphthaleneacetic acid (NAA), the formation of globular structures being stimulated with 0.1 mg/l zeatin riboside and 1 mg/l 2,4-dichlorophenoxyacetic acid (2,4-D). Embryo growth was enhanced by addition of soluble starch to the hormone-free medium used to dilute the protoplast cultures. Embryos which had reached the size of mature alfalfa seeds germinated within 24 h after transfer to a filter paper substratum saturated with fresh starchless hormone-free medium. Other workers have reported modifications of the Kao and Michayluk (1980) procedure, in terms of media composition and the times of dilution, for embryogenesis from mesophyll protoplasts of Regen S clones of alfalfa (Johnson et al. 1981). Similarly, Lu et al. (1983) also modified the original procedure for leaf protoplasts of European lucerne, but reported starch-containing medium to be of little benefit in embryogenesis. However, as indicated by Johnson et al. (1981), the auxin 2,4-D appears to play a significant role in the induction of embryogenesis is alfalfa, consistent with the ability of this growth regulator to induce embryogenesis in cultures of non-leguminous plants (Ammirato 1984).

The results of studies using explant-derived callus should be applicable to protoplast systems, particularly in the case of poorly responding genotypes. Thus, addition of amino acids to Schenk and Hildebrandt (1972) medium caused a threefold increase in embryogenesis in callus of *M. sativa* clone RA-3 (Stuart and Strickland 1984a). Proline was most stimulatory, although its effect was dependent on the ammonium concentration in the medium, regeneration being optimal with 100 mM proline and 25 mM ammonium (Stuart and Stickland 1984b). The same workers also investigated the effect of a range of carbohydrates on embryogenesis, with maltose and maltotriose improving embryo morphology and the development of embryos to plants. In agreement with the results of Kao and Michayluk (1980), soluble starch stimulated embryogenesis in explant-derived callus (Strickland et al. 1986).

The source tissues used for protoplast isolation and the culture method also determine the period required for embryogenesis. Lu et al. (1983) reported that in European lucerne the culture period was 60–70 days for leaf protoplasts, 40–50 days for green and etiolated seedling cotyledon protoplasts, and 30–35 days for root protoplasts in liquid medium. When protoplast-derived colonies were transferred from liquid to agar-solidified regeneration media, the culture periods required for embryogenesis in lucerne were extended to 90–100, 60–70 and 40–50 days for leaf, cotyledon and root protoplasts respectively (Lu et al. 1983). The media employed, and culture periods on these media were critical for embryogenesis and organogenesis in *M. sativa* (Santos et al. 1980). Protoplast-derived colonies formed friable yellow callus which developed green areas on its surface on UM-based medium containing 2 mg/l 2,4-D and 0.25 mg/l kinetin. Such green areas produced somatic embryos and/or shoots when the callus was subcultured to MS medium, containing 0.05 mg/l NAA and 0.5 mg/l benzylaminopurine for a maximum period of 4 weeks. An intermediate step in this regeneration procedure, where protoplast-derived colonies were transferred to agar solidified (0.6% w/v; Sigma) cell culture medium (Kao 1977) for 14–21 days before subculture to UM medium, promoted growth of colonies from cotyledon protoplasts of wild *Medicago* species into small callus tissues (Gilmour et al. 1987 a). A 4-week induction period on UM medium has been used successfully to stimulate embryogenesis in *M. coerulea* and *M. glutinosa* (Arcioni et al. 1982), and in *M. difalcata, M. falcata, M. glutinosa, M. hemicycla* and *M. varia* (Gilmour et al. 1987 a).

Using liquid medium, Kao and Michayluk (1980) reported that in the best preparations, over 50% of alfalfa cv. Canadian No. 1 mesophyll protoplasts developed into compact cellular clumps which gave rise directly to globular embryoids. However, in *M. sativa* cv. Europe (Lu et al. 1983) and in *M. falcata, M. glutinosa* and *M. hemicycla* (Gilmour et al. 1987 a), only a limited number of protoplast-derived cell clumps developed directly into green, free-floating globular structures. Higher rates of regeneration occurred through an intervening callus phase. Similarly, in a comparative study of the development of *M. sativa* mesophyll protoplasts, Dijak and Brown (1987) also reported callus development to be a prerequisite for somatic embryogenesis in the cultivars Rambler and Regen S, but protoplasts of the cultivar Rangelander underwent direct embryogenesis. Interestingly, exposure of protoplasts to low voltage electric fields altered the pattern of development and induced protoplasts of Regen S to follow the Rangelander pattern of direct embryogenesis (Dijak et al. 1986). Histological studies have shown that where callus development occurs, embryos arose from groups of meristematic cells situated at the periphery of cell clumps in liquid medium (Lu et al. 1983). Likewise, in *Medicago* callus cultured on agar solidified medium, embryos also developed from groups of meristematic cells (Santos et al. 1983).

Plantlet formation from embryos may occur whilst these structures are still attached to the parent tissue, or following transfer of individual embryos to hormone-free medium with 3.0% w/v sucrose (MSO). The emergence of a unifoliate leaf usually precedes the development of trifoliate leaves (Fig. 4), with extensive root formation occurring on plantlets 2–4 weeks after subculture to MSO agar medium (Fig. 5). Further multiplication often takes place by secondary embryogenesis from cotyledons, hypocotyls, and roots of existing embryos, and from hypocotyls of regenerated plantlets. Such adventitious embryos differ from protoplast-derived embryos in their origin from epidermal cells of the parent structures (Santos et al. 1983).

Subsequent establishment of in vitro-grown plants under glasshouse (Fig. 6) or field conditions can be readily achieved by transferring plantlets, 8–10 cm in height with well-developed root systems, to soil-less compost. Such potted plants are maintained in a controlled environment cabinet (18 h daylight, 5000 lx; 18 °C day; 16 °C night) for 3–4 weeks prior to transfer to the glasshouse or field plot. Desiccation, the major cause of plant loss during the early stages of transfer, can be prevented by covering pots with polythene bags to maintain humidity. Alternatively, plants can be transferred to a mist propagation unit immediately after potting, but under these conditions close attention must be paid to disease control and watering regimes.

5 Fusion of Protoplasts and Culture of Heterokaryons

Protoplast fusion has the potential to create novel plants because of the extensive nuclear-cytoplasmic combinations which may arise (Davey and Kumar 1983). Indeed, somatic hybridization should be of particular value in overcoming sexual incompatiblity between many *Medicago* species. Prior to 1986, selectable markers had not been introduced into *Medicago*, which precluded the use of biochemical selection to recover somatic hybrid cells resulting from protoplast fusion. However, fusion of green chloroplast-containing mesophyll and cotyledon protoplasts with chloroplast-free protoplasts from cell suspensions, seedling roots or etiolated cotyledons has facilitated visual identification of heterokaryons, enabling the latter to be isolated using a micromanipulator. Heterokaryon identification has been confirmed by labelling the colourless protoplast partner with fluorescein diacetate (FDA; 50 μl of a 5 mg/ml stock solution in acetone, added to 10 ml of enzyme solution prior to protoplast isolation). Following fusion, heterokaryons are identified under UV illumination by the presence of dual red and green fluorescence from the chlorophyll-containing and fluorochrome-labelled protoplast partners respectively.

Medicago protoplasts have been fused chemically, using large and small-scale techniques, and electrically. In large-scale chemical fusion, the density of isolated protoplasts is adjusted to 1×10^5/ml in CPW13M solution, and 4.0 ml aliquots of each partner mixed in 16×120 mm screw-capped tubes and pelleted by centrifugation (80 *g*, 5 min). The supernatant is removed and the protoplasts resuspended in 0.5 ml of CPW13M. Two ml of PEG solution (30% w/v PEG MW 6000 Koch Light; 4% w/v sucrose, 0.147% w/v $CaCl_2 \cdot 2H_2O$, sterilized by autoclaving) is added, and the mixture incubated (22 °C, 10–15 min). As much PEG is removed as possible and 4 ml of high pH/Ca^{2+} buffer, based on the formulation of Keller and Melchers (1973), is added (0.375% w/v glycine, 0.74% w/v $CaCl_2 \cdot 2H_2O$, 4% mannitol, pH 10.4, filter sterilized). After 10 min the volume is made to 15 ml with CPCW13M (CPW13M to which 0.74% w/v $CaCl_2 \cdot 2H_2O$

is added) and the tube contents mixed. Protoplasts are pelleted (80 *g*, 5 min), washed with CPCW13M, counted, and plated in K8P culture medium.

Small-scale fusion is useful when only a limited number of protoplasts are available. The density of each preparation is adjusted to 5×10^5/ml in CPW13M, and 200 µl of each solution mixed in a droplet placed in the centre of a 56×14 mm Nunclon dish. After 20 min, 3×50 µl of PEG are added to the edge of the droplet and allowed to coalesce. As much of the fusogen as possible is removed after 20–30 min, and replaced by 0.5 ml of high pH/Ca^{2+} solution. Aggregated protoplasts are washed with CPCW13M (10 min), followed by K8P culture medium (three changes, 120 min). Finally, 4 ml of K8P is added to the dish.

For electrofusion, freshly isolated *Medicago* protoplasts are washed by centrifugation through a solution of 13% w/v mannitol containing 1 mM $CaCl_2$ (80 *g*, 6 min; two changes), the protoplast density adjusted to 5×10^4/ml, and equal volumes of each protoplast partner mixed. One ml aliquots of the protoplast mixture are placed in the centre nine wells of 10 cm square Sterilin Petri dishes (the latter are divided into 25, 2 cm^2 chambers). The electrofusion apparatus, based on that described by Watts and King (1984) has a transferrable type of electrode consisting of five copper plates (18×30 cm), separated by 0.4×1.8×2.0 cm plastic spacers, connected to a pulse generator. The electrode is sterilized by rinsing with 70% ethanol before placing in a well containing the protoplast mixture. Protoplasts are aligned to form pearl chains in an AC electric field (500 K Hz, 40 V/cm) applied to the sample for 1–4 min (Fig. 7). The AC field is reduced to almost zero, and a DC pulse of 600 V/cm applied to the chamber for 2 ms to fuse the aligned protoplasts (Figs. 9–13). After delivery of the DC pulse, the electrode is transferred to another well and the procedure repeated. Two to three percent heterokaryon formation is achieved, protoplasts which coalesce taking up to 20 min to regain a spherical shape. One ml of protoplast culture medium (Kao 1977) is added to each chamber following electrofusion, and the dishes incubated overnight in the dark at 25 °C. Twenty four hours later, the protoplasts are transferred to 56×14 mm Nunclon dishes, and an additional 1.0 ml of culture medium added. The percentage of protoplast fusion depends on the protoplast systems and the procedure employed, for example, in fusing chlorophyll-containing seed-

Fig. 7. Alignment of *Medicago* chlorophyll-containing with etiolated seedling cotyledon protoplasts in an AC electric field. ×123

Fig. 8–13. A 20-min time sequence showing the fusion of chlorophyll-containing with etiolated seedling cotyledon protoplasts following application of a DC pulse. ×275

Fig. 14. Variation in morphology of leaves from four regenerated plants (**a–d**) of *M. sativa* cv. Vertus compared to the morphology of leaves of the parental plant used for protoplast isolation (**e**). ×0.5

Fig. 15. Heterokaryons resulting from the electrofusion of chlorophyll-containing with etiolated *Medicago* seedling cotyledon protoplasts. The etiolated protoplasts were FDA-labelled prior to fusion, and heterokaryons selected from unfused parental protoplasts and homokaryons using a Coulter Epics 541 flow cytometer. ×200

Fig. 16. Flow sorted *Medicago* heterokaryons viewed under UV illumination. Heterokaryons show dual red fluorescence (from chlorophyll of the green partner) and green fluorescence of the FDA-labelled partner. ×200

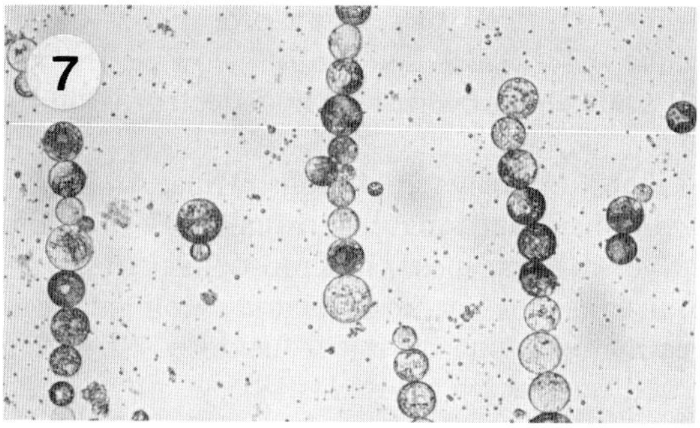
7

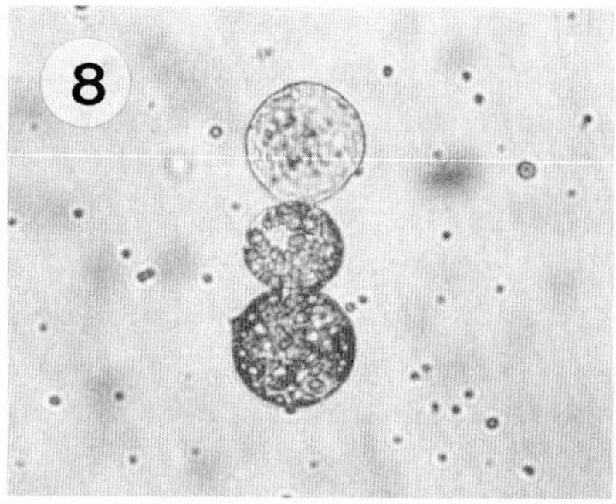
8

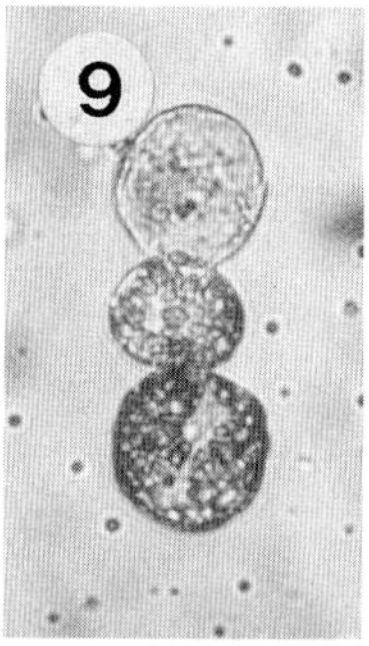
9

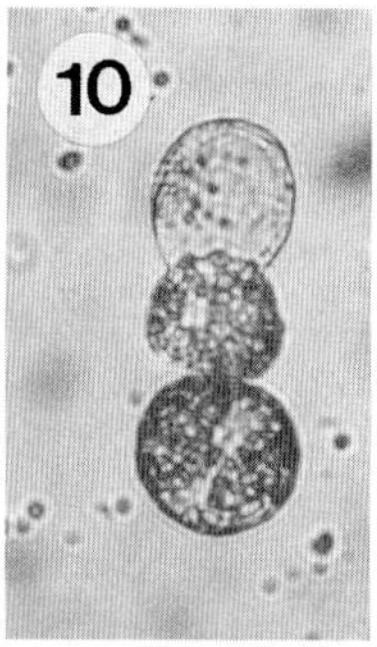
10

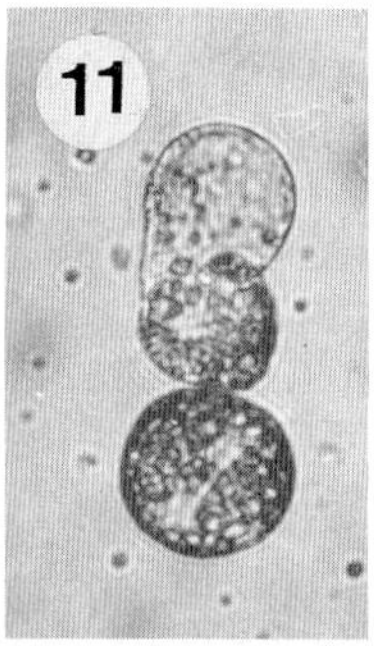
11

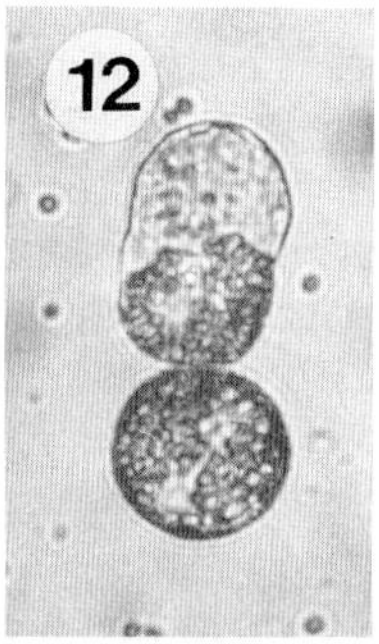
12

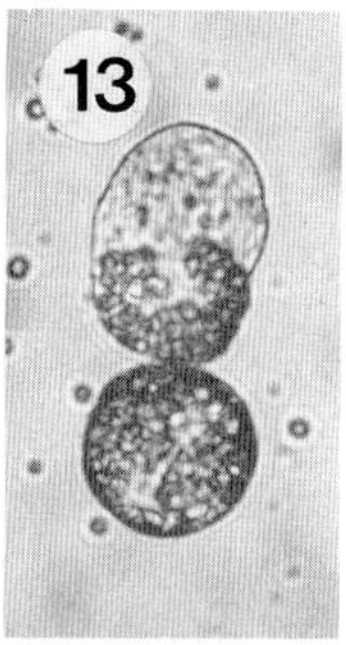
13

14
a
b
c
d
e

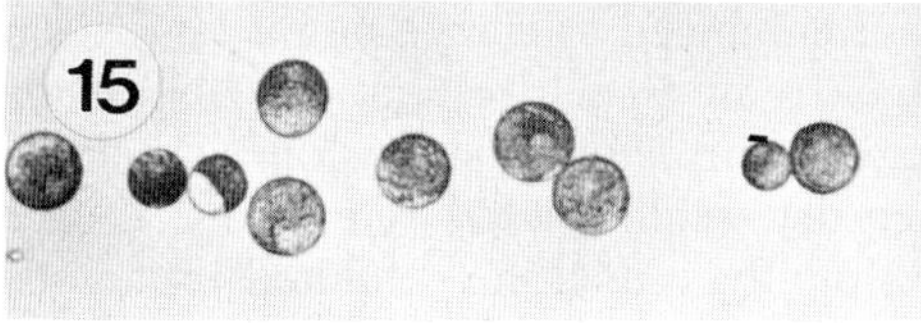
15

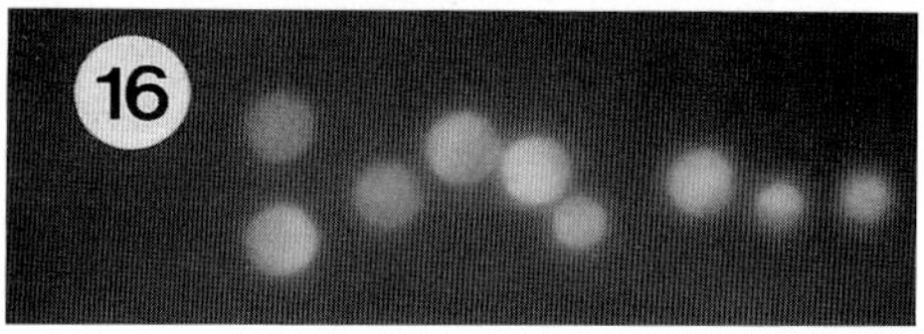
16

ling cotyledon protoplasts of *M. sativa* with cell suspension protoplasts of *M. falcata*, Gilmour (1986) obtained values of 1.5, 3.3 and 2.6% following the use of large-scale chemical fusion, small-scale chemical fusion, and electrofusion respectively.

There are reports of the successful culture of heterokaryons. Thus, Arcioni et al. (1986) reported somatic hybrid colonies following electrofusion of root and cell suspension protoplasts of *M. sativa* with leaf mesophyll protoplasts of *M. arborea*, while Gilmour et al. (1987b) used isoenzyme analysis and cytological evidence to confirm the hybridity of callus following chemical fusion of *M. sativa* seedling cotyledon protoplasts with cell suspension protoplasts of *M. falcata* and cotyledon protoplasts of *M. quasifalcata.*

To date, there is only one report of the production of a somatic hybrid plant by fusing protoplasts from a cytoplasmic male sterile line of *M. sativa* with a male fertile line of *M. falcata* (Téoulé 1983), these species being sexually compatible. In this case, mesophyll protoplasts from the two partners were fused using high pH/Ca^{2+} (Keller and Melchers 1973), and protoplast-derived embryos regenerated to plants. Of 59 embryos selected at random from the culture dishes, 28 formed plants, one of which was a somatic hybrid, as confirmed by its morphological characteristics (intermediate leaf shape, flower form and colour compared to the parents), and the electrophoretic banding patterns of peroxidase and β-amylase isoenzymes. The hybrid had a chromosome complement of 55 compared to 32 of each of the parents.

6 Miscellaneous Studies

Many workers have observed that plants regenerated from tissue cultures of several crops, including *Medicago* (Fig. 14) differ both morphologically and cytologically, offering the breeder a novel source of genetic variability (Larkin and Scowcroft 1981). Historically, such variants were described as "calliclones" if derived from callus, or "protoclones" when originating from protoplasts. Currently, the blanket term "somaclonal variation" is more widely applicable (Evans et al. 1984). In a study of somaclonal variation in *M. sativa*, plants regenerated from mesophyll protoplasts were highly tolerant to the wilt fungus *Verticillium albo-atrum* compared to the parental clone (Latunde-Dada and Lucas 1983). Since cytological analysis revealed the variant plants to have a ploidy level higher than the normal (4n = 32), it was concluded that enhanced tolerance might have been associated with the increase in gene dosage. In another investigation, field trials of mesophyll protoplast-derived alfalfa regenerants showed somaclones superior, equal to and inferior to their parents when dry weight yields were recorded over four harvests (Johnson et al. 1984). Additionally, one clone exhibited resistance to *Fusarium solani* and/or *F. roseum* and *Rhizoctonia solani.* Physiological assessments of protoplast-derived plants of *M. sativa* should determine whether such somaclonal variants exhibit improved photosynthesis and nitrogen fixation (Golds, Davey, Sheehy and Minchin unpubl.).

Recently, there has been interest in developing transformation systems for *Medicago*. Indeed, alfalfa protoplasts have been shown to be suitable for microinjection (Reich et al. 1986a). The intracellular compartments of cell suspension protoplasts were distinguished using fluorescent stains, enabling small droplets of olive oil (10^{-17}l) to be injected into the nuclei. This work preceded the injection of *Agrobacterium tumefaciens* Ti plasmid DNA and subsequent transformation of alfalfa protoplasts (Reich et al. 1986b). When nopaline synthase activity was used as a biochemical marker for transformation, 15–26% of the callus tissues recovered from microinjected protoplasts were nopaline-positive.

7 Conclusions and Prospects

Prospects for the production of a range of somatic hybrid and cybrid plants in the genus *Medicago* now appear promising, particularly with the identification of high regenerating cultivars of *Medicago* (Brown and Atanassov 1985) and the regeneration of several wild *Medicago* species from protoplasts (Arcioni et al. 1982; Mariotti et al. 1984; Gilmour et al. 1987a). Improved methods of protoplast culture using agarose solidified media (Holbrook et al. 1985; Gilmour et al. 1987a) or a nurse culture technique (Gilmour et al. 1987a) have increased the plating efficiency and colony formation from protoplasts. Recently, the frequency of chemically induced fusion of *Medicago* protoplasts has been elevated by the use of deionized polyethylene glycol (Kao and Saleem 1986). Similarly, high rates of heterokaryon formation (2–3%) have been obtained by electrofusion of *Medicago* protoplasts (this laboratory, unpubl.).

Introduction of plasmid DNA into the nucleus of alfalfa protoplasts by microinjection (Reich et al. 1986b) or transformation of *Medicago* cells by *Agrobacterium*-mediated gene transfer (Deak et al. 1986) may serve to introduce selectable markers for somatic hybridization, or other resistance genes, such as those for herbicide or pesticide tolerance.

Despite improvements in culture, fusion and plant regeneration from *Medicago* protoplasts, the difficulty of developing biochemical systems and the inability to select large heterokaryon populations by micromanipulation both impose limitations on the throughput of hybrid and cybrid protoplast-derived tissues. A technique likely to be applicable in the production of somatic hybrid plants in *Medicago*, is the use of fluorescence-activated cell sorting to isolate heterokaryons from unfused parental protoplasts following fusion (Afonso et al. 1985). Indeed, this approach has already been used to isolate fusion products resulting from the electrofusion of green seedling cotyledon protoplasts with etiolated, FDA-labelled cotyledon protoplasts of *M. sativa* (Figs. 15, 16) and will be extended to heterokaryons between protoplasts of *M. sativa* and those wild *Medicago* species habouring agronomically useful characteristics.

8 Protocol for the Isolation and Fusion of *Medicago* Seedling Cotyledon Protoplasts (See text for details of media composition)

Protoplast isolation

(1) Surface sterilize seeds by immersion in 0.1% w/v mercuric chloride and 0.1% w/v sodium lauryl sulphate (5 m) followed by 20% v/v Domestos bleach (30 min). Wash with five changes of sterile tap water.

(2) Sow seeds on 50 ml aliquots of agar solidified hormone-free MS medium with 2% w/v sucrose (MSO). Incubate at 25 °C under continuous fluorescent light (2000 lx). Incubate seeds in the dark to obtain etiolated seedlings.

(3) Excise cotyledons from 4–7-day-old seedlings. Cut transversely into 1 mm strips. Plasmolyze in CPW13M solution (1 h).

(4) Incubate cotyledon slices from 30 seedlings in 10 ml of 2% w/v Rhozyme HP150, 2% w/v Meicelase and 0.03% w/v Macerozyme R10 in CPW13M solution (16 h, 25 °C, 30 rpm). The enzyme mixture for etiolated cotyledon slices should contain FDA at 25 µg/ml.

(5) Gently squeeze the digested tissue to release the protoplasts, Filter the enzyme-protoplast suspension through a 64 µm mesh nylon sieve. Centrifuge (80 *g*, 5 min).

(6) Remove supernatant; resuspend protoplasts in CPW13M. Centrifuge (80 *g*, 5 min). Repeat this washing using two changes of K8P medium.

(7) Resuspend protoplasts in a known volume of K8P medium. Count the protoplasts.

Protoplast fusion (large-scale fusion of chlorophyll-containing protoplasts, with etiolated seedling cotyledon protoplasts).

(1) Adjust the density of the two protoplast suspensions to 1×10^5/ml in CPW13M solution.

(2) Mix 4 ml aliquots of each protoplast suspension in a 16×120 mm screw-capped tube. Centrifuge (80 *g*, 5 min).

(3) Remove the supernatant. Resuspend the protoplast pellet in 0.5 ml of CPW13M solution.

(4) Add 2 ml of PEG 6000 solution. Mix. Incubate at 22 °C for 10–15 min.

(5) Remove as much PEG as possible using a Pasteur pipette. Add 4 ml of high pH/Ca^{2+} buffer. Incubate for 10 min.

(6) Adjust the volume to 15 ml with CPCW13M. Mix the tube contents by gentle inversion. Pellet the protoplasts (80 *g*, 5 min). Wash the protoplasts in CPCW13M solution.

(7) Identify heterokaryons under UV illumination by their dual fluorescence (red chlorophyll autofluorescence; green FDA fluorescence).

(9) Isolate heterokaryons either (a) manually, using a micromanipulator, or (b) using a flow cytometer.

Culture of Heterokaryons

The method of culture will depend on the number of heterokaryons available. A small-scale technique using membrane chambers is ideal for limited numbers of manually isolated heterokaryons.

(1) Prepare chambers by forming 3×0.75 cm strips of 12 µm pore size Sartorius filter membrane into cylinders and sealing the ends with acetone.

(2) Autoclave the chambers and wash (three times) with K8P medium.

(3) Mix equal volumes of 1.2% w/v Sea Plaque agarose at 40 °C with double strength K8P medium. Dispense 2 ml volumes into 56×14 mm Nunclon dishes. Stand a cylinder vertically in the medium in the centre of each dish. Allow the medium to set.

(4) Place 2 ml of liquid K8P medium containing nurse protoplasts, e.g. those from seedling cotyledons, at a density of 5×10^4/ml over the agarose medium around the outside of each chamber.

(5) Incubate for 1–2 days at 25 °C in the dark.

(6) Introduce the heterokaryons into the chamber in 0.2 ml of liquid K8P medium.

(7) Gradually reduce the osmolality of the liquid layer after 14, 21 and 28 days of growth of the nurse protoplasts by adding 1 ml aliquots of K8P:K8 media. The latter should be mixed before use in the ratios of 2:1, 1:1, and 0:1 respectively.

(8) Transfer heterokaryon-derived cell colonies, when approx 1 mm in diameter, to agarose solidified K8 medium for further growth.

References

Afonso CL, Harkins KR, Thomas-Compton MA, Krejci AE, Galbraith DW (1985) Selection of somatic hybrid plants in *Nicotiana* through fluorescence-activated sorting of protoplasts. Bio/Technology 3:811–816

Ammirato PV (1984) Embryogenesis. In: Evans DA, Sharp WR, Ammirato PV, Yamada Y (eds) Handbook of plant cell culture, Vol. 1, Macmillan, New York, pp 82–123

Arcioni S, Davey MR, Santos AVP dos, Cocking EC (1982) Somatic embryogenesis in tissues from mesophyll and cell suspension protoplasts of *Medicago coerulea* and *M. glutinosa*. Z Pflanzenphysiol 106:105–110

Arcioni S, Damiani F, Pezzotti M (1986). Protoplast electrofusion between *Medicago sativa* and *Medicago aborea*. In: Somers DA, Gengenbach BG, Biesboer DD, Hackett WP, Green CE (eds) Abstr VI Int Congr Plant Tissue Cell Cult Univ Minnesota, Mineapolis, p 86

Atanassov A, Brown DCW (1984) Plant regeneration from suspension culture and mesophyll protoplasts of *Medicago sativa*. Plant Cell Tissue Organ Cult 3:149–162

Barnes DK, Ratcliffe RH (1969). Evaluation of annual species of *Medicago* as sources of alfalfa weevil resistance. Crop Sci 9:640–642

Barnes DK, Bingham ET, Murphy RP, Hunt OJ, Beard DF, Skrdla WH, Teuber LR (1977). Alfalfa germplasm in the United States: genetic vulnerability, use, improvement and maintenance. USDA Tech Bull 1571:1–21

Bingham ET, Hurley LV, Kaatz DM, Saunders JW (1975) Breeding alfalfa which regenerates from callus tissue in culture. Crop Sci 15:719–721

Brown DCW, Atanassov A (1985). Role of genetic background in somatic embryogenesis in *Medicago*. Plant Cell, Tissue Organ Cult 4:111–122

Davey MR, Kumar A (1983) Higher plant protoplasts–retrospect and prospect. Int Rev Cytol Suppl 16:219–299

Deak M, Kiss GB, Koncz C, Dudits D (1986) Transformation of *Medicago* by *Agrobacterium*: mediated gene transfer. Plant Cell Rep 5:97–100

Dijak M, Brown DCW (1987) Patterns of direct and indirect embryogenesis from mesophyll protoplasts of *Medicago sativa*. Plant Cell, Tissue Organ Cult 9:121–130

Dijak M, Smith DL, Wilson TJ, Brown DCW (1986) Stimulation of direct embryogenesis from mesophyll protoplasts of *Medicago sativa*. Plant Cell Rep 5:468–470

Evans DA, Sharp WR, Medina-Fillic HP (1984) Somaclonal and gametoclonal variation. Am J Bot 71:759–774

Frearson EM, Power JB, Cocking EC (1973) The isolation, culture and regeneration of *Petunia* leaf protoplasts. Dev Biol 33:130–137

Gilmour DM (1986) In vitro Genetic Manipulation of *Medicago*. Ph. D. Thesis, University of Nottingham

Gilmour DM, Davey MR, Cocking EC (1987a) Plant regeneration from cotyledon protoplasts of wild *Medicago* species. Plant Sci 48:107–112

Gilmour DM, Davey MR, Cocking EC (1987b) Isolation and culture of heterokaryons following fusion of protoplasts from sexually compatible and sexually incompatible *Medicago* species. Plant Sci 53:263–270

Gilmour DM, Davey MR, Cocking EC, Pental D (1987c) Culture of low numbers of forage legume protoplasts in membrane chambers. J Plant Physiol 126:457–465

Holbrook LA, Reich TJ, Iyer VN, Haffner M, Miki BL (1985) Induction of efficient cell division in alfalfa protoplasts. Plant Cell Rep 4:229–232

Johnson LB, Stuteville DL, Higgins RK, Skinner DZ (1981) Regeneration of alfalfa plants from protoplasts of selected Regen-S clones. Plant Sci Lett 20:297–304

Johnson LB, Stuteville DL, Schlarbaum SE, Skinner DZ (1984) Variation in phenotype and chromosome number in alfalfa protoclones regenerated from nonmutagenized calli. Crop Sci 24:948–951

Kao KN (1977) Chromosomal behaviour in somatic hybrids of soybean-*Nicotiana glauca*. Mol Gen Genet 150:225–230

Kao KN, Michayluk MR (1975) Nutrient requirements for growth of *Vicia hajastana* cells and protoplasts at very low population density in liquid medium. Planta 126:105–110

Kao KN, Michayluk MR (1980) Plant regeneration from mesophyll protoplasts of alfalfa. Z Pflanzenphysiol 96:135–141

Kao KN, Michayluk MR (1981) Embryoid formation in alfalfa cell suspension cultures from different plants. In vitro 17:645–648

Kao KN, Saleem M (1986) Improved fusion of mesophyll and cotyledon protoplasts with PEG and high pH-Ca^{2+} solutions. J Plant Physiol 122:217–225

Kao KN, Wetter LR (1977) Advances in techniques of plant protoplast fusion and culture of heterokaryons. In: Brinkley BR, Porter K (eds) International cell biology 1976–1977. Rockefeller University Press, New York, p 216

Keller WA, Melchers G (1973) The effect of high pH and calcium on tobacco leaf protoplast fusion. Z Naturforsch 28C:737–741

Larkin PJ, Scowcroft WR (1981) Somaclonal variation – a novel source of variability from cell cultures for plant improvement. Theor Appl Genet 60:197–214

Latunde-Dada AO, Lucas JA (1983) Somaclonal variation and reaction to *Verticillium* wilt in *Medicago sativa* plants regenerated from protoplasts. Plant Sci Lett 32:205–211

Lu, DY, Davey MR, Pental D, Cocking EC (1982) Forage legume protoplasts: somatic embryogenesis from protoplasts of seedling cotyledons and roots of *Medicago sativa*. In: Fujiwara A (ed) Plant tissue culture 1982. Maruzen, Tokyo, pp 597–598

Lu DY, Davey MR, Cocking EC (1983) A comparison of the cultural behaviour of protoplasts from leaves, cotyledons and roots of *Medicago sativa*. Plant Sci Lett 31:87–99

Mariotti D, Arcioni S, Pezzotti M (1984) Regeneration of *Medicago arborea* L. plants from tissue and protoplast cultures of different organ origin. Plant Sci Lett 37:149–156

McCoy TJ, Smith LY (1984) Uneven ploidy levels and a reproductive mutant required for interspecific hybridization of *Medicago sativa* L.×*Medicago dzhawakhetica* Bordz. Can J Genet Cytol 26:511–518

Mitten DH, Sato SJ, Skokut TA (1984) In vitro regenerative potential of alfalfa germplasm sources. Crop Sci 24:943–945

Murashige T, Skoog F (1962) A revised medium for rapid growth and bioassays with tobacco tissue cultures. Physiol Plant 15:473–497

Nagarajan P, McKenzie JS, Walton PD (1986) Embryogenesis and plant regeneration of *Medicago* spp. in tissue culture. Plant Cell Rep 5:77–80

Pezzoti M, Arcioni S, Mariotti D (1984) Plant regeneration from mesophyll, root and cell suspension protoplasts of *Medicago sativa* cv. 'Adriana'. Genet Agrar 38:195–208

Reich TJ, Iyer VN, Scobie B, Miki BL (1986a) A detailed procedure for the intranuclear microinjection of plant protoplasts. Can J Bot 64:1255–1258

Reich TJ, Iyer VN, Miki BL (1986b) Efficient transformation of alfalfa protoplasts by the intranuclear microinjection of Ti plasmids. Bio Technology 4:1001–1004

Santos AVP dos, Outka DE, Cocking EC, Davey MR (1980) Organogenesis and somatic embryogenesis in tissues derived from leaf protoplasts and leaf explants of *Medicago sativa*. Z Pflanzenphysiol 99:261–270

Santos AVP dos, Cutter EG, Davey MR (1983) Origin and development of somatic embryos in *Medicago sativa*. Protoplasma 117:107–115

Schenk RV, Hildebrandt AC (1972) Medium and techniques for induction and growth of monocotyledonous and dicotyledonous plant cell cultures. Can J Bot 50:199–204

Strickland SG, McCall CM, Nichol JW, Stuart DA (1986) Enhanced somatic embryogenesis in *Medicago sativa* by addition of maltose and its higher homologs to the culture medium. In: Somers DA, Gengenbach BG, Biesboer DD, Hackett WP, Green CE (eds) Abstr VI Int Congr Plant Tissue Cell Cult. University of Minnesota, Minneapolis, p 188

Stuart DA, Strickland SG, (1984a) Somatic embrygenesis from cell cultures of *Medicago sativa*. I. The role of amino acid additions to the regeneration medium. Plant Sci Lett 34:165–174

Stuart DA, Strickland SG (1984b) Somatic embryogenesis from cell cultures of *Medicago sativa* II. The interaction of amino acids with ammonium. Plant Sci Lett 34:175–181

Téoulé E (1983) Hybridation somatique entre *Medicago sativa* et *Medicago falcata*. CR Acad Sci Paris. 297 Ser 111:13–16

Uchimiya, H, Murashige T (1974) Evaluation of parameters in the isolation of viable protoplasts from cultured tobacco cells. Plant Physiol 54:936–944

Watts JW, King JM (1984) A simple method for large-scale electrofusion and culture of plant protoplasts. Biosci Rep 4:335–342

Xu ZH, Davey MR, Cocking EC (1982) Organogenesis from root protoplasts of the forage legumes *Medicago sativa* and *Trigonella foenum-graecum*. Z Pflanzenphysiol 107:231–235

III.8 Somatic Hybridization Between *Moricandia* and *Brassica*

K. TORIYAMA[1], T. KAMEYA[2], and K. HINATA[1]

1 Introduction

Brassica crops have been utilized in a variety of forms such as vegetables, oil crops, fodder, condiments and ornamentals. These species are widely distributed in the temperate regions of the world. The genus *Brassica* is classified as a member of the subtribe Brassicinae of the tribe Brassiceae in the family Cruciferae (Schulz 1936).

As early as 1978, intergeneric somatic hybrids in Cruciferae were reported between *Arabidopsis thaliana* and *Brassica campestris* (Gleba and Hoffmann 1978). This was the first case of intertribal hybridization of flowering plants. However, most of the somatic hybrids were abnormal, probably because these parental species were distantly related. Asymmetric hybrids were also obtained in this study and they were more regular in morphology and bore flowers in vitro (Gleba and Hoffmann 1980; Hoffmann and Adachi 1981). Recently, we succeeded in producing intergeneric somatic hybrids between *B. oleracea* and *Sinapis turgida*, and between *B. nigra* and *S. turgida* (Toriyama et al. 1987b).

Regarding interspecific somatic hybridization in the same genus, artificial synthesis of *B. napus* from *B. oleracea* and *B. campestris* has been reported. These somatic hybrids showed hybrid vigour, and this trait was used for the selection of the hybrids (Schenck and Röbbelen 1982; Taguchi and Kameya 1986). Other selection methods, which were independent of a specific trait, were also applied for the synthesis of *B. napus*, for example, mechanical isolation by micromanipulator (Sundberg and Glimelius 1986) and by fluorescence-activated cell sorter (Glimelius et al. 1986), and inactivation of protoplasts by iodoacetamide (Terada et al. 1987). Further, production of cytoplasmic hybrid plants has been extensively studied in this genus, especially in *B. napus* (Pelletier et al. 1983; Yarrow et al. 1986; Barsby et al. 1987; Menczel et al. 1987). In these studies, protoplast fusion was used to combine cytoplasmic atrazine resistance (encoded in the chloroplast DNA) and cytoplasmic male sterility (probably encoded in the mitochondrial DNA).

The tribe Brassiceae includes xerophylous wild plants such as the subtribes Moricandiinae, Zillinae and Vellinae (Schulz 1936). *Moricandia arvensis* is a wild species distributed from the Sahara desert to the Mediterranean region, and it has

[1] Faculty of Agriculture, Tohoku University, Sendai 980, Japan
[2] Institute for Agricultural Research, Tohoku University, Sendai 980, Japan

Biotechnology in Agriculture and Forestry, Vol. 8
Plant Protoplasts and Genetic Engineering I (ed. by Y. P. S. Bajaj)

been identified as being intermediate between C_3 and C_4 plants with respect to low CO_2 compensation point and low photorespiratory activity (Holaday et al. 1981; Holaday and Chollet 1984). No other C_4-like species has been reported in the familiy Cruciferae. Genetic analysis of these photosynthetic systems may be of interest for the study of evolution of photosynthetic systems as well as for crop improvement in view of yield and adaptation to drought.

Although a number of cases ordinary inter-species hybridization in *Brassica* and some others have been carried out extensively for the study of genomic relationships, inter-subtribe crosses have not been widely made except between Raphaninae and Brassicinae (Karpechenko 1924, see review Prakash and Hinata 1980). Recently Apel et al. (1984) were successful in the hybridization between *Moricandia arvensis* and *Brassica alboglabra*. They could obtain only one hybrid plant from 2000 crossed flowers. The difficulty in hybridization by the ordinary method may be due to their remote relationships. Protoplast fusion can be considered to be applied for obtaining hybrid plants in such cases.

In a preliminary survey on tissue culture for about 50 species in *Brassica* and its wild allies, we found that *Moricandia arvensis* delivered good cell suspension (Toriyama et al. 1987c). It was also observed that the suspended cells easily produce abundant protoplasts. We have made novel hybrid plants between *Moricandia* and *Brassica* by protoplast fusion (Toriyama et al. 1987a).

2 Isolation of Protoplasts

A strain of *M. arvensis* DC. was collected near Biskra, Algeria, in 1975 and preserved in our laboratory (accession No. 14). Callus culture was initiated by inoculating the leaf segments on MS medium containing 1 mg/l 2,4-D and 0.8% agar. The callus was then transferred to a liquid medium containing the same component without agar, and subcultured every 2 weeks as described for rice culture by Toriyama and Hinata (1985). The *Moricandia* calli subcultured for more than a year, having lost their regenerating ability, were used as one of the partners (Fig. 1).

The other partners used were red cabbage (*B. oleracea* L. cv. Ruby Ball) and cauliflower (*B. oleracea* L. cv. Nozakiwase). Seeds were surface sterilized with a 75% ethanol solution for a few seconds and soaked in a 2% sodium hypochlorite solution for 20 min. The seeds were germinated in the dark at 25 °C on 1.5% agar containing 0.5% $CaCl_2 \cdot 2H_2O$. Five- or six-day-old hypocotyls were cut transversely into 0.5–1 mm segments to make protoplasts in enzyme solution.

The enzyme solution for releasing protoplasts contained 2% Cellulase Onozuka R10, 0.2% Macerozyme R10, 0.1% $CaCl_2 \cdot 2H_2O$ and 10% mannitol at pH 5.5. After overnight incubation at 25 °C, the solution with respective materials was gently agitated for 1 h on a rotary shaker. The isolated protoplasts were filtered through nylon net and washed three times, with a washing solution containing 0.1% $CaCl_2 \cdot 2H_2O$ and 10% mannitol at pH 5.5, by centrifugation. A half bottle (20 ml) of *Moricandia* callus or 50 hypocotyls were enough for obtaining protoplasts in one Petri dish for fusion. To purify the protoplasts, the protoplast solution was overlaid on 20% sucrose solution and centrifuged. The undamaged

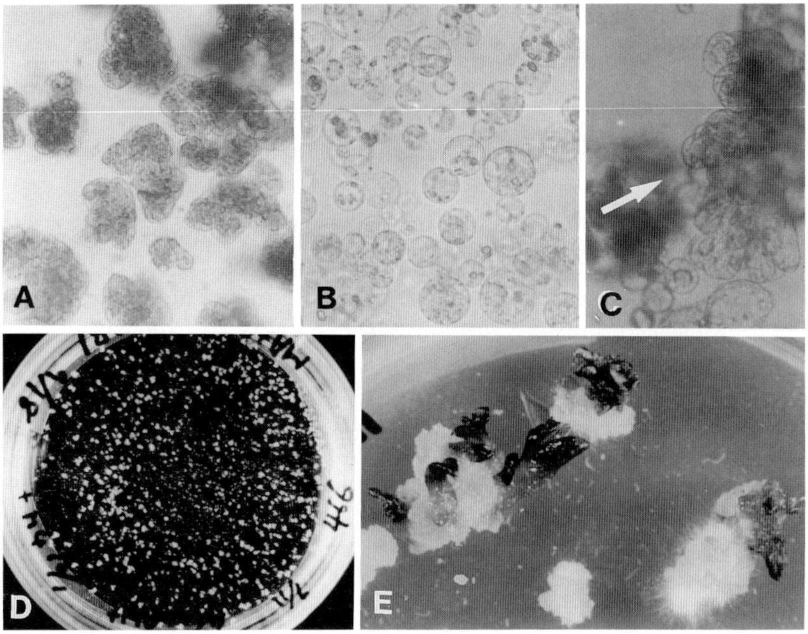

Fig. 1. A cell suspensions of *Moricandia arvensis*. **B** freshly isolated protoplast of *M. arvensis*. **C** division of microcallus in DBN medium 2 weeks after fusion treatment between *M. arvensis* and red cabbage. **D** cell colonies in D2B1 medium after 1 month of culture. **E** regeneration of shoots from calli. (**D** and **E** after Toriyama et al. 1987a)

protoplasts floating on the surface of the solution were collected and washed again with the washing solution.

3 Culture of Protoplasts

Two kinds of culture media were used. Both of them were composed of B5 medium supplemented with 8% mannitol, but they differed in hormone content; DBN medium contained 0.25 mg/l 2,4-D, 0.1 mg/l BAP and 0.5 mg/l NAA, and D2B1 medium contained 2 mg/l 2,4-D, and 1 mg/l BAP. After 2–3 weeks of culture, the equivalent fresh medium (1 ml) containing 4% mannitol was added to each Petri dish, and at the 4th week the fresh medium (1 ml) without mannitol was again added. At the 6th day of culture, cell division was observed. Plating efficiency after 1 month of culture was 1% in the DBN medium and 0.1% in the D2B1 medium (Fig. 1).

Five or six weeks after the fusion treatment, the developing calli were transferred to MS medium containing 1 mg/l NAA, 1 mg/l BAP, 100 mg/l casamino acids and 0.8% agar. The calli were subcultured every 3 weeks on the same medium. The callus, once it had regenerated shoots, continued to proliferate, producing a number of shoots. The regenerated shoots were transplanted to hormone-free MS or B5 medium containing 1% agar for rooting. These plantlets were then transferred to soil in pots.

4 Fusion of Protoplasts and Somatic Hybridization

Protoplast fusion was carried out as described by Kameya (1983) with modifications. The details are given in the protocol section.

The rate of protoplast fusion was not determined in this experiment, because the protoplasts of partners could not be distinguished from each other. According

Table 1. Data of protoplast culture after protoplast fusion between *M. arvensis* and red cabbage (MA + RC) and between *M. arvensis* and cauliflower (MA + CA). (Toriyama et al. 1987a)

Exp. no.	Combination of materials in fusion	Medium for protoplast culture	Number of calli		
			transplt. regen. m.	produced shoots	produced plants
466	MA + RC	D2B1	112	15	8
466	MA + RC	DBN	ca. 200	0	0
464	MA + CA	DBN	ca. 500	4	2

Fig. 2. Young plants of *M. arvensis* (*MA*), somatic hybrid (466-6) and red cabbage (*RC*) (*upper, left to right*), and isozyme patterns of acid phosphatase in leaves. The patterns of 466-6 were a combination of the bands of both parents (MA and RC). (Upper photograph after Toriyama et al. 1987a)

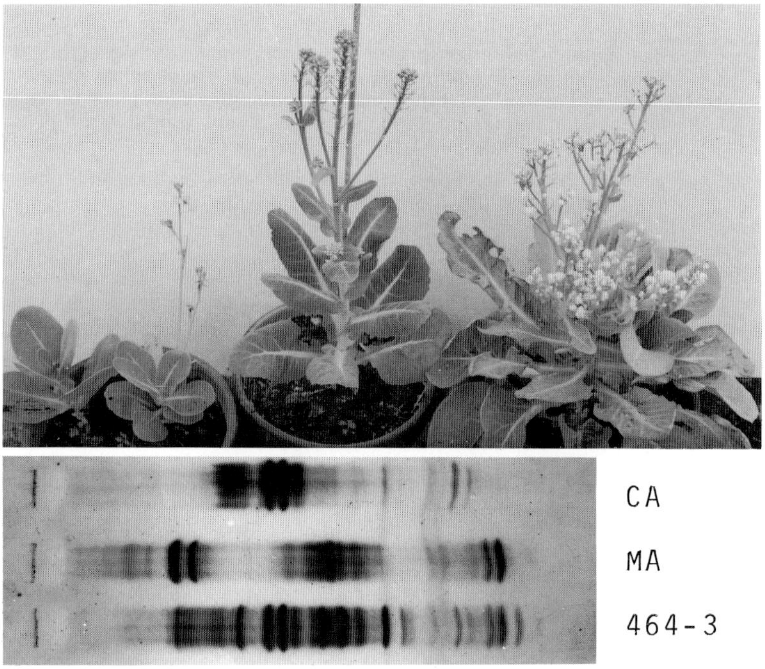

Fig. 3. Mature plants of *M. arvensis* (*MA*), somatic hybrid (464-3) and cauliflower (*CA*) (*upper, left to right*), and isozyme patterns of acid phosphatase in leaves. The patterns of 464-3 are a combination of the bands of both parents (*CA* and *MA*)

to other studies using the dextran method, the rate was estimated to be about 10% (Toriyama et al. 1987b).

Out of 112 calli cultured in D2B1 medium, 15 calli produced shoots (Fig. 1). Finally, more than 100 plants which had originated from 8 calli were grown in pots in the fusion between *M. arvensis* and red cabbage. No plants regenerated from the callus cultured in the DBN medium (Table 1).

In the fusion between *M. arvensis* and cauliflower, we used only DBN medium and two hybrid plants were obtained from 500 calli that were transferred to the regeneration medium (Table 1).

All the regenerated plants were determined to be hybrids because of their morphology. The regenerated plants showed combined morphological characteristics of both parental species (Figs. 2 and 3). Stalked leaves in young plants were typical features of red cabbage and cauliflower. Greyish green-coloured succulent leaves with almost entire margins were like *M. arvensis. B. oleracea* has yellowish petals and *M. arvensis* has violet-veined petals. The petals of many hybrid plants were creamy purple with violet veins, and those of some hybrids were white. Many hybrids showed doubled flowers (Fig. 4). The size of flowers was generally larger than both parental species. This may be caused by their amphidiploidal nature. The hybrids could be multiplied by cutting.

The haploid chromosome number (n = 23) of some plants was determined to be the composition of those of the parental species (n = 14 in *M. arvensis* and

Fig. 4. Flowers and flowering plants of red cabbage, somatic hybrid (466-3) and *M. arvensis* (*left to right*). Many strains had doubled floweres as 466-3

n = 9 in *B. oleracea*). They formed normal bivalents in general with some exceptional cases where one or two tetravalents were observed. Anthers were stunted with sterile pollen grains. Therefore, it was difficult to obtain seeds, but when the hybrids were pollinated with the pollen of *B. oleracea* a few seeds were obtained. Observation of the descendants is now proceeding with more precise analysis of chromosomal behaviour.

Isozyme analysis for the leaves on acid phosphatase showed that these plants had a combination of bands of the parental species (Figs. 2, 3). As for aspartate aminotransferase band profiles in the hybrids, an additional band was found adding to the bands which were the composition of the parental bands. These isozyme patterns were identical in all the hybrids using cabbage and in all those using cauliflower.

All the data indicated that they were novel hybrid plants between *B. oleracea* and *M. arvensis*, and these plants could be called "*Brassicomoricandia*". We expected hypocotyl protoplasts of *B. oleracea* to have regenerative ability (Glimelius 1984), but no plant of *B. oleracea* was found. Probably the high regenerative ability of the hybrids was due to the complementary action of the parental species and hybrid vigour. The present result is a case where selection was unnecessary for obtaining hybrid plants.

5 Protocol for Protoplast Fusion by the Dextran Method

(1) Prepare protoplasts with enzyme solution and purify by repeated washing and centrifugation. The washing solution is composed of 10% mannitol and 0.1% $CaCl_2 \cdot 2H_2O$ at pH 5.5.
(2) Protoplasts in the washing solution are mixed together in equal quantities. Then, the density of protoplasts is adjusted to be more than 10^6/ml by using centrifugation.
(3) Place 0.2 ml dextran solution (15% dextran, MW 500000 in 6% KCl solution) in the centre of a Petri dish, 60 mm in diameter.
(4) Add the 0.2 ml mixed protoplast solution onto the contents of dextran solution.
(5) Add 0.6 ml dextran solution again. Then stir the contents of Petri dish gently to mix the protoplasts and dextran solution. Keep in the Petri dish for 15 min.
(6) Confirm that protoplasts are aggregating, under the microscope.
(7) Add one to two drops of KOH-KCl solution (containing 0.5% KOH and 6% KCl) to the protoplast-dextran mixture and mix them by stirring. Keep the Petri dish at 30 °C for 10 min.
(8) Drop 10 ml of elution medium (containing 5% mannitol and 2% $CaCl_2 \cdot 2H_2O$, pH 5.5) along the wall of the Petri dish gently and keep it for 10 min.
(9) Stir gently and mix the solutions gradually. Keep the Petri dish at room temperature for 1 h. Protoplasts fuse by this procedure.
(10) Keep the Petri dish in a refrigerator over night.
(11) Transfer the protoplast solution into the test tube and centrifuge at 100 *g* for 3 min for protoplast collection.
(12) Add culture medium (B5 containing 2 mg/l 2,4-D, 1 mg BAP and 8% mannitol). Protoplast density should be 2×10^5/ml.
(13) Generally, 3 ml of protoplast solution is cultured in a Petri dish 60 mm in diameter. Protoplasts begin to divide within 1 week.
(14) After 2–3 weeks, decrease the concentration of mannitol gradually.
(15) Five or six weeks after the fusion treatment, transfer the developing calli to MS medium containing 1 mg/l NAA, 1 mg/l BAP, and 100 mg/l casamino acids.
(16) When regenerated shoots are observed, transplant the shoots to hormone-free MS or B5 medium containing 1% agar for rooting.
(17) After rooting, transplant the plantlets to soil in pots in a greenhouse. The plants can be multiplied by cutting.

References

Apel P, Bauwe H, Ohle H (1984) Hybrids between *Brassica alboglabra* and *Moricandia arvensis* and their photosynthetic properties. Biochem Physiol Pflanz 179:793–797

Barsby TL, Chuong PV, Yarrow SA, Wu SC, Coumans M, Kemble RJ, Powell AD, Beveradorf WD, Pauls KP (1987) The combination of Polima cms and cytoplasmic triazine resistance in *Brassica napus*. Theor Appl Genet 73:809–814

Gleba YY, Hoffmann F (1978) Hybrid cell lines *Arabidopsis thaliana* + *Brassica campestris*: No evidence for specific chromosome elimination. Mol Gen Genet 165:257–264

Gleba YY, Hoffmann F (1980) "Arabidobrassica": A novel plant obtained by protoplast fusion. Planta 149:112–117

Glimelius K (1984) High growth rate and regeneration capacity of hypocotyl protoplasts in some Brassicaceae. Physiol Plant 61:38–44

Glimelius K, Djupsjobacka M, Fellner-Feldegg H (1986) Selection and enrichment of plant protoplast heterokaryons of Brassicaceae by flow sorting. Plant Sci 45:133–141

Hoffmann F, Adachi T (1981) "Arabidobrassica": Chromosomal recombination and morphogenesis in asymmetric intergeneric hybrid cells. Planta 153:586–593

Holaday AS, Chollet R (1984) Photosynthetic/photorespiratory characteristics of C_3-C_4 intermediate species. Photosynth Res 5:307–323

Holaday AS, Shieth YJ, Lee KW, Chollet R (1981) Anatomical, ultrastructural and enzymic studies of leaves of *Moricandia arvensis*, a C_3-C_4 intermediate species. Biochem Biophys Acta 637:334–341

Kameya T (1983) Studies on plant cell fusion by dextran: Effects of pH, inorganic salts and electrical stimulus. Cytologia 48:873–878

Karpechenko GD (1924) Hybrids of *Raphanus sativus* L. × *Brassica oleracea* L. J Genet 15:375–396

Menczel L, Morgan A, Brown S, Maliga P (1987) Fusion-mediated combination of Ogura-type cytoplasmic male sterility with *Brassica napus* plastids using X-irradiated CMS protoplasts. Plant Cell Rep 6:98–101

Pelletier G, Primard C, Vedel F, Chetrit P, Remy R, Rousselle P, Renard M (1983) Intergeneric cytoplasmic hybridization in Cruciferae by protoplast fusion. Mol Gen Genet 191:244–250

Prakash S, Hinata K (1980) Taxonomy, cytogenetics and origin of crop Brassicas, a review. Opera Bot 55:1–57

Schenck HR, Röbbelen G (1982) Somatic hybrids by fusion of protoplasts from *Brassica oleracea* and *B. campestris*. Z Pflanzenzücht 89:278–288

Schulz OE (1936) Cruciferae. In: Engler A, Harms H (eds) Die natürlichen Pflanzenfamilien 17b (2nd edn 1960). Duncker & Humblot, Berlin, pp 227–658

Sundberg E, Glimelius K (1986) A method for production of interspecific hybrids within Brassiceae via somatic hybridization, using resynthesis of *Brassica napus* as a model. Plant Sci 43:155–162

Taguchi T, Kameya T (1986) Production of somatic hybrid plants between cabbage and Chinese cabbage through protoplast fusion. Jpn J Breed 36:185–189

Tereda R, Yamashita Y, Nishibayashi S, Shimamoto K (1987) Somatic hybrids between *Brassica oleracea* and *B. campestris*: selection by the use of iodoacetamide inactivation and regeneration ability. Theor Appl Genet 73:379–384

Toriyama K, Hinata K (1985) Cell suspension and protoplast culture in rice. Plant Sci 41:179–183

Toriyama K, Hinata K, Kameya T (1987a) Production of somatic hybrid plants, 'Brassicomoricandia', through protoplast fusion between *Moricandia arvensis* and *Brassica oleracea*. Plant Sci 48:123–128

Toriyama K, Kameya T, Hinata K (1987b) Selection of a universal hybridizer in *Sinapis turgida* Del. and regeneration of plantlets from somatic hybrids with *Brassica* species. Planta 170:308–313

Toriyama K, Kameya T, Hinata K (1987c) Ability of callus growth and shoot regeneration in the wild species of Brassicaceae. Plant Tissue Cult Lett 4:75–78

Yarrow SA, Wu SC, Barsby TL, Kemble RJ, Shepard JF (1986) The introduction of CMS mitochondria to triazine tolerant *Brassica napus* L., var. Regent, by micromanipulation of individual heterokaryons. Plant Cell Rep 5:415–418

III.9 Somatic Hybridization in *Trifolium*

J. HONKANEN and P. RYÖPPY[1]

1 Introduction

The clovers (*Trifolium* spp.) belong to the family Leguminoseae. There are some 300 species within the genus, distributed all over the world, but mostly in the northern temperate and subtropical zones. They are grown for hay, pasture or soil improvement, usually in mixtures with grasses except when intended for seed production. Species belong to numerous sections and subsections between which hybrid barriers are very strong (Taylor 1980). Very few natural hybrids between species are known, although species within sections can be hybridized by applying the embryo rescue technique (Merker 1984).

Hybrid barriers between species may have a number of different geographical, physiological or genetic reasons (Stebbins 1950). Barriers between the clovers appear to be physiological and/or genetic. Many clovers have a one-locus multiple-allele incompatibility system which prohibits selfing on a gametophytic level (Taylor and Smith 1979). This mechanism secures cross-pollination within species, but should not be functional in an interspecific cross. We believe that interspecific barriers arise due to failure of fertilization or possibly hybrid abortion at an early stage of embryo formation (Kivinen 1987).

Red clover (*T. pratense*) and alsike clover (*T. hybridum*) are both cultivated species that have undergone considerable plant breeding, which has resulted in a number of diploid and artificially produced autotetraploid cultivars. Tetraploid cultivars generally show good growth and usually become larger plants than their diploid ancestors. However, the tetraploids often suffer from partial sterility and thus seed production is badly hampered. Sterility may be caused by meiotic irregularities, but in the case of red clover it also seems to be the morphology of the tetraploid flower which causes difficulties for pollinating insects. It is hoped that, at least in the allotetraploid between the two species, meiotic regularity would be retained.

Red clover and alsike clover belong to two different sections within the genus, sections *Trifolium* and *Lotoidea* respectively (Table 1). It can be seen from Table 1, that red clover has a basic chromosome number of $x = 7$ while that of alsike clover is $x = 8$, a more common basic number in the genus.

We have attempted to overcome the hybrid barrier between the two clover species by (1) using the embryo rescue method (Kivinen 1987), and (2) applying the

[1] Department of Plant Breeding, University of Helsinki, SF-00710 Helsinki, Finland

Biotechnology in Agriculture and Forestry, Vol. 8
Plant Protoplasts and Genetic Engineering I (ed. by Y.P.S. Bajaj)

Table 1. Taxonomic status and chromosome number of the most common clover species within the two sections relevant to the hybridization of red and alsike clovers. (Simon 1983)

Section	Species	Chromosome number (x)
Trifolium	*alpestre* L.	8
	diffusum Ehrn.	8
	medium L.	Aneuploid
	pallidum Waldst. & Kit.	8
	pratense L.	7
	sarosiense Hazl.	24
Lotoidea	*ambiguum* Bieb.	8
	hybridum L.	8
	nigrescens Viv.	8
	repens L.	16

protoplast fusion method (Honkanen et al. 1986). The embryo rescue method posed a number of problems and hybrid plants have not yet been obtained. Protoplast fusion between the two clover species may result in tetra-, hexa- and octoploid plants, depending on the initial ploidy level of the protoplasts. In addition, variation may arise in the cytoplasmic constitution, giving rise to new "cybrid" plants. At the moment the protoplast fusion method looks more promising, but the utilization of this technique depends on a well-developed protoplast regeneration system.

Although plants have been regenerated from protoplasts of the Fabaceae, so far, no somatic hybrids have been reported between *Trifolium* species. Table 2 summarizes the protoplast work done on *Trifolium* protoplasts. As can be seen, regeneration of whole plants from *Trifolium* species is still quite rare; see also the review article on tissue culture of red clover and other forage legumes by Phillips and Collins (1984).

2 Isolation of Protoplasts

2.1 Plant Material

Protoplasts of *Trifolium* species have usually been isolated from mesophyll tissue or from suspension cultures (see Table 2). In our experiments four diploid and three tetraploid alsike clover, and six diploid and two tetraploid red clover cultivars served as the source of protoplasts which were isolated from leaves and from suspension cultures. The material was selected from 28 alsike clover and 32 red clover cultivars according to the results of preliminary studies (Honkanen et al. 1986, 1988a).

Plants were maintained as aseptic shoot cultures on L2 medium (Phillips and Collins 1979) or on half strength MS medium (Murashige and Skoog 1962) without hormones at 24±2 °C under 16-h photoperiod. Cell suspension cultures were

Table 2. Growth responses of *Trifolium* protoplast cultures

Plant species	Origin of protoplasts	Growth response	Reference
T. arvense	Suspension	Callus	White and Bhojwani (1981)
T. hybridum	Mesophyll Root tissue	Shoots	Webb et al. (1986)
T. hybridum	Suspension	Plants	Honkanen and Ryöppy, (this chapter)
T. hybridum	Mesophyll Suspension	Callus Roots	Honkanen et al. (1986)
T. hybridum	Mesophyll Suspension	Shoots	Honkanen et al. (1988b)
T. hybridum	Mesophyll	Callus	Oelck et al. (1983)
T. incarnatum	Root tip	Callus	Rybczynski, J.J. (1984)
T. pratense	Mesophyll Suspension	Callus Roots	Honkanen et al. (1986)
T. pratense	Suspension	Embryolike structures	Rakousky et al. (1984)
T. pratense	Mesophyll	Callus	Oelck et al. (1983)
T. repens	Mesophyll	Shoots Roots	Bhojwani and White (1982)
T. repens	Mesophyll	Plants	White (1983)
T. repens		Shoots	Bhojwani et al. (1984)
T. repens	Mesophyll	Plants	Ahuja et al. (1983)
T. repens	Suspension	Plants	Gresshoff (1980)
T. repens	Mesophyll Root tissue	Shoots	Webb et al. (1986)
T. repens	Mesophyll	Shoot-like structures	Oelck et al. (1983)
T. resupinatum	Mesophyll	Callus	Oelck et al. (1982)
T. rubens	Mesophyll Suspension	Plants	Grosser and Collins (1984)

first established on B5 medium (Gamborg et al. 1968) supplemented with 2 mg/l 2,4-D and 0.5 mg/l kinetin and then maintained on B5 or SL medium (Phillips and Collins 1980) with 0.06 mg/l 2,4-D and 0.1 mg/l BAP. Sucrose contents of all media were 20 g/l and pH was adjusted to 5.8. Suspensions were cultured at 130 rpm on a platform shaker and subcultured once a week. Cultures were screened for best-growning cell lines and only these were used as donor material for protoplast isolation.

2.2 Enzymes

The leaves were sliced into strips about 2 mm thick, which were incubated overnight in an enzyme solution consisting of 0.5% Macerozyme R-10 (Yakult Co.), 1% Cellulase Onozuka R-10 (Yakult Co.) and 1% Cellulysin (Calbiochem) in 0.5 M sucrose with 5 mM $CaCl_2 \cdot 2H_2O$. The pH was adjusted to 5.6 and the solution was filter-sterilized. After enzyme treatment the released protoplasts were filtered through 100-μm stainless steel sieves and centrifuged for 10 min at 100 *g*. The floating protoplast layer was further washed with 0.35 M sucrose or with W5 salts (154 mM NaCl, 125 mM $CaCl_2 \cdot 2H_2O$, 5 mM KCl, 5 mM glucose). Protoplasts (Fig. 1) were either used for fusion studies or they were diluted with culture medium to a density of 10^5 protoplasts/ml and used for regeneration studies.

Protoplasts were also isolated from suspension cultures by incubating actively growing cell clusters with enzyme solution. Enzyme solution, purification and washing procedures were the same as with mesophyll protoplasts.

2.3 Factors Influencing Yield

Physiological condition of the donor plant was the most important factor influencing protoplast yield and quality. Bhojwani and White (1982) had best results when donor plants were cultured in controlled environment in a growth cabinet. Shoot cultures gave low protoplast yields in their studies. This was in contrast to our studies, in which expanded leaves of aseptically grown shoot cultures gave best results; donor plants had to be well adapted to in vitro culture and the age of the plants was critical.

3 Culture of Protoplasts

Protoplasts were cultured as liquid cultures and by using bead-type technique (Shillito et al. 1983) in which protoplasts were plated into agarose (type VII, Sigma)-containing culture medium. After gelling, sections of agarose were transferred into liquid medium. Culture media used were a KM8p (Kao and Michayluk 1975) and B5 medium supplemented with 250 mg/l casein hydrolysate, 250 mg/l NH_4NO_3, 630 mg/l $CaCl_2 \cdot 2H_2O$, 90 g/l glucose and 2 mg/l each of 2,4-D, NAA and kinetin and 90 g/l glucose (modified from Bhojwani and White 1982). Media (pH 5.8) were filtered-sterilized.

Cultures were kept in darkness at +24 °C for the first week after isolation and then cultured under fluorescent light (Osram L36 W/30 Warmton Warm White, photoperiod 16 h light/8 h dark) at 24±2 °C. After protoplasts had started to divide, osmolarity and auxin concentration of the culture medium were reduced stepwise by adding fresh culture medium with lower osmolarity. Developing colonies were transferred to various agar-solidified media for differentiation.

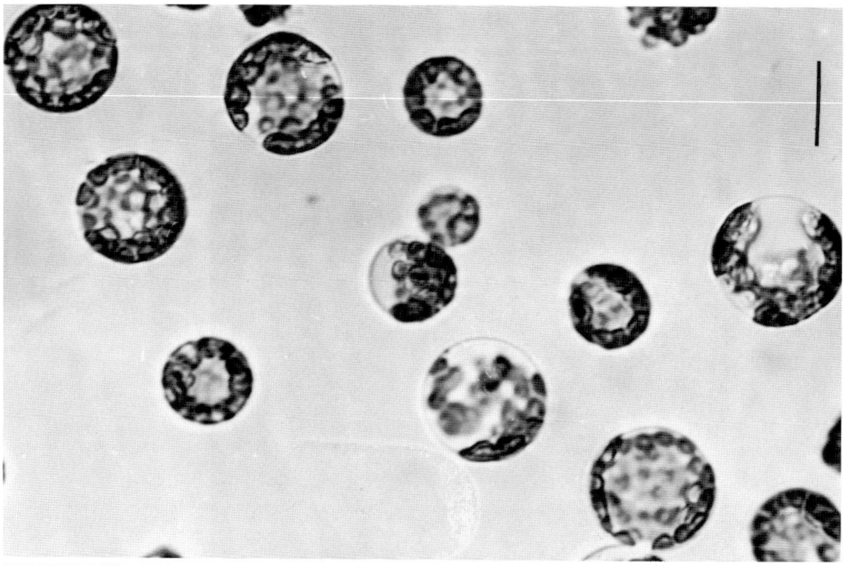

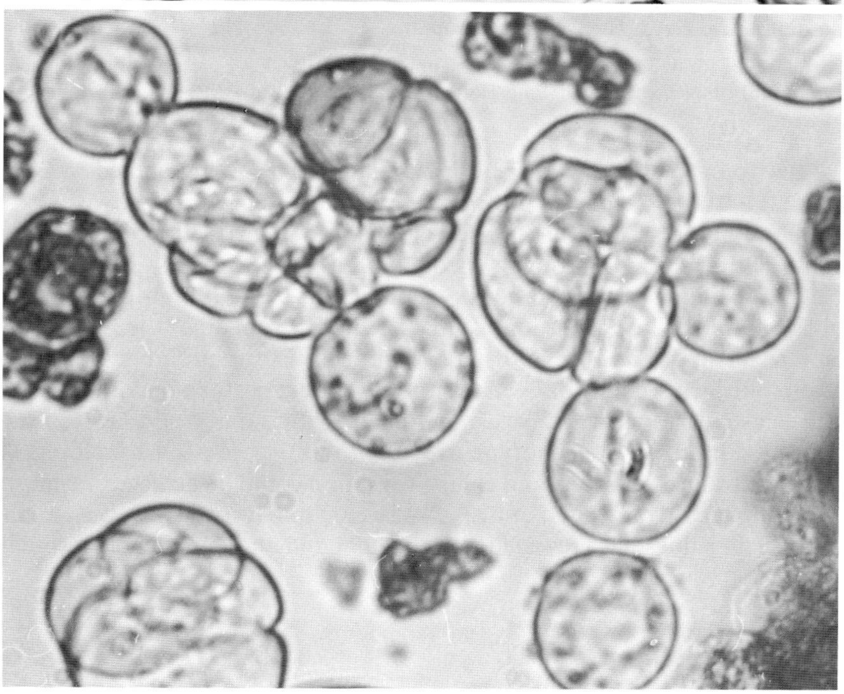

Fig. 1. Freshly isolated protoplasts of red clover. Bar = 50 μm

Fig. 2. Dividing protoplast of alsike clover 2 weeks after isolation in B5 medium supplemented 2 mg/l each of 2,4-D, NAA and kinetin

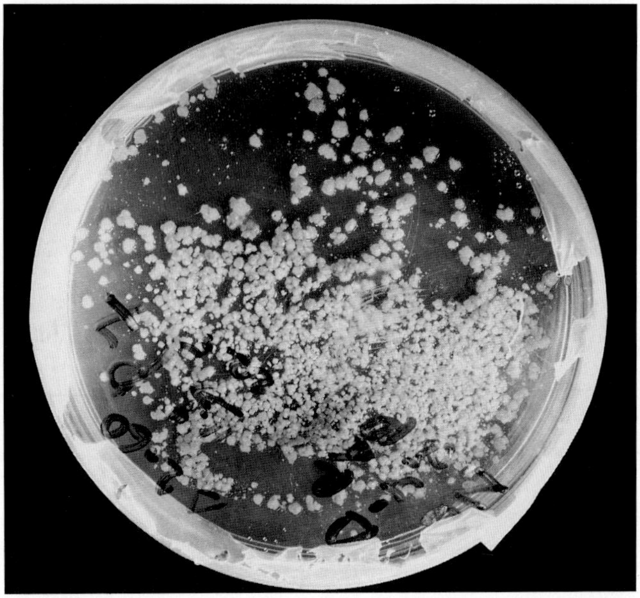

Fig. 3. Callus formation (8 weeks after isolation) from alsike clover protoplasts on N6 medium (Chu et al. 1975) supplemented with 1 mg/l of 2,4-D and 0.2 mg/l of BAP

Best protoplast growth of both parental species was obtained in B5 medium (Fig. 2) and 1–5% of the protoplasts started to divide. First divisions of alsike clover could be seen within 1 week (sometimes 3 days), and developing colonies could be transferred onto agar media within 1 to 2 months. Also the bead-type culture technique proved to be very suitable for *Trifolium* protoplasts. Sustained divisions and callus formation could be achieved from all of the seven tested alsike clover cultivars (Fig. 3). No differences in callus formation between diploid and tetraploid forms could be seen. With red clover, profuse budding was observed and it usually took over 2 weeks before first divisions could be seen. Divisions depended more on the genotype used than in the case of alsike clover. With red clover, sustained divisions and callus formation could be achieved from only three of the eight cultivars tested and all three were diploid forms. Chromosome studies of the callus lines have revealed both normal and aneuploid chromosome counts.

4 Regeneration of Plants

Shoot development (Fig. 4) from protoplast-derived calli has been induced in three diploid alsike clover cultivars. Of the three regenerating cultivars, one (callus was regenerated from protoplasts isolated from suspension culture) produced shoots which could be easily rooted and whole normal-looking plants (Fig. 5) could be regenerated and transferred to the greenhouse. However, about 90% of these plants died due to damping-off. The other two cultivars produced only

Fig. 4. Shoots regenerating from 1-year-old protoplast derived calli of alsike clover on MS medium supplemented with 0.5 mg/l kinetin and 0.1 mg/l NAA

shoots which were abnormal-looking and which have so far not been able to be rooted. Webb et al. (1986) have reported similar difficulties with alsike and white clover, and Bhojwani and White (1982) and Ahuja et al. (1983) with white clover. Media which induced shoot formation were MS medium with 1 mg/l BAP and B5 media supplemented with 0.5 mg/l BAP and 0.2 mg/l IAA or 0.5 mg/l kinetin, and 0.1 mg/l NAA. Red clover produced roots on all protoplast-derived callus lines but shoot regeneration has so far failed. However, in our previous experiments with callus cultures shoot formation was achieved and somatic embryos were formed in suspension cultures. Grosser (1985) has regenerated shoots and somatic embryos from protoplast-derived calli of one selected genotype of red clover.

5 Fusion of Protoplasts

5.1 Chemical PEG-Fusion

PEG solution contained 40% PEG (MW 1540), 100 mmol/l $CaCl_2$ in 0.3 M glucose. Two parts of mesophyll protoplasts and one part of suspension culture-derived protoplasts in W5 medium were mixed. The larger proportion of mesophyll protoplasts is due to their high frequency of deterioration when PEG-treated. Fusions were made in plastic Petri dishes by adding PEG fusion solution onto the droplets of mixture so that the final PEG concentration was about 20%. Treatment time was 5 to 15 min, after which PEG was removed by washing protoplasts

Fig. 5. Well-rooted alsike clover plants ready to be transferred to soil. Bar = 10 mm

carefully with W5 medium or by gradually lowering PEG concentration by adding a solution with lower concentration of PEG. Final wash was made with culture medium.

5.2 Electrofusion

Fusions were made in 0.35 M sucrose using a High Voltage Cell Processor (Model FPH-01001, D.E.P. Systems, Inc., 1186 Combermere, Troy, Michigan 48083). Equal parts of mesophyll and suspension culture-derived protoplasts were mixed so that the final protoplast density was at least 2×10 protoplasts/ml. Volumes of 0.3 to 0.5 ml of protoplast mixture were placed in the fusion chamber with 3-mm spacing between the electrodes (Models HCO6031 and HCO12125, D.E.P. Inc.). An electric impulse of 200 to 600 V with pulse duration of 30 μs was given to the protoplasts after a 1 to 3 s RF-field in the frequency range of 10-500 Mhz (amplitude 100–200 VAC).

5.3 Comparison of Fusion Methods

Of the two fusion methods we have tested, electrofusion seems clearly more promising, although we have not been able to produce callus and plants from electric field-treated protoplasts as we have done with PEG-treated protoplasts. In our preliminary studies with electrofusion, the fusion frequency was at least tenfold compared to PEG method. With our chemical fusion method the frequencies were usually below 1%, whereas 5% is optimal. After electrofusion, protoplasts float freely in liquid and they are thus easier to handle than protoplasts fused with PEG, which tend to adhere to the bottom of the the fusion dish (Honkanen et al. 1988 b).

5.4 Selection and Culture of Fused Protoplasts

Fusion products between mesophyll protoplasts and colourless protoplasts isolated from suspension cultures were visually identified by the presence of green chloroplasts originating from mesophyll protoplasts and cytoplasmic strands originating from suspension protoplasts. These heterokaryons were picked out into microcapillaries by using a micromanipulator. Grosser (1985) has used a similar method for isolating heterokaryons between *Trifolium rubens* and *Trifolium pratense.* Heterokaryons could still be easily identified 36 h after fusion and they

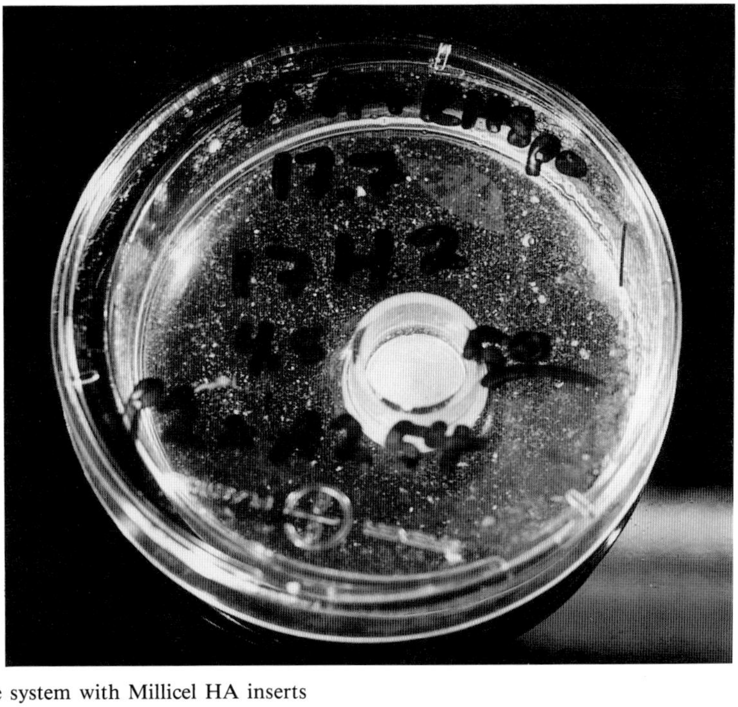

Fig. 6. Nurse culture system with Millicel HA inserts

could be picked out into microcapillaries. Rate of the picking was about 50–100 hybrids per one hour. Heterokaryon frequencies in fusion dishes were about 10% after 24 h from fusion. Unselected fusion products were also cultured directly on the fusion plates with unfused protoplasts.

Selected protoplasts were cultured in Cuprak (Costar) dishes and in 10-μl multivial dishes (Nunc) or they were cultured on Millicell HA (Millipore) plates (Fig. 6) which were inserted in small Petri dishes with 1.5 ml of *Trifolium* protoplast suspension which served as nurse cells. Densities in microdrops were 1000–10000 protoplasts/ml and number of selected protoplasts was 100–200 per one Millicell HA plate.

5.5 Regeneration of Fusion Products

Heterokaryons, produced with either PEG or electrofusion, and cultured in Cuprak and multivial plates, were observed to undergo few divisions, usually only up to the 20-cell stage (Fig. 7). Sustained divisions could not be achieved, mainly due to excessive phenol output of protoplasts. The preliminary experiments with a new nurse culture technique of Millicell HA inserts are promising. Phenol problems have been slightly reduced and heterokaryons fused with PEG have divided to macroscopically visible colonies and small calli.

From the unselected fusion plates a large number of plantlets have been regenerated. Most of them resemble *T. hybridum* morphologically but there are two plants which differ from the other regenerated plants and from the parents morphologically (i.e. leaf shape, hairiness). They have not flowered so their meiotic behaviour and fertility is not known. These two plants were regenerated from the fastest-growing calli. Further studies of these plantlets failed to confirm their nature. Plantlets died in greenhouse.

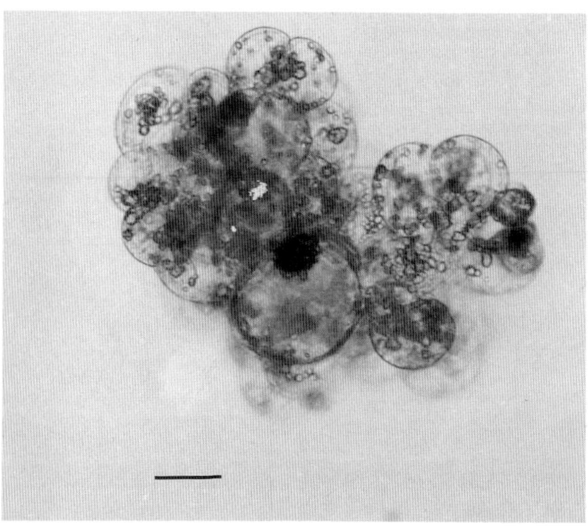

Fig. 7. Colony formation from a selected heterokaryon in microdrop culture in B5 medium. Cultivars used for fusion were diploid alsike clover (Hankkija 869) and diploid red clover (Björn). Bar = 50 μm

6 Conclusions and Prospects

Most interspecific protoplast fusion work has been done within the families Solanaceae and Brassicaceae (Negrutiu et al. 1984; Glimelius et al. 1986). It has become clear through these experiments that organelle "cybrids" may a priori be of more immediate use to the plant breeder than full, fledged nuclear hybrids. The reasoning behind this is that the large nuclear genome is the result of a very long genetic adaptation process within the species and that the breaking up of such coadapted complexes through nuclear fusion would result in unbalanced genomes. It would take a long time for the breeder to restore balance in such cases.

Organelle cybrids, however, do affect critical steps in anther development and pollen formation through the mitochondrial DNA information, which can be utilized by the breeder in cytoplasmic male sterility mechanisms. Particularly the genetic information in the chloroplast DNA may have direct applications for the breeder. Some of the important genes of the ribulose-1,5-bisphosphate-carboxylase (RuBPC) enzyme and the photosystems I and II that can be used in manipulating toxin or herbicide resistance are located in the chloroplast DNA. Here are also genes affecting photorespiration rates connected with the functioning of the RuBPC enzyme.

7 Protocol

For isolation of mesophyll protoplasts of *Trifolium*, leaves of 4–6 months old, well-rooted sterile seedlings maintained on L2 medium without hormones, are used as donor material. Protoplasts are released with overnight incubation in enzyme solution containing Macerozyme R-10 (0.5%), Cellulase Onozuka R-10 (1.0%), Cellulysin (1.0%) and 5 mM $CaCl_2$ dissolved in 0.5 M sucrose. The same enzyme solution is also suitable for the isolation of protoplasts from actively growing *Trifolium* suspensions. Protoplasts are cultured on modified B5 medium (see text) with 2 mg/l each of 2,4-D, NAA and kinetin. Mesophyll and suspension culture-derived protoplasts are fused in 0.35 M sucrose with an electric impulse of 200–600 V (pulse duration of 30 μs) after 1 to 2 s treatment in RF-field (frequency 10–500 Mhz, amplitude 100–200 VAC).

References

Ahuja PS, Lu DY, Cocking EC, Davey MR (1983) An assessment of the cultural capabilities of *Trifolium repens* L. and *Onobrychis viciifolia* Scop. mesophyll protoplasts. Plant Cell Rep 2(5):269–272

Bhojwani SS, White DRW (1982) Mesophyll protoplasts of white clover: Isolation, culture and organogenesis. Plant Sci Lett 26:2–3

Bhojwani SS, Mullins K, Cohen D (1984) Intravarietal variation for in-vitro plant regeneration in the genus *Trifolium*. Euphytica 33 (3):915–922

Chu CC, Wang CC, Sun CS (1975) Establishment of an effective medium for anther culture of rice through comparative experiments on nitrogen sources. Sci Sin 18 (5):659–668

Gamborg OL, Miller RA, Ojima K (1968) Nutrient requirements of suspension cultures of soybean root cells. Exp Cell Res 50:151–158

Glimelius K, Fahlesson J, Sjödin C, Sundberg E, Djupsjöbacka M (1986) Somatic hybridization and cybridization as potential methods for widening of the genepools of crops within Brassicaceae and Solanaceae. In: Horn W, Jensen CJ, Odenbach W, Shieder O (eds) Genetic manipulation in plant breeding: Proc Int Symp. EUCARPIA, Sept. 8–13, 1985, Berlin (West), Germany, de Gruyter, Berlin, pp 663–682

Gresshoff PM (1980) In-vitro culture of white clover: callus, suspension, protoplast culture and plant regeneration. Bot Gaz 141 (2):157–164

Grosser JW (1985) Approaches to interspecific hybridization of red clover with perennial *Trifolium* species via embryo rescue and somatic cell fusion. Dissertation Abstr Int Vol 45, No 09

Grosser JW, Collins GB (1984) Isolation and culture of *Trifolium rubens* protoplasts with whole plant regeneration. Plant Sci Lett 37:1–2

Honkanen J, Ryöppy P, Tigerstedt PMA (1986) Protoplast culture and fusion of red and alsike clover. In: Horn W, Jensen CJ, Odenbach W, Schieder O (eds) Genetic manipulation in plant breeding: Proc Int Symp. EUCARPIA, Sept. 8–13, 1985, Berlin (West), Germany, de Gruyter, Berlin, pp 685–687

Honkanen J, Ryöppy P, Tigerstedt PMA (1988a) Towards somatic hybridization in clovers. Proc Int Symp Genetic Manipulation in Crops Oct. 22–26, 1984. Beijing, China (IRRI), pp 413

Honkanen J, Ryöppy P, Tigerstedt PMA (1988b): Experiments on protoplast fusion in *Trifolium*. In: Proc Int Biotech Workshop January 12–15, 1987, ICRISAT, Hyderabad, India

Kao KN, Michayluk MR (1975) Nutritional requirements for growth of *Vicia hajastana* cells and protoplasts at a very low population density in liquid media. Planta 126:105–110

Kivinen M (1987) Lajien välinen hybridisaatio: Puna-ja alsikeapilan risteyttäminen alkioviljelytekniikkaa apuna käyttäen. (In Finnish). Master's thesis, Dept. of Plant Breeding, University of Helsinki

Merker A (1984) Hybrids between *Trifolium medium* and *Trifolium pratense*. Hereditas 101:267–268

Murashige T, Skoog F (1962) A revised medium for rapid growth and bioassay with tobacco tissue cultures. Physiol Plant 15:473–497

Negrutiu I, Jacob M, Caboche M (1984) Advances in somatic cell genetics of higher plants – the protoplast approach in basic studies on mutagenesis and isolation of biochemical mutants. Theor Appl Gen 67(4):298–304

Oelck MM, Bapat VA, Schieder O (1982) Protoplast culture of three legumes: *Arachis hypogaea, Melilotus officinalis, Trifolium resupinatum.* Z Pflanzenphysiol 106:173–177

Oelck MM, Rao PS, Ozias-Akins P, Schieder O (1983) Protoplast regeneration of some legume species. In: Potrykus I, Harms CT, Hinnen A, Hutter R, King PJ, Shillito RD (eds) Protoplasts 1983, Poster Proc. 6th Int Protoplast Symp Basel, August 12–16 1983. Birkhäuser, Basel, pp 50–51

Phillips GC, Collins GB (1979) In vitro tissue culture of selected legumes and plant regeneration from callus cultures of red clover. Crop 19:59–64

Phillips GC, Collins GB (1980) Somatic embryogenesis from cell suspension cultures of red clover. Crop Sci 20:323–326

Philips GC, Collins GB (1984) Red clover and other forage legumes. In: Sharp WR, Evans DA, Ammirato PV, Yamada Y (eds) Handbook of plant cell culture, Vol 2. Crop species, Macmillan, New York, pp 169–210

Rakousky S, Nasinek V, Nemcova B (1984) Protoplast cultures of *Trifolium pratense* L. In: Novak FJ, Havel L, Dolezel J (eds) Proc Int Symp of Plant Tissue and Cell Culture Application to Crop Improvement, Olomouc, Czechoslovakia, 24–29 Sept. 1984, Czechoslovak Academy of Sciences, Prague, pp 403–404

Rybczynski JJ (1984) The in vitro culture of selected Leguminosae – plant regeneration and somatic embryogenesis. In: Novak FJ, Havel L, Dolezel J (eds) Proc Int Symp Plant Tissue and Cell Culture Application to Crop Improvement, Olomouc, Czechoslovakia, 24–29 Sept. 1984, Czechoslovak Academy of Sciences, Prague, pp 129–130

Shillito RD, Paszkowski J, Potrykus I (1983) Culture in agarose improves protoplast plating and permits division in otherwise unresponsive systems. In: Potrykus I, Harms CT, Hinnen A, Hutter R, King PJ, Shillito RD (eds) Protoplasts 1983: Poster Proc 6th Int Protoplast Symp, Basel, Aug 12–16, 1983, Birkhäuser, Basel

Simon U (1983) Limitations to the wide crosses in clover and lucerne breeding. Reinhold, New York

Stebbins GL (1950) Variation and evolution in plants. Columbia University Press, New York

Taylor NL (1980) Clovers. In: Fehr WR, Hadley HH (eds) Hybridization of crop plants. Madison, USA: Am Soc Agr and Crop Sci Soc Am, pp 261–272
Taylor NL, Smith RR (1979) Red clover breeding and genetics. Adv Agric 31:125–154
Webb KJ, Woodcock S, Pike LS, Dale PJ (1986) Plant regeneration in the forage legumes. In: Withers LA, Alderson PA (eds) Plant tissue culture and its agricultural applications. Butterworths, London, pp 99–104
White DWR (1983) Plant regeneration from mesophyll protoplasts of white clover (*Trifolium repens*). In: Potrykus I, Harms CT, Hinnen A, Hutter R, King PJ, Shillito RD (eds) Protoplasts 1983, Poster Proc 6th Int Protoplast Symp Basel, August 12.–16. 1983. Birkhäuser, Basel, pp 60–61
White DRW, Bhojwani SS (1981) Callus formation from *Trifolium arvense* protoplast-derived cells plated at low densities. Z Pflanzenphysiol 102:257–261

III.10 Somatic Hybridization in Rice × Soybean

M. NIIZEKI[1]

1 Introduction

Rice, *Oryza sativa* L., is the world's most important cereal food crop in terms of caloric intake. It contains about 72% glucoside and 10% protein (Standard Tables of Food Composition, Agency of Science and Technology, Japan 1986). In terms of total world grain production it is exceeded only by wheat. Rice is a warm area crop and is grown most extensively in the humid tropical and subtropical regions of the world. In temperate climates, and at high elevations, rice is grown as a summer crop. About 90% of the world's rice is grown in the People's Republic of China, India, Japan, Korea and southeastern Asia, outside of Asia, less than 10% is produced in the USA, Brazil, Africa and southern Europe (FAO 1985). Soybean, *Glycine max* (L.) Merr., is the world's most important crop as a source of plant protein and oil. It contains about 35% protein, 19% lipid and 24% of glucoside (Standard Tables of Food Composition, Agency of Science and Technology, Japan 1986). The USA now accounts for about 48% of the world production of soybeans. China and Brazil account for another 33% of the world's total production. The remaining production is scattered among various countries in Asia and South America (FAO 1985).

The application of tissue, cell and protoplast culture involving genetic manipulation for crop improvement has been investigated on these crops all over the world. The recently developed technique of rice protoplast culture has led to successful callus formation and the regeneration of whole plants (Deka and Sen 1976; Tsai et al. 1978; Wakasa et al. 1984; Toriyama and Hinata 1985; Fujimura et al. 1985; Yamada et al. 1985, 1986; Coulibaly and Demarly 1986; Abdullah et al. 1986; Kyozuka et al. 1987). In the case of soybean, there has been successful division of the protoplasts in culture with the subsequent formation of calli (Kao et al. 1970; Zieg and Outka 1980; Niizeki and Kita 1981; Xu et al. 1982; Lu et al. 1983; Chowhury and Widholm 1985). However, there has been no success in the regeneration of whole plants from protoplasts, although adventitious embryogenesis and whole plant regeneration is now available from tissue cultures of soybean (Gamborg et al. 1983a; Barwale et al. 1986; Wright et al. 1986).

Cell division of somatic hybrids originating from fused protoplasts has been observed in some combinations of plant species between different families (Kao

[1] Plant Breeding Laboratory, Faculty of Agriculture, Hirosaki University, Hirosaki, Aomori-ken 036, Japan

Biotechnology in Agriculture and Forestry, Vol. 8
Plant Protoplasts and Genetic Engineering I (ed. by Y. P. S. Bajaj)

et al. 1974; Kartha et al. 1974; Dudits et al. 1976; Constabel et al. 1976; Binding and Nehls 1978). Furthermore, callus formation has been reported in somatic hybrids between species of different families, such as *Glycine max* and *Nicotiana glauca* (Kao 1977), *Glycine max* and *N. tabacum* (Chien et al. 1982), and *N. tabacum* and *Triticum durum* (Imamura 1984). Niizeki and Kita (1981) reported cell division of a somatic hybrid of *Oryza sativa* and *Glycine max*, and recently Niizeki et al. (1985) have reported callus induction and selection of the somatic hybrid callus and some characteristics of this wide combination.

Such a somatic hybridization of distantly related plant species is the most interesting investigation with regard to the reconstruction of higher plant cells. In the hybrid cells new and impossible genetic combinations can be constructed in the nucleus, chloroplast and mitochondrial genomes. Genetic complementation may reveal physiological and morphological traits which are absent in both the parental cells. Also the hybrid cells may give new data concerning the mechanisms of various cellular activities. Moreover, the hybridization of distantly related species by protoplast fusion has a practical tool for removing the incompatibility barriers in sexual crossing of agriculturally important plant species (see also Chaps. II.1 and II.5, this Vol. on rice and soybean protoplasts).

2 Isolation of Protoplasts

2.1 Rice Protoplasts

Seeds of 20 cultivars and four strains of rice were cultured aseptically on the medium of Murashige and Skoog (1962) (MS medium) containing 2 mg/l 2,4-dichlorophenoxyacetic acid (2,4-D) (M-1 medium). Calli were subsequently obtained from a portion of an embryo after 1 month in the dark at 25 °C (Niizeki and Kita 1981; Niizeki et al. 1982). These calli were divided and subcultured every 4 weeks. About 1 – 2 g of fresh calli were used for the isolation of protoplasts by 20 ml of enzyme solution containing 2% Cellulase Onozuka R-10, 0.4% Macerozyme R-10, 1% driselase and 0.7 M mannitol (pH 5.8). The enzyme solution containing callus was incubated in the dark at 28 °C for 14 – 17 h in a shaker bath. The protoplasts were separated from indigested cellular material by using four sheets of gauze (Iwatsuki Co., Ltd.) followed by centrifugation at 80 *g* for 4 min and washing four times with 0.7 M mannitol. Protoplast release clearly reflected the differences in genotypes (cultivars or strains) of the callus tissue used (Table 1). The percentage of isolated protoplasts in most of the cultivars and strains was less than 20%. Only two strains, A-58 MS (a cytoplasmic male sterile line) and A-58 Nor (a normal line) showed satisfactory results with about 94% production of protoplasts. A recent result, however, showed that for the isolation of protoplasts 1 h of incubation was sufficient when an improved enzyme solution was used which contains 4% Cellulase Onozuka RS, 1% Macerozyme R-10 and 0.7 M mannitol (pH 5.8).

Table 1. Isolation of protoplasts from rice calli. (Niizeki and Kita 1981)

Cultivar or strain	Protoplast isolation (%)[a]
Hōryu	5.30
Fukuyuki	6.41
Yūkara	8.05
Sorachi	4.32
Tominishiki	33.09
Tomoemasari	6.31
Teine	56.41
Iwakogane	14.45
Shinsetsu	13.29
Eikō	7.94
Himehonami	16.58
Toyohikari	6.82
Hokusetsu	14.29
Sasahonami	2.54
Sachiho	2.03
Yūnami	14.71
Shiokari	13.85
Nōrin 20	3.70
Yukimochi	3.10
Sōhōmochi	8.18
A-58 MS[b]	94.00
A-58 Nor.[c]	93.82
H-126 MS[b]	4.75
H-126 Nor.[c]	6.25

[a] $\frac{\text{Number of protoplasts}}{\text{Number of protoplasts + number of cells with indigested cell wall}} \times 100.$

[b] Cytoplasmic male sterile line.

[c] Normal line.

2.2 Soybean Protoplasts

Soybean (cv. Harosoy) seeds were germinated on the medium of Nitsch and Nitsch (1969) without growth regulators, and about 1 cm of hypocotyls from 7-day-old seedlings was transferred onto M-1 medium (Niizeki and Kita 1981; Niizeki et al. 1982). The callus which resulted was subcultured every 3 weeks. About 1–2 g of fresh calli were used for the isolation of protoplasts. An enzyme solution containing 4% Cellulase Onozuka R-10, 0.4% Macerozyme R-10, and 0.7 M mannitol (pH 5.8) was used and the procedure of incubation and protoplast isolation was the same as in the case of rice. More recently, it has been found that about 2.5 h of incubation in an enzyme solution containing 2% Cellulase Onozuka R-10, 0.2% Pectolyase Y-23 and 0.7 M mannitol (pH 5.8) was more effective in the isolation of protoplasts. Soybean cells from suspension culture for protoplast isolation were used by Kao and Michayluk (1974) and Kao (1977). The suspension culture was maintained by subculture every 2 to 3 days. Forty-hour cultured suspension cells were used for protoplast isolation. In this case, the capacity for cell division in the protoplast culture was very high and the initial cell division occurred in less than 24 h. In the case of calli, the isolated protoplasts were less active in cell division than the suspension-cultured cells and required 2 to 3 days for the initial cell division to occur. The use of calli, however, makes it possible to easily manipulate and obtain a large number of protoplasts when compared with suspension cultured cells. In the two cases of rice and soybean, the activity of cell division can be improved by using younger calli after transplantation to a fresh medium, because they have larger amounts of meristematic tissue from which protoplasts may be released and which are highly active in cell division.

3 Culture of Protoplasts

3.1 Rice Protoplasts

The isolated protoplasts of the two rice strains A-58 Nor. and A-58 MS have opaque dense cytoplasms (Niizeki and Kita 1981; Niizeki et al. 1982). These protoplasts were cultured in MS medium containing 29 mM sucrose, 0.7 M mannitol, 4 mg/l 2,4-D and 2 mg/l benzyladenine (BA). A newly formed cell wall could sometimes be observed within 24 h after the isolation of the protoplasts by plasmolyzing the protoplasm. After a few days, cell division occurred and formed five to ten cell clusters within 2 weeks. At 1 month, after the dilution of MS medium containing 0.4 M mannitol, the cell clusters grew into visible colonies of 1 to 2 mm in diameter. They were transplanted onto a solid medium with agar and subsequently formed calli of 4 to 5 mm diameter after 1 month. Reproducibility of these calli by this method, however, was so low that it was necessary to improve the method of protoplast culture.

Recently, reproducible protoplast cultures of rice resulting in callus formation and plant regeneration have been reported by many investigators, as mentioned in the introduction (see Table 5). They are usually obtained from protoplasts having totipotency from immature embryos and seed calli or from suspension cultured cells in very short-term cultures which are subcultured as frequently as every 5 days. The capacity of callus formation and plant regeneration in these cases varied, depending on the genotype of the rice strain or cultivar. In a recent study, Kyozuka et al. (1987), using a novel culture method with agarose-bead type culture, reported high frequency plant regeneration.

3.2 Soybean Protoplasts

Niizeki and Kita (1981) and Niizeki et al. (1982) reported callus formation from soybean protoplasts. The isolated protoplasts from hypocotyl-derived calli showed transparent features and changed their shape gradually from round to oval in modified 8p medium (Kao and Michayluk 1975) containing BA instead of zeatin and without coconut water. About 32% of the protoplasts divided into two to three cells within 3 days. After 1 week some cells which had divided developed into small cell clusters, and later formed cell colonies in successive liquid cultures. When the colonies were transferred onto M-1 medium, callus formation occurred vigorously but no somatic embryogenesis or organogenesis occurred.

Barwale et al. (1986) succeeded in regeneration of viable plants by using calli derived from immature embryos. The regeneration occurred via embryogenesis or via organogenesis. The embryogenesis resulted when the immature embryos were cultured on an MS medium containing 43 μM α-naphthaleneacetic acid (NAA). Organogenesis also occurred when a medium containing 13.3 μM BA, 0.2μM NAA and four times the normal concentration of MS minor salts were used. These results were obtained from 54 genotypes tested in this manner, and may perhaps be applicable to protoplast-derived callus.

4 Fusion of Protoplasts

4.1 Fusion Treatment

The fusion of rice (A-58 Nor.) and soybean (cv. Harosoy) protoplasts was performed by using a solution containing 50% (w/v) polyethyleneglycol 1540 (PEG), 10 mM $CaCl_2$, 0.7 mM KH_2PO_4 and 0.1 M glucose (Niizeki et al. 1985). The mixed protoplasts of rice and soybean were incubated with the PEG solution for 15 – 20 min at 25 °C. After the incubation, the protoplasts were washed four times with a solution containing 0.6 M mannitol and 50 mM $CaCl_2$. They were then cultured in a liquid modified 8p medium (Kao and Michayluk 1975) containing BA instead of zeatin and without coconut water. The rice (A-58 Nor.) callus had a blackish purple pigment of flavonoids, which was under the control of nuclear genes (Fig. 1 A). The pigmentation of the callus, however, was very unstable, so that a white callus was sometimes obtained on the M-1 medium. The rice callus derived from A-58 Nor. was very hard but friable. The soybean callus was usually yellow and very soft (Fig. 1 B). The yellow colour of the callus sometimes turned into a green colour on the M-1 medium. When rice protoplasts with the flavonoid pigment and dense cytoplasmic strands, and soybean protoplasts with a transparent cytoplasm were used for protoplast fusion, the difference between the parental cytoplasms was suitable for easy recognition of the fusion products of rice and soybean protoplast (Fig. 1 C). The cytoplasm of the fused protoplasts gradually became mixed when the PEG was washed out (Fig. 1 D). This method for identification of the hybrid protoplasts was a reasonably good as well as commonly used

Fig. 1. A A blackish purple rice callus of strain A-58 which is hard but friable after 14 days of culture. **B** Yellow and very soft soybean callus of cv. Harosoy after 14 days of culture. **C** Fusion of rice and soybean protoplasts by PEG. In fused protoplasts the blackish part is rice and the whitish part is soybean protoplast. **D** Cytoplasmic mixing of the rice and soybean protoplasts after washing out the PEG. **E** Hybrid calli derived from a colony after 30 days of culture. A hybrid callus which is blackish purple is a character of rice calli, but being very soft, which is a character of soybean calli. **F** Lipid granules (indicated by an *arrow*) observed in the hybrid callus cell. **G** Lipid granules observed in rice callus cells. **H** No lipid granule observed in soybean callus cells. All calli of **A, B, E, F, G, H** were cultured on the medium of Murashige and Skoog (1962) containing 2 mg/l 2,4-D. (Niizeki et al. 1985)

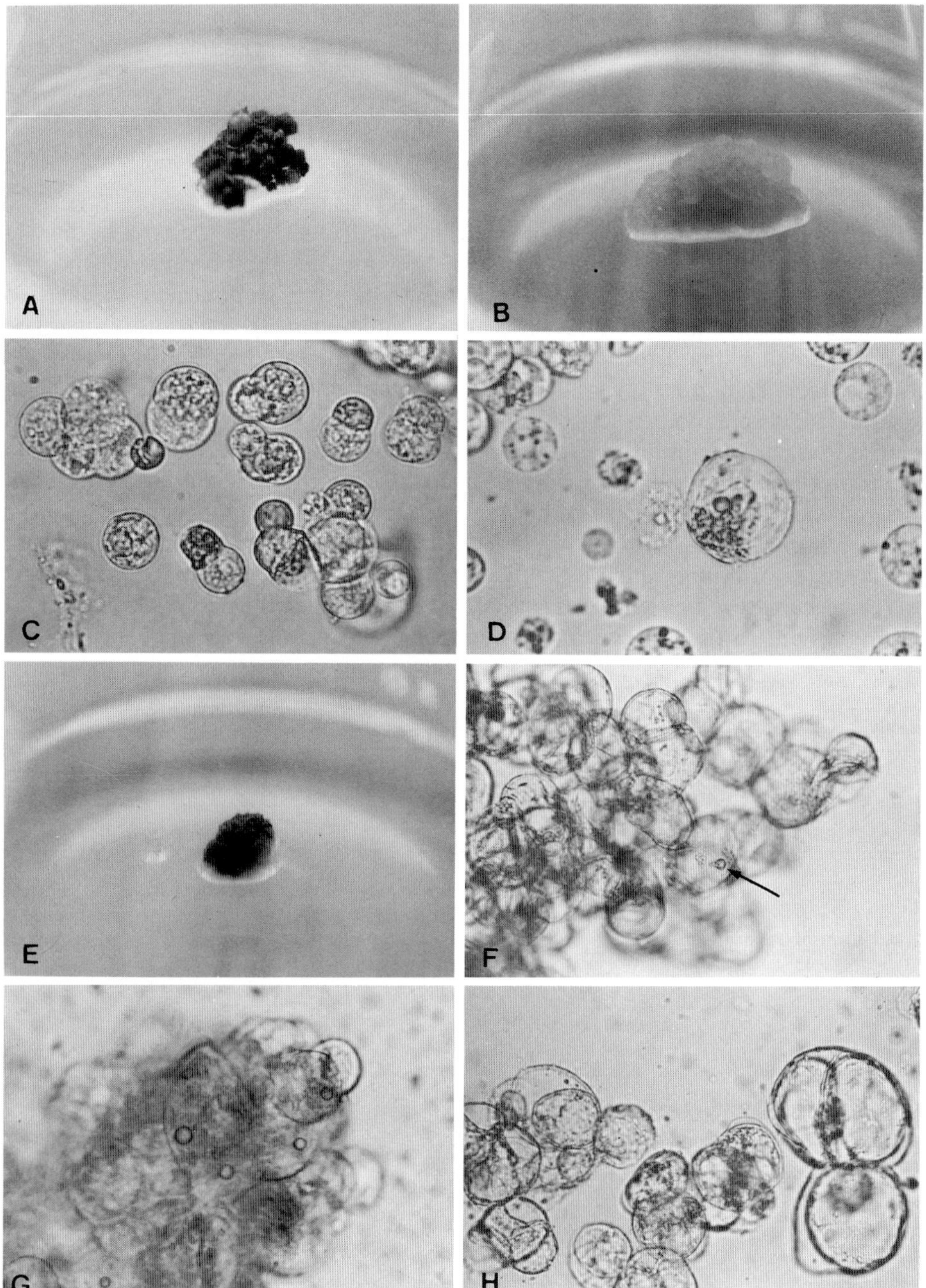
A
B
C
D
E
F
G
H

method in which green protoplasts from leaf mesophyll of one parent and non-green protoplasts from cultured tissue or cells of another parent are fused (Bajaj 1983).

4.2 Culture of Protoplasts After the Fusion Treatment

Table 2 indicates the frequencies of the total calli formed and the selected hybrid calli (Niizeki et al. 1985). In an attempt to eliminate the chromosomes of one parent, 1 to 9 kR of X-irradiation (60 kV, 3 mA) was given to donor protoplasts of rice or soybean, and then fused with non-irradiated recipient protoplasts. The non-irradiated parent protoplast mixtures were also treated with a PEG solution as a control. They were then cultured in the liquid modified 8p medium. Falcon plastic dishes were used for the cultures. The dishes were sealed with Parafilm and placed in a plastic humid box in the dark at 25 °C. Cells which were formed from the protoplasts were exposed to light (2000 lx) after they divided a few times in 7–10 days. When soybean protoplasts were irradiated by high dosages of X-rays, colony formation was completely inhibited, while colony formation occurred at low dosages of X-rays, such as 1 kR. On the other hand, when rice protoplasts were irradiated even by high dosages of X-rays, colonies formed vigorously. This result was due to the fact that the rice protoplast itself was not capable of colony

Table 2. Callus formation of the somatic hybrid between rice and soybean. (Niizeki et al. 1985)

Dosage of X-rays (kR)	No. of colonies explanted on agar-solidified medium[a]	No. of calli formed	No. of hybrid calli
0 (3)	116	54	3
Irradiation of soybean protoplasts			
1 (3)	70	5	0
4 (3)	0	–	–
6 (1)	0	–	–
Irradiation of rice protoplasts			
1 (2)	80	57	0
3 (2)	90	71	0
6 (2)	40	14	0
9 (2)	129	82	1

[a] Basic medium of MS containing 2 mg/l 2,4-D.
( ) Test frequency.

formation in the modified 8p medium. while soybean protoplasts had the capacity to form colonies. Therefore, high dosages of X-rays caused such severe damage to the nuclear genome of soybean protoplasts that both the soybean protoplast itself and soybean protoplasts that fused with rice protoplasts could not survive. This resulted in loss of ability to form colonies. In the case of soybean protoplasts irradiated by 1 kR of X-rays, the damage to the genome was not so severe as to inhibit the formation of colonies, although the frequency of callus formation was very low.

4.3 Selection of Hybrid Callus

In many cases in the production of cell hybrids, direct visual observation of the hybrids is possible because morphologically different parental cell types, generally callus and leaf mesophyll protoplasts, are fused. Early dividing fusion cells were mechanically isolated and cultured in droplets of conditioned medium (Kao 1977). In most cases, however, the observations were limited to the first few divisions of the fusion products, and not many attempts were made to isolate the hybrid cell lines. It is sometimes also possible to select hybrid cell lines by observing the combined markers of the parent. Indeed, hybrid callus of rice and soybean could be selected by observing the phenotypic characters of blackish purple flavonoid pigment and the softness of the calli (Niizeki et al. 1985) (Fig. 1 E). This pigmentation is a character of the rice callus of strain A-58 and the soft callus is that of soybean. The possibility that these selected calli originated from rice protoplasts was excluded because the rice protoplasts could not form the colonies in the modified 8p medium. On the contrary, they were confirmed to be hybrids by the fact that the lipid granules observed only in rice callus cells could also be detected in the selected callus cells (Fig. 1 F, G, H). Thus, four hybrid calli were obtained, of which three were from the control in which both parent protoplasts were not irradiated and only one was from the combination in which irradiated rice protoplasts fused with non-irradiated soybean protoplasts. No hybrid calli were obtained in cases where soybean protoplasts were irradiated by X-rays. The blackish purple pigment of the hybrid calli, however, disappeared rapidly in succeeding subcultures and resulted in a brown or brownish yellow colour. Also the lipid granules could no longer be observed in the hybrid calli after a few subcultures. Therefore, it could be estimated that the rice chromosomes were rapidly eliminated from the hybrid callus. From microscope observations it was not possible to detect any rice chromosomes in the hybrid calli. X-irradiation of the rice protoplasts was also considered to enhance the elimination of the rice chromosomes. Therefore, the phenotypic marker for the selection of hybrid callus might be lost in the early stages of cell division and the hybrid callus might not be able to be identified. This might be a cause of the extreme inability to select hybrid callus when irradiated rice protoplasts fused with non-irradiated soybean protoplasts.

Kao (1977) observed that as the chromosomal behaviour of soybean and *Nicotiana glauca* in the hybrids was not synchronous in the first few cell generations and that the chromosomes of *N. glauca* had a tendency to stick together and break into pieces, only a few individual chromosomes of *N. glauca* were observable. Also many cases of abnormal chromosome behaviour, such as chromosomal bridges, extremely long chromosomes, ring chromosomes, chromosomal fragments and multiconstrictional chromosomes were observed. After 6 months of culturing, most of the hybrid lines lost the chromosomes of *N. glauca* completely. Chien et al. (1982) also reported the loss of tobacco chromosomes in the hybrids of tobacco and soybean. These chromosome losses in hybrid cells are estimated from the observation of premature chromatin condensation, as found generally in mammalian hybrid cultured cells.

4.4 Isozyme Analysis

The peroxidase and esterase isozyme band patterns of the hybrid calli of rice and soybean after several subcultures indicated only that of the soybean callus (Fig. 2), and all four hybrid callus lines showed the same band pattern (Niizeki et al.

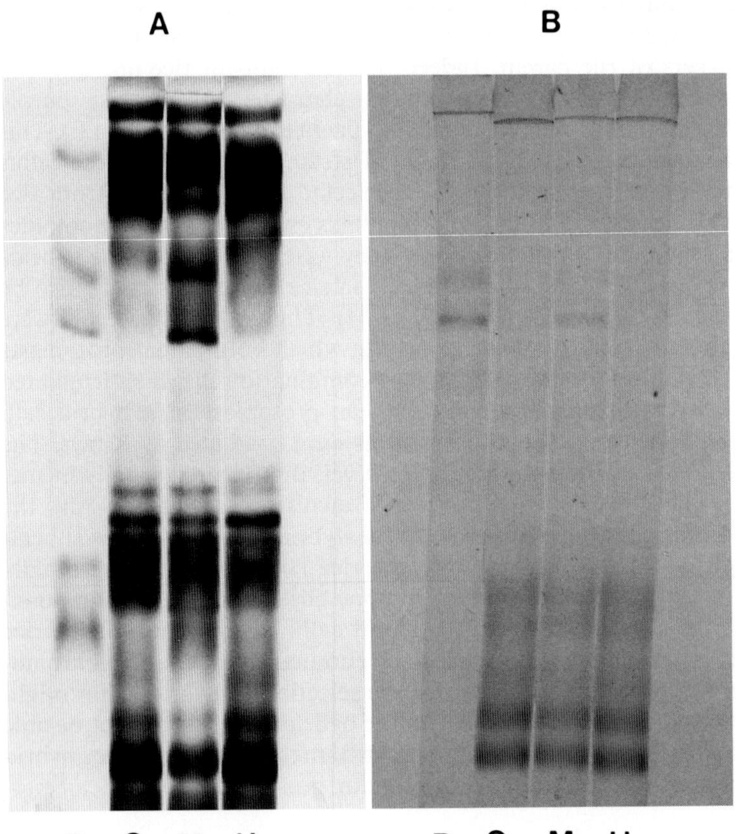

Fig. 2. Electrophoretic patterns of peroxidase (**A**) and esterase (**B**) obtained from rice calli (*R*), soybean calli (*S*), a mixture of rice and soybean calli (*M*), and hybrid calli of rice and soybean (*H*)

1985). This result supports the fact that the rice chromosomes are eliminated from the hybrid callus and that the soybean nuclear genome may be retained after several subcultures. Wetter (1977) studied the isozyme patterns of aspartate aminotransferase and alcohol dehydrogenase from soybean-*Nicotiana glauca* hybrid calli that had been cultured from 2 to 8 months. The study confirmed the presence of specific isozyme bands for both *N. glauca* and soybean in all of the cell lines at the early stages of culture (2–4 months). However the specific enzyme bands of *N. glauca* were no longer detected in the majority of the cultures after 8 months. This is in good agreement with the data of chromosome elimination reported by Kao (1977). In this case, chromosome elimination of the *N. glauca* genome is not so drastic as compared with the hybrids of rice and soybean.

4.5 Segregation of Different Callus Colours from Hybrids

Table 3 indicates the colour of hybrid calli of rice and soybean on two different media. On medium M-1 (basic MS medium containing 2 mg/l 2,4-D), rice calli were blackish purple or whitish yellow and soybean calli were yellow or green. Calli of somatic hybrids were usually brown or brownish yellow. However, yellow, green and white calli have been segregated from three hybrid calli during subcultures. When the parental calli, the hybrid calli and the segregated calli were transferred onto medium M-2 (basic MS medium containing 4 mg/l NAA and 2.5 mg/l kinetin) yellow and green soybean calli or yellow and green calli segregated from

Table 3. Colour of calli of rice, soybean, their somatic hybrids and of calli segregated from somatic hybrids on two types of media. (Niizeki et al. 1985)

Materials	Medium	
	M-1	M-2
Rice	Blackish purple	Blackish purple
(strain A-58)	Whitish yellow	Whitish yellow
Soybean	Yellow	Green
(cv. Harosoy)	Green	Green
Somatic hybrid		
III-1	Brownish yellow	Pale green
	Green[a]	Green
III-2	Brownish yellow	Green
	Green[a]	Green
	White[a]	Whitish yellow
III-3	Brown	Pale green
	Yellow[a]	Green
	Green[a]	Green
	White[a]	Whitish yellow
IV-1	Brownish yellow	Pale green

M-1 Basic medium of MS containing 2 mg/l 2,4-D.
M-2 Basic medium of MS containing 4 mg/l NAA and 2.5 mg/l kinetin.
[a] Colour of calli segregated from hybrid calli during subcultures.

the hybrids and brown or brownish yellow hybrid calli became green or pale green, or otherwise maintained their green colour. Whitish yellow callus of rice and white callus segregated from the hybrids, but did not turn green, even though the peroxidase and esterase isozyme band patterns of the segregated white calli indicated a soybean callus. During long-term culture of over 1 month, the segregated white calli, however, produced green parts in a mosaic pattern. Therefore, it might be concluded that the yellow and green calli which segregate from the hybrids have mainly the soybean nuclear genome and the soybean chloroplasts, and the white calli which segregate from the hybrids have mainly the soybean nuclear genome and the rice and soybean chloroplasts in a mosaic pattern. When white parts of calli were selected from a white-green mosaic callus for four to seven subcultures, the selected white calli became stable and never entered the mosaic state.

4.6 Fraction I Protein

Fraction I protein (ribulose-1,5-bisphosphate carboxylase/oxygenase) from the green callus segregated from the hybrids was isolated and analyzed by the method described by Hirai (1982). In the case of the white-green mosaic callus, 15 g fresh weight of callus was ground in 40 ml of 100 mM Tris-HCl (pH 7.8), 5 mM DTT, 10 mM $MgCl_2$, 1 mM EDTA and 12.5% (v/v) glycerol. The callus extract was centrifuged and solid ammonium sulfate was added to the supernatant at a final

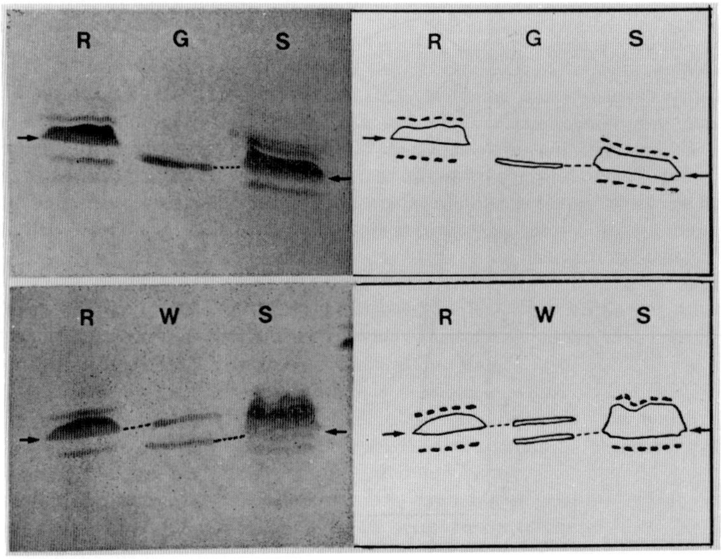

Fig. 3. Gels after isoelectrofocusing of large subunits of Fraction I protein from rice, strain A-58 (*R*), soybean, cv. Harosoy (*S*), and green callus (*G*) and white callus (*W*) segregated from their somatic hybrid callus (III-3). Green calli have only soybean large subunits whereas white calli have both rice and soybean subunits. *Arrows* indicate positions of each large subunit polypeptide

concentration of 0.75 saturation. The solution was centrifuged and the protein fraction in the pellet was dissolved in 2 ml of 10 mM Tris-HCl (pH 7.8), 1 mM DTT, 1 mM $MgCl_2$, 1 mM EDTA and 12.5% (v/v) glycerol and desalted by Sephadex G-25. About 2 to 3 ml of the void volume fraction was incubated with 1/20 volume of tobacco anti-Fraction I protein. Precipitated Fraction I protein was analyzed as described by Hirai (1982). The result showed that the green callus which segregated from the hybrids had soybean chloroplast genomes and the mosaic callus had both rice and soybean chloroplast genomes (Fig. 3). This indicates that the soybean and rice chloroplast genomes were sorting out into green calli and white calli selected from the white-green mosaic callus, respectively. The small subunit of Fraction I protein was not detected in this study.

5 Response to Streptomycin

Four months after the induction of primary somatic hybrid calli, the parent calli, the hybrid calli and those produced by the sorting-out of chloroplasts were cut into 20–30-mg pieces and cultured in MS medium containing 0, 125, 250, 500, 750 and 1000 ppm of streptomycin (Niizeki et al. 1986). The medium also contained 2 mg/l of 2,4-D. This experiment was performed under light (ca. 3000 lx) and dark conditions at 25 °C. Table 4 shows the growth index of each callus under light and dark conditions. The growth index expresses the value of the fresh callus weight on the medium with streptomycin, divided by the fresh callus weight on the streptomycin-free medium after 11 weeks of callus culturing. The results showed that all four heteroplasmic hybrid calli with both parent chloroplasts were extremely sensitive to streptomycin under both light and dark conditions. On the other hand, the parent calli and the green and white calli with sorted-out chloroplast genomes were resistant as compared with the heteroplasmic hybrid calli under light conditions. However, the parent calli were almost as sensitive as the hybrid calli under dark conditions. The green and white calli showed slightly higher resistance levels than the hybrid and parent calli under dark conditions, but they were extremely sensitive as compared with those under light conditions. From these results the following assumptions can be drawn. The differences in sensitivity to streptomycin under light and dark conditions in the parent calli or the calli with sorted-out chloroplast genomes suggest that the response of callus to streptomycin is presumably related to the chloroplast genomes which require light for their function. Furthermore, the recovery of streptomycin resistance in the calli produced by the sorting-out of chloroplast genomes indicates that the chloroplast genomes may be related to the streptomycin resistance or sensitivity. These data support the results of experiments carried out on the SR-1 chloroplast mutant for streptomycin resistance in *Nicotiana tabacum* by Menczel et al. (1981); Medgyesy et al. (1980); Maliga et al. (1982), and Fluhr et al. (1983). However, the possibility that the streptomycin resistance resides in the other organelles cannot be completely ruled out in this study. Also, it should be noted that the cause of the heteroplasmic state of the callus, which is extremely sensitive to streptomycin even under light conditions, remains to be determined. In the case of SR-1 of

Table 4. Response of callus growth to streptomycin for 11 weeks of culture under light and dark conditions (Niizeki et al. 1986)

Kind of callus	Growth index[a] Concentration of streptomycin (ppm)				
	125	250	500	750	1000
Light conditions					
Rice	0.780	0.808	0.374	0.275	0.166
Soybean	0.717	0.602	0.519	0.375	0.253
Hybrid callus line					
III-1	0.169	0.029	0.016	0.011	0.010
III-2	0.249	0.087	0.145	0.102	0.112
III-3	0.160	0.108	0.132	0.128	0.150
IV-1	0.057	0.009	0.034	0.018	0.014
Segregated callus[b]					
Green callus	0.883	0.811	0.536	0.439	0.432
White callus	1.093	1.052	0.936	0.537	0.396
Dark conditions					
Rice	0.372	0.106	0.022	0.020	0.022
Soybean	0.237	0.098	0.095	0.101	0.065
Hybrid callus line					
III-1	0.017	0.015	0.013	0.014	0.012
III-2	0.255	0.087	0.081	0.050	0.050
III-3	0.037	0.089	0.012	0.026	0.028
IV-1	0.029	0.036	0.023	0.030	0.021
Segregated callus[b]					
Green callus	0.754	0.391	0.052	0.058	0.074
White callus	0.970	0.600	0.220	0.117	0.031

[a] Fresh callus weight on the medium with streptomycin/Fresh callus weight on the medium without streptomycin.

[b] Callus segregated from the hybrid callus by sorting-out of chloroplasts. Green callus has soybean nucleus and soybean chloroplast and white callus has soybean nucleus and rice chloroplasts.

N. tabacum, the resistance to streptomycin is a dominant character in the heteroplasmic state of the callus having both resistant and sensitive chloroplasts (Maliga et al. 1982).

After 1 year, the growth ability of excised small pieces of the hybrid calli maintained by subcultures was examined on the medium containing 500 ppm of streptomycin under light conditions (Fig. 4) (Niizeki et al. 1986). The results clearly showed that the number of streptomycin-resistant calli increased as compared with the number of calli subcultured for a period of 4 months. The colour of the streptomycin-resistant calli was usually pale brown and the highly resistant calli were green, yellow or white. These findings suggest that the sorting-out of chloroplast genomes took place, and that the number of callus parts or cells containing one of the parent chloroplast genomes increased. Thus, many callus parts or cells had acquired the streptomycin-resistant trait.

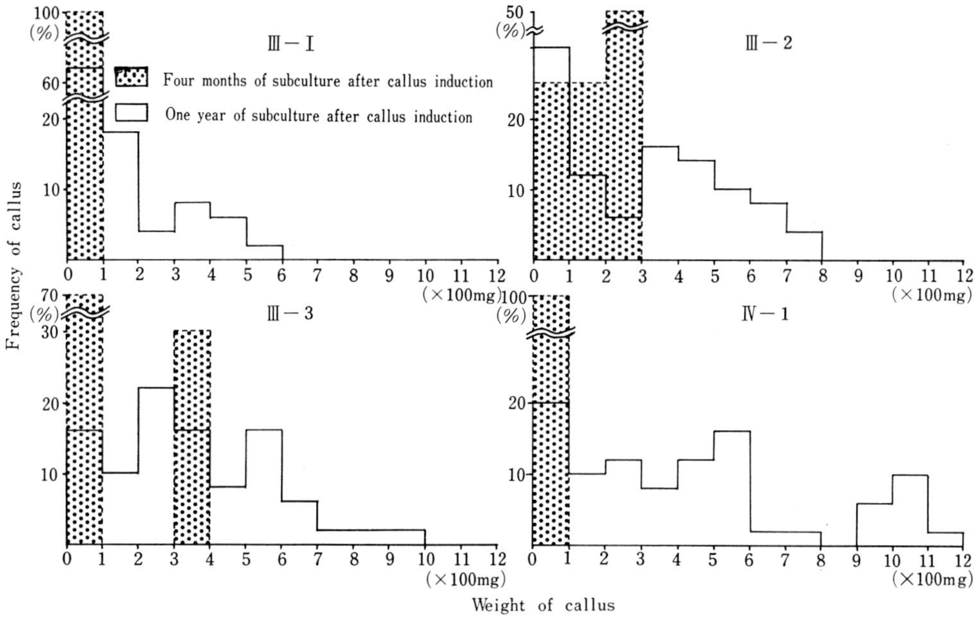

Fig. 4. Growth of the excised 20−30 mg pieces of somatic hybrid callus during 11 weeks on the medium containing 500 ppm streptomycin under light conditions. *III-1, III-2, III-3* and *IV-1* indicate the somatic hybrid callus lines

6 Conclusion and Prospects

In this study on the fusion of protoplasts of rice x soybean the segregation of the colour of the calli from the somatic hybrid cells was due to the segregation of chloroplast genomes of rice and soybean, because the nuclear genome of the somatic hybrids was assumed to be mainly that of soybean. This hypothesis supports the result of Akada et al. (1983), who analyzed the Fraction I protein from the somatic hybrid calli of *Nicotiana glauca* and *N. langsdorffii.* In the early stage of the hybrid calli, 70% of the calli contained two kinds of chloroplast genomes. However, the selection of a small piece of the callus followed by further growth and subculturing led to the conclusion that the callus was already a mosaic of cells, where elimination of one of the two kinds of chloroplast genomes had occurred in some cells. In the case of the somatic hybrid calli of rice and soybean, segregation of chloroplast genomes of the two parents was represented in the callus colour segregation into green and white. However, white calli turned into green-white mosaic calli after long-term culture, but after selection of white parts of the calli for four to seven subcultures, only white calli were obtained. From the results of the Fraction I protein analyses it may be presumed that the calli with chloroplasts of rice or soybean are isolated independently. In the genus *Nicotiana*, chloroplast-substituted plants were induced by the somatic hybrids in which one of the parent protoplasts was enucleated by centrifugation or X-irradiation

Table 5. Methods of protoplast isolation and culture of rice and soybean

Materials	Enzyme and conditions for protoplast isolation	Medium and conditions for protoplast culture	Medium and conditions for plant regeneration	Reference
A. Rice: *Oryza sativa* L. Mesophyll protoplasts of soft whitish leaf-sheaths	Enzyme solution 2% pectinase (Serva) 0.45 M mannitol (pH 5.4) Incubation Vacuum for 10 min and then under normal pressure for 1 h at 30 °C with shaking ↓ Transfer the cells 3% Cellulase Onozuka SS 0.45 M mannitol (pH 5.4) 2 h at 30 °C with shaking	Medium Medium of Deka and Sen (1976) Protoplast density $2.0-2.5\times10^4$/ml Culture condition 26 °C under the continuous light Obtained product Calli	Medium SH medium with 0.5 mg/l 2,4-D 0.5 mg/l IAA 0.25 mg/l NAA 0.1 mg/l kinetin Culture condition 26 °C and 12 h photoperiod Obtained product Roots	Deka and Sen (1976)
Callus	Enzyme solution 3% Cellulase Onozuka SS 2% pectinase 0.45 M mannitol (pH 5.4) Incubation 4 h at 30 °C with shaking	ditto	ditto	
Suspension-cultured cells derived from pollen-callus of cv. Lianchiang Mitsao	Enzyme solution 4% Cellulase Onozuka R-10 0.6% Macerozyme R-10 0.2% hemicellulase (Koch Light Laboratories Co., Ltd) 0.2% potassium dextran sulphate 1 mM $CaCl_2\cdot2H_2O$	Medium Medium of Tsai et al. (1978) Protoplast density 4×10^6/ml Culture condition 1. 10 h photoperiod at 28 °C for 14 days then transfer to 26 °C in dark 2. 26 °C in dark	–	Tsai et al. (1978)

	0.8 M mannitol (pH 5.4) Incubation 0.5 h at room temperature and 3.5 h at 34 °C without shaking	Obtained product Calli		
Nitrate reductase-deficient callus of cv. Norin 8	Enzyme solution 4% Cellulase Onozuka RS 1% Macerozyme R-10 0.3 M mannitol (pH 5.8) Incubation 40 min at 28 °C without shaking	Medium AA medium with 2 mg/l 2,4-D 0,1 or 2 mg/l zeatin 0.3 M glucose and/or sucrose Protoplast density $1\times10^5-2\times10^6$/ml Obtained product Cell colonies	–	Wakasa et al. (1984)
Suspension-cultured cells of cv. Taichung 65 (a cytoplasmic male sterile line)	Enzyme solution 4% Cellulase Onozuka R-10 0.2% Macerozyme R-10 6.5% mannitol 0.1% $CaCl_2 \cdot 2H_2O$ Incubation 1.5 h at 25 °C with shaking and then 1.5 h at 30 °C without shaking	Medium B5 medium modified as following: 1. B5 medium with 2500 mg/l KNO_3 134 mg/l $(NH_4)_2SO_4$ 2. B5 medium with 2500 mg/l KNO_3 3. B5 medium with 2500 mg/l KNO_3 134 mg/l $(NH_4)_2SO_4$ ↓ Transfer the cells B5 medium with Amino acid of AA medium All media containing 1 mg/l 2,4-D 0.2 mg/l kinetin 0.1 mg/l GA_3 Protoplast density 2×10^5/ml	Medium B5 medium with 0.2 mg/l NAA 1 mg/l kinetin Obtained product Green spots Roots	Toriyama and Hinata (1985)

Table 5 (continued)

Materials	Enzyme and conditions for protoplast isolation	Medium and conditions for protoplast culture	Medium and conditions for plant regeneration	Reference
		Culture condition 12 h photoperiod Obtained product Calli		
Suspension-cultured cells of cv. Nipponbare and Sasanishiki	Enzyme solution 4% Cellulase Onozuka RS 1% Macerozyme R-10 0.4 M mannitol 0.5% $CaCl_2 \cdot 2H_2O$ 0.5% potassium dextran sulfate (pH 5.5) Incubation 3 h with shaking and then 3 h without shaking at 27 °C	Medium R2 Medium with vitamins of B5 medium 0.4 M sucrose 2 mg/l 2,4-D Protoplast density 1×10^6/ml Culture condition 25 °C in the dark Obtained product Calli	Medium N6 medium with 3% sucrose 2 mg/l 2,4-D 0.5 mg/l BA ↓ Transfer the calli N6 medium without growth regulators Culture condition Under continuous light Obtained product Plants	Fujimura et al. (1985)
Suspension-cultured cells of strain A-58 MS (a cytoplasmic male sterile line) and cv. Fujiminori	Enzyme solution 2% Cellulase Onozuka RS 2% Macerozyme R-10 0.1% Pectoryase Y-23 1% Driselase 0.8% calf serum 80 mM $CaCl_2 \cdot 2H_2O$ 0.125 mM $MgCl_2$ 0.5 mM MES LS medium with 0.3 M glucose but without growth regulators (pH 5.6) Incubation 9 h in the dark without shaking	Medium RY-2 medium Protoplast density 2.5×10^5/ml Culture condition 25 °C in the dark Obtained product Calli	Medium N6 liquid medium with 8% sucrose but without growth regulators ↓ Transfer the calli LS agar medium with 8% sucrose 4×10^{-6} M BA (pH 5.6) Culture condition 25 °C under the continuous light Obtained product Plants	Yamada et al. (1985, 1986)

Callus of cv. Moroberekan	Enzyme solution 1% Cellulase Onozuka R-10 1% Macerozyme R-10 0.6% Driselase 0.5% Pectolyase Y-23 5.5% mannitol 5.5% sorbitol (pH 5.4) Incubation 20 h at 25 °C with shaking	Medium 8p medium with 10^{-4} M FeEDTA 2 mg/l 2,4-D 0.2 mg/l BA ↓ Transfer the protoplasts MS medium with 10^{-4} M FeEDTA vitamins of Morel and Wetmore (1951) 2 mg/l 2,4-D 0.1 mg/l BA (pH 6.0) Protoplast density 3×10^4/ml Culture condition 16 h photoperiod at 28 °C Obtained product Calli	Medium MS medium with 10^{-4} M FeEDTA vitamins of Morel and Wetmore (1951) 4% sucrose 0.7% agar (pH 6.0) Obtained product Plants	Coulibaly and Demarly (1986)
Suspension-cultured cells of cv. Taipei 309 and Fujisaka 5	Enzyme solution 1% Cellulase Onozuka RS 0.1% Pectolyase Y-23 5 mM MES, CPW salt (Frearson et al. 1973) 13% mannitol (pH 5.6)	Medium KpR medium with 0.5 mg/l 2,4-D 10% glucose (pH 5.6) Protoplast density 3×10^5/ml Culture condition 27 °C in the dark Obtained product Calli	Medium N6 or MS medium without growth regulators containing 8% sucrose, 0.8% agar (pH 5.8) in the dark at 27 °C Transfer the ↓embryoids MS basal medium with 0.8% agar in continuous light (2500 lx) Obtained product Plants	Abdullah et al. (1986)
Seed callus and suspension-cultured cells of cv. Nipponbare, Iwaimochi, Norin 14, Fujisaka 5 and Koshihikari	Enzyme solution 4% Cellulase Onozuka RS 1% Macerozyme R-10 0.4 M mannitol	Medium R2 medium with 5.6 mg/l $FeSO_4$ 7.5 mg/l Na_2EDTA	Medium N6 medium with 0–10 mg/l kinetin, zeatin or BAP	Kyozuka et al. (1987)

Table 5 (continued)

Materials	Enzyme and conditions for protoplast isolation	Medium and conditions for protoplast culture	Medium and conditions for plant regeneration	Reference
	Incubation 5 – 7 h for seed callus and 3 – 4 h for suspension-cultured cells at 30 °C without shaking	vitamins of MS medium 2 mg/l 2,4-D 0.4 M sucrose The same volume of 2.5% agarose as the protoplast suspension (pH 5.6) ↓ Transfer the colonies N6 medium with 2 mg/l 2,4-D 6% sucrose 0.25% agarose Protoplast density 1×10^6/ml Culture condition Agarose-bead method (Shillito et al. 1983) with nurse cells Shaking slowly at 25 °C in the dark, then under the light (3000 lx) after transfer the colonies Obtained products Calli	6% sucrose 1% agarose Obtained product Plants	
Glycine max (L.) Merr. Suspension-cultured cells	Enzyme solution Cellulase Onozuka P1500 Pectinase (Sigma Chemical Co., Ltd.) B5 medium with 0.1 M sucrose 0.3 M sorbitol 0.005 M citric acid (pH 5.5)	Medium B5 medium with 0.13 M sorbitol 0.05 M sucrose 0.001 M $CaH_4(PO_4)_2 \cdot 2H_2O$ (pH 5.5) Obtained product Cell culsters	–	Kao et al. (1970)

	Incubation 4 – 6 h with shaking for a few seconds each hours			
Pod tissue of strain T219	Enzyme solution 1. 3% Macerozyme R-10 0.6 M mannitol (pH 4.5) 2 h at 30 °C with shaking Transfer the pod ↓ pieces 5% Cellulase Onozuka R-10 0.6 M mannitol 8 mM $CaCl_2$ (pH 5.2) 2. 3% Macerozyme R-10 5% Cellulase Onozuka R-10 Mannitol-$CaCl_2$ solution (pH 5.2) Incubation In the dark	Medium Medium of Kao (1977) Protoplast density 1×10^5/ml Culture condition Keep at 27 °C in the dark for 3 days, then culture medium was added ↓ Transfer the entire contents B5 medium with 1% coconut water The suspension cultures were maintained at 29 °C on a reciprocal shaker Obtained product Calli	Medium MS medium with 1 mg/l 2,4-D 1 mg/l kinetin ↓ Transfer the calli HRM medium Culture condition 25 °C in the dark or under 2300 lx of illumination (16 h day length) Obtained product Roots	Zieg and Outka (1980)
Callus of cv. Harosoy	Enzyme solution 4% Cellulase Onozuka R-10 0.4% Macerozyme R-10 0.7 M Mannitol (pH 5.8) Incubation 14 – 17 h in the dark at 28 °C with shaking	Medium 8p medium with BA instead of zeatin and without coconut water Protoplast density $1 \times 10^4 - 1 \times 10^5$/ml Culture condition 25 °C Obtained product Calli	–	Niizeki and Kita (1981)
Root tip of cv. Fiskeby 5	Plasmolysis treatment 1 h in CPW 13 M solution which containing 13% mannitol 27.2 mg/l KH_2PO_4 101 mg/l KNO_3 1480 mg/l $CaCl_2 \cdot 2H_2O$	Medium 1. B5P medium with 3% sucrose 7% mannitol 1 mg/l *p*-chlorophenoxy-acetic acid 1 mg/l BAP	Medium MS medium with 1. 2 mg/l IAA 1 mg/l BAP 2. 0.05 mg/l NAA 0.5 mg/l BAP Culture condition	Xu et al. (1982)

Table 5 (continued)

Materials	Enzyme and conditions for protoplast isolation	Medium and conditions for protoplast culture	Medium and conditions for plant regeneration	Reference
	240 mg/l $MgSO_4 \cdot 7H_2O$ 0.16 mg/l KI 0.025 mg/l $CuSO_4 \cdot 5H_2O$ (pH 5.6) Enzyme solution 2% Rhozyme 4% Meicelase R-10 0.3% Macerozyme R-10 CPW 13 M solution Incubation 16 h at 25 °C with shaking	2. B5 medium with Vitamins of B5P 1% glucose 3.5% sorbitol 3.5% mannitol 200 mg/l casein hydrolysate 0.25 or 2.0 mg/l 2,4-D 0.5 mg/l NAA 0.1 or 0.5 mg/l BAP 3. KM8P + KM8 (2:1, v/v) Protoplast density 2.5×10^4/ml Culture condition 23 °C in the dark Obtained product Calli	Under diffuse light (700 lx) Obtained product Roots	
Cotyledons of cv. Fiskeby 5, AP-10, HP 20-20, HP 2530 and HP 4800	Enzyme solution EM-1 0.2% Cellulase Onozuka R-10 0.1% Pectolyase Y-23 EM-2 0.2% Cellulase Onozuka R-10 1% Rhozyme 0.1% Pectolyase Y-23 EM-3 0.2% Cellulase Onozuka RS 0.1% Pectolyase Y-23 EM-4 0.2% Cellulase Onozuka RS 1% Rhozyme 0.1% Pectolyase Y-23	Medium Medium of Kao (1977) Protoplast density 5×10^4/ml Culture condition In the dark at 23 °C for 8 – 10 days then transfer to 2000 lx continuous illumination at 23 °C Colonies were transferred to the medium of Kao (1977) solidified with 0.6% agar Obtained products Calli	–	Lu et al. (1983)

	EM-1 was dissolved in 0.12% $CaCl_2 \cdot 2H_2O$ and 8% mannitol (pH 5.8) EM-2 – 4 were dissolved in CPW salts and 9% mannitol (pH 5.8) Incubation 8 – 10 h at 25 °C with shaking			
Suspension-cultured photoautotrophic and heterotrophic cell lines derived from cotyledons or hypocotyl cv. Corsoy, Asgrow and A 3127	Enzyme solution 1% Cellulase Onozuka R-10 0.2% Pectolyase Y-23 2% Driselase IS-1 protoplast wash medium (Gamborg et al. 1983b) (pH 5.6) Incubation 2 to 3 h at 28 °C with shaking in the dark	Medium Medium of Chowhury and Widholm (1985) On day 8 and 12, KT-1 medium was added Transfer the cell ↓ aggregates KT-1 medium solidified with 0.6% agarose Protoplast density 1×10^6/ml Culture condition Under the continuous light (fluorescent, 300 lx) Obtained products Calli	–	Chowhury and Widholm (1985)

SH medium: Schenk and Hildebrandt (1972), AA medium: Müller and Grafe (1978), B5 medium: Gamborg et al. (1968), R2 medium: Ohira et al. (1973), N6 medium: Chu et al. (1975), LS medium: Linsmaier and Skoog (1965), RY-2 medium: Yamada et al. (1986), KM8P (8P) and KM8 medium: Kao and Michayluk (1975), KpR medium: Kao (1977), MS medium: Murashige and Skoog (1962), HRM medium: Evans et al. (1977), B5P medium: Donn (1978), KT-1 medium: Horn et al. (1983), Cellulase Onozuka SS: All Japan Biochemicals Co., Ltd., Cellulase Onozuka P1500: All Japan Chemicals Co., Ltd., Cellulase Onozuka R-10: Yakult Pharmaceutical Industry Co., Ltd., Cellulase Onozuka RS: Yakult Pharmaceutical Industry Co., Ltd., Rhozyme: Rohm and Haas Co., Ltd., Meicelase: Meiji Seika Co., Ltd., Macerozyme R-10: Yakult Pharmaceutical Industry Co., Ltd., Pectolyase Y-23: Seishin Pharmaceutical Co., Ltd., Driselase: Kyowa Hakko Kogyo Co., Ltd.

(Maliga et al. 1982; Aviv et al. 1984). If regeneration of plants is possible in the white calli produced in this study, it may be possible to obtain chloroplast-substituted plants which have soybean nuclei and rice chloroplasts. Unfortunately, the regeneration of the plants has not been successful so far. By recent advances in protoplast culture of rice, it may be possible to obtain a plant with rice nuclei and soybean chloroplasts from enucleated soybean protoplasts. It may also be possible to apply this chloroplast substitution technique in distantly related species to improve some plants with high photosynthetic ability and also plants having useful genes coded in the chloroplast genome.

Streptomycin resistance was transferred via protoplast fusion from *Nicotiana tabacum* (SR-1) to other *Nicotiana* species (Menczel et al. 1981; Medgyesy et al. 1980; Maliga et al. 1982). Fluhr et al. (1983) indicated from the results mentioned above, and their experimental results, that streptomycin resistance may reside in the chloroplast genome. The hybrid calli of rice and soybean were very sensitive to streptomycin when compared with parent calli, and the green and white calli segregated from the hybrids. If the response of these calli to streptomycin is controlled by the chloroplast genome, the hybrid calli have as much resistance as their parents. The mechanism of streptomycin resistance and sensitivity in hybrid calli composed of distantly related species may be substantially different from the reaction in closely related species. Resolving the different mechanism between them will clarify novel genetic and physiological phenomena and contribute to a new somatic hybrid breeding programme.

7 Protocol

The protocols for the isolation, culture of rice and soybean protoplasts adopted by various workers are shown in Table 5 (also see Chap II.1, and II.5, this Vol).

References

Abdullah R, Cocking EC, Thompson JA (1986) Efficient plant regeneration from rice protoplasts through somatic embryogenesis. Bio/Technology 4:1087–1090

Akada S, Hirai A, Uchimiya H (1983) Studies on mode of separation of chloroplast genomes in parasexual hybrid calli. 1. Fraction I protein composition in unseparated hybrid callus. Plant Sci Lett 31:223–230

Aviv D, Bleishman S, Arzee-Gonen P, Galun E (1984) Intersectional cytoplasmic hybrids in *Nicotiana*: Identification of plastomes and chondriomes in *N. sylvestris* + *N. rustica* cybrids having *N. sylvestris* nuclear genomes. Theor Appl Genet 67:499–504

Bajaj YPS (1983) Somatic hybridization and cryopreservation studies on rice x pea and wheat x pea protoplasts. In: Potrykus I, Harms CT, Hinnen A, Hutter R, King PJ, Shillito RD (eds) Protoplasts 1983, Birkhäuser, Basel, pp 248–249

Barwale UB, Kerns HR, Widholm JM (1986) Plant regeneration from callus cultures of several soybean genotypes via embryogenesis and organogenesis. Planta 167:473–481

Binding H, Nehls R (1978) Somatic cell hybridization of *Vicia faba* + *Petunia hybrida.* Mol Gen Genet 164:137–143

Chien YC, Kao KN, Wetter LR (1982) Chromosomal and isozyme studies of *Nicotiana tabacum-Glycine max* hybrid cell lines. Theor Appl Genet 62:301–304

Chowhury VK, Widholm JM (1985) Callus production from photoautotrophic soybean cell culture protoplasts. Plant Cell Rep 4:289–292

Chu ZC, Wang CC, Sun CS, Xu Z, Zhu ZY, Yin GC, Bi FY (1975) Attempt at establishing a better medium for anther culture of rice by comparison between different nitrogen-sources. Sci Sin 2:484–490

Constabel F, Weber G, Kirkpatrick JW, Pahl K (1976) Cell division of intergeneric protoplast fusion products. Z Pflanzenphysiol 79:1–7

Coulibaly MY, Demarly Y (1986) Regeneration of plantlets from protoplasts of rice, *Oryza sativa* L. Z Pflanzenzücht 96:79–81

Deka PC, Sen SK (1976) Differentiation in calli originated from isolated protoplasts of rice (*Oryza sativa* L.) through plating technique. Mol Gen Genet 145:239–243

Donn G (1978) Cell division and callus regeneration from leaf protoplasts of *Vicia narbenensis*. Z Pflanzenphysiol 86:65–76

Dudits D, Kao KN, Constabel F, Gamborg OL (1976) Fusion of carrot and barley protoplasts and division of heterokaryocytes. Can J Genet Cytol 18:263–269

Evans DA, Sharp WR, Paddock EF (1977) Variation in callus proliferation and root morphogenesis in leaf tissue cultures of *Glycine max* strain T219. Phytomorphology 26:379–384

FAO (1985) Vol. 39 Production Year book

Fluhr R, Aviv D, Edelman M, Galun E (1983) Cybrids containing mixed and sorted-out chloroplasts following interspecific somatic fusions in *Nicotiana*. Theor Appl Genet 65:289–294

Frearson EM, Power JB, Cocking EC (1973) The isolation, culture and regeneration of *Petunia* leaf protoplasts. Dev Biol 33:130–137

Fujimura T, Sakurai M, Akagi H, Negishi T, Hirose A (1985) Regeneration of rice plants from protoplasts. Plant Tissue Cult Lett 2:74–75

Gamborg OL, Miller RA, Ojima K (1968) Nutrient requirements of suspension cultures of soybean root cells. Exp Cell Res 50:151–158

Gamborg OL, Davis BP, Stahlhut RW (1983 a) Somatic embryogenesis in cell cultures of *Glycine* species. Plant Cell Rep 2:209–212

Gamborg OL, Davis BP, Stahlhut RW (1983 b) Cell division and differentiation in protoplasts from cell culture of *Glycine* species and leaf tissue of soybean. Plant Cell Rep 2:213–215

Hirai A (1982) Isoelectrofocusing of non-carboxymethylated fraction I protein from green callus. Plant Sci Lett 25:37–41

Horn ME, Sherrard JH, Widholm JM (1983) Photoautotrophic growth of soybean cells in suspension culture. Plant Physiol 72:426–429

Imamura J (1984) Callus formation from tobacco (nia 115) + *Triticum durum* fusion hybrid. Plant Tissue Cult Lett 1:60–61

Kao KN (1977) Chromosomal behaviour in somatic hybrids of soybean-*Nicotiana glauca*. Mol Gen Genet 150:225–230

Kao KN, Michayluk MR (1974) A method for high-frequency intergeneric fusion of plant protoplasts. Planta 115:355–367

Kao KN, Michayluk M (1975) Nutritional requirements for growth of *Vicia hajastana* cells and protoplasts at a very low population density in liquid media. Planta 126:105–110

Kao KN, Keller WA, Miller RA (1970) Cell division in newly formed cells from protoplasts of soybean. Exp Cell Res 62:338–340

Kao KN, Constabel F, Michayluk MR, Gamborg OL (1974) Plant protoplast fusion and growth of intergeneric hybrid cells. Planta 120:215–227

Kartha KK, Gamborg OL, Constabel F, Kao KN (1974) Fusion of rape seed and soybean protoplasts and subsequent division of heterokaryocytes. Can J Bot 52:2435–2436

Kyozuka J, Hayashi Y, Shimamoto K (1987) High frequency plant regeneration from rice protoplasts by novel nurse culture methods. Theor Appl Genet 206:408–413

Linsmaier EF, Skoog F (1965) Organic growth factor requirements of tobacco tissue culture. Physiol Plant 18:100–127

Lu DY, Cooper-Bland S, Pental D, Cocking EC, Davey RM (1983) Isolation and sustained division of protoplasts from cotyledons of seedlings and immature seeds of *Glycine max* L. Z Pflanzenphysiol 111:389–394

Maliga P, Lörz H, Lazak G, Nagy F (1982) Cytoplast-protoplast fusion for interspecific chloroplast transfer in *Nicotiana*. Mol Gen Genet 185:211–215

Medgyesy P, Menczel L, Maliga P (1980) The use of cytoplasmic streptomycin resistance. Chloroplast transfer from *N. tabacum* into *N. sylvestris* and isolation of their somatic hybrids. Mol Gen Genet 179:693–698

Menczel L, Nagy F, Kiss ZR, Maliga P (1981) Streptomycin resistant and sensitive somatic hybrids of *N. tabacum* + *N. knightiana*. Correlation of resistance to *N. tabacum* plastids. Theor Appl Genet 59:191–195

Morel G, Wetmore RH (1951) Fern tissue culture. Am J Bot 38:141–143

Müller AJ, Grafe R (1978) Isolation and characterization of cell lines of *Nicotiana tabacum* lacking nitrate reductase. Mol Gen Genet 161:67–76

Murashige T, Skoog F (1962) A revised medium for rapid growth and bioassays with tobacco tissue cultures. Physiol Plant 15:473–497

Niizeki M, Kita F (1981) Cell division of rice and soybean and their fused protoplasts. Jpn J Breed 31:161–167

Niizeki M, Kita F, Takahashi M (1982) Cell division in fused protoplasts of rice and soybean and their selection system. In: Fujiwara A (ed) Plant tissue culture 1982. Maruzen, Tokyo, pp 629–630

Niizeki M, Tanaka M, Akada S, Hirai A, Saito K (1985) Callus formation of somatic hybrid of rice and soybean and characteristics of the hybrid callus. Jpn J Genet 60:81–92

Niizeki M, Tanaka M, Saito K (1986) Response of somatic hybrid callus between rice and soybean to streptomycin. Jpn J Breed 36:75–79

Nitsch JP, Nitsch C (1969) Haploid plants from pollen grains. Science 163:85–87

Ohira K, Ojima K, Fujiwara A (1973) Studies on the nutrition of rice cell culture I. A simple, defined medium for rapid growth in suspension culture. Plant Cell Physiol 14:1113–1121

Schenk RU, Hildebrandt AC (1972) Medium and technique for induction and growth of monocotyledonous and dicotyledonous plant cell cultures. Can J Bot 50:199–204

Shillito RD, Paszkowski J, Potrykus I (1983) Agarose plating and a bead type culture technique enable and stimulate development of protoplast-derived colonies in a number of plant species. Plant Cell Rep 2:244–247

Toriyama K, Hinata K (1985) Cell suspension and protoplast culture in rice. Plant Sci 41:179–183

Tsai CK, Chien YC, Chou YL, Wu SX (1978) A further study on the isolation and culture of rice (*Oryza sativa* L.) protoplast. In: Proc Symp Plant Tissue Cult. Science Press, Peking, pp 317–324

Wakasa K, Kobayashi K, Kanda H (1984) Colony formation from protoplasts of nitrate reductase deficient rice cell lines. J Plant Physiol 117:223–231

Wetter LR (1977) Isozyme patterns in soybean-*Nicotiana* somatic hybrid cell lines. Mol Gen Genet 150:231–235

Wright MS, Koehler SM, Hinchee MA, Carnes MG (1986) Plant regeneration by organogenesis in *Glycine max*. Plant Cell Rep 5:150–154

Xu ZH, Davey MR, Cocking EC (1982) Callus formation from root protoplasts of *Glycine max* (soybean). Plant Sci Lett 24:111–115

Yamada Y, Yang ZQ, Tang DT (1985) Regeneration of rice plants from protoplasts. Rice Genet Newslett 2:94

Yamada Y, Yang ZQ, Tang DT (1986) Plant regeneration from protoplast-derived callus of rice (*Oryza sativa* L.). Plant Cell Rep 5:85–88

Zieg RG, Outka DE (1980) The isolation, culture and callus formation of soybean pod protoplasts. Plant Sci Lett 18:105–114

Subject Index